Mechanical Technology

5.

Mechanical Technology

A Two-Year Course

G. D. Redford
C.Eng., M.I.Mech.E., A.M.I.E.D.

J. G. Rimmer
B.Sc., C.Eng., M.I.Mech.E.

D. Titherington
B.Eng., C.Eng., M.I.Mech.E.

SECOND EDITION

M

First edition 1969
Second edition 1971
Reprinted 1977, 1979, 1981

Published by
THE MACMILLAN PRESS LTD
London and Basingstoke
Companies and representatives
throughout the world

ISBN 0 333 00760 3

Printed in Hong Kong

Preface

This text book is intended primarily to meet the requirements of the Mechanical Technology syllabus of the new Higher National Certificate course in engineering, but it is hoped that it will have a wider appeal because it combines much of the subject matter taught under the separate headings of Strength of Materials, Theory of Machines, Applied Thermodynamics and Mechanics of Fluids. In this respect it should prove useful to those students who are taking Higher National Diploma and degree courses.

The imminence of metrication makes it imperative that the engineer should become familiar with the International System of units. In the authors' view the conversion from one system of units to another is unnecessary and time consuming and for these reasons S.I. units (Système International d'Unités) have been used throughout the book. The student should relate the new units directly to the corresponding physical entities. He should, for example, know that the modulus of elasticity for steel is approximately 200 GN/m^2, that energy in all its forms has only one unit, the joule, and that the power output of an internal combustion engine is measured in kilowatts.

Inevitably, in a transitional period such as this, there is much discussion and criticism of the form which the S.I. system should take. It would be wrong at this stage to think that all the details have been settled once and for all, but the principles are now clear.

The contents of the book cover both years of the course. Worked examples are included and problems involving numerical calculations are provided for the student to solve. In addition "tutorial" problems are given at the end of each chapter. These should prove useful for group discussion, and are intended to direct the student's interest to the more peripheral areas, as well as to encourage him to examine the fundamental aspects of the subject.

The authors wish to express their appreciation of the excellent work of Mrs. M. Robinson in the typing of the manuscript and of Miss G. Tyldesley in the tracing of the diagrams. They also thank the publishers for their help and advice during the preparation of the book.

G. D. Redford
J. G. Rimmer
D. Titherington

Preface to the Second Edition

The general arrangement of the book remains unaltered but changes in the notation have been made in accordance with recent British Standard and S.I. requirements. These include the use of sigma (σ) and tau (τ) for normal and shear stresses, epsilon (ε) and gamma (γ) for direct and shear strains, and nu (ν) for Poisson's ratio. In addition, the 'kelvin' (K) now replaces the 'degree kelvin' and the practice of distinguishing between temperature and temperature interval by using the symbol ° and the abbreviation 'deg' has been abandoned. Thus °K and degK are replaced by K, while deg C is replaced by °C. Certain derived units are therefore also affected, for example, kJ/kg degK becomes kJ/kg K. In additional several minor alterations and corrections have been made to bring the book fully up to date.

G. D. Redford
J. G. Rimmer
D. Titherington

Contents

viii

CHAPTER 1

Kinematics and Dynamics of a Rigid Body

1.1 MOTION

There is no such thing as absolute rest or movement but only relative motion. The study of relative motion in engineering is termed kinematics while the study of the forces involved in the cause or the effect of motion is called kinetics. Kinetics, together with statics which is the study of bodies at rest relative to each other under the action of forces, form dynamics.

Motion may be linear, angular or a combination of both and is described by three quantities

(i) displacement, which means distance;
(ii) velocity, which is rate of change of displacement;
(iii) acceleration, which is rate of change of velocity.

Thus, if s represents linear displacement and t the time, then linear velocity $v = \dfrac{ds}{dt}$ and linear acceleration $a = \dfrac{dv}{dt}$. For angular motion, if θ represents angular displacement, angular velocity $\omega = \dfrac{d\theta}{dt}$ and angular acceleration, $\alpha = \dfrac{d\omega}{dt}$.

Equations of motion

The fundamental kinematic relationships are

$$s = \tfrac{1}{2}(u + v)t \qquad \theta = \tfrac{1}{2}(\omega_1 + \omega_2)t$$
$$s = ut + \tfrac{1}{2}at^2 \qquad \theta = \omega_1 t + \tfrac{1}{2}\alpha t^2$$
$$v = u + at \qquad \omega_2 = \omega_1 + \alpha t$$
$$v^2 = u^2 + 2as \qquad \omega_2{}^2 = \omega_1{}^2 + 2\alpha\theta$$

Linear and angular motion are connected by

$$s = \theta r$$
$$v = \omega r$$
$$a = \alpha r$$

In these formulae, s and θ refer to linear and angular displacement respectively, u_1 and ω_1 to initial linear and angular velocity, v_2 and ω_2 to final linear and angular velocity, a and α to linear and angular acceleration and r is the radius of curvature of the path of the body.

Reference frames

A frame of reference is necessary to describe the motions of a body.

Vector representation

Linear motion. Suppose a body moves horizontally from A to B through a distance of s metres (Fig. 1.1 (*a*)). The body when at B is said to have been displaced, relative to A, through s metres horizontally to the right. This displacement can be represented by a vector drawn from a to b (Fig. 1.1(*b*)). The length of the vector represents the magnitude,

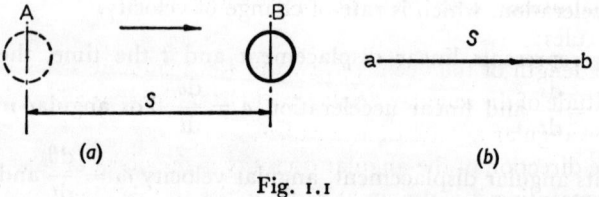

(*a*)   (*b*)

Fig. 1.1

s metres, the horizontal direction the direction of the displacement and the arrow pointing from a to b, the sense of the displacement. The vector is written as $\overrightarrow{ab}$. Conversely, $\overrightarrow{ba}$ would denote the displacement of a body at A relative to B.

Example. *A body moves vertically upwards from* A *to* B *with a velocity of 2 m/s for 3 seconds, then horizontally from* B *to* C *to the right at 3 m/s for 3 seconds and finally at 1 m/s for 2 seconds vertically downwards. What is the resultant displacement of the body?*

Multiplying velocity by time in each case, the component displacements are 6, 9 and 2 m respectively in the directions stated.

From the displacement diagram (Fig. 1.2) drawn to some suitable scale, the resultant displacement of D relative to A, $\overrightarrow{ad} = 9.85$ m at 23° 58′ to the horizontal.

Velocity and acceleration are also vector qualities and can be represented in a manner analogous to that for displacement since they are simply the first and second differentials of displacement with respect to time. Thus in Fig. 1.3, $\overrightarrow{pn}$ could represent the velocity of a point N relative to a point P as 10 m/s at 30° to the horizontal, the sense being from p to n. Alternatively, $\overrightarrow{np}$ represents the velocity of P relative to N as having the same magnitude and direction but being of opposite sense.

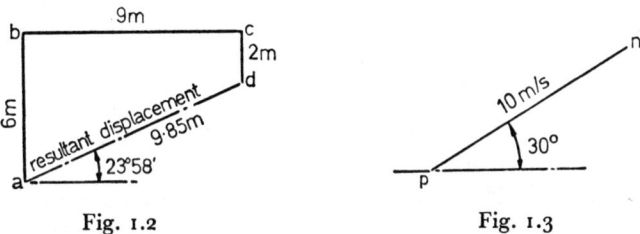

Fig. 1.2   Fig. 1.3

Angular motion. Angular displacement, velocity and acceleration are also vector quantities. They can be represented in accordance with the following rules:

(*a*) The length of the vector, drawn to some suitable scale, represents the magnitude of the angular displacement, θ radians, or angular velocity, ω rad/s, or angular acceleration α rad/s².

(*b*) The direction of the angular quantity is the plane of rotation. The vector representing it is drawn at right angles to this plane, in the direction of the axis of rotation.

(*c*) The sense, whether clockwise or anticlockwise, is given by lettering the vector to represent the way in which a right hand screw would advance.

Thus, the angular velocity of a body A (Fig. 1.4 (*a*)) rotating relative to some frame of reference O at 10 rad/s in a vertical plane, clockwise as looked at in the direction of the arrow N, would be represented by $\overrightarrow{oa}$ drawn horizontally to some suitable scale. Fig. 1.4 (*b*) shows the body with the opposite sense of rotation and its appropriate vector representation. Conversely, $\overrightarrow{ao}$ represents in each case, the velocity of the reference frame O *relative to the body* A.

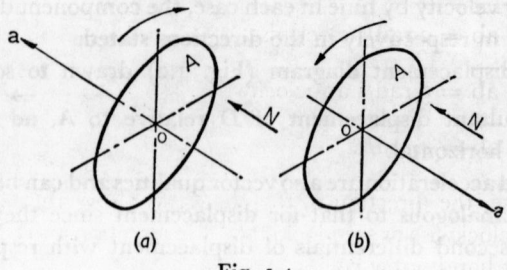

Fig. 1.4

Example. *A disc rotates about a horizontal axis XX. Represent vectorially the change in angular velocity in each of the following cases:*

   (i) *From 2 rad/s clockwise to 5 rad/s clockwise.*
   (ii) *From 2 rad/s clockwise to 3 rad/s anticlockwise.*
   (iii) *Maintaining constant magnitude of 4 rad/s clockwise while the disc rotates anticlockwise through an angle of 60 degrees about vertical axis YY.*

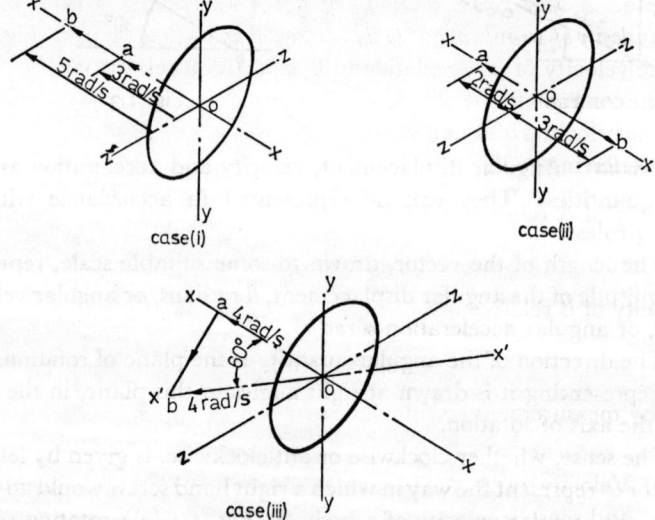

Fig. 1.5

Let o represent the frame of reference, a the body before the change of velocity and b the body after the change. Then in each case $\overrightarrow{oa}$ is the initial velocity, $\overrightarrow{ob}$ the final velocity and $\overrightarrow{ab}$ the change in velocity, that is, the velocity of b relative to a. From Fig. 1.5:

Case (i) $\overrightarrow{ab}$ = 2 rad/s clockwise.

Case (ii) $\overrightarrow{ab}$ = 5 rad/s anticlockwise.

Case (iii) $\overrightarrow{ab}$ = 4 rad/s.

In each case, the direction of the change is along ab and the sense of the change is clockwise as seen looking in the direction a to b in accordance with the right hand screw rule.

1.2 RELATIVE VELOCITY AND ACCELERATION

The velocity of one body relative to another can be found quite readily by the vector method.

Example. *A body A has a velocity of 5 m/s horizontally to the right and a body B a velocity of 10 m/s at 30° to the horizontal upwards to the right* (Fig. 1.6). Find the velocity of (i) A relative to B, and (ii) B relative to A.

Let the common frame of reference against which the velocities of A and B are measured be O. Referring to Fig. 1.6 (*b*).

Velocity of A relative to O,

$$\overrightarrow{oa} = 5 \text{ m/s}$$

Velocity of B relative to O,

$$\overrightarrow{ob} = 10 \text{ m/s}$$

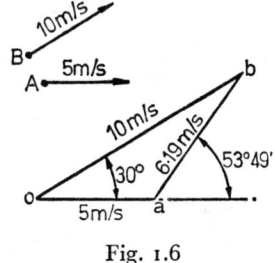

Fig. 1.6

Then, by measurement or calculation:

(i) Velocity of A relative to B, $\overrightarrow{ba}$ = 6·19 m/s with direction ba and sense b to a.

(ii) Velocity of B relative to A, $\overrightarrow{ab}$ = 6·19 m/s with direction as before but with sense a to b.

Example. *The bodies A and B in the last example have the following acceleration at an instant. A is accelerating at 6 m/s², is B decelerating at 4 m/s². What is the acceleration of B relative to A?*

Using letters with subscripts to distinguish acceleration from velocity vectors, the acceleration diagram will be as shown in Fig. 1.7, the values being drawn to some suitable scale, then:

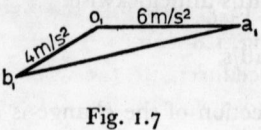

Fig. 1.7

Acceleration of B relative to A is given by $\overrightarrow{a_1b_1}$ in magnitude, direction and sense.

1.3 GRAPHS RELATING DISPLACEMENT, VELOCITY AND ACCELERATION WITH TIME

Velocity-time and acceleration-time graphs can be prepared provided that it is possible to plot a displacement curve.

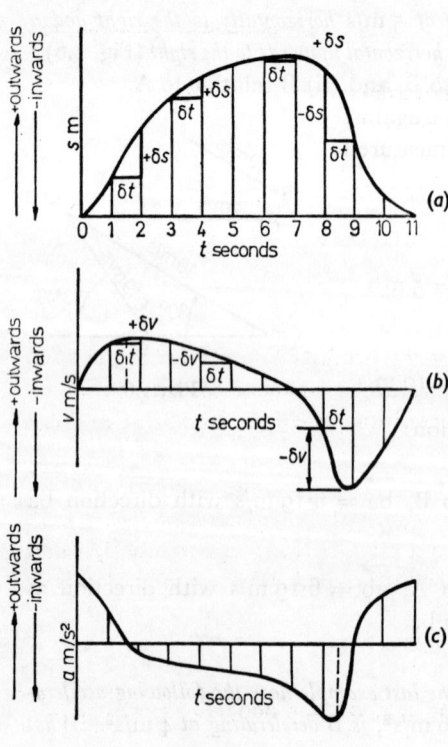

Fig. 1.8

Graphical differentiation

The displacement-time curve shown in Fig. 1.8 (*a*) represents the forward and return movement *s* metres of a machine part during time *t* seconds. The outward displacement of *s* metres is made in 7 seconds and the return movement in 4 seconds. It is assumed that displacement, velocity and acceleration are positive when their sense is outwards from the starting point. Time intervals δt are positive throughout. Velocity-time and acceleration-time curves can then be found by graphical differentiation.

Ordinates are erected at suitable time points, the interval between each two adjacent ordinates being δt seconds. The corresponding increments or decrements in displacement are set off and measured as $\pm \delta s$ metres. The average velocity during each time interval is now obtained from $v = \delta s/\delta t$, the values being positive or negative as s is positive or negative. These average velocity values are next marked off as mid-ordinates of the time interval over which they occur to give points through which the velocity-time graph can be drawn (Fig. 1.8 (*b*)).

By repeating the procedure with the velocity-time curve, acceleration values are given by $\delta v/\delta t$ in each case. These are then plotted as mid-ordinates to obtain the acceleration-time graph Fig. 1.9 (*c*). Since $\delta s/\delta t$ is a measure of the *average* velocity and $\delta v/\delta t$ is a measure of the *average* acceleration during interval δt, the smaller the increments the greater the accuracy.

The reverse of differentiation is integration and since:

$$\delta v = a\delta t$$

then $v = \int a\, dt$ which is the area under acceleration—time curve.

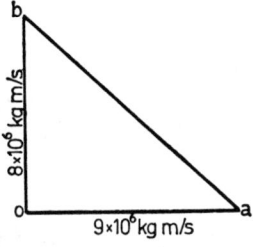

Fig. 1.9

Similarly $\delta s = v\delta t$

so that $s = \int v\, dt$ which is the area under the velocity-time curve.

1.4 MOMENTUM

Linear Momentum

The linear momentum of a body is the product of the mass of the body and its linear velocity. Since the product involves velocity (a vector quantity), momentum will have magnitude, direction and sense and can be represented vectorially.

Example. *A rocket with mass 3000 kg has its velocity changed from 3 km/s horizontally to the right when at A to 4 km/s vertically upwards at B, during which time its mass is reduced to 2000 kg by the burning of fuel. Determine its total change of momentum.*

Let O represent the reference plane. Then

$$\text{Momentum at A, } \overrightarrow{oa} = 3000 \times 3000 = 9\ 000\ 000 \text{ kg m/s}$$
$$\text{Momentum at B, } \overrightarrow{ob} = 2000 \times 4000 = 8\ 000\ 000 \text{ kg m/s}$$

From the momentum diagram (Fig. 1.9), the change of momentum is the momentum of B relative to momentum at A or $\overrightarrow{ab}$ where

$$\overrightarrow{ab} = 10^6\sqrt{(8^2 + 9^2)} = 12\cdot04 \times 10^6 \text{ kg m/s}$$

Angular momentum

Angular momentum is analogously, the product of the moment of inertia of the body and its angular velocity, and is also a vector quantity. In other words,

$$\text{Angular momentum} = I\omega = mk^2\omega \text{ kg m}^2/\text{s}$$

where m is the mass of the body in kilograms and k is the radius of gyration in metres. The vector representation of angular momentum is essentially the same as that for angular velocity.

Example. *What is the angular momentum of a disc having a mass of 5 kg and radius of gyration of 0·5 m if its angular velocity is 10 rad/s in a vertical plane in a clockwise sense as seen from the right? What is the rate of change of angular momentum if the disc is now made to rotate at 5 rad/s about a vertical axis, anticlockwise when seen from above.*

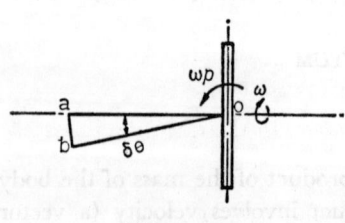

Fig. 1.10

Fig. 1.10 shows the disc edge on as seen from above. Let O represent the "fixed" frame of reference and A the body when beginning to rotate at 5 rad/s about its vertical axis, then:

Angular momentum,

$$mk^2\omega = 5 \times 0\cdot5^2 \times 10 = 12\cdot5 \text{ kg m}^2/\text{s}$$

and is represented by $\overrightarrow{oa}$.

Let disc rotate at 5 rad/s about its vertical axis displacing itself through angle $\delta\theta$ in time δt to position B. Then the body's momentum after time δt will be represented by $\overrightarrow{ob}$, so that

$$\text{Change in angular momentum, } \overrightarrow{ab} \simeq I\omega\delta\theta$$

and

$$\text{Rate of change of momentum} = \frac{\text{Change of momentum}}{\text{Time for change}} = I\omega\frac{\delta\theta}{\delta t}$$

In the limit, as $\delta t \to$ zero,

$$\text{Rate of change of momentum} = I\omega \frac{d\theta}{dt} = I\omega\omega_\rho$$

where ω_ρ = Angular velocity of rotation about the vertical axis.

$$\text{Rate of change of momentum} = mk^2\omega\omega_\rho = 5 \times 0.5^2 \times 10 \times 5$$
$$= 62.5 \text{ kg m/s}^2$$

Conservation of momentum

The principle of the conservation of momentum is that the sum of the momenta of a system of bodies before impact or separation is equal to the sum of the momenta after impact or separation.

Example. *A body A with a mass 2 kg is approaching a second body B with a velocity of 10 m/s. B has a mass of 3 kg and is at rest. A third body C of mass 4 kg is also approaching B with a velocity of 1 m/s but from the opposite direction. If the bodies collide with each other simultaneously and remain together, what is now the velocity of the composite mass?*

Consider A's velocity as positive, so that its momentum will also be positive. By the same convention, C's momentum will be negative. Applying the principle of the conservation of momentum:

$$\text{Sum of momenta before impact} = \text{Sum of momenta after impact}$$

$$(2 \times 10) + (3 \times 0) - (4 \times 1) = (2 + 3 + 4)v$$

so that
$$v = \frac{+16}{9} = +1.77 \text{ m/s}$$

The positive sign indicates that the mass moves in the direction in which A was moving before impact.

Example. *A skater rotates on the tip of one skate with an angular velocity of 2π rad/s with arms and legs as close as possible to the vertical axis of spin. In this position the radius of gyration of the skater is 0.14 m. What will be the speed of rotation when the limbs of the skater are spread to give a radius of gyration of 0.3 m? Neglect friction.*

$$\text{Momentum in the closed position} = \text{Momentum in the open position}$$

$$I_1\omega_1 = I_2\omega_2$$
$$mk_1{}^2\omega_1 = mk_2{}^2\omega_2$$
$$m \times 0.14^2 \times 2\pi = m \times 0.3^2 \times \omega_2$$
$$\therefore \omega_2 = \frac{0.14^2 \times 2\pi}{0.3^2} = 0.1365 \text{ rad/s}$$

1.5 THE LAWS OF MOTION

The laws which relate force, mass and acceleration were formulated by Sir Isaac Newton. Their validity rests upon observation backed by empirical verification.

Newton's first law. A body will remain at rest or continue to move linearly (or angularly) with uniform motion unless acted upon by some resultant external force (or torque).

Newton's second law. The rate of change of linear or angular momentum of a body is proportional to the applied resultant external force or torque and will take place in the direction of the force or torque.

Newton's third law. To every action there is an equal and opposite reaction.

The second law enables force and torque to be defined and evaluated. It can be restated as follows:

$$\text{Force} \propto \text{Rate of change of momentum}$$
$$\propto \text{Mass} \times \text{Rate of change of velocity}$$
$$\propto \text{Mass} \times \text{Acceleration}$$

In other words,

$$\text{Force} = \text{Constant} \times \text{Mass} \times \text{Acceleration}$$

The unit of force is the *Newton* (N) and is that force which, if allowed to act on a unit mass of one kilogram (kg), will produce an acceleration of one metre per second^2 (m/s^2). With these units,

$$\text{Force (N)} = \text{Mass (kg)} \times \text{Acceleration (m/s}^2\text{)}$$

i.e. $$F = M \times a \qquad (1.1)$$

Analogously for angular motion,

$$\text{Torque} \propto \text{Rate of change of angular momentum}$$
$$\propto \text{Moment of inertia} \times \text{Rate of change of angular velocity}$$
$$\propto I\alpha$$

or $$\text{Torque} = \text{Constant} \times I\alpha$$

Unit moment of inertia can be represented by unit mass concentrated at a point at unit radius of gyration k. Applying unit torque, i.e. unit force at unit radius, to this mass will produce a linear acceleration of 1 m/s^2 in the mass. But

$$\alpha = \frac{a}{r} = \frac{1 \text{ m/s}^2}{1 \text{ m}} = 1 \text{ rad/s}^2$$

Therefore, unit torque in Newton metres (Nm) applied to unit moment of inertia (kg m²) produces unit angular acceleration (rad/s²). Therefore the constant is unity and

$$\text{Torque (Nm)} = I(\text{kg m}^2) \times \text{Angular acceleration (rad/s}^2)$$

i.e. $\qquad\qquad T = I\alpha \qquad\qquad\qquad\qquad (1.2)$

1.6 ACCELERATED MOTION

Falling bodies and weight

A body of mass m falling freely at the earth's surface in the vicinity of London is accelerated by the pull of the earth at 9·81 m/s². Since the earth's pull is the *weight* of the body at that particular geographical location, then in accordance with Newton's second law

$$\text{Force (N)} = \text{Mass (kg)} \times \text{Acceleration (m/s}^2)$$

i.e. in the vicinity of London:

$$\text{Weight of a body (N)} = 9\cdot81 \times \text{Mass of the body (kg)}$$

or more generally,

$$W = gm$$

where g is the gravitational constant at the location to which the body is referred.

Example. *A flywheel* (Fig. 1.11) *of mass 100 kg and radius of gyration 0·25 m is keyed to a horizontal shaft 0·1 m in diameter carried in bearings. Around the shaft is wound a cord and from its free end is suspended a mass of 10 kg. Calculate the resultant angular acceleration of the flywheel.*

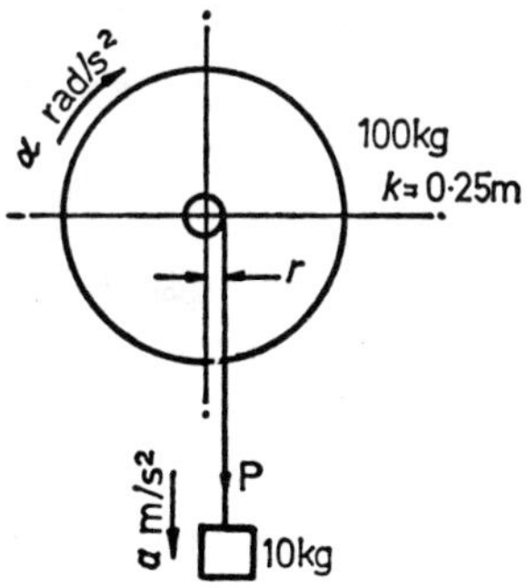

Fig. 1.11

Let α be the angular acceleration of the flywheel.

a the associated linear acceleration of the mass on the cord

Then $\alpha = a/r$ where $r =$ shaft radius.

Tension in the cord $P =$ (Force to support the 10 kg mass)
$\qquad\qquad\qquad\qquad$ − (Force to accelerate 10 kg mass)

$$P = (9\cdot81 \times 10) - (10 \times a)$$

Then since   Torque $= I\alpha$

$$Pr = mk^2\alpha$$

i.e.

$$[(9\cdot81 \times 10) - (10 \times a)]0\cdot05 = 100 \times 0\cdot25^2 \times \frac{a}{0\cdot05}$$

or $4\cdot905 - 0\cdot5a = 125a$

∴ $a = \dfrac{4\cdot905}{125\cdot5} = 0\cdot0391$ m/s^2

But

$$\alpha = \frac{a}{r} = \frac{0\cdot0391}{0\cdot05} = 0\cdot782 \text{ rad/s}^2$$

Bodies moving in curved paths

Referring to Fig. 1.12 (*a*), let body B move in a curved path of radius *r* metres about a fixed point O with uniform angular velocity ω rad/s. At any instant the tangential velocity of the body will be V m/s where $V = \omega r$.

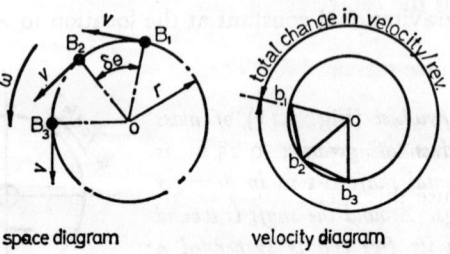

space diagram   velocity diagram

Fig. 1.12

Let the body move, in succession, through positions B_1, B_2, B_3, etc. undergoing a small angular displacement $\delta\theta$ radians between adjacent points in time δt seconds. Then the tangential velocity will change in direction accordingly. Referring to Fig. 1.12 (*b*), $\overrightarrow{ob_1}$ represents velocity of body when at B_1, $\overrightarrow{ob_2}$ represents velocity of body when at B_2 and so on for succeeding positions, while $\overrightarrow{b_1b_2}$, $\overrightarrow{b_2b_3}$, etc., represent the changes in velocity from position to position. Considering infinitely small time intervals dt and correspondingly small angular displacements dδ then, for one complete rotation of B,

$$\text{Total velocity change} = \overrightarrow{b_1b_2} + \overrightarrow{b_2b_3} + \overrightarrow{b_3b_4} + \text{etc.}$$

$$= \text{Circumference of velocity diagram circle}$$

$$= 2\pi v \text{ m/s}$$

$$\text{Time for the change} = \text{Time for one revolution of B}$$

$$= \frac{2\pi r}{v}$$

$$\therefore \text{ Rate of change of velocity} = \frac{\text{Change in velocity}}{\text{Time for the change}}$$

$$= \frac{2\pi v^2}{2\pi r} = \frac{v^2}{r} \text{ m/s}^2$$

From the velocity diagram, it is clear that the change in velocity during any interval dt, when referred to the mid-position of the corresponding angular displacement on the space diagram Fig. 1.12 (*a*), is directed inwards towards the centre of rotation O. That is,

$$\text{Centripetal (centre seeking) acceleration} = \frac{v^2}{r} = \omega^2 r \text{ (m/s)}$$

From Newton's second law of motion,

$$\text{Force to cause this acceleration} = \text{Mass of body} \times \text{Acceleration}$$

$$= m\frac{v^2}{r} \text{ (N)} = m\omega^2 r \text{ (N)}$$

The body reacts to the centripetal force with an inertial force which is equal and opposite. This radially outward force is termed the Centrifugal (Centre fleeing) force.

Example. *Bodies contained within a rocket in "free fall" are usually weightless. Determine the velocity of spin about a longitudinal axis which must be given to a rocket 40 metres in diameter if bodies within the rocket at maximum distance from the longitudinal axis are to "weigh" as much as they would if at rest on the earth's surface in London.*

The "weight" of the body will result from the centripetal acceleration imparted by the spin of the rocket. If m is the mass of the body in kilograms:

$$\text{Weight of body} = m \times 9 \cdot 81 = m\omega^2 r$$

$$\therefore \; \omega^2 = \frac{9 \cdot 81}{20} = 0 \cdot 4905$$

$$\therefore \; \omega = 0 \cdot 7 \text{ rad/s} = \frac{60}{2\pi} \times 0 \cdot 7 = 6 \cdot 68 \text{ rev/min}$$

Example. *With what force is an aircraft pilot held against his seat at the top of "looping the loop" at a speed of* 240 km/h *if the radius of the loop is* 300 m *and the mass of the pilot is* 80 kg?

$$\text{Tangential velocity } v = 240 \text{ km/h} = \frac{240 \times 1000}{3,600} = 66 \cdot 7 \text{ m/s}$$

Centripetal force applied through seat to the pilot to maintain his circular path

$$= m\frac{v^2}{r} = \frac{80 \times 66 \cdot 7^2}{300} = 1185 \text{ N}$$

$$\text{Weight of the pilot} = 80 \times 9 \cdot 81 = 784 \cdot 8 \text{ N}$$

$\therefore$ Resultant force between pilot and seat

$$= 1185 - 784 \cdot 8 = 400 \cdot 2 \text{ N}$$

Acceleration of a body subjected to combined radial and angular velocity

The body P (Fig. 1.13 (*a*)) is moving radially outwards with instantaneous linear velocity along and relative to a rod OX, while OX itself is rotating about O with instantaneous angular velocity ω. Let OX rotate from position OX_1 to position OX_2 through angle $\delta\theta$ in a small interval of time δt. Let the body be at P_1 radius r from O at the commencement of time δt, and at position P_2 radius $r + \delta r$ from O at end of time δt. Let Y be a point on the link OX which is "coincident" with P during time δt. Then the body P will experience an acceleration relative to Y at right angles to OX. This acceleration can be described in two stages as follows:

Stage 1: Referring to Fig. 1.13 (*b*)

Let $\overrightarrow{\text{yp}_1}$ represent the velocity of P relative to Y *along the rod* at the beginning of time δt.

Let $\overrightarrow{\text{yp}_2}$ represent the velocity of P relative to Y *along the rod* at the end of time δt.

$$\text{Change in velocity} = \overrightarrow{\text{p}_1\text{p}_2} \simeq v\delta\theta \text{ (when } \delta\theta \text{ is small)}$$

and is at right angles to OX. Then

$$\text{Acceleration of P relative to Y} = \frac{\text{Change in velocity}}{\text{Time for change}}$$

$$= v\frac{\delta\theta}{\delta t} = v\omega$$

Stage 2: Referring to Figs. 1.13 (*a*) and (*c*):
Tangential velocity of P relative to O when at radius r and coincident with Y is equal to ωr and is represented by $\overrightarrow{\text{op}_1}$.

Tangential velocity of P relative to O when at radius $(r + \delta r)$ is equal to $\omega(r + \delta r)$ and represented by $\overrightarrow{\text{op}_2}$.
Change in tangential velocity of P relative to O (and relative to Y)

$= \omega\delta r$ and is at right angles to OX

$= \omega v\delta t$

Acceleration of P relative to Y

$$= \frac{\text{Change in velocity}}{\text{Time for change}}$$

$$= \frac{\omega v\delta t}{\delta t} = \omega v$$

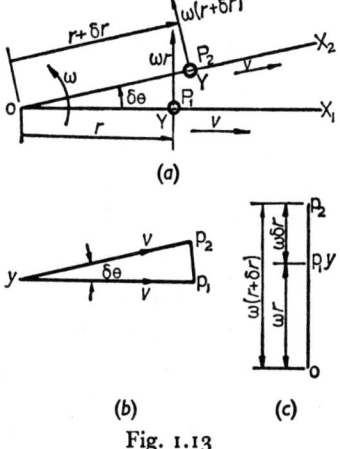

(a)

(b)   (c)

Fig. 1.13

$\therefore$ Total acceleration of P relative to Y

$= 2\omega v$ and is always at right angles to OX

The sense of P's acceleration will be reversed if the sense of either ω or v is reversed (Fig. 1.14). The above acceleration of P relative to Y is known as the Coriolis component of acceleration of P (see Chapter III) and exists whether or not ω and v are uniform.

In addition to the Coriolis component of acceleration, P may have a component of sliding f_s acceleration *along* OX relative to Y, so that these two components must be added vectorially to give the *total* acceleration of P relative to Y. In addition, the point Y, itself fixed on OX_1, will

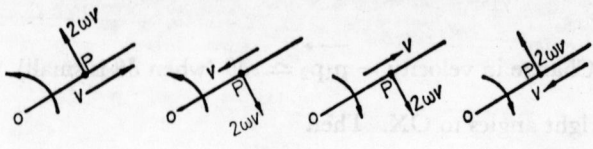

Fig. 1.14

have a centripetal acceleration relative to O equal to $\omega^2 r$, and if OX has an angular acceleration α, will also have a tangential acceleration relative to O equal to αr. Then:

Total acceleration of P relative O

$$= \begin{bmatrix} \text{Acceleration of} \\ \text{P relative to Y} \end{bmatrix} + \begin{bmatrix} \text{Acceleration of} \\ \text{Y relative to O} \end{bmatrix}$$

$$= \begin{bmatrix} \text{Coriolis component of acceleration} & + & \text{Acceleration of sliding} \\ \text{of P relative to Y} & & \text{of P relative to Y} \end{bmatrix}$$

$$+ \begin{bmatrix} \text{Centripetal acceleration} & + & \text{Tangential acceleration} \\ \text{of Y relative to O} & & \text{of Y relative to O} \end{bmatrix}$$

$$= [2\omega v + f_s] + [\omega^2 r + \alpha r]$$

Example. *A link AB is rotating about end A and a block S is sliding along the link. Determine for the following instantaneous conditions the total acceleration of S relative to A when S is coincident with a point Y on AB situated 2 m from A.*

    (i) Link, at 30° to the horizontal (Fig. 1.15), rotating clockwise with uniform angular velocity of 5 rad/s and the block sliding uniformly towards end B at 10 m/s.

    (ii) Link at 30° to the horizontal (Fig. 1.17) rotating anticlockwise at 5 rad/s and accelerating clockwise at 3 rad/s² while S is moving towards A at 10 m/s and accelerating towards B at 40 m/s².

Case (i). Refer to Fig. 1.16.

Centripetal acceleration of Y relative to A,

$$\overrightarrow{a_1 y_1} = \omega^2 r = 5^2 \times 2 = 50 \text{ m/s}^2$$

Because the link has uniform angular velocity,

$$\text{Tangential acceleration of Y relative to A} = 0$$

Coriolis component of acceleration of S relative to Y,

$$\overrightarrow{y_1s_1} = 2\omega v = 2 \times 5 \times 10 = 100 \text{ m/s}^2$$

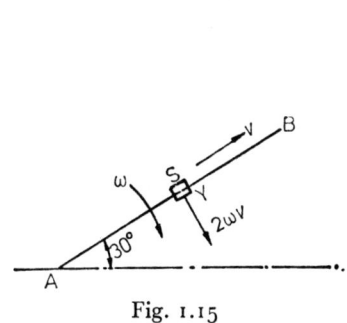

Fig. 1.15

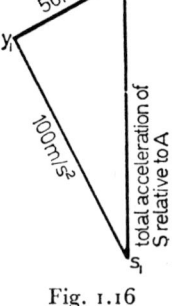

Fig. 1.16

Because S has uniform motion along link,

$$\text{Acceleration of sliding of S relative to Y} = 0$$

Fig. 1.16 shows the required acceleration diagram in which the total acceleration of S relative to A is given by $\overrightarrow{a_1s_1}$ and equals 112 m/s².

Case (ii). Refer to Fig. 1.18.

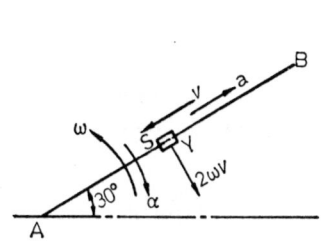

Fig. 1.17

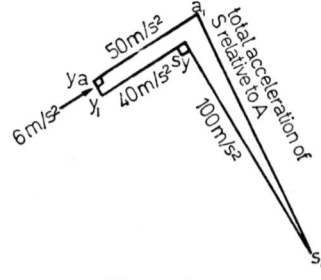

Fig. 1.18

Centripetal acceleration of Y relative to A,

$$\overrightarrow{a_1y_a} = \omega^2 f = 5^2 \times 2 = 50 \text{ m/s}^2$$

Tangential acceleration of Y relative to A,

$$\overrightarrow{y_ay_1} = \alpha r = 3 \times 2 = 6 \text{ m/s}^2$$

Acceleration of sliding of S relative to Y,

$$\overrightarrow{y_1s_y} = 40 \text{ m/s}^2$$

Coriolis component of acceleration of S relative to Y,

$$\overrightarrow{s_ys_1} = 100 \text{ m/s}^2 \text{ (as before)}$$

Fig. 1.18 shows the acceleration diagram. The total acceleration of S relative to A is again given by $\overrightarrow{a_1s_1}$ and is equal to 106·5 m/s².

Note. The component accelerations of Y relative to A are always paired at right angles. So too are the component accelerations of S relative to Y. The junction points of the component accelerations are denoted by using the two relevant letters with one as a subfix to the other.

1.7 WORK, ENERGY AND POWER

Work is the overcoming of a resistance through a distance and is measured either as:

Work done by a force = Applied force × the distance moved by the point of application in the sense and direction of the force

= F (newtons) × s (metres)

or = Fs joules

Work done by a torque = Applied torque × the angle turned through by the torque arm in the sense and direction of the torque

= T (newton metres) × θ (radians)

= $T\theta$ joules

Energy

Energy is the ability to do work. The unit is the joule.

Potential energy. Potential energy is associated with state or position relative to some conventional standard. The most obvious example of this is the potential energy of a body due to its height above the surface of the earth.

Let a mass of m kilogrammes be raised with uniform velocity through height of h metres:

Force required to lift the mass = Weight of the mass (W) = gm (N)

Potential energy given to the mass = Work done in lifting the mass

$$= \text{Force} \times \text{Distance lifted}$$

$$= Wh = gmh \text{ joules}$$

Kinetic energy. Kinetic energy is the energy possessed by virtue of motion.

(i) Bodies with linear motion. If a body of mass m kilogrammes initially at rest is acted upon by a force of F newtons causing an acceleration of a metres per second², then the work done on the body during its displacement of s metres will be stored in the body as kinetic energy. For a body under uniform acceleration

$$v^2 = u^2 + 2aS$$

or
$$s = \frac{v^2}{2a} \quad \text{when } u = 0$$

But also,
$$\text{Force} = \text{Mass} \times \text{Acceleration}$$

i.e.
$$F = ma$$

$$\text{Energy stored in body} = \text{Work done}$$

$$= F \times s = ma \times \frac{v^2}{2a}$$

i.e.
$$\text{Kinetic energy} = \tfrac{1}{2}mv^2 \text{ joules}$$

(ii) Bodies with angular motion. If a mass of m rotating at ω rad/s is thought of as concentrated at its radius of gyration k and a torque is applied to it in the form of a force F newtons at a radius of k metres, then it can be shown in a manner analogous to that for linear motion above that

$$\text{Kinetic energy of rotating mass} = \tfrac{1}{2}mk^2\omega^2 = \tfrac{1}{2}I\omega^2 \text{ joules}$$

(iii) Bodies with both linear and angular motion.

$$\text{Total kinetic energy} = \text{KE of translation} + \text{KE of rotation}$$
$$= \tfrac{1}{2}mv^2 + \tfrac{1}{2}I\omega^2 \text{ joules}$$

The principle of the conservation of energy. Energy within a system of bodies in motion may undergo conversion to potential, kinetic, heat, electrical or chemical energy. No system can be entirely isolated from its surrounding environment so that there is usually a loss or gain in total energy.

However, when referring to the entire universe as a system, the principle of the conservation of energy may be said to hold good. This principle is that:

Energy can neither be created nor destroyed but only changed in form, so that the total amount of energy in the universe is constant.

The energy principle. If the forces within a system consist only of forces which can convert potential energy into kinetic energy—gravitational forces, for example—then there is a restricted form of the principle of the conservation of energy which is known as the energy principle and which takes the form:

The sum of the potential and kinetic energies of a system is constant.

The energy principle does not apply when friction is involved, for energy is then converted into heat. The principle applies to impact only when the bodies concerned are perfectly elastic. Although the total momentum is always conserved on impact, the sum of the potential and kinetic energies may be decreased if some energy is converted into heat.

Power

Power is the rate of doing work. In the S.I. system of units, the unit of power is the watt and is that power which enables work to be done at the rate of one joule per second. Thus

$$\text{Power (watts)} = \text{Work done (joules)}/\text{time (seconds)}$$

Example. *A flywheel of mass 5000 kg is in the form of a disc 1·5 metres in diameter. The flywheel is mounted in bearings with its shaft horizontal and is coupled directly to the shaft of an electric motor.* Neglecting friction:

(i) Find the torque which the motor must exert in order to give the flywheel an angular acceleration of 5 rad/second/second.

(ii) What power will the motor be transmitting when the flywheel has reached a speed of 500 revolutions per minute if the latter is still being accelerated at 5 radians per second?

(iii) When the flywheel speed reaches 600 revolutions per minute the motor is disconnected and the flywheel shaft geared to a stationery drum to which is fastened a rope. From the free end of the rope is suspended a mass of 2000 kg. Determine the height to which the mass will be raised by the flywheel before the latter is brought to rest.

(i) For a disc $k^2 = \dfrac{r^2}{2} = \dfrac{0 \cdot 75^2}{2} = 0 \cdot 282$ m²

$$\text{Moment of inertia of flywheel} = mk^2 = 5{,}000 \times 0 \cdot 282$$
$$= 1410 \text{ kg m}^2$$
$$\text{Accelerating torque} = I\alpha = 1{,}410 \times 5$$
$$= 7050 \text{ Nm}$$

(ii) Angular velocity of flywheel $= \dfrac{2\pi500}{60} = \dfrac{100\pi}{6}$ rad/s

Power $=$ Torque $\times$ Angle turned through/second

$$= 7050 \times \frac{100\pi}{6} = 117500\pi \text{ W}$$

$$= 368 \text{ kW}$$

(iii) $\qquad 600 \text{ rev/minute} = \dfrac{2\pi600}{60} = 20\pi$ rad/s

$$\text{Kinetic energy in flywheel} = \tfrac{1}{2}I\omega^2$$
$$= \tfrac{1}{2} \times 1{,}410 \times 20^2\pi^2$$
$$= 2\ 780\ 000 \text{ joules}$$

Let h be the height through which the mass is lifted.
Gain in potential energy of the mass

$$= \text{Loss in kinetic energy of the flywheel}$$
$$gmh = 2\ 780\ 000$$
$$9 \cdot 81 \times 2{,}000 \times h = 2\ 780\ 000$$

so that $\qquad\qquad\qquad h = 142$ m

PROBLEMS

For tutorials

1. What is meant by "absolute motion" and "absolute rest"? Do these phrases have meaning?

2. A body such as the earth bulges at its equator due to its angular motion causing centrifugal force. Since this motion is about the body's *own* axis, why should the amount of "bulge" not be measured in order to ascertain the body's absolute angular velocity?

3. A rocket is orbiting the earth beyond the atmosphere at a speed of 480 km/s in "free fall" with engines shut off. Its position at any instant can be determined from the ground. If, when in position A the rocket explodes, fragmenting in all directions, what will be the position, relative to A, of the centre of gravity of the system of fragments four seconds after the explosion? Give a reasoned explanation.

4. A body is moving in a circular path on a frictionless, horizontal plane surface. It is constrained to move in this manner by being attached to a cord which passes downwards through a hole in the surface at the centre of the body's path and which has a mass hanging freely from its lower end which is sufficient to balance the centrifugal force of the body. What will happen if:

(a) the suspended mass is pulled down a little and then released?

(b) the suspended mass is increased?

(c) the suspended mass is decreased?

5. A body is being rotated rapidly at the end of a cord. The cord breaks, due to centrifugal force. Does the body now move, (a) tangentially to the circular path in which it previously rotated. (b) radially outwards, that is, in the direction opposite to the pull previously exerted by the cord, or (c) in a direction which is a combination of (a) and (b)?

6. A person does not see, feel, hear, smell or taste forces, but only their effects. What then *is* a force and wherein lies a person's justification for creating or analysing systems of forces as he does?

7. Refer to the worked example involving the momentum of a skater. As friction is to be neglected there will, apparently, be no loss or gain in energy. Why then should not the Principle of Conservation of Energy be used as an alternative to the Principle of the Conservation of Momentum in solving this and similar problems?

General

1. An aircraft carrier A lies 10 km due north of a second carrier B. Aircraft leave the two carriers simultaneously, the one from A flying at 300 km/h on a course 30° east of south and the one from B flying at 320 km/h on a course due east. If these speeds, relative to the ocean, are maintained determine (a) the velocity of the aeroplane from B relative to the aeroplane from A, (b) the time from launching to when the planes are nearest to each other.

Ans. (a) 310·4 km/h, 33° 11′ E of N

(b) 1·617 minutes

2. A hollow sphere, internal diameter one metre (Fig. 1.19), contains several small ball bearings. A series of holes slightly larger than the ball bearings is drilled in the shell of the sphere at a vertical height of 0·25 metres above the base of the sphere. Circumferential grooves on the inner surface of the sphere extend from these holes to the base and form guides for the ball bearings. The sphere stands upon a horizontal surface XX and is rotated so that the bearings rise within it along the grooves until they are able to slip through the holes. Assuming

that the depth of the grooves is such that the centres of the balls lie in the plane of the inner surface of the sphere, determine (i) the minimum speed of rotation at which the balls are released, and (ii) the distance from the points at which the balls first make contact with XX after being released to the point of intersection of the vertical axis of spin of the sphere with surface XX. Neglect friction and effects of shell thickness.

Ans.(i) 6·25 rad/s
(ii) 0·749 m

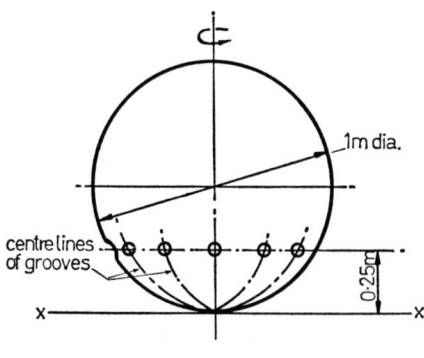

Fig. 1.19

3. A piece of fairground equipment takes the form of a large open topped cylinder of timber which can rotate about its vertical longitudinal axis. The floor can be moved upwards or downwards. People enter through a door in the wall and stand against the latter. The cylinder is rotated and the floor lowered, leaving people firmly held against the wall and clear of the floor.

If the cylinder diameter is 8 metres and the minimum coefficient of friction between a person's clothing and the wall is 0·30 determine:

(*a*) the minimum speed of rotation necessary to hold a person flat against the wall (assume the centre of gravity of the person to be in the plane of the wall),

(*b*) the new minimum speed and the angle relative to the wall at which a person would have to lean at this speed in order to "stand" with his feet on the wall and his head towards the cylinder axis if his centre of gravity was 3·25 metres from the axis of rotation. Assume edge contact between feet and wall.

Ans. (*a*) 27·5 rev/min
(*b*) 30·2 rev/min. 73° 14′

CHAPTER 2

Complex Plane Mechanisms—1

2.1 The Structure of a Machine

Machines

A machine is an assemblage of parts capable of moving in relation to each other to transmit power and convert energy into useful work. The machine parts which have relative motion are called links. Levers, pivot pins, pulleys, belts, gear wheels, power screws, shafts, hydraulic fluids, connecting rods and pistons are all examples of links. Machine frames are links too for they move in relation to the parts they support even though they are usually at rest in relation to the earth. A link may be as simple as a lever made from a single piece of metal, or as complex as a shaft to which several gear wheels are keyed. The shaft, gear wheels and keys form one link since they rotate together without relative motion.

Constraint. The motion of a machine must be predictable and hence ordered, which can be achieved only if the links impose some degree of control on each other. This mutual control is referred to as constraint.

Kinematic pairs

Two links which interact directly with, and mutually constrain, each other form a *kinematic pair*. The constraint will be exerted at the point, line or area of contact and the links involved are referred to as a lower or higher pair according to the nature of the contact between them and their relative motion. Lower pairs usually involve contact over an area, with surfaces sliding over each other. Shafts in plain bearings, pin-jointed links, slide bars and slide blocks, screw and nut assemblies are all forms of lower pairs. In higher pairing, contact is usually along a line or at a point, and the motion is that of, or equivalent to, rolling—roller and ball bearings, gear wheel teeth in mesh, cam and follower and belt drives, for example. In certain cases, *complete* constraint is achieved only by applying external forces in the forms of gravitational pulls, spring loads, belt tensions etc.

Kinematic chains

A simple machine may involve only one kinematic pair, as in the case of a lever and fulcrum, but more complex machines have several parts arranged so that each link forms one element in each of two kinematic pairs. Such an arrangement of links is called a kinematic chain. The connecting rod and crankshaft of a reciprocating engine form what is called a turning pair, the crankshaft and the frame (with bearings) a second, the frame (with cylinder attached) and the piston a sliding pair and the piston (with gudgeon pin) and the connecting rod a third turning pair.

Mechanisms. For a kinematic chain to transmit motion, one link must be fixed relative to the earth or to the machine as a whole. With constraints, movement of one link will now produce predictable relative movements of the others.

A complete assemblage of links, considered simply as a means of transferring or transmitting motion, is called a *mechanism*. When the parts are proportioned so as to be capable of transmitting forces or torques and hence of converting energy into work, the mechanism becomes a *machine*. A mechanism is, in effect, the skeletal form of a machine and can usually be shown as a simple line diagram where each line represents either the line of centroids of the cross sections of each link or a simplified version of such a line.

2.2 BASIC KINEMATIC CHAINS

The designs of many machines are based upon two important kinematic chains—the four-bar or quadric cycle chain and the slider crank chain.

The four-bar chain consists essentially of four bars, links or elements connected to give four turning pairs as shown in Fig. 2.1. Various combinations of turning members (cranks) and oscillating members (levers) can be obtained by altering the relative lengths of the links.

If AD is fixed (Fig. 2.1), the relative lengths of the links will permit links AB and CD to oscillate but not to rotate, giving a double lever mechanism. By shortening CD in relation to AB (Fig. 2.2), CD can rotate and AB can oscillate forming a lever-crank mechanism. Fig. 2.3

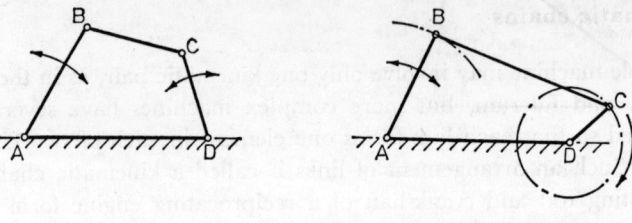

Fig. 2.1   Fig. 2.2

shows AB and CD as having equal lengths and both able to rotate, providing a special case of the double crank mechanism well known in the arrangement for coupling the wheels of locomotives. Shortening AB in relation to CD (Fig. 2.4) produces a second lever-crank mechanism.

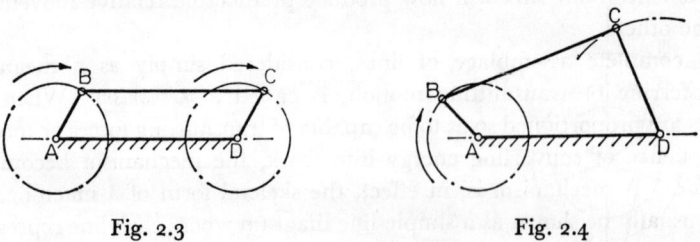

Fig. 2.3   Fig. 2.4

Inversions. With a kinematic chain of suitably proportioned links (Fig. 2.5), the above combinations can be obtained by fixing each link in turn so that:

    (i) with AB fixed, BC can rotate and AD oscillate (lever-crank mechanism);

    (ii) with BC fixed, *either* AB *or* CD can rotate (double-crank mechanism);

    (iii) with CD fixed, BC can rotate and AD oscillate (lever-crank mechanism);

    (iv) with AD fixed, AB and CD can oscillate (double-lever mechanism).

The process of fixing a different link to produce a different motion is known as *inversion*. In general, the number of changes in operation obtained by inversion is equal to the number of links in the chain.

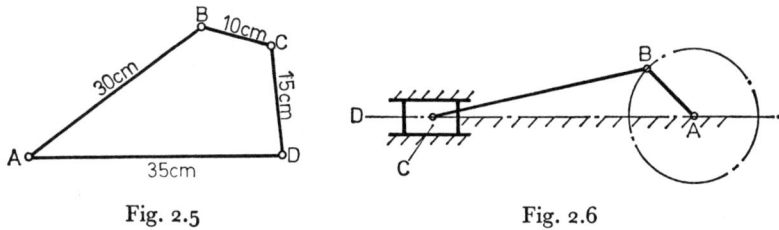

Fig. 2.5   Fig. 2.6

The slider-crank chain consists of three turning pairs and one sliding pair and (Fig. 2.6) forms the kinematic arrangement of the reciprocating engine mechanism. Here AB represents the crank, BC the connecting rod and DA the frame, cylinder and crank shaft bearings. DA is the fixed member. Inversion of the slider crank chain produces different mechanisms, some of these are given below and later in this and the next chapter.

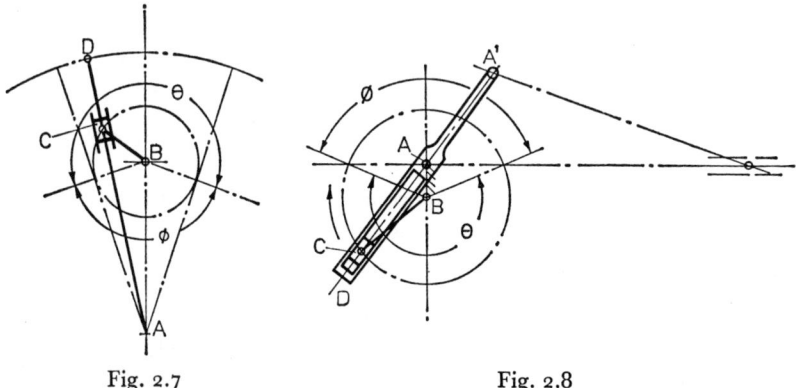

Fig. 2.7   Fig. 2.8

Inversions of the slider-crank chain

Quick-return mechanisms. (i) If the crank AB of Fig. 2.6 is made longer in relation to the other links and fixed and link BC made the driving and rotating member, the result is the quick-return mechanism used in most shaping machines (Fig. 2.7). In this case link AD will oscillate.

(ii) By keeping link AB short and again fixing it, a second form of quick-return mechanism known as the Whitworth quick-return motion (Fig. 2.8) is obtained. Link BC will be the driving member and, with link AD, can rotate. In both cases the cutting stroke occurs while the driving member rotates through angle θ and the return stroke during rotation through angle ϕ.

Oscillating cylinder mechanism. By fixing link BC and allowing link AB to rotate about end B, the oscillating cylinder mechanism is obtained (Fig. 2.9). Link AD, embodying the cylinder, piston and piston rod, now oscillates in trunnions situated at C. This is a relatively compact mechanism.

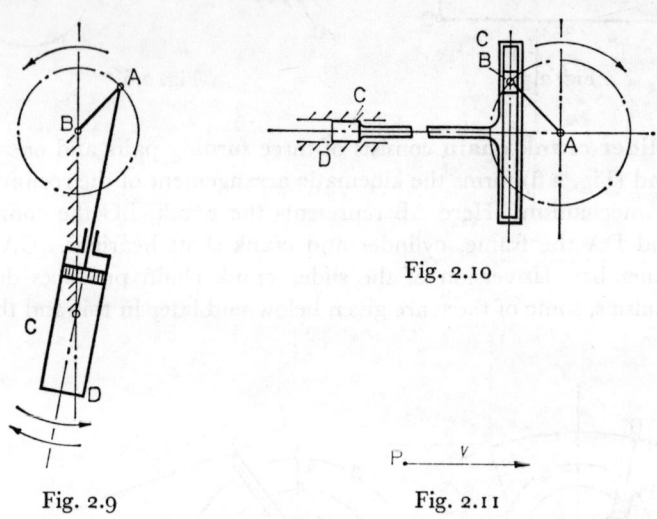

Fig. 2.10

Fig. 2.9

Fig. 2.11

Double slider crank mechanism. If the connecting rod BC in the slider crank mechanism (Fig. 2.6) could be made infinitely long, the piston C would have simple harmonic motion (see Section 2.6). In practice, alternative means must be used to obtain simple harmonic motion from rotary motion. One way is to use a double slider-crank chain in which there are two turning pairs and two sliding pairs, each two similar pairs being adjacent to each other (Fig. 2.10).

2.3 Instantaneous Centre of Velocity

For a single particle

Consider a particle P moving with velocity v (Fig. 2.11). Since the figure represents an *instantaneous* state of unconstrained motion, there is no information about the past or the future motion of P. The particle may therefore be travelling along any one of the infinite number of paths to which the specified velocity is tangential. Five such possible paths are shown in Fig. 2.12. Whatever the paths, the centre of curvature I at the instant under consideration must lie somewhere on the line YY drawn through the point P at right angles to the tangential velocity. But where along this line does the instantaneous centre of curvature lie? Everything depends on the paths along which P is travelling, and only when this

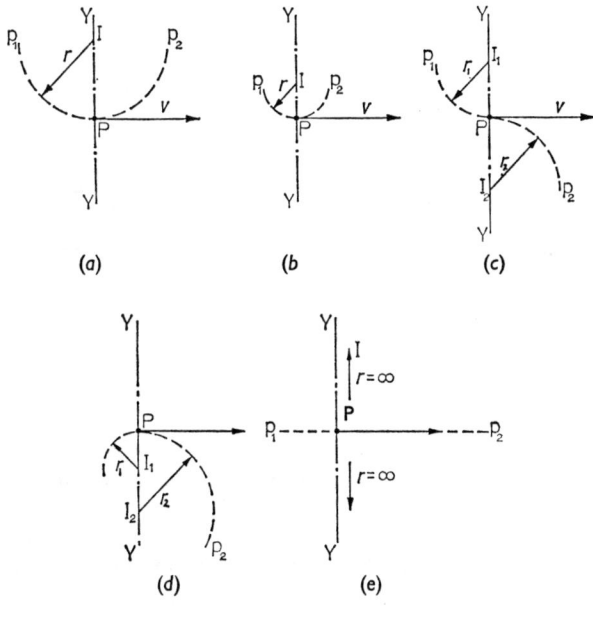

Fig. 2.12

instantaneous centre of curvature of the paths is known will it be possible to fix the centre of the velocity of P somewhere along YY. (See Fig. 2.12 (*a*) to (*e*).)

For a link

Consider next a rigid link AB which is part of a kinematic chain (Fig. 2.13). Suppose that at an instant end A has velocity v_A and end B velocity v_B. The instantaneous centre of rotation of A must lie on Y_1Y_1 drawn at right angles to v_A and the instantaneous centre of rotation of B must lie on Y_2Y_2 drawn at right angles to v_B. Because the link is rigid, the distance from A to B must always be the same which means that A and B must have a common centre of rotation I lying at the intersection of Y_1Y_1 and Y_2Y_2. But because the link is rigid the same argument must apply to all points *within* it. So that I is the instantaneous centre for the link as a whole. As the position of the link AB alters from instant to instant, so will the position of I.

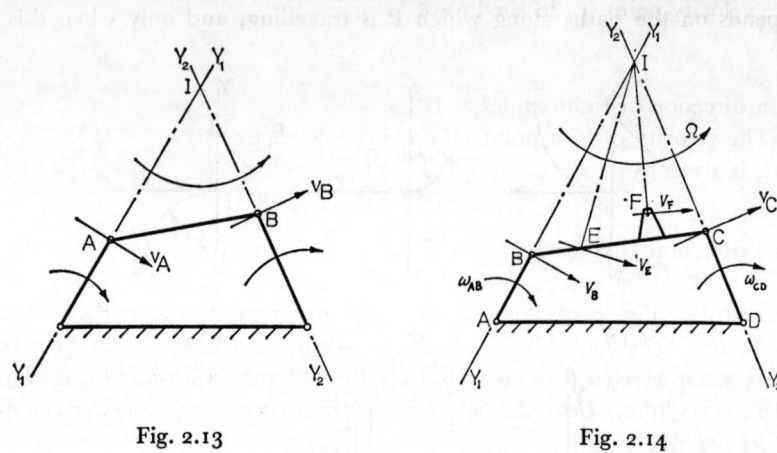

Fig. 2.13   Fig. 2.14

Determination of velocities by the instantaneous centre method

Fig. 2.14 shows an instantaneous configuration of a four-bar chain in which AD is the fixed link. Link AB is assuumed to have an instantaneous angular velocity ω_{AB} clockwise about A and end B a tangential velocity $v_B = \omega_{AB}$. AB as shown. Similarly end C of link CD will have a tangential velocity v_c as shown. But v_B and v_c are also the velocities of the extremities of link BC, so that if Y_1Y_1 and Y_2Y_2 are drawn perpendicular to v_B and v_C respectively, their intersection I is the instantaneous centre of rotation of the link *as a whole*.

Let Ω be the instantaneous angular velocity of BC about I. Then

$$v_B = \omega_{AB} . AB = \Omega . IB$$

$$\therefore \ \Omega = \omega_{AB} . (AB/IB) \qquad (2.1)$$

In the same way, $v_c = \Omega . IC$, and substituting for Ω from Eq. (2.1),

$$v_c = \omega_{AB} . (AB/IB)IC \qquad (2.2)$$

Thus if ω_{AB} is known, v_c can be found from Eq. (2.2). Also, if ω_{CD} is the instantaneous angular velocity of CD then,

$$\omega_{CD} = v_c/CD \qquad (2.3)$$

When using Eq. (2.2), the lengths of IB and IC must be measured to the same scale as that to which the kinematic chain ABCD is drawn. Similarly, the velocity point, E, in the link BC is given by

$$v_E = \Omega \,.\, IE$$

in a direction at right angles to IE.

The velocity of F, a point offset from but rigidly attached to the link BC, is given by

$$v_F = \Omega \,.\, IF$$

in a direction at right angles to IF.

Example. *Fig. 2.15 shows an instantaneous configuration of a slider crank mechanism. The crank AB = 50 mm, the connecting rod BP = 150 mm. The crank makes an angle θ = 30° with the inner dead centre position and is rotating at 1200 rev/min. Determine the velocity of the piston P and the angular velocity of the link BP.*

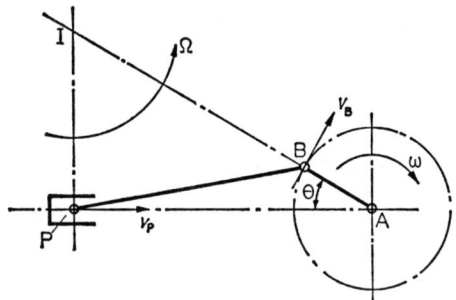

Fig. 2.15

Tangential velocity of B,

$$v_B = \omega_{AB} \,.\, AB = (2\pi 1200/60)(50/1000) = 2\pi \text{ m/s}$$

Angular velocity of BP,

$$\Omega = v_B/IB = 2\pi/(172\cdot 5/1000) = 36\cdot 5 \text{ rad/s}$$

Also $\Omega = v_P/IP$,

$$\therefore v_P = \Omega \,.\, IP = 36\cdot 5(112\cdot 5/1000) = 3\cdot 24 \text{ m/s}$$

2.4 THE TRANSMISSION OF POWER THROUGH A MECHANISM

In the slider crank mechanism shown in Fig. 2.16, AC is fixed and a force F_P is applied to the piston P causing it to move with velocity v_{PA} in the direction of the force. The force results in a thrust Q along the connecting rod PB which is exerted at the crank pin B. Force Q may be thought of as having components F_n and F_r at right angles to and along the crank BA respectively. Because the crank BA is rigid, the radial component

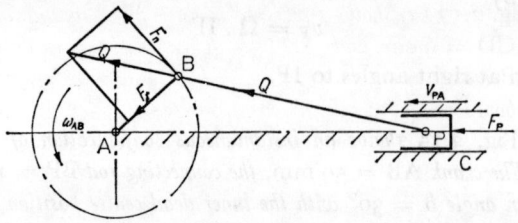

Fig. 2.16

F_r will do no work on the crank. The crank pin B has however a tangential velocity v_{BA} in the direction of F_n so that

$$\text{Power being supplied to the piston} = F_P \cdot v_{PA}$$

$$\text{Power being supplied to the crank} = F_n \cdot v_{BA}$$

If friction and the inertial effects of the links are neglected the power supplied to the piston is equal to the power transmitted to the crank, so that

$$F_P v_{PA} = F_n v_{BA}$$

Taking friction into account,

$$F_n v_{BA} = \eta F_P v_{PA}$$

where η is the efficiency of the system. But $v_{BA} = \omega_{AB} \cdot AB$, so that,

$$F_n \omega_{AB} \cdot AB = \eta F_P v_{PA}$$

or $$T_{AB} \omega_{AB} = \eta F_P v_{PA} \qquad (2.4)$$

where T_{AB} is the torque exerted at the crank AB.

2.5 Velocity Diagrams

The linear and angular velocities of points and links in mechanisms can be obtained by drawing vector diagrams in the manner explained in Sections 1.2 and 1.3. The fixed link is always the frame relative to which other velocities are measured.

Velocity diagrams for quadric-cycle and slider-crank mechanisms

Example. Fig. 2.17 (*a*) *shows a four-bar chain* ABCD *in which* AB = 6 cm, BC = 9 cm, CD = 9 cm *and* AD (*the fixed link*) = 12 cm. *If, at the instant when* AB *makes an angle of* 60° *with* AD, *it is rotating at* 180 rev/min *anticlockwise about end* A, *determine the linear velocity of* C *and the angular velocity of link* BC. *If* AB *is being turned by a torque of* 15 *newton metres applied at* A, *what will be the torque at* D *if the efficiency of the mechanism is* 85 *per cent?*

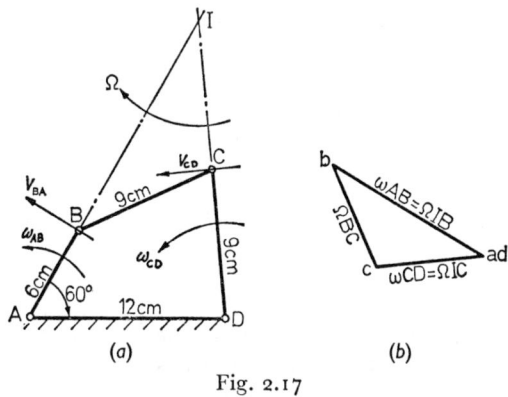

Fig. 2.17

Angular velocity of AB,

$$\omega_{AB} = 2\pi 180/60 = 6\pi \text{ rad/s}$$

Velocity of B relative to A,

$$v_{BA} = \omega_{AB} \cdot AB = 6\pi 6 = 113 \text{ cm/s}$$

The velocity of B relative to A has the sense and direction shown in Fig. 2.17 (*a*) and is represented in the velocity diagram by vector ab $\overrightarrow{(ab)}$ drawn to some suitable scale (Fig. 2.17 (*b*)). (A vector which represents the velocity of one end of a link relative to the other end is referred to as the *velocity image* of that link.) The velocity of C relative to B must be at right angles to BC, but only its direction is known. In the same way, the velocity of C relative to D will be at right angles to DC but only its direction is known.

These two tangential velocities are now drawn in the directions stated from b and d respectively. (Note that d is coincident with a since AD is fixed and therefore D has zero velocity relative to A.)

The vectors bc and dc intersect at a point which fixes C since C lies on both *bc* and *dc*. These vectors are the velocity images of links BC and CD respectively.

If $\overrightarrow{dc}$ is now measured to the same scale as $\overrightarrow{ab}$, the magnitude of the velocity of C relative to D will be found and its sense will be from d to c.

From the velocity diagram, the velocity of C is given by

$$v_{CD} = \overrightarrow{dc} = 70 \text{ cm/s}$$

Also, by the instantaneous centre method,
Angular velocity of link BC,

$$\Omega = v_{BA}/BI = 113/14 \cdot 55 = 7 \cdot 78 \text{ rad/s}$$

Also, from Section 2.4,

$$T_{AB}\omega_{AB} \cdot \eta = T_{CD}\omega_{CD}$$

where $\omega_{CD} = v_{CD}/CD = 70/9 = 7 \cdot 78 \text{ rad/s}$

$$\therefore \ T_{CD} = 15 \cdot 6\pi 0 \cdot 85/7 \cdot 78$$

$$= 30 \cdot 9 \text{ newton metres}$$

Alternative method of determining the angular velocity of a link. In the last example (Fig. 2.17 (*a*)) I is the instantaneous centre of rotation and Ω is the instantaneous angular velocity of link BC. Then

$$\overrightarrow{ab} = v_{BA} = \Omega \cdot IB$$

and will be at right angles and proportional to IB for a given value of Ω. Also,

$$\overrightarrow{dc} = v_{CD} = \Omega \cdot IC$$

and will be at right angles to and proportional to IC. The velocity triangle abc and the space triangle IBC are therefore similar, so that

$$\overrightarrow{bc} = \Omega \cdot BC \quad \text{or} \quad \Omega = \overrightarrow{bc}/BC$$

That is, the angular velocity of a link may be found without resorting to the instantaneous centre method from the relationship

Angular velocity of a link

$$= \frac{\text{Tangential velocity of one end of the link relative to the other}}{\text{Length of the link}} \quad (2.5)$$

In the last example, $\overrightarrow{bc}$ scales 70 cm/s so that

$$\text{angular velocity of BC} = \overrightarrow{bc}/\text{BC}$$
$$= 70/9 = 7 \cdot 78 \text{ rad/s}$$

as before.

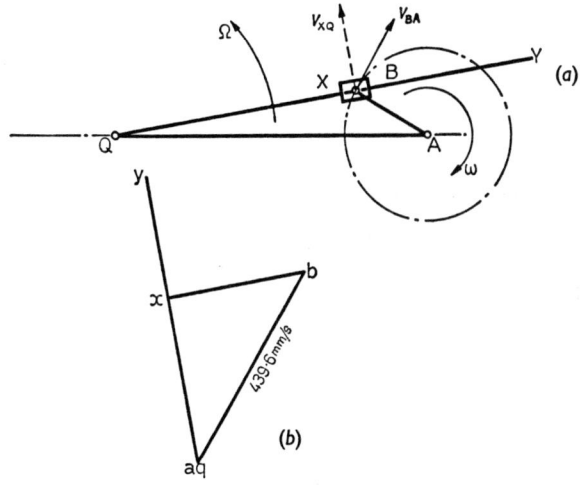

Fig. 2.18

Velocity of sliding at the surface of a connecting pin. The pin connecting two links is usually an interference fit in one of the links, rotating with it. Its angular motion relative to the other link (in which it turns freely) will be that of the relative angular motion of the two links. If, ω_1 and ω_2 are the angular velocities of the two links relative to some reference plane, then the angular velocity of one link (and the pin) relative to the other link $= \omega_1 \pm \omega_2$. The positive sign applies when the links are turning with opposite senses and the negative sign when the senses are the same. It follows that if r is the radius of the pin,

Velocity of sliding of the pin surface with mating surface of free link

$$= (\omega_1 \pm \omega_2)r \quad (2.6)$$

This velocity of sliding is sometimes referred to as the velocity of rubbing.

Problem. *The crank AB of the slider crank mechanism in Fig. 2.18 (a) rotates clockwise at 60 rev/min about A and carries a slide block at end B. The link QY is pivoted at Q and passes through the block B so that B is able to slide along QY as crank AB rotates. Determine, for the position shown, (i) the linear velocity of B relative to the link QY; (ii) the linear velocity of Y, and (iii) the angular velocity Ω of the link QY. The lengths of the links are QY = 350 mm, AB = 70 mm, QA = 260 mm. The velocity diagram is shown as Fig. 2.18 (b).*

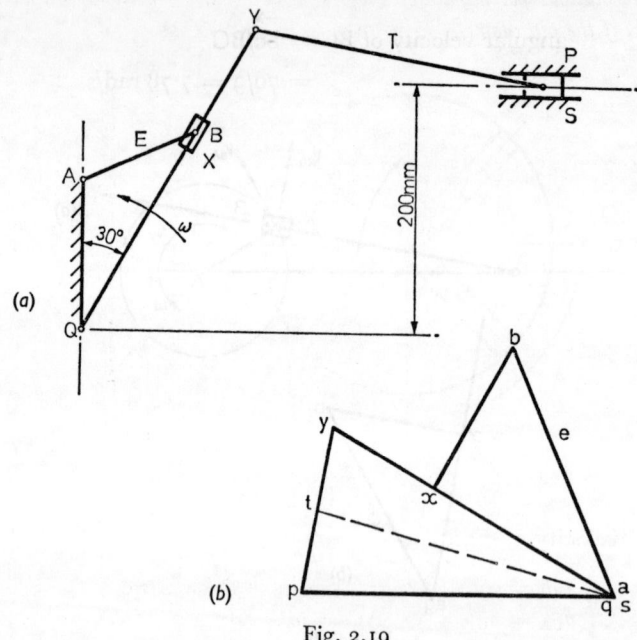

Fig. 2.19

Problem. *Fig. 2.19 (a) shows an instantaneous configuration of a slider crank mechanism in which the link QY is rotating about end Q with angular velocity 180 rev/min anti-clockwise. Slide block B is free to move along QY but is attached to end B of link AB which is pivoted at A. AQ is a fixed link. If AQ = 120 mm, AB = 100 mm, QY = 280 mm, YP = 240 mm, AE = 60 mm and angle AQY 30°, determine;*

    (i) *The velocity of sliding of B along link QY.*
   (ii) *The angular velocity of link AB.*
  (iii) *The velocity of Y.*
  (iv) *The velocity of point E on link AB.*
   (v) *The velocity of block P, which is constrained to move horizontally in slide bars S which are at rest relative to A and Q.*
  (vi) *The velocity of T and the mid-point of YP.*

The velocity diagram is shown as Fig. 2.19 (b).

Example. *In the double slider-crank mechanism in* Fig. 2.20 (*a*) A *and* B *are parallel shafts with axes* 100 mm *apart. An arm* CD, 200 mm *long is keyed to shaft* A *so that* AC = AD. *Keyed to shaft* B *is a disc with slots cut at* 90° *to each other. The arm* CD *has blocks pinned at each of its ends, the blocks sliding in the slots so that motion is transmitted from shaft* A *to shaft* B. *Shaft* A *is the driver and rotates uniformly at* 210 rev/min *anticlockwise. Draw the velocity diagram for the system when angle* BAD = 90° *and hence determine the angular velocity in revolutions per minute of shaft* B. *Also find the velocity of rubbing of the pin at* C *if the pin diameter is* 20 mm.

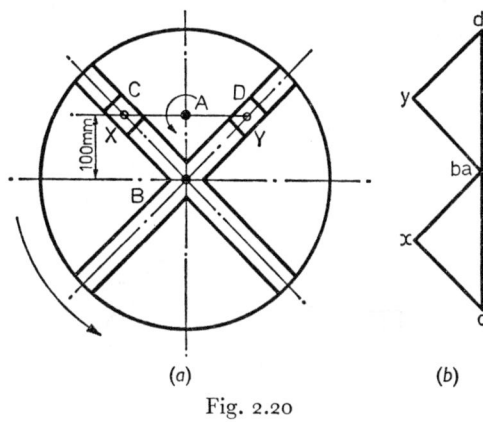

(*a*)   (*b*)

Fig. 2.20

Angular velocity of shaft A,

$$\omega = 2\pi 210/60 = 22 \text{ rad/s anticlockwise}$$
$$\therefore \ v_{CA} = v_{DA} = \omega \cdot AC = \omega \cdot AD$$
$$= 22 \times 100 = 2200 \text{ mm/s}$$

These velocities are both at right angles to CD but are opposite in sense. If X and Y are points on the slotted disc instantaneously coincident with C and D respectively, the velocity diagram can be drawn (Fig. 2.20 (*b*)) so that $\overrightarrow{ac}$ represents v_{CA} and $\overrightarrow{ad}$ represents v_{DA}. Points a and b will coincide, since shafts A and B have zero relative velocity. $\overrightarrow{xc}$ represents the velocity of sliding of C relative to X (known only in direction) and $\overrightarrow{bx}$ represents the velocity of X relative to B (known only in direction). Point x is fixed by the intersection of $\overrightarrow{xc}$ and $\overrightarrow{bx}$. Similarly $\overrightarrow{yd}$ represents the velocity of sliding of D relative to Y and $\overrightarrow{by}$ the velocity of Y relative to B. Therefore point y is fixed by the intersection of $\overrightarrow{yd}$ and $\overrightarrow{by}$.

Angular velocity of shaft B

$$= \text{(Tangential velocity of X relative to B)}/BX$$
$$= \overrightarrow{bx}/BX$$

From the velocity diagram $\overrightarrow{bx}$ scales 1555 mm/s.

From the space diagram, BX scales 140 mm:

$$\therefore \text{Angular velocity of B} = 1555/140$$
$$= 11\cdot1 \text{ rad/s} = 106 \text{ rev/min}$$

Angular velocity of end C of link DC $= \omega_1 = 22$ rad/s anticlockwise

Angular velocity of block C $=$ Angular velocity of B
$$= 11\cdot1 \text{ rad/s anticlockwise}$$
$$= \omega_2$$

Velocity of rubbing of the pin at C $= (\omega_1 - \omega_2)r$
$$= (22 - 11\cdot1)10$$
$$= 109 \text{ mm/s}$$

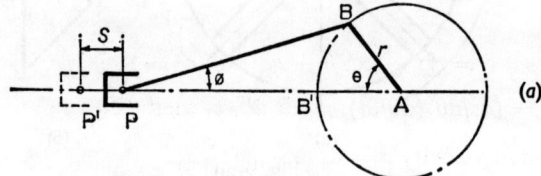

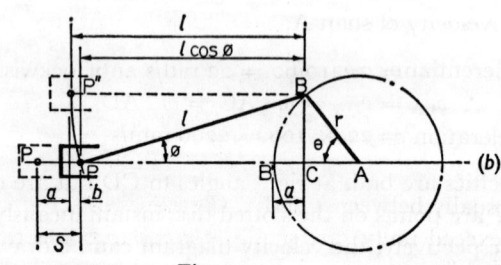

Fig. 2.21

2.6 MATHEMATICAL DETERMINATION OF PISTON DISPLACEMENT, VELOCITY AND ACCELERATION

Fig. 2.21 (*a*) shows a reciprocating engine mechanism with crank AB displaced through angle θ from the inner dead centre position AB'. Because of this angular displacement the piston P has undergone a corresponding linear displacement along the line of stroke equal to P'P $= S$. Fig. 2.21 (*b*) shows how the displacement S can be considered as made up of two displacements a and b where a is the displacement due to the left to right movement B′C of the crank pin B while the connecting rod remains parallel with the line of stroke, and b is the displacement due to the angle of obliquity ϕ of the connecting rod to the line of stroke.

Let r be the crank radius and l the length of the connecting rod.

Piston displacement
$$S = a + b$$
$$= (r - r\cos\theta) + (l - l\cos\phi)$$
$$= r(1 - \cos\theta) + l(1 - \cos\phi)$$
$$= r(1 - \cos\theta) + l[1 - \sqrt{(1 - \sin^2\phi)}]$$

but
$$\sin\phi = BC/l = r\sin\theta/l = \sin\theta/n$$

where $\qquad n = l/r$

$$\therefore \; S = r(1 - \cos\theta) + rn[1 - \sqrt{(1 - \sin^2\phi)}]$$

$$= r(1 - \cos\theta) + rn\left[1 - \frac{1}{n}\sqrt{(n^2 - \sin^2\theta)}\right]$$

$$= r[1 - \cos\theta + n - \sqrt{(n^2 - \sin^2\theta)}] \tag{2.7}$$

Since $ds/dt = (ds/d\theta)(d\theta/dt) = (ds/d\theta)\omega$, then differentiating

Piston velocity
$$v = ds/dt$$
$$= \omega r[\sin\theta + \sin 2\theta/2\sqrt{(n^2 - \sin^2\theta)}] \tag{2.8}$$

Further, $\qquad dv/dt = (dv/d\theta)(d\theta/dt) = (dv/d\theta)\omega$

so that, differentiating again,

Piston acceleration $\quad a = \omega^2 r \dfrac{\cos\theta + (n^2\cos 2\theta + \sin^4\theta)}{(n^2 - \sin^2\theta)^{3/2}}$ $\qquad$ (2.9)

Since n is usually between 4 and 5, the values of $\sin^2\theta$ and $\sin^4\theta$ (which can never exceed unity) are small by comparison with n^2. Therefore if $\sin^2\theta$ and $\sin^4\theta$ are neglected, Eqs. (2.8) and (2.9) reduce to the following expressions:

Piston velocity $\quad v = \omega r(\sin\theta + \sin 2\theta/2n)$ $\qquad$ (2.10)

Piston acceleration $\quad a = \omega^2 r(\cos\theta + \cos 2\theta/n)$ $\qquad$ (2.11)

Note. The piston velocity and acceleration will be positive when their sense is from P to A and negative when from A to P. (Fig. 2.21 (*b*)).

Example. *A reciprocating engine has a stroke of 60 mm and a connecting rod length of 120 mm. The mass of the reciprocating parts is 2 kg. If the engine speed* *is* 3,000 rev/min, *determine the velocity, acceleration and inertia force of these parts when the crank has turned through* 120° *from the inner dead centre.*

$$\omega = 2\pi 3{,}000/60 = 100\pi \text{ rad/s}$$
$$r = 30 \text{ mm} \qquad = 0{\cdot}03 \text{ m}$$
$$n = l/r \qquad = 120/30 \qquad = 4$$
$$\sin \theta = \sin 120° \qquad = +\sin 60° = +0{\cdot}8660$$
$$\cos \theta = \cos 120° \qquad = -\cos 60° = -0{\cdot}5000$$
$$\sin 2\theta = \sin 240° \qquad = -\sin 60° = -0{\cdot}8660$$
$$\cos 2\theta = \cos 240° \qquad = -\cos 60° = -0{\cdot}5000$$

Then

$$\text{Piston velocity } v = 100\pi 0{\cdot}03(0{\cdot}8660 - 0{\cdot}8660/8)$$
$$= 10\ 07 0{\cdot}03 \cdot 0{\cdot}7577$$
$$= +7{\cdot}138 \text{ m/s}$$
$$\text{Piston acceleration } a = (100\pi)^2 0{\cdot}03(-0{\cdot}5000 - 0{\cdot}5000/4)$$
$$= 10\ 000\pi^2 0{\cdot}03(-0{\cdot}625)$$
$$= -1848{\cdot}7 \text{ m/s}^2$$
$$\text{Inertia force } F = ma$$
$$= 2 \times 1848{\cdot}7$$
$$= 3697{\cdot}4 \text{ N}$$
$$= 3{\cdot}6974 \text{ kN}$$

This force will act along the line of stroke and its sense will be opposed to that of the acceleration.

PROBLEMS

For tutorials

1. When a body moves with simple harmonic motion, it is said that the body's acceleration *is* a maximum when its velocity *is* zero, at the end of each stroke. How can this be so, since acceleration is change in velocity with respect to time? Could it be that it is incorrect to use the word "is" which is part of the verb *to be*, which means *to exist*; implying passage of time. For how long *is* the body at rest? Discuss this apparent contradiction and rephrase the statement to convey the correct interpretation of the body's motion.

2. Explain carefully what is meant by the magnitude, direction and sense of a body's displacement and velocity when the body is moving (a) linearly, (b) angularly.

3. (a) Describe, with the aid of sketches if necessary, as many forms of constraint as seem possible which would produce the following relative movements:

    (i) Linear motion without rotation.
    (ii) Rotation without linear motion.
    (iii) Rotation with a proportional linear movement.

(*b*) How are the following components constrained in their movements and why?

    (i) The table of a planing machine.

    (ii) The valves of an internal combustion engine.

    (iii) A multi-collar thrust bearing.

    (iv) The races of a ball or roller bearing.

    (v) A belt on a "flat" pulley.

    (vi) A belt on a grooved pulley.

    (vii) A piston in its cylinder.

    (viii) A drill spindle.

4. Where is the instantaneous centre for each wheel of a locomotive travelling along a horizontal track? Assuming that the wheels are each 1 metre in diameter and that the locomotive has a speed of 60 kilometres per hour, determine, by drawing the necessary velocity diagrams, the magnitude, sense and direction of the instantaneous velocities of (*a*) the wheel's geometric centre and (*b*) all those points on the wheel's rim lying at 30° intervals, when one such point is in contact with the track.

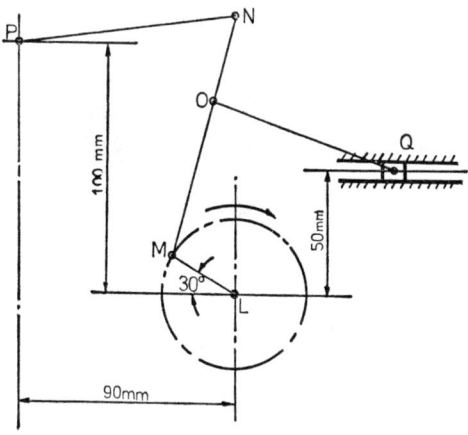

Fig. 2.22

General

1. Fig. 2.22 shows a kinematic chain in which L and P are fixed points. The crank ML rotates at 480 rev/min clockwise. The lengths of the links are ML = 30 mm, MN = 100 mm, PN = 90 mm, OQ = 80 mm and point O is 65 mm from M. For the given instantaneous position when ML is at 30° to the horizontal, find the linear velocity of the slider Q and the angular velocities of OQ and PN.

     Ans. 0·385 m/s; 19·4 rad/s clockwise; 17·4 rad/s anticlockwise

2. In the slider-crank mechanism shown in Fig. 2.23 the pivots T and M are fixed. The crank TS rotates uniformly in an anticlockwise sense at 90 rev/min.

The block S is free to slide along link MN. For the position shown, determine
(a) the velocity of sliding of S, relative to MN, and (b) the angular velocity of MN.

Ans. 2·5 m/s; 3·46 rad/s clockwise

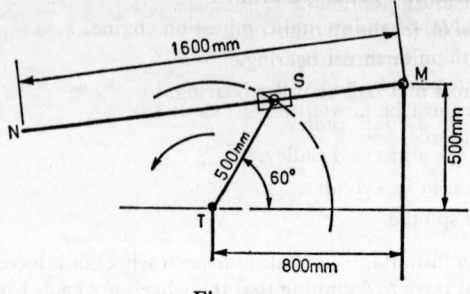

Fig. 2.23

3. Two cranks OL and OY (Fig. 2.24) are attached to co-incident shafts O.
Crank OL rotates clockwise at 360 rev/min and crank OY rotates clockwise at
720 rev/min. Slide blocks A and Z are connected to L and Y respectively by
rods AL and ZY. A slotted link AC is pin-jointed at A and is supported by a
block X over which it is free to slide. The block X is mounted in trunnions on
rod YZ at a fixed distance from Z. The blocks A and Z are each constrained to

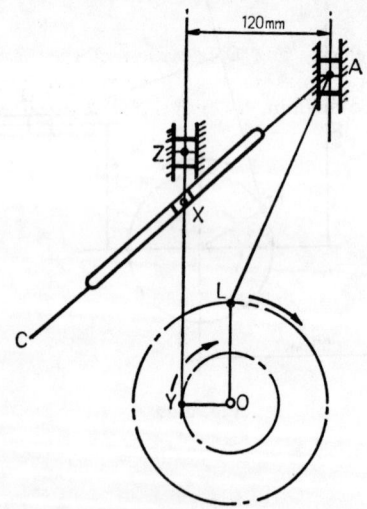

Fig. 2.24

move vertically in slides. The lengths of the links are, OL = 80 mm, OY =
40 mm, AL = 200 mm, YZ = 200 mm and AC = 320 mm. The distance of X
from Z is 40 mm.

In the position shown crank OY is 90° behind crank OL (which is vertical)
and Z is vertically above Y. Determine for this position (a) the velocity of sliding
of X relative to AC, and (b) the linear velocity of C.

Ans. (a) 1·1 m/s towards A   (b) 3·7 m/s

4. A lever-crank mechanism is shown in Fig. 2.25. The links have the following dimensions: DB = 20 mm, DC = 120 mm, AC = 60 mm and AB = 90 mm. AB is fixed. If the crank DB rotates at a uniform speed of 60 rev/min clockwise determine for the given position (*a*) the angular velocity of the connecting rod CD, (*b*) the velocity of C, (*c*) the angular velocity of AC. Use the instantaneous centre method to determine these values.

If, in the given position a force *F* of 1,000 N is being applied at C as shown, (*d*) what torque must be exerted on the shaft B to which DB is keyed, in order to overcome this force, and (*e*) what is the power required at B at this instant?

Ans. (*a*) 2·13 rad/s (*b*) 0·223 m/s (*c*) 3·72 rad/s
(*d*) 35·5 Nm (*e*) 222·9 W

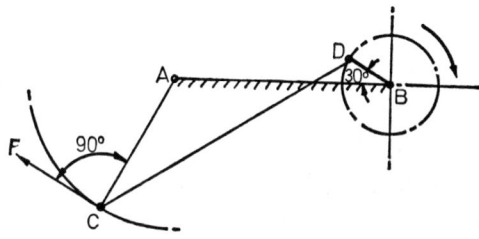

Fig. 2.25

5. Determine the velocity and acceleration of the piston of a slider crank mechanism when the crank makes an angle of 210° with its inner dead centre and the engine speed is 2400 rev/min. Length of crank 50 mm, length of connecting rod 200 mm.

Ans. −4·92 m/s; −2338 m/s²

CHAPTER 3

Complex Plane Mechanisms—2

3.1 ACCELERATION DIAGRAMS

The preparation of acceleration diagrams usually involves, as a necessary preliminary, space and velocity diagrams.

The vector characteristics of an acceleration diagram are essentially similar to those of velocity diagrams, the chief difference being that points on links may have two components of acceleration—tangential and centripetal. Such components must be added vectorially to give the total acceleration relative to the frame of reference. Sometimes, of course, there is only one component to be considered. Points which rotate with uniform motion experience only centripetal acceleration and points which move in straight paths have only tangential acceleration. As well as these relatively simple cases, certain points in a mechanism may move with combined radial and angular velocities when two additional components are involved—the Coriolis component of acceleration and the radial component of acceleration (see Section 3.3).

As with velocity diagrams, the first step is to consider those links and points about which most information is known or obtainable. The structure of the acceleration diagram is then extended to the remainder of the mechanism by the intersection of vectors to give new points of reference.

Acceleration diagrams for a single link

Link AB in Fig. 3.1 (*a*) is assumed to be rotating clockwise about A with *uniform* angular velocity ω rad/s. End B, therefore, has a centripetal acceleration directed towards A equal to $\omega^2 . AB$. This is the only and therefore the total acceleration of B relative to A. The acceleration diagram for the link can now be drawn (Fig. 3.1 (*b*)) using the conventions employed in Section 1.3. Thus the total acceleration of B relative to A is represented by $\overrightarrow{a_1b_1}$.

In general, the vector representing the total acceleration of one end of a link relative to the other is called the *acceleration image* of the link and is denoted by lower case letters with strokes as subfixes.

If link AB at the instant under consideration has an angular acceleration α rad/s² about A as well as an angular velocity ω rad/s (Fig. 3.2 (*a*)), then end B relative to A will have a second component of acceleration (a tangential component) equal to α . AB directed at right angles to AB. Its sense can be derived by observation from the sense of the angular acceleration. The components are drawn out as in Fig. 3.2 (*b*) and are represented as follows:

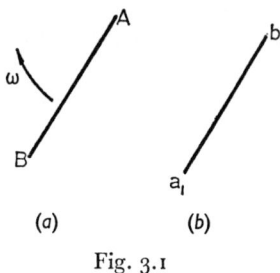

(*a*)   (*b*)

Fig. 3.1

$$\text{Centripetal acceleration of B relative to A} = \overrightarrow{a_1b_a}$$

$$\text{Tangential acceleration of B relative to A} = \overrightarrow{b_ab_1}$$

The two components will always be at right angles to each other, one tangential to B's path and the other directed radially inwards towards A. It will be seen that in most problems, the centripetal component can be determined in magnitude, direction and sense, but that the tangential component is known only in direction. It is therefore usual to start by

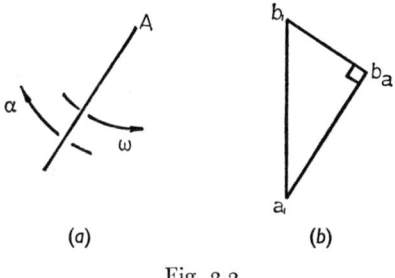

(*a*)   (*b*)

Fig. 3.2

drawing the centripetal components and adding vectors at right angles to denote the directions of the tangential components, the magnitudes of which are determined by their intersection with other vectors. The junction of a centripetal and a tangential component is indicated by the two letters relating to the particular link, one as a subfix. As before, the total acceleration of B relative to A in Fig. 3.2 (*b*) is given by $\overrightarrow{a_1b_1}$ and is the acceleration image of the link AB.

Acceleration of in-line and offset points. As for velocities, the total acceleration of points within or offset from a link can be obtained by similarity.

Example. Fig. 3.3 (*a*) *shows a link EF of length* 200 mm *with angular velocity* 2 rad/s *and a angular acceleration* 3 rad/s^2 *about E. Point S lies on* EF 150 mm *from E. Two other points T and U are rigidly connected to the link, T on* EF *continued and U offset from the link. Find the total accelerations of F, S, T and U relative to E.*

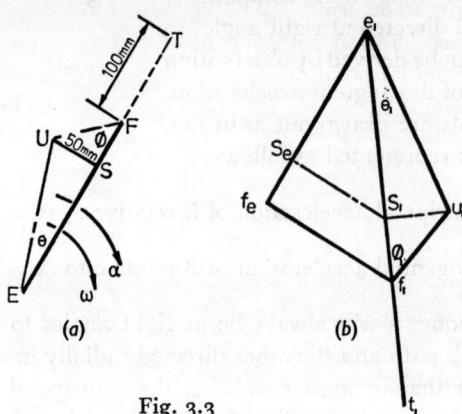

Fig. 3.3

Centripetal acceleration of F relative to E $= \omega^2$. FE

$$= 2^2 \times 200 = 800 \text{ mm/s}^2$$

$$= \overrightarrow{e_1 f_e}$$

Tangential acceleration of F relative to E $= \alpha$. FE

$$= 3 \times 200 = 600 \text{ mm/s}^2$$

$$= \overrightarrow{f_e f_1}$$

Then, from the acceleration diagram Fig. 3.3 (*b*),

Total acceleration of F relative to E $= \overrightarrow{e_1 f_1}$

$$= 1000 \text{ mm/s}^2$$

The centripetal and tangential accelerations of S relative to E are given by ω^2 . SE and α . SE respectively and are represented by $\overrightarrow{e_1 s_e}$ and $\overrightarrow{s_e s_1}$. If these are superimposed on the existing acceleration diagram for the link FE, commencing at e_1 they will be found to give a total acceleration of S relative to $E = \overrightarrow{e_1s_1}$ where $\overrightarrow{e_1s_1}/\overrightarrow{e_1f_1} = ES/EF$. It follows that

$$\overrightarrow{e_1s_1} = (ES/EF)\overrightarrow{e_1f_1}$$
$$= (150/200)1000 = 750 \text{ mm/s}^2$$

When finding the total acceleration of an inset point such as S, it is not necessary therefore to determine its component accelerations but sufficient to fix s_1 on the acceleration image of the link in a position similar to that occupied by the actual point on the actual link.

Similarly for point t_1,

$$\overrightarrow{e_1t_1}/\overrightarrow{e_1f_1} = ET/EF$$

Thus $\overrightarrow{e_1t_1} = (300/200)1000 = 1500 \text{ mm/s}$

Similarly, for the offset point u_1, where angle $\theta_1 = \theta$ and angle $\phi_1 = \phi$, so that the vector triangle $e_1u_1f_1$ is similar to the space triangle EUF. Then

$$\text{Total acceleration of U relative to } E = \overrightarrow{e_1u_1}$$
$$= 800 \text{ mm/s}^2$$

Acceleration diagrams for mechanisms

The following examples illustrate this general method of determining the acceleration of points and links in mechanisms.

Example. *Fig. 3.4 (a) is an instantaneous configuration of a slider crank mechanism. The crank AB rotates clockwise at 180 rev/min about the fixed centre A, the slide block E is driven through link DE which is attached at the mid-point D of the connecting rod BC. The slide block C reciprocates horizontally and the slide block E vertically. AB = 90 mm, BC = 270 mm and DE = 200 mm.*

Determine the linear velocity and acceleration of E and the angular velocity and acceleration of the link DE.

$$\text{Angular velocity } \omega \text{ of link AB} = 2\pi180/60$$
$$= 6\pi \text{ rad/s}$$
$$\text{Velocity of B relative to A} = \omega . \text{AB}$$
$$= 6\pi90/1000$$
$$= 1 \cdot 69 \text{ m/s}$$

The velocity diagram (Fig. 3.4 (*b*)) gives (by scaling):

$$\text{Velocity of C relative to B} = \overrightarrow{bc} = 1 \cdot 55 \text{ m/s}$$
$$\text{Velocity of E relative to D} = \overrightarrow{de} = 1 \cdot 17 \text{ m/s}$$

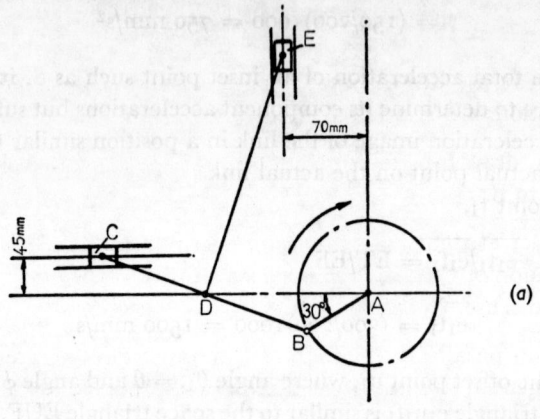

(*a*)

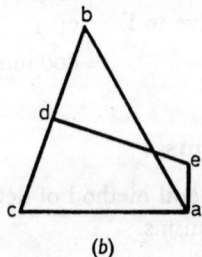

(*b*)

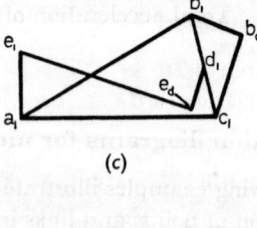

(*c*)

Fig. 3.4

The centripetal components of acceleration of each link can now be calculated.

$$\text{Centripetal (total) acceleration of B relative to A} = \overrightarrow{(ab)^2}/AB$$
$$= 1 \cdot 69^2/0 \cdot 09$$
$$= 31 \cdot 8 \text{ m/s}^2$$

$$\text{Centripetal acceleration of C relative to B} = \overrightarrow{(bc)^2}/CB$$
$$= 1 \cdot 55^2/0 \cdot 27$$
$$= 89 \text{ m/s}^2$$

$$\text{Centripetal acceleration of E relative to D} = \overrightarrow{(de)^2}/ED$$
$$= 1 \cdot 17^2/0 \cdot 2$$
$$= 6 \cdot 85 \text{ m/s}^2$$

These accelerations are directed radially inwards along their respective links. The corresponding tangential accelerations are unknown in magnitude and sense, but they must occur at right angles to their respective links. The slide blocks C and E, which move along straight paths, will have only tangential (total) accelerations relative to the fixed point A in known directions. The acceleration diagram can be drawn as follows (Fig. 3.4 (c)).

(i) Starting at a_1, draw $\overrightarrow{a_1b_1}$ to represent the centripetal (total) acceleration of B relative to A.

(ii) From b_1, draw $\overrightarrow{b_1b_c}$ to represent the centripetal acceleration of C relative to B.

(iii) Draw $\overrightarrow{b_cc_1}$ at right angles to link BC to represent the tangential acceleration of C relative to B. Because it is unknown in magnitude and sense, C_1 cannot yet be fixed.

(iv) Draw $\overrightarrow{a_1c_1}$ parallel to the path of C to represent the direction of the tangential (total) acceleration of C relative to A. The position of c_1 is now located at the point of intersection of $\overrightarrow{b_cc_1}$ with $\overrightarrow{a_1c_1}$.

(v) Join b_1 to c_1. Then $\overrightarrow{b_1c_1}$ represents the total acceleration of C relative to B and is the acceleration image of link BC.

(vi) The position of d_1 on $\overrightarrow{b_1c_1}$ can now be fixed because $\overrightarrow{b_1d_1}/\overrightarrow{b_1c_1} =$ BD/BC.

(vii) From d_1, draw $\overrightarrow{d_1e_d}$ to represent the centripetal acceleration of E relative to D.

(viii) From e_d, draw $\overrightarrow{e_de_1}$ at right angles to $\overrightarrow{d_1e_d}$ to represent the tangential acceleration of E relative to D. Because this acceleration is known only in direction, the position of e_1 cannot yet be located.

(ix) Draw $\overrightarrow{a_1e_1}$ parallel to the path of E to represent the direction of the tangential (total) acceleration of E relative to A. The position of e_1 is now fixed by the intersection of $\overrightarrow{e_de_1}$ and $\overrightarrow{a_1e_1}$.

From the velocity diagram:

$$\text{Velocity of E (relative to A)} = \overrightarrow{ae} = 0.35 \text{ m/s}$$

with sense and direction a to e.

Angular velocity of link DE $= \dfrac{\text{Tangential velocity of E relative to D}}{\text{ED}}$

$= \overrightarrow{de}/\text{ED}$

$= 1\cdot15/0\cdot20 = 5\cdot75$ rad/s clockwise

From the acceleration diagram:

Total acceleration of E (relative to A) $= \overrightarrow{a_1 e_1}$

$= 10\cdot8$ m/s^2

with sense and direction a_1 to e_1.

Angular acceleration of link DE

$= \dfrac{\text{Tangential acceleration of E relative to D}}{\text{ED}}$

$= \overrightarrow{e_d e_1}/\text{ED}$

$= 29/0\cdot20 = 145$ rad/s^2 anticlockwise

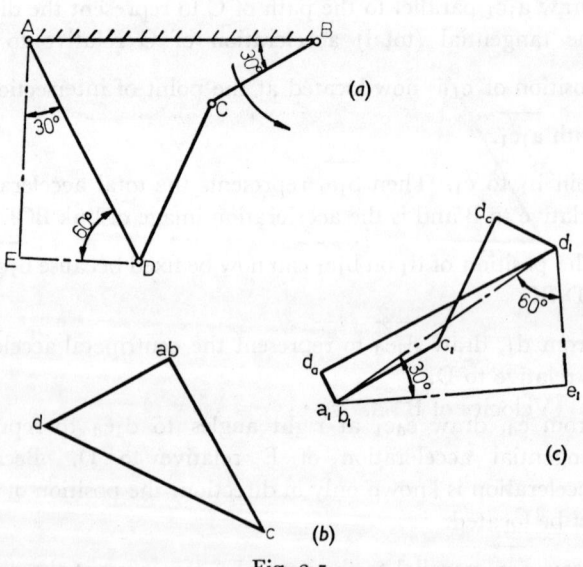

Fig. 3.5

Problem. Fig. 3.5 (*a*) *shows a four-bar chain. Link CB is rotating about the fixed end B with a uniform angular velocity of* 300 *rev/min anticlockwise. Point E is rigidly connected to link AD in the position shown.* AB (*fixed*) $= 2\cdot4$ m, CB $= 1\cdot0$ m, CD $= 1\cdot4$ m, AD $= 2\cdot0$ m. *Determine the acceleration of* E *and the angular acceleration of link AD. Figs.* 3.5 (*b*) *and* (*c*) *are the velocity and acceleration diagrams.*

The Slider Crank Chain (Fig. 2.6) forms the kinematic linkage of the reciprocating mechanism in most steam and internal combustion engines. The velocity and acceleration diagrams can be prepared by the general method, although more direct methods are sometimes preferable (see Section 3.2).

Example. *In the slider crank mechanism Fig. 3.6 (a), crank AB is at 30° to the inner dead centre and turning uniformly clockwise at 240 rev/min. The lengths of the links are AB = 60 mm, BC = 210 mm. The piston C reciprocates horizontally along the line of stroke AD. Determine the velocity and acceleration of the piston C relative to A.*

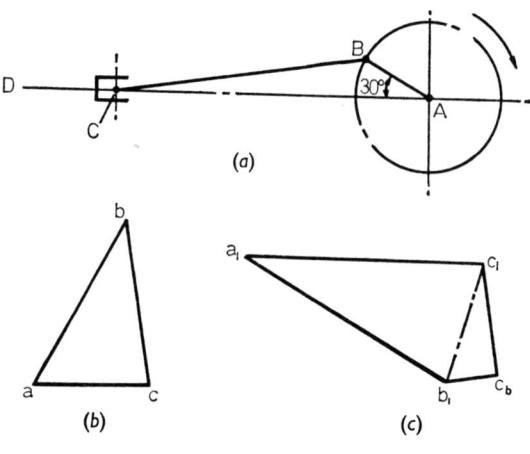

Fig. 3.6

Velocity of B relative to A $= \omega \cdot AB$

$$= (2\pi 240/60)(60/1000)$$

$$= 1\cdot507 \text{ m/s}$$

$$\overrightarrow{= ab}$$

The velocity diagram Fig. 3.6 (b) can now be drawn and the following values obtained:

Velocity of C relative to B, $\overrightarrow{bc} = 1\cdot325$ m/s

Velocity of C relative to A, $\overrightarrow{ac} = 0\cdot950$ m/s

Then, centripetal (total) acceleration of B relative to A $= v_{BA}^2/AB$
$$= 1 \cdot 507^2/(6/100)$$
$$= 37 \cdot 85 \text{ m/s}^2$$

Centripetal acceleration of C relative to B $= v_{CB}^2/CB$
$$= 1 \cdot 325^2/(21/100)$$
$$= 8 \cdot 35 \text{ m/s}^2$$

The acceleration diagram (Fig. 3.6 (*c*)) is drawn as follows:

(i) Draw $\overrightarrow{a_1 b_1} = 37 \cdot 85$ m/s² to represent the centripetal (total) acceleration of B relative to A.

(ii) Draw $\overrightarrow{b_1 c_b} = 8 \cdot 35$ m/s² to represent the centripetal acceleration of C relative to B.

(iii) Draw $\overrightarrow{c_b c_1}$ at right angles to CB to represent the tangential acceleration of C relative to B (magnitude not known).

(iv) Draw $\overrightarrow{a_1 c_1}$ parallel to the line of stroke to represent the tangential (total) acceleration of C relative to A (magnitude not known).

The intersection of $\overrightarrow{a_1 c_1}$ and $\overrightarrow{c_b c_1}$ fixes C_1.

From the diagram:

$$\text{Acceleration of C relative to A} = \overrightarrow{a_1 c_1}$$
$$= 38 \text{ m/s}^2$$

3.2 Reciprocating Engine Mechanism

In the design of reciprocating engine mechanisms it is sometimes necessary to obtain velocity and acceleration diagrams for several crank angle positions. These can be obtained most conveniently by drawing them directly onto the configuration diagrams.

Velocity diagrams

Fig. 3.7 is a reproduction of the slider crank mechanism shown in Fig. 3.6 (*a*) for which the velocity and acceleration diagrams have already been drawn. To obtain the velocity diagram, let the connecting rod CB (produced if necessary) intersect a perpendicular to the line of· stroke through A in point D. *Then triangle ABC represents the velocity diagram for that particular configuration.*

To prove this, the instantaneous centre I of the connecting rod is determined (see Section 2.3):

$$v_{CA} = v_{BA} \cdot (IC/IB) \tag{3.1}$$

But ICB and ADB are similar triangles (all three angles equal).

$$\therefore \ IC/IB = AD/AB$$

Substituting in Eq. (3.1),

$$v_{CA} = v_{BA} \cdot (AD/AB)$$

but
$$v_{BA} = \omega \cdot AB$$

$$\therefore \ v_{CA} = \omega \cdot AB \cdot (AD/AB)$$

$$= \omega \cdot AD$$

This means that the piston velocity v_{CA} can be obtained by measuring to scale the intercept AD and then multiplying by ω, the angular velocity of the crank.

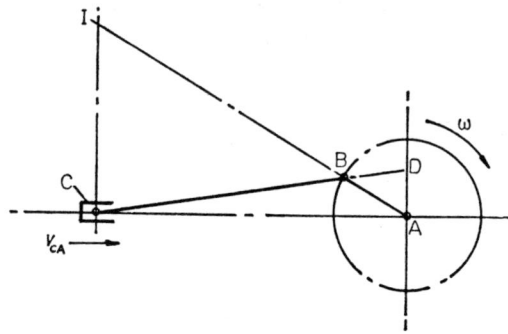

Fig. 3.7

The triangle BDA of Fig. 3.7 and the triangle of velocities bca of Fig. 3.6 (*b*) are similar since the sides AB, BD and DA are at right angles to the sides ab, bc and ac respectively, which means that the angles of the two triangles are respectively equal. Therefore, because

$$v_{CA} = \overrightarrow{ac} = \omega \cdot AD$$

it follows that
$$\overrightarrow{ab} = \omega \cdot AB$$

and
$$\overrightarrow{bc} = \omega \cdot BD$$

The triangle ABD may therefore be used to represent the velocity diagram provided that each side (measured on the scale to which the mechanism is drawn) is multiplied by ω.

Acceleration diagrams

Two of the better known constructions are given below. Their proofs (omitted) are based upon the similarity of the diagrams obtained by the "direct" and the "general" methods.

Klein's construction. Figs. 3.8 (*a*) and 3.9 (*a*) show a reciprocating engine mechanism with its crank in two different positions. The crank shaft AB is assumed to have uniform angular velocity. Klein's construction, valid for any position of the crank, is as follows.

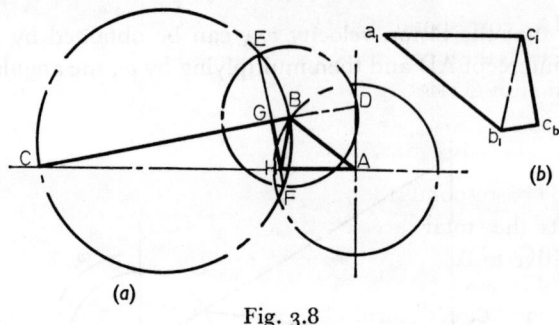

Fig. 3.8

With the mid-point of the connecting rod BC as centre, describe a circle with BC as diameter.

Let the connecting rod BC (produced if necessary) meet a vertical through A in D.

With B as centre and radius BD, describe circle DEF to intersect the previous circle CEF in E and F.

Join E to F and produce if necessary to meet the line of stroke CA in H.

Let G be the point where EF intersects BC. Then the quadrilaterals ABGH represent the acceleration diagrams of the mechanism for the two positions shown.

Figs. 3.8 (*b*) and 3.9 (*b*) show the diagrams produced by the "general" method. Comparison of the two forms of diagram will indicate their similarity, but it is left to the student to prove that they are similar.

It will be seen that side a_1b_1 of the diagrams produced by the general method corresponds with side BA of the "direct" diagram. But $\overrightarrow{a_1b_1}$ represents the centripetal (total) acceleration of B relative to A. Moreover the centripetal acceleration of B relative to A is given by $\omega^2 \cdot BA$, that is, $\omega^2 . \overrightarrow{BA}$ corresponds to a_1b_1 and represents the centripetal (total) acceleration of B relative to A, and, since the diagrams are similar, it follows that:

$\omega^2 . \overrightarrow{BG}$ corresponds to b_1c_b and represents the centripetal acceleration of C relative to B.

$\omega^2 . \overrightarrow{GH}$ corresponds to c_bc_1 and represents the tangential acceleration of C relative to B.

$\omega^2 . \overrightarrow{HA}$ corresponds to a_1c_1 and represents the tangential (total) acceleration of C relative to A.

Similarly,

$\omega^2 . \overrightarrow{BH}$ corresponds to b_1c_1 and represents the total acceleration of C relative to B.

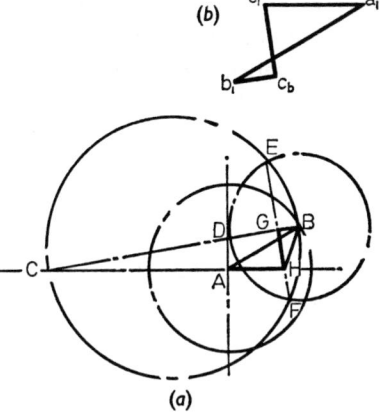

Fig. 3.9

In using the "direct" form of diagram, the sides of the figure must be measured to the same scale as that used for drawing the mechanism.

Note. Klein's construction is applicable only when the crank can be assumed to be rotating with *uniform* angular velocity. This condition however includes maximum speeds when the inertia forces on the links are greatest, so that the construction is most suitable for design purposes.

As the crank assumes different positions, the radius BD of circle DEF will vary from a maximum when it is equal to BA in both dead centre positions to zero when the crank angle is 90°. In the latter position B, D, E, F and G will coincide and the chord EF (produced) of circle CEBF becomes, in this limiting case, a tangent at point B to circle CEBF and is drawn as before to meet CA in H.

The relative positions of the links at which the piston C has zero acceleration are not easily determined geometrically. Klein's construction shows however that when the crank and connecting rod are at right angles to each other, the piston acceleration is *almost* zero. For most practical purposes this position is assumed to give zero piston acceleration.

Ritterhaus's construction. This construction gives acceleration diagrams identical with those obtained by Klein's method, but the construction, although simple, is not as convenient for certain crank angles. In Fig. 3.10:

Let the connecting rod CB (produced if necessary) intersect a perpendicular to the line of stroke through A in D.

Draw DF parallel to the line of stroke to meet crank AB, produced if necessary.

From F draw FG at right angles to the line of stroke to meet BC in G.

From G draw GH at right angles to BC to meet the line of stroke in H.

The quadrilateral ABGH again represents the acceleration diagram

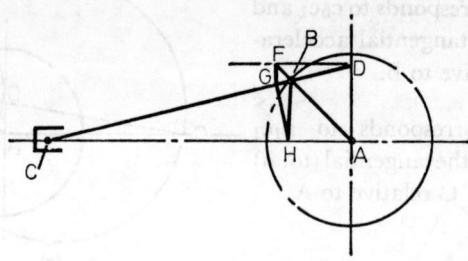

Fig. 3.10

for the mechanism. As with Klein's construction, each side must be measured on the scale to which the mechanism is drawn and then multiplied by ω^2 to give the required acceleration.

Ritterhaus's construction, like Klein's applies only when the crank has uniform angular velocity.

3.3 COMBINED RADIAL AND ANGULAR VELOCITIES

From Section 1.11, a body with both angular and radial motions suffers the Coriolis component of acceleration which must be added vectorially to whatever centripetal and tangential accelerations there may be. If the radial motion is not uniform, a fourth component called the acceleration of sliding or radial acceleration must also be taken into account.

Problem. Fig. 3.11 (a) shows a quick return mechanism in which crank OS rotates about O so as to cause a slotted link AB to rotate about Q. End A is joined by a connecting rod AC to a slider C. The slider C reciprocates in fixed guides. The lengths of the links are, OS = 200 mm, AB = 525 mm, AC =

600 mm *and* AQ = 200 mm. *The crank* OS *rotates clockwise at* 120 rev/min. *Determine the velocity and acceleration of C and the angular velocity and acceleration of link* AC. *Velocity and acceleration diagrams are shown in* Figs. 3.11 (*b*) *and* (*c*).

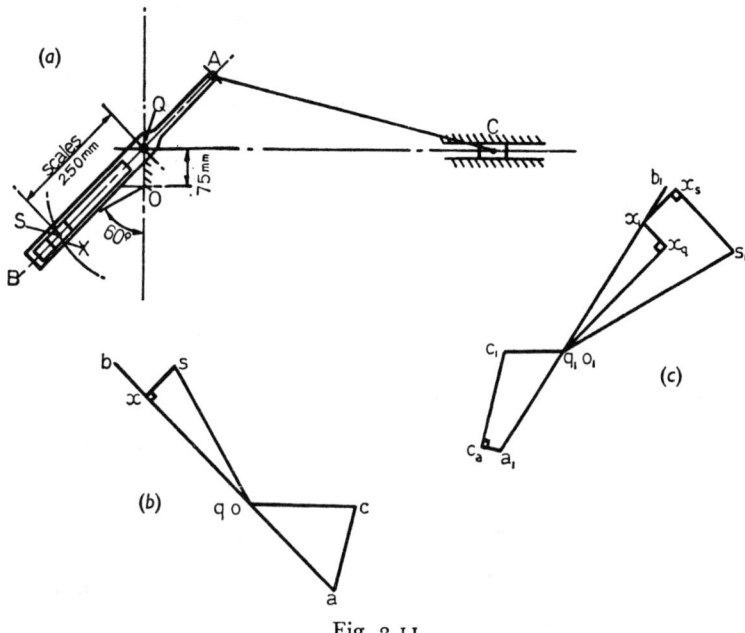

Fig. 3.11

Example. *Two parallel shafts* A *and* B (Fig. 3.12 (*a*)) *are set with their axes* 20 cm *apart. Shaft* A *carries an arm* CD 40 cm *long and* AC = AD. *Shaft* B *carries a slotted disc, the slots being at* 90° *to each other. The arm* CD *has blocks pinned at each end and these blocks slide in the slots so that motion is transmitted from* A *to* B. *Shaft* A *is the driver and rotates at* 210 rev/min *clockwise.*

Draw velocity and acceleration diagrams for the system at the instant when angle BAD = 45° *and hence determine the angular velocity in revolutions per minute and the angular acceleration of the shaft* B.

This is an example of the double slider crank mechanism. Let X and Y be points instantaneously coincident with C and D.

$$\text{Angular velocity of shaft A} = \omega = 2\pi 210/60$$
$$= 22 \text{ rad/s}$$
$$\text{Velocity of C relative to A} = \omega \cdot \text{CA} = 22(20/100)$$
$$= 4\cdot 4 \text{ m/s}$$
$$\text{Velocity of D relative to A} = \omega \cdot \text{DA} = 22(20/100)$$
$$= 4\cdot 4 \text{ m/s}$$

The velocity diagram (Fig. 3.12 (*b*)) can now be drawn and gives:

$$\text{Velocity of X relative to B} = v_{XB} = \overrightarrow{bx} = 4 \text{ m/s}$$

$$\text{Velocity of Y relative to B} = v_{YB} = \overrightarrow{by} = 1\cdot8 \text{ m/s}$$

$$\text{Velocity of sliding of C relative to X} = \overrightarrow{xc} = 1\cdot8 \text{ m/s}$$

$$\text{Velocity of sliding of D relative to Y} = \overrightarrow{yd} = 4 \text{ m/s}$$

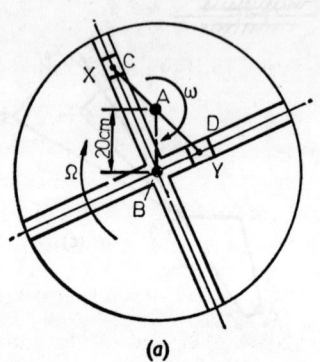

(*a*)

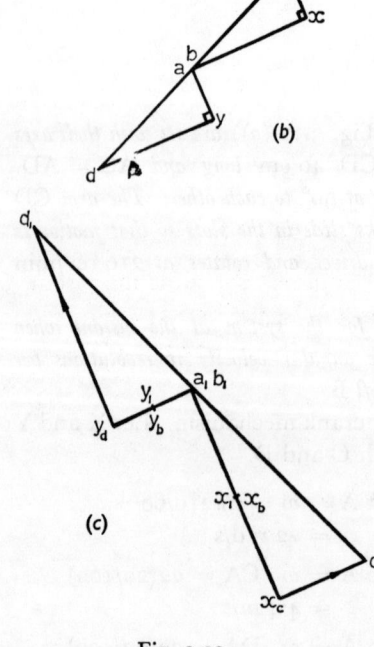

(*b*)

(*c*)

Fig. 3.12

On the space diagram

> BY scales 16 cm
> BX scales 37 cm

Centripetal (total) acceleration of

$$\text{C relative to A} = \overrightarrow{a_1c_1}$$
$$= (ac)^2/CA$$
$$= 4\cdot4^2/(20/100)$$
$$= 96\cdot8 \text{ m/s}^2$$

Centripetal (total) acceleration of

$$\text{D relative to A} = \overrightarrow{a_1d_1}$$
$$= (ad)^2/DA$$
$$= 4\cdot4^2/(20/100)$$
$$= 96\cdot8 \text{ m/s}^2$$

Centripetal acceleration of

$$\text{X relative to B} = \overrightarrow{b_1x_b}$$
$$= (bx)^2/XB$$
$$= 4^2/(37/100)$$
$$= 43\cdot2 \text{ m/s}^2$$

Centripetal acceleration of

$$\text{Y relative to B} = \overrightarrow{b_1y_b}$$
$$= (by)^2/YB$$
$$= 1\cdot8^2/(16/100)$$
$$= 20\cdot3 \text{ m/s}^2$$

Angular velocity

$$\text{of slotted disc} = \Omega = v_{XB}/XB$$
$$= 4\cdot06/(37/100)$$
$$= 11 \text{ rad/s}$$

Coriolis component of C relative to $X = \overrightarrow{x_c c_1}$
$$= 2\Omega v_{CX}$$
$$= 2 \times 10{\cdot}8 \times 1{\cdot}8$$
$$= 38{\cdot}8 \text{ m/s}^2$$

Coriolis component of D relative to $Y = \overrightarrow{y_d d_1}$
$$= 2\Omega v_{DY}$$
$$= 2 \times 10{\cdot}8 \times 4$$
$$= 86{\cdot}4 \text{ m/s}^2$$

The tangential accelerations of X and Y relative to B are represented in the acceleration diagram (Fig. 3.12 (*c*)) by $\overrightarrow{x_b x_1}$ and $\overrightarrow{y_b y_1}$ respectively, and are paired with the corresponding centripetal accelerations. Points x_1 and x_b will be found to coincide, as will points y_1 and y_b, indicating that both these tangential accelerations are, in fact, zero. The sliding accelerations of C relative to X and D relative to Y and known only in direction. They are represented in the acceleration diagram by $\overrightarrow{x_1 x_c}$ and $\overrightarrow{y_1 y_d}$. The Coriolis components of C relative to X and of D relative to Y are drawn in backwards from c_1 and d_1 respectively and paired at right angles with the corresponding accelerations of sliding.

Angular velocity of shaft B, $\Omega = 11$ rad/s clockwise

Angular acceleration of shaft B

$$= \frac{\text{Tangential acceleration of X relative to B}}{XB}$$

$$= \overrightarrow{x_b x_1}/XB$$
$$= 0(37/100)$$
$$= 0$$

The mechanism is therefore transmitting uniform motion at the instant considered, and it can be shown that this will be the case at *any* instant during the complete rotation of shaft A.

3.4 THE DYNAMICAL PROPERTIES OF A LINK

Acceleration of a link due to applied forces

For the link PQ with centre of gravity at G (Fig. 3.13), let

$m =$ the mass,
$k =$ the radius of gyration of about G, and
$I =$ the moment of inertia of the link about G.

Suppose that at an instant, the link is acted upon by forces S and T exerted by other links which can be represented by a single resultant force F. It is unlikely that the line of action of F will pass through G. If equal and opposite forces of magnitude F are imagined to be applied at G (Fig. 3.14), the *additional* resultant force is zero but the three force system is now equivalent to

(*a*) a force F at G causing a linear acceleration a where

$$F = ma \qquad (3.2)$$

(*b*) a couple or torque $T = Fb$ about G causing an angular acceleration α, where

$$T = I\alpha$$
$$= mk^2\alpha \qquad (3.3)$$

The force F and the resultant accelerations a and α are shown in Fig. 3.15.

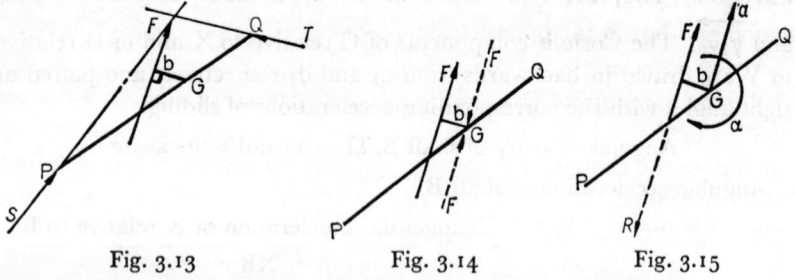

Fig. 3.13   Fig. 3.14   Fig. 3.15

Inertial reaction of a link

The link PQ will resist the change of motion imposed upon it by a reaction R equal to F in magnitude but opposite in sense (Fig. 3.15) which is called the *inertial reaction of the link*.

Example. *A link AB of length* 2 m *is acted upon by forces of* 3 kN *and* 4 kN *at ends A and B respectively in the directions shown* (Fig. 3.16). *Link centre of gravity is G,* 0·6 m *from A. If the mass of the link is* 300 kg *and its radius of gyration about G is* 0·4 m, *determine the inertia reaction and the linear and angular accelerations of the link.*

The lines of action of the applied forces meet at o and from the parallelogram of forces opqr, the resultant force F is found to be 5 kN. The line of action of F passes between G and B and is found by measurement to be at a perpendicular distance of 0·25 m from G. Hence $b = 0·25$ m.

From Eq. (3.2):

Linear acceleration of

$$G = a = F/m$$
$$= 5000/300$$
$$= 16\cdot66 \text{ m/s}^2$$

(in the direction of F)

From Eq. (3.3):

Angular acceleration

$$\alpha = T/I$$
$$= Fb/mk^2$$
$$= 5000 \times 0\cdot25/300 \times 0\cdot4^2$$
$$= 26\cdot04 \text{ rad/s}^2$$

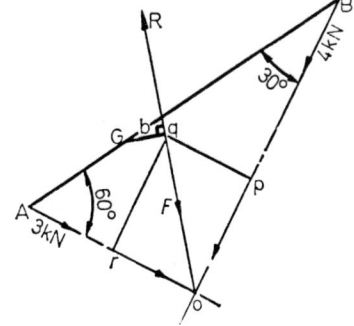

Fig. 3.16

(clockwise in accordance with the sense of the torque Fb).
The inertial reaction R of the link will be 5 kN acting along the line of
action of F but in the opposite sense (see Fig. 3.16).

D'Alembert's principle. The reaction R of the link PQ (Fig. 3.15) is not
applied to the link by some external agent but is simply the reaction of
the link to the change of its motion. The link, moving as it is with non-
uniform motion, cannot therefore normally be treated as a body in
static equilibrium. D'Alembert pointed out, however, that if the inertial
reactions of a body are *imagined to be* externally applied forces and torques
equal and opposite to the actual forces and torques which cause acceleration (their equilibrants), then the body can be treated as though it were
in static equilibrium under the action of the system of forces. In Fig. 3.16,
the two external forces actually applied at A and B (or their resultant F)
and the reaction R imagined as an externally applied force can be considered as a sytem of forces putting the link PQ in static equilibrium.

Example. *A link* ST *of uniform cross section (Fig. 3.17 (a)) has its ends
pin-jointed to blocks which slide within guides so as to constrain end* S *to move
vertically and end* T *horizontally. The link has a length of 3 metres, a mass of
400 kg and a radius of gyration about its centre of gravity of* 0·9 *metres.*

*At a certain instant the link is at an angle of 40° to the horizontal as shown
and end* S *has a velocity of 2·5 m/s and an acceleration of 20 m/s² vertically
upwards. Neglecting friction at the slides, determine the vertical force* P *required
at* S *to produce this motion:*

(a) when the weight of the link is neglected,

(b) *when the weight of the link is taken into account and a resisting force* Q
of 1000 N *is applied horizontally at* T.
What will be the reactions at the slide blocks in case (b)?

Since the link is of uniform cross section, its centre of gravity G will
lie at the mid-point of ST. Let o represent the frame of reference then,

Velocity of S = $\overrightarrow{os}$ = 2·5 m/s

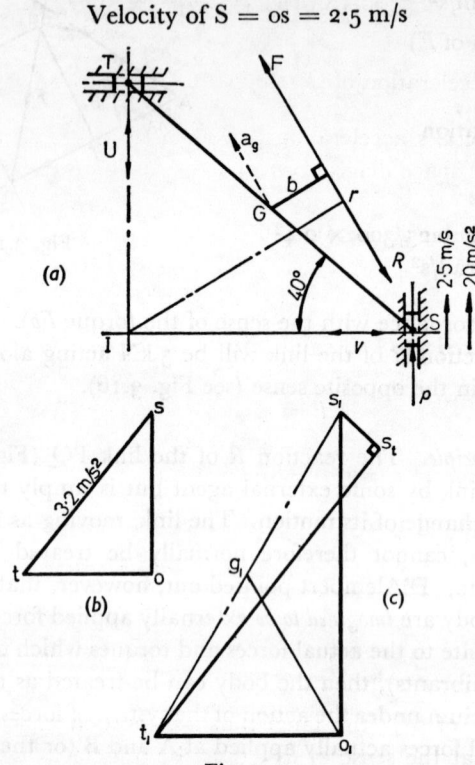

Fig. 3.17

The velocity diagram can now be drawn as in Fig. 3.17 (b), and gives the

Velocity of T relative to S = $\overrightarrow{st}$ = 3·2 m/s

The acceleration diagram is drawn using the following values.

Tangential (total) acceleration of S relative to O = 20 m/s²

$$= \overrightarrow{o_1 s_1}$$

Centripetal acceleration of T relative to S = (st)²/ST

$$= 3·2^2/3$$

$$= 3·41 \text{ m/s}^2$$

$$= \overrightarrow{s_1 s_t}$$

The tangential accelerations of T relative to S and T relative to O (total) are known in direction. Referring to the acceleration diagram Fig. 3.17 (c), $\overrightarrow{s_1t_1}$ will be the acceleration image of link TS and g_1 will lie at its mid-point such that:

$$\overrightarrow{t_1g_1}/\overrightarrow{t_1s_1} = TG/TS$$

then,

$$\text{Total acceleration of G relative to O} = \overrightarrow{o_1g_1} = 11 \cdot 6 \text{ m/s}^2$$

The direction of G's acceleration is from o_1 to g_1. This acceleration is indicated on the space diagram as a_g.

$$\text{Angular acceleration of TS} = \alpha_{TS} = \overrightarrow{s_tt_1}/ST$$
$$= 23/3$$
$$= 7 \cdot 66 \text{ rad/s}^2$$

The force required at G to cause the linear acceleration a_g is, from Eq. (3.2):

$$F = ma_g$$
$$= 400 \times 11 \cdot 6$$
$$= 4640 \text{ N}$$

The torque required to cause the angular acceleration α_{TS} is, from Eq. (3.3):

$$Fb = mk^2\alpha_{TS}$$
$$\therefore b = (400 \times 0 \cdot 9^2 \times 7 \cdot 66)/4640$$
$$= 0 \cdot 534 \text{ m}$$

From the sense of $\overrightarrow{s_tt_1}$ (the tangential acceleration of T relative to S) it is obvious that the sense of α_{TS} must be anticlockwise, therefore the line of action of F must pass to the right of G at a perpendicular distance of $b = 0 \cdot 534$ m from G. Hence the inertial reaction R of link TS will be equal in magnitude to F but opposite in sense, as shown in Fig. 3.17 (a) where F (and R) are drawn parallel to a_g.

When the weight of the link is neglected (Case (a)) there are the following forces:

(i) The force P causing motion.
(ii) A force U at T normal to the slides and arising from their reaction.
(iii) A force V at S normal to the slides and due to their reaction.

If the inertial reaction R is considered as an external force applied to the link, then the latter may be treated as though it were in static equilibrium and the moments of these forces about any point in their plane must balance.

By taking moments about I the instantaneous centre of the link (Fig. 3.17 (a)) the moments of U and V become zero since the lines of action of these forces pass through I, then:

$$P \times \text{IS} = R \times \text{Ir}$$

IS is found to be 2·298 m and Ir 2 m so that:

$$P = 4640 \times 2/2 \cdot 298$$
$$= 4030 \text{ N}$$

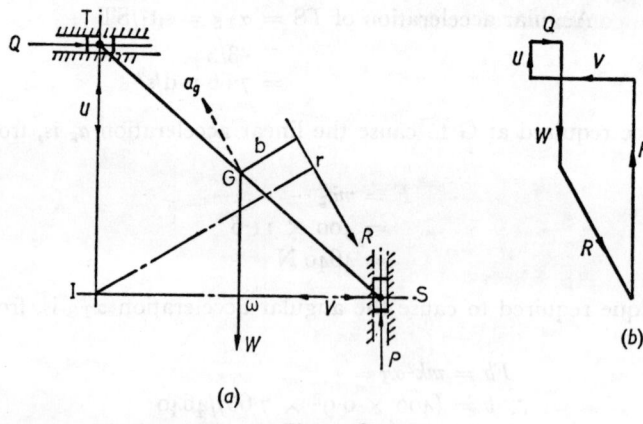

Fig. 3.18

For part (b) there will be, in addition to forces P, R, U and V, the resisting force $Q = 1{,}000$ N and the weight W of the link where,

$$W = mg = 400 \times 9 \cdot 81 = 3924 \text{ N}$$

The system of forces is shown in Fig. 3.18 (a). Taking moments about I:

$$P \times \text{IS} = (R \times \text{Ir}) + (W \times \text{Iw}) + (Q \times \text{IT})$$

or

$$P \times 2 \cdot 298 = (4{,}640 \times 2) + (3{,}924 \times 1 \cdot 149) + (1{,}000 \times 1 \cdot 93)$$
$$P \times 2 \cdot 298 = 9280 + 4509 + 1930$$
$$P = 15\ 719/2 \cdot 298$$
$$= 6880 \text{ N}$$

The slide block reactions U and V can be found by drawing the force diagrams. The magnitudes, directions and senses of Q, W, R and P

are known. Forces U and V are known in direction and form the closing forces (Fig. 3.18 (b)). From the figure it is found that $V = 3400$ N and $U = 1200$ N in the senses indicated by the arrows on the vectors in the diagram.

Two-mass dynamically equivalent systems

The inertial reaction R of a link is the resultant of the inertial reactions of every particle of matter in it. Thus, for a given acceleration and position, R depends only upon the quantity of matter and its geometrical disposition, or on the mass of the link, the position of its centre of mass and the radius of gyration of the link about that centre. Systems which, under the same conditions of acceleration, produce inertial reactions R which

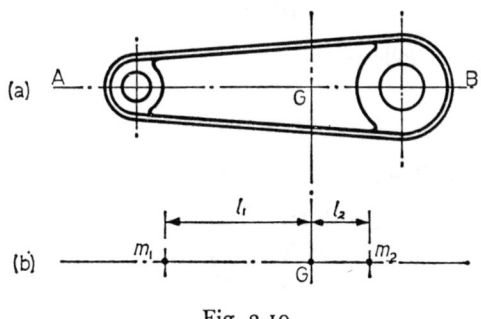

Fig. 3.19

are identical in magnitude, direction, sense and position are known as dynamically equivalent systems. It is sometimes useful at the design stage to replace an *actual* link by a dynamically equivalent *theoretical* system comprising two point masses rigidly connected together. This is known as a two-mass dynamically equivalent system and can be obtained as follows:

Fig. 3.19 (a) shows an "actual" link AB with centre of mass at G. Fig. 3.19 (b) represents the theoretical two-mass dynamically equivalent system, the centre of gravity of which is also at G. If

m is the mass of the actual link AB,

k is the radius of gyration about an axis through G perpendicular to the plane of motion, and

m_1 and m_2 the two point masses at distances l_1 and l_2 from G respectively;

then for the two systems to be dynamically equivalent, the following conditions must exist.

1. The total masses of the two systems must be equal:

$$m_1 + m_2 = m \qquad (3.4)$$

2. The centres of gravity of the two systems must occupy the same position relative to the ends of the link:

$$m_1 l_1 = m_2 l_2 \qquad (3.5)$$

3. The radii of gyration (and hence the moments of inertia) of the two systems about the centre of mass must be equal:

$$m_1 l_1{}^2 + m_2 l_2{}^2 = mk^2 \qquad (3.6)$$

From these relationships can be derived a fourth:

$$l_1 l_2 = k^2 \qquad (3.7)$$

which is obtained as follows.

From Eq. (3.5)

$$m_1 = m_2 l_2 / l_1$$

Substituting for m_1 in Eq. (3.4)

$$m_2 l_2 / l_1 + m_2 = m$$
$$m_2 l_2 + m_2 l_1 = m l_1$$
$$m_2 = m l_1 / (l_1 + l_2) \qquad (a)$$

Substituting for $m_1 l_1 = m_2 l_2{}^2$ in Eq. (3.6):

$$m_2 l_2 l_1 + m_2 l_2{}^2 = mk^2$$
$$m_2 l_2 (l_1 + l_2) = mk^2$$

or equating (a) and (b) $m_2 = mk^2 / l_2(l_1 + l_2) \qquad (b)$

$$m l_1 / (l_1 + l_2) = mk^2 / l_2(l_1 + l_2)$$

from which,

$$l_1 l_2 = k^2 \text{ as above}$$

It will be seen from Eq. (3.7) that *either* m_1 or m_2 can be located relative to G where most convenient, *but not both* since the product of l_1 and l_2 must equal k^2. Thus if l_1 is fixed arbitrarily then $l_2 = k^2 / l_1$.

The following example illustrates the manner in which the two—mass system is used.

Example. *By using a two-mass dynamically equivalent system in place of the link ST in Fig. 3.17 (a), determine the inertial reaction R of the link for the given position. Use the velocity and acceleration given in the worked example on page 62.*

Fig. 3.20 (*a*) shows the essentials of the link mechanism. The velocity and acceleration diagrams must be drawn as before, but only the acceleration diagram is repeated here (Fig. 3.20 (*b*)).

In the two-mass system, mass m_1 may be located anywhere along the line TS but, for convenience, has been placed at T. Therefore:

$$l_1 = TG = 1 \cdot 5 \text{ m}$$

Therefore, from Eq. (3.7)

$$l_2 = k^2/l_1 = 0 \cdot 9^2/1 \cdot 5 = 0 \cdot 54 \text{ m}$$

that is, m_2 is located on TS $0 \cdot 54$ m from G on the opposite side of G from m_1.

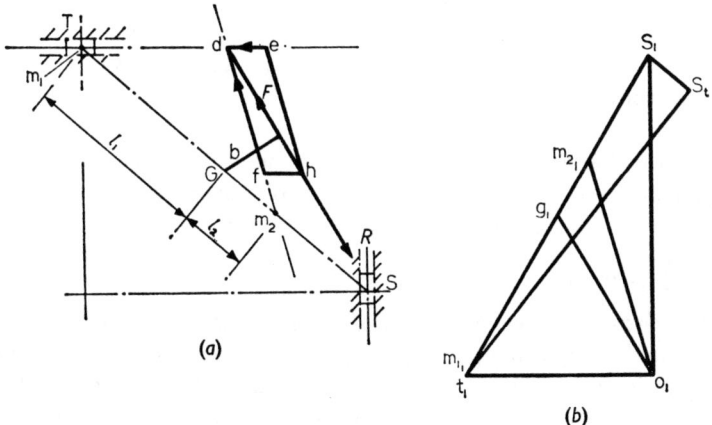

(*a*)   (*b*)

Fig. 3.20

The mass of the link $ST = m = 400$ kg (given).
Therefore, by Eq. (3.5)

$$m_1 l_1 = m_2 l_2$$
$$(400 - m_2) l_1 = m_2 l_2$$
$$(400 - m_2) 1 \cdot 5 = m_2 \times 0 \cdot 54$$
$$m_2 = 294 \text{ kg}$$

and

$$m_1 = 400 - 294$$
$$= 106 \text{ kg}$$

The acceleration of m_2 is given by $\overrightarrow{o_1 m_{2_1}}$ where m_{2_1} is located on the acceleration image $\overrightarrow{s_1 t_1}$ of the link ST as follows:

$$\overrightarrow{s_1 m_{2_1}}/\overrightarrow{s_1 t_1} = Sm_2/ST$$

or

$$\overrightarrow{s_1 m_{2_1}} = (0 \cdot 96/3)23 \cdot 5$$
$$= 7 \cdot 52 \text{ m/s}^2$$

then, from the acceleration diagram:

$$\text{Acceleration of } m_2 = \overrightarrow{o_1 m_{2_1}} = 14 \text{ m/s}^2$$

$$\text{Acceleration of } m_1 = \overrightarrow{o_1 m_{1_1}} = \overrightarrow{o_1 t_1} = 12 \text{ m/s}^2$$

Therefore

$$\text{Force to accelerate } m_1 = 106 \times 12$$
$$= 1272 \text{ N}$$

$$\text{Force to accelerate } m_2 = 294 \times 14$$
$$= 4116 \text{ N}$$

Since m_1 and m_2 are assumed to be concentrated at points, the lines of action of these forces must pass through m_1 and m_2 and must be parallel to the accelerations of m_1 and m_2 respectively. These lines of action are drawn as shown and intersect at point d. Drawing the forces to scale from point d these are represented by ed = 1272 N and fd = 4116 N. By the parallelogram of forces their resultant hd is found to be F = 4640 N at a perpendicular distance $b = 0.534$ m as obtained by the first method.

The inertia reaction R of the two-mass system is equal in magnitude but opposite in sense to F, as before.

Graphical determination of the position of m_2 given the position of m_1. It has been pointed out that the position of one mass (say m_1) can be fixed quite arbitrarily in relation to G but that the position of m_2 must then be in accordance with the relationship $l_1 l_2 = k^2$ [Eq. (3.7)]. The position of m_2 can be found by calculation as shown above or geometrically.

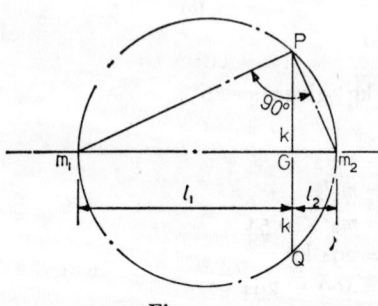

Let G be the centre of gravity of the two-mass system (Fig. 3.21), and let m_1 be located arbitrarily at distance l_1 from G. Let k be the radius of gyration of the actual link about an axis through G.

Fig. 3.21

Construction

Erect a perpendicular GP to m_1G to represent k to the same scale. Join m_1P. Draw Pm_2 at right angles to m_1P to meet m_1G (produced) in m_2. This is the position of the second mass. This construction is valid because, by symmetry, GQ also equals k and from the properties of intersecting chords:

$$l_1 l_2 = k^2$$

Inertial reaction of a connecting rod (geometrical method)

By replacing the connecting rod of a slider crank mechanism by a dynamically equivalent two-mass system, it is possible to use an entirely geometrical method for finding the inertial reaction R of the rod.

Example. *Determine the magnitude, direction, sense and position of the inertial reaction R for the connecting rod of the slider-crank mechanism shown in* Fig. 3.22. *Crank* AB = 90 mm. *Connecting rod* BC = 270 mm. *Crank angle* = 45° *from inner dead centre. Mass of connecting rod* = 10 kg. *Centre of gravity of rod* (G^1) = 90 mm *from* B. *Speed of crankshaft* 2400 rev/min. *Radius of gyration of rod* = 80 mm.

The slider-crank mechanism must first be drawn to scale. Next, using Klein's construction, the quadrilateral ABGH representing the acceleration diagram for the mechanism is obtained. Then:

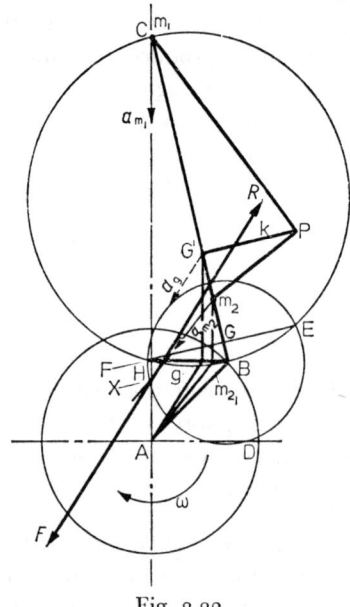

Fig. 3.22

$\omega^2 \cdot$ BA = centripetal (total) acceleration of B relative to A;

$\omega^2 \cdot$ HA = tangential (total) acceleration of C relative to A;

$\omega^2 \cdot$ BH = total acceleration of B relative to C, i.e. the acceleration image for the connecting rod BC.

One of the masses (m_1) is located arbitrarily at the gudgeon pin C. Its acceleration is thus along the line of stroke CA (a useful simplification). By the geometrical construction previously described, m_2 is located as shown. Draw G^1g parallel to the line of stroke to meet HB in g. Then, by simple proportion,

ω^2gA = acceleration of G^1, its direction being parallel to gA and its sense from g to A. This acceleration is denoted by a_g in the figure.

Similarly, draw $m_2 m_{2_1}$ to meet HB in m_{2_1} then proportionally,

$\omega^2 m_{2_1}$A = acceleration of m_2 its direction being parallel to m_{2_1}A and its sense from m_{2_1} to A.

The acceleration of m_1 is given by ω^2. HA and takes place along the line of stroke with sense from H to A. The acceleration of m_1 and m_2 are denoted by a_{m_1} and a_{m_2} in the figure hence the forces required to give m_1 and m_2 these accelerations must act in the directions of a_{m_1} and a_{m_2} and will therefore intersect on the line of stroke. Therefore the resultant (F) of these forces must pass through point X. But, since the resultant force must produce the linear acceleration of the centre of mass G^1 it must act through X in a direction parallel to a_g (the acceleration of G^1) with sense g to A.

By Newton's second law

$$\text{Resultant } F = \text{Mass} \times \text{Acceleration}$$
$$= 10(\omega^2 . \text{gA})$$
$$= 10(2\pi2400/60)^2(78/1000)$$
$$= 10 \times 6400\pi^2 \times 0\cdot078$$
$$= 49\ 300\ \text{N}$$
$$= 49\cdot3\ \text{kN}$$

Hence the magnitude, direction, sense and line of action of the resultant accelerating force on the connecting rod BC have been determined. The inertial reaction of BC will therefore be equal but opposite to F (R in the figure).

Turning moment at the crank shaft. Once the inertia reaction of the connecting rod has been determined it is possible to find the turning moment applied at the crank shaft for that particular position of the rod.

Example. Fig. 3.23 *is a reproduction of the slider-crank mechanism shown in* Fig. 3.22. *The crank angle position, lengths of the links and mass of connecting rod BC are therefore the same as before. Additional details are: piston diameter* 120 mm, *mass of reciprocating parts (gudgeon pin and piston)* 8 kg, *gas pressure* 6000 kN/m². *For the same engine speed (2400 rev/min), determine the turning moment at the crank shaft. Neglect friction.*

In accordance with D'Alembert's principle, the connecting rod BC can be assumed to be in static equilibrium under the following forces and inertial reactions.

   (i) The gas force on the piston acting inwards at C along the line of stroke.

  (ii) The gravitational pull (W_r) on the piston and gudgeon pin acting inwards at C along the line of stroke (for a vertical engine).

 (iii) The inertial reaction of piston and gudgeon pin acting outwards at C along the line of stroke in opposition to the acceleration of the piston, etc.

(iv) The inertial reaction R of the connecting rod (previous example).

(v) The reaction N of the cylinder on the piston at C, at right angles to the cylinder walls.

(vi) The gravitational pull W on the connecting rod acting through G^1, the centre of gravity of the rod.

(vii) A force applied by the crank pin to the connecting rod at B which can be considered as having rectangular components S and Q acting along the crank shaft and at right angles to it respectively.

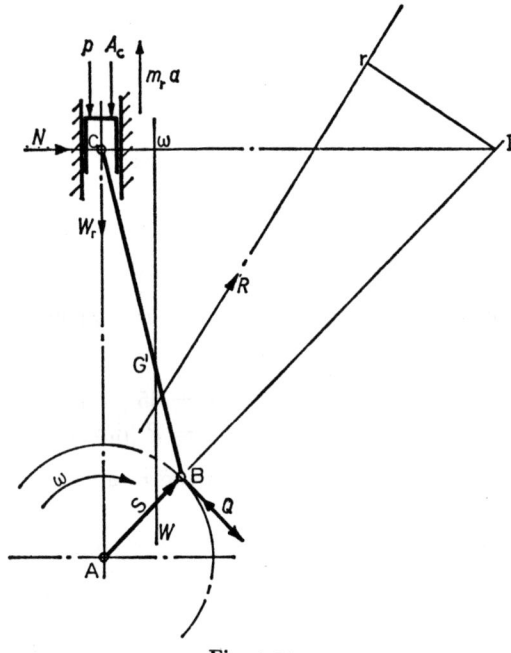

Fig. 3.23

The magnitudes of these forces are:

(i) Gas force on piston = Pressure × Area of piston
$$= (6000 \times 10^3)(0 \cdot 7854 \times 120^2/10^6)$$
$$= 67 \cdot 858 \text{ kN (downwards)}$$

(ii) Gravitational pull W_r on piston, etc. $m_r g$
$$= 8 \times 9 \cdot 81$$
$$= 78 \cdot 48 \text{ N}$$
$$= 0 \cdot 07848 \text{ kN (downwards)}$$

Inertial reaction of piston, etc. $= m_r a$

where a the piston acceleration is given by $\omega^2 r (\cos \theta + \cos 2\theta/n)$.

In this case,

$$\omega = 2\pi 2{,}400/60 = 80\pi \text{ rad/s}$$
$$r = 90 \text{ mm} = 0{\cdot}09 \text{ m}$$
$$n = 1/r = 27/9 = 3$$
$$\cos \theta = \cos 45^\circ = 0{\cdot}7071$$
$$\cos 2\theta = \cos 90^\circ = 0$$

so that

$$a = (80\pi)^2 0{\cdot}09(0{\cdot}7071 + 0)$$
$$= 4100 \text{ m/s}^2$$

Alternatively, a can be obtained by scaling vector HA = 0·0636 m in Fig. 3.22 when

$$a = \omega^2 \cdot \text{HA}$$
$$= (80\pi)^2 \times 0{\cdot}0636 = 4100 \text{ m/s}^2 \text{ as before.}$$

Then the inertial reaction of pistons, etc.,

$$= 8 \times 4{,}100$$
$$= 32\,800 \text{ N}$$
$$= 32{\cdot}8 \text{ kN (upwards)}$$

The net force on the connecting rod at C along the line of stroke is given by:

$$67{\cdot}858 + 0{\cdot}07848 - 32{\cdot}8 = 35{\cdot}13648$$

say $35{\cdot}13 \text{ kN} = P$ (downwards)

(iv) Inertial reaction R of connecting rod = 49·3 kN (determined previously)

(v) The magnitude of the reaction N's not known.

(vi) Gravitational pull W on the connecting rod

$$= mg$$
$$= 10 \times 9{\cdot}81$$
$$= 98{\cdot}1 \text{ N}$$
$$= 0{\cdot}0981 \text{ kN (downwards)}$$

(vii) Forces S and Q are not known in magnitude or sense.

For equilibrium, the algebraic sum of the moments of the forces about any point in the plane of the forces must equal zero. By taking moments about I, the instantaneous centre of the connecting rod the moments of the forces N and S are zero. Then, *assuming* that the sense of Q is outwards away from the rod, scaling perpendicular distances from the lines of action of the forces to I in metres and taking clockwise moments as positive:

$$-(P \times \text{CI}) - (W \times \omega\text{I}) + (R \times \text{rI}) - (Q \times \text{BI}) = 0$$

that is,

$$-(35 \cdot 13 \times 0 \cdot 328) - (0 \cdot 0981 \times 0 \cdot 284) + (49 \cdot 3 \times 0 \cdot 128)$$
$$- (Q \times 0 \cdot 372) = 0$$
$$-11 \cdot 53 - 0 \cdot 0278 + 6 \cdot 31 - 0 \cdot 372Q = 0$$

from which

$$Q = -5 \cdot 2478/0 \cdot 372$$
$$= -14 \cdot 1 \text{ kN approximately.}$$

The negative sign indicates that the assumption that Q acts outwards is incorrect. Thus the force Q applied to the connecting rod at B is 14·1 kN inwards, towards the rod and by Newton's third law, the rod must react by applying an equal but opposite thrust Q on the crank pin tending to cause the crank shaft to rotate in a clockwise direction. The turning moment on crank shaft

$$= \text{Tangential thrust } Q \times \text{Crank radius}$$
$$= 14 \cdot 1 \, (90/1000)$$
$$= 1 \cdot 269 \text{ kNm (clockwise)}$$

If required, the unknown forces N and S can be found by drawing out all the forces acting on the connecting rod to form a polygon of forces in the manner shown in a previous example in this section.

PROBLEMS

For tutorials

1. The link AB shown in Fig. 3.24 is slotted so that a slider S can travel with uniform velocity v from A to B. During the movement of S the rod rotates with uniform angular velocity ω about the centre O. How does the magnitude, direction and sense of the Coriolis component of acceleration of the slider S vary, if at all, during the movement from A through O to B?

2. Fig. 3.25 (a), (b), (c) and (d) shows four slotted links each rotating about a fixed centre O with uniform angular velocity ω and containing a slider S within its slot. If, in each case, S is moving with uniform linear velocity v relative to a co-incident point X on the link, discuss the possible existence of a Coriolis component of acceleration ($2\omega v$) in each case. If such a component is found to exist, in which direction and with what sense is it acting? (*Hint:* In Fig. 3.25 (d)) the total linear velocity of S relative to O is given by ($\omega_r + v$).

Is the Coriolis component involved with the centripetal acceleration in any of the cases shown?

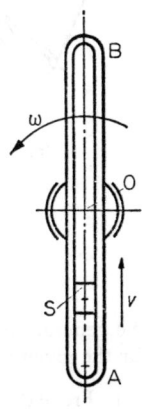

Fig. 3.24

3. Referring to case (*d*) in Fig. 3.25. If, in addition to the previous motion, the circular link is made to precess about axis aa with angular velocity ω_p anti-clockwise when viewed from the right in the direction of arrow E, what will be

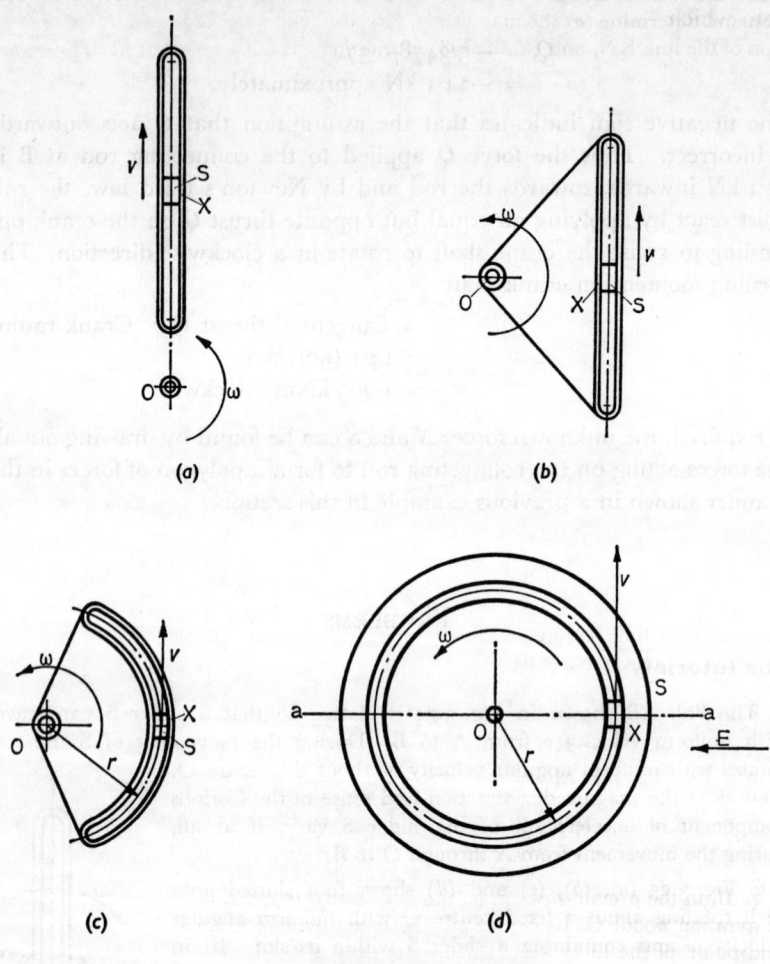

(*a*)   (*b*)

(*c*)   (*d*)

Fig. 3.25

the general direction and sense of the *resultant* Coriolis component of acceleration for the given instantaneous position of the slider S?

4. Prove the similarity of the quadrilaterals ABGH and $a_1b_1c_bc_1$ in Fig. 3.8 (*a*) and (*b*) and so verify Klein's construction.

General

1. In the four-bar chain shown in Fig. 3.26, M and N are fixed pivots with M vertically above N and crank LM is rotating clockwise uniformly at 720 rev/min. Draw the velocity and acceleration diagrams for the given instantaneous configuration and determine (*a*) the magnitude, direction and sense of the angular acceleration of the link KN, and (*b*) the torque necessary at the crankshaft M to overcome

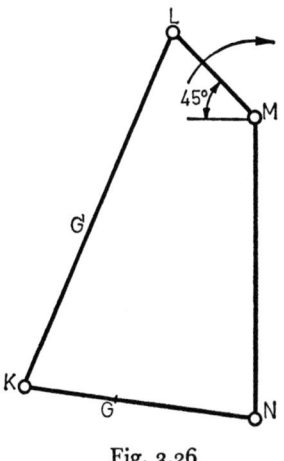

Fig. 3.26

the inertia of link KN given that the mass of link KN is 0·5 kg and its radius of gyration about G, its centre of gravity, is 40 mm.

$$LM = 60 \text{ mm} \qquad LK = 190 \text{ mm}$$
$$KN = 120 \text{ mm} \qquad GN = 70 \text{ mm}$$
$$MN = 150 \text{ mm}$$

Ans. (*a*) 1,080 rad/s² in the plane of the links, anticlockwise
(*b*) 1·722 Nm anticlockwise

2. If, in the previous example, the mass of the link KL is 0·75 kg and its radius of gyration about G' its centre of gravity is 60 mm (where G' is situated at the mid-point of the link) what will be the torque required at the crankshaft M to overcome the combined inertia of links KN and KL.

Ans. 4·95 Nm anticlockwise

3. Assuming that the mass of the link LM in Fig. 3.26 is 0·25 kg and its centre of gravity is at the mid-point of LM what must be the torque required at the crankshaft M to overcome the total inertia reaction of all three links and the gravitational pulls on these links? Assume all other relevant information is as given in, or found from the solution of, the two previous examples.

Ans. 4·26 Nm anticlockwise

4. Fig. 3.27 shows an instantaneous configuration of a mechanism in which A, D and G are fixed pivots. AB is a lever which oscillates through a certain angle so causing levers DE and FG to have corresponding oscillations. When AB is at 45° above the horizontal as shown and rotating clockwise C is moving horizontally and directly towards A and the links ED and FG are both vertical. The dimensions of the links are as follows: AB = 30 mm, BC = 60 mm, DE = 90 mm, DC = 60 mm, EF = 60 mm and FG = 160 mm.

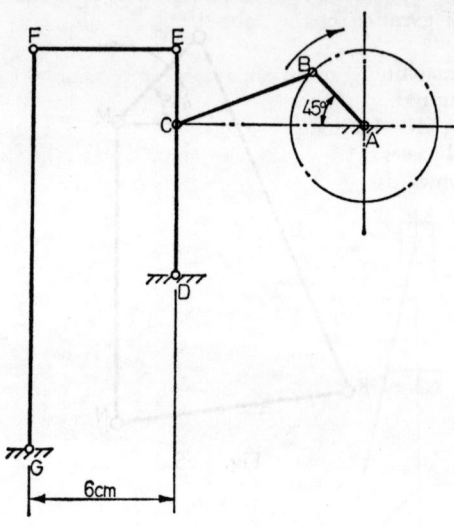

Fig. 3.27

For the given configuration, assuming that AB has a uniform velocity of 8π rad/s clockwise determine (a) the total acceleration of F relative to E, (b) the angular acceleration of the link FE and (c) the total acceleration of C relative to B.

> *Ans.* (a) 5·92 m/s² vertically upwards
> (b) 98·7 rad/s² clockwise
> (c) 5·8 m/s²

5. A slider-crank mechanism has a crank AB = 40 mm and connecting rod BC = 150 mm. The crank is rotating at a uniform speed of 1800 rev/min. Determine the velocity and acceleration of the piston C when the crank is 135° past the inner dead centre. (a) Obtain the required values by drawing velocity and acceleration diagrams using Kleins and Ritterhaus's constructions. (b) Check these results by using the equations for piston velocity and acceleration derived in Section 2.6.

> *Ans.* Piston velocity 4·18 m/s; piston acceleration 1005 m/s²

6. A reciprocating engine has a crank AB of 80 mm and a connecting rod BC of 280 mm length. The centre of gravity of the connecting rod G lies at a point 80 mm from the big end centre. The mass of the rod if 5 kg and its radius of gyration about its centre of gravity is 0·06 m. If the speed of the engine is 300

rev/min determine for a crank angle of 120° past inner dead centre, the magnitude of the inertia force of the connecting rod and the perpendicular distance of its line of action from G.

Ans. 326 N; 13·8 mm (the line of action lying between G and B)

7. Fig. 3.28 shows the slider-crank mechanism of a vertical engine, the details of which are as follows: Crank AB = 90 mm, connecting rod BC = 270 mm, crank angle = 135° from inner dead centre, mass of connecting rod = 10 kg, centre of gravity of connecting rod is 90 mm, from B, speed of crankshaft = 2400 rev/min, radius of gyration of the connecting rod about its centre of gravity = 80 mm.

Determine the magnitude, direction, sense and position of the inertia reaction *R* of the connecting rod.

Ans. *R* = 49·3 kN. Its line of action lies at a perpendicular distance from G of 20 mm and passes between G and B at 35° to the line of stroke, the general sense being downwards.

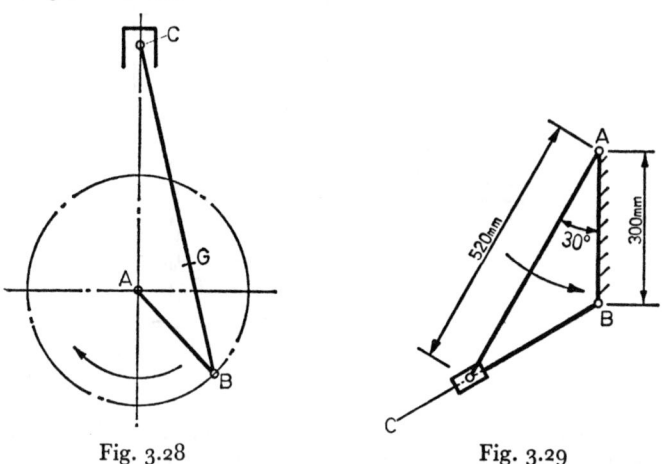

Fig. 3.28   Fig. 3.29

8. With reference to the slider-crank mechanism of the previous question: assume the following additional information is available and determine, for the same engine speed and crank angle position, the turning moment on the crank shaft. Piston diameter 120 mm, mass of reciprocating parts (gudgeon pin and piston) 8 kg, gas pressure 2000 kN/m². Neglect friction.

Ans. 4·69 kNm clockwise

9. If the slider-crank mechanism of the previous question formed part of a *horizontal* engine in what way would the calculations relating to the previous question need to be altered in order to determine the turning moment on the crank shaft?

10. In Fig. 3.29 A and B are fixed pivots. The link AS, which is 520 mm in length, carries a slider at S through which passes the link BC. For the given position, assuming that link AS is uniformly rotating anticlockwise at 10 rad/s, determine (*a*) the velocity of sliding of S along CB, (*b*) the angular acceleration of the link CB.

Ans. (*a*) 2·6 m/s towards B
(*b*) 173·3 rad/s² anticlockwise

11. In Fig. 3.30 A, B and E are fixed pivots. The crank BC carries a slide block at C through which passes the link AD, the latter being pivoted at A. At end D of link AD a slide block is fixed through which passes link EF which is pivoted at E. Three positions of the crank BC and the corresponding positions of the other links are shown, one in full and the other two in dotted lines. When in the position shown by the full lines AD is horizontal and EF vertical. The dimensions of the links are, BC = 30 mm, AD = 140 mm and EF = 200 mm.

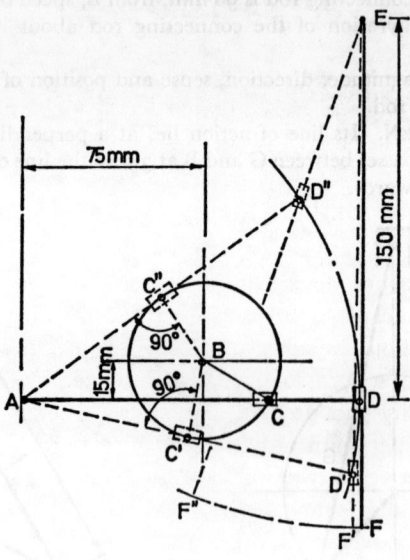

Fig. 3.30

If the crank BC rotates about B with a uniform angular velocity of 180 rev/min, determine the values of the Coriolis components of acceleration of C and D relative to the co-incident points on the engaging links for the three given instantaneous configurations.

Ans. At C 2·55 m/s² upwards; C′ o m/s²; C″ o m/s²; D o m/s²; D′ o m/s²; D″ o m/s².

CHAPTER 4

The Gyroscope

4.1 BEHAVIOUR OF GYROSCOPES

The gyroscope was one device (another being the pendulum) by means of which Léon Foucault, a French physicist, was able to prove that the earth is rotating about its axis.

The behaviour of a gyroscope will be illustrated by the disc in Fig. 4.1, which is universally mounted and therefore free to rotate about any of three mutually perpendicular axes xx, yy and zz. To begin with, the disc has a clockwise velocity of spin ω about axis xx viewed in the direction of arrow X. If the disc is now made to precess* about axis yy with angular velocity ω_p, clockwise viewed in the direction of arrow Y, it will immediately start to rotate anticlockwise about axis zz as viewed in the direction of arrow Z. The movement will continue, unless resisted, until the disc has undergone an angular displacement to 90° about axis zz and until its axis of spin is co-incident with axis yy. The rotation may be prevented by applying a torque T to the disc about axis zz in a clockwise sense, and this will result in an equal but opposite inertial reaction (the gyroscopic torque) by the disc. The torque T required to prevent rotation of the disc about axis zz depends on the rates of spin and precession and on the moment of inertia of the disc about the axis of spin and will be shown later to be $T = I\omega\omega_p$ where I is the polar moment of inertia of the disc. Conversely, if the disc is acted upon by an externally applied torque

* To precess means, in this context, to rotate. Precess refers generally to precedence in time or order. It is associated here with the astronomical phenomenon of the earlier occurrences of the equinoxes in successive sidereal years caused by the slow "wobble" of the earth's axis as it responds to the gravitational pulls of the other planets. The earth itself a spinning body, is behaving as a gyroscope. The wobble is the precession of its axis.

T about **axis zz** while spinning about **axis xx**, it will precess about **axis yy** with angular velocity ω_p. If the sense of either ω or ω_p is reversed, the sense of *T* must be correspondingly reversed. Likewise, if the sense of *T* or ω is reversed the sense of ω_p will be changed.

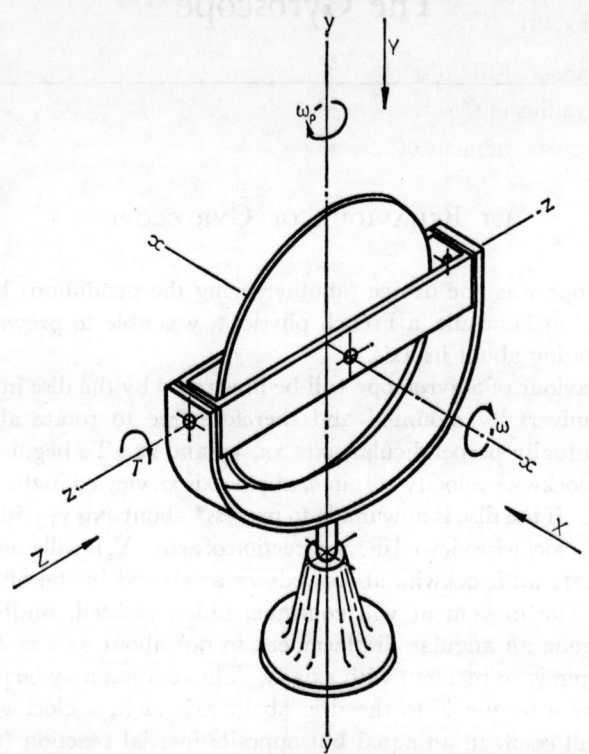

Fig. 4.1

This apparently odd behaviour is common to all spinning bodies, so that gyroscopic torque appears whenever the direction of the axis of spin of a body is changed. Engineers meet gyroscopic torque with rotating parts aboard ship, in aircraft, automobiles and locomotives; in fact, wherever the members supporting a rotating part are able to move angularly. Applications of gyroscopic principles are to be found in ships' stabilizers, roll and pitch recorders, gyro-compasses, etc.

4.2 RELATIONSHIP BETWEEN TORQUE, SPIN AND PRECESSION

From consideration of change of angular momentum

Fig. 4.2 shows a disc spinning clockwise with angular velocity ω rad/s about axis xx. If

$m =$ the mass of the disc (kg),

$k =$ the radius of gyration about xx (m), and

$I =$ the polar moment of inertia (kg m²), then

$I = mk^2$

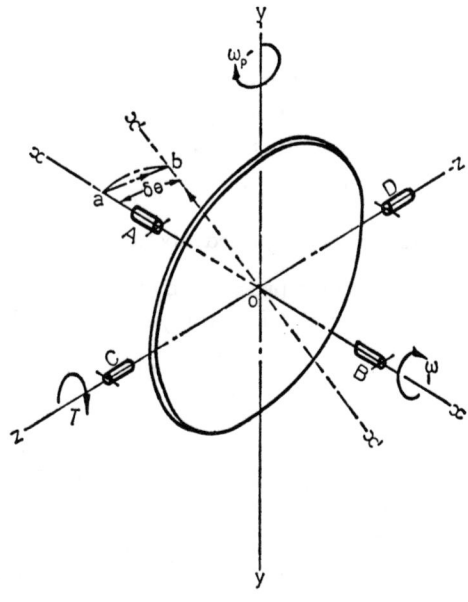

Fig. 4.2

From Section 1.6,

Angular momentum of disc $= I\omega = mk^2\omega$ kg m²/s

and, in accordance with the right hand screw rule (Sections 1.1 and 1.4) can be represented by the vector $\overrightarrow{oa}$ in Fig. 4.2. If the disc is now caused to precess clockwise about axis yy with angular velocity ω_p radians/s, xx will move through angle $\delta\theta$ radians to position x¹x¹ in time δt seconds.

At the end of this time the angular momentum of the disc is represented by $\overrightarrow{ob}$, so that the

$$\text{Change in angular momentum} = \overrightarrow{ab} \backsimeq I\omega\delta\theta$$

when $\delta\theta$ is a small angle.

$$\therefore \text{ Rate of change of angular momentum} = \frac{\text{Change in momentum}}{\text{Time for change}}$$

$$\backsimeq I\omega\delta\theta/\delta t$$

In the limit as $t \to$ zero,

$$\text{Rate of change of angular momentum} = I\omega \, d\theta/dt$$
$$= I\omega\omega_p$$

Now   Torque $\propto$ Rate of change of angular momentum

$$= \text{Constant} \times I\omega\omega_p$$

As shown in Section 1.5, when the torque is measured in Newton metres, the moment of inertia of the disc in kg m² and the angular velocities in rad/s, the constant becomes unity, so that,

$$\text{Torque } (T) = I\omega\omega_p \qquad\qquad (4.1)$$

The torque T is that which, applied about axis zz, will produce the change in angular momentum represented by $\overrightarrow{ab}$ and hence the precession. The sense of the torque is obtained by applying the right hand screw rule to $\overrightarrow{ab}$, i.e. movement from a to b would be produced by rotating a right hand screw clockwise about $\overrightarrow{ab}$, viewed from the left of the diagram. Thus the external torque T must be applied in this sense about axis zz. The *gyroscopic torque* which forms the inertial reaction of the disc, is equal in magnitude but opposite in sense to T and is therefore anticlockwise. This is the sense in which the disc would heel if the external torque T were not applied to keep it upright and precession was brought about by other means. In short, an external clockwise torque about zz will cause the disc to precess clockwise about yy and to react with an anticlockwise torque about zz. Conversely, if the disc is made to precess clockwise about yy by movement of the gyro frame, it will exert an anticlockwise inertial torque pressing downwards on bearing A and upwards on bearing B. For stability, the bearing A must exert an equal and upward force and the bearing B an equal but downward force, so producing an external torque equal and opposite to the gyroscopic torque. Reversal of the sense of ω or ω_p will alter the sense of T, and reversal of T or ω will alter the sense of ω_p.

From consideration of the coriolis component of acceleration

The above derivation of the relationship $T = I\omega\omega_p$ is sometimes confusing, and an alternative method which does not involve angular momentum directly may be more easily understood.

In Fig. 4.3, S represents one particle of the disc material at a radius r_s from the centre. Orthographic views of the disc are seen in the direction of arrows X, Y and Z (Figs. 4.4 (a), (b) and (c)). From Fig. 4.4 (a),

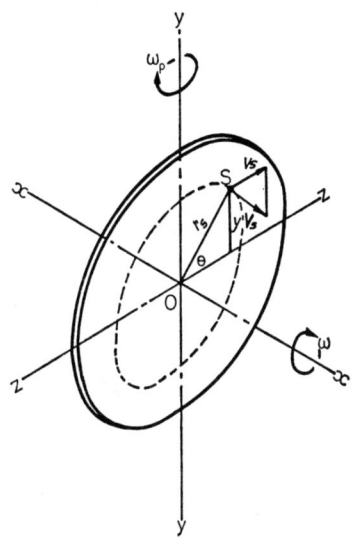

Tangential velocity of P,

$$V_s = \omega r_s$$

Component of P's velocity parallel to zz,

$$v_s = V_s \sin \theta$$

No matter what other component velocities S may have, it has both a radial velocity v_s and an angular velocity ω_p relative to axis yy Fig. 4.4 (b). It follows from Section 1.11 that S must have a Coriolis component of acceleration relative to zz which is equal to $2\omega_p v_s$ at right angles to the disc.

Fig. 4.3

For the given sense of spin, all particles such as S which are situated momentarily in the upper half of the disc (above zz) will have component velocities v_s directed towards the right (Fig. 4.4 (b)) while particles in the lower half, will have corresponding component velocities towards the left. Therefore, all particles in the upper half have a Coriolis component of acceleration *towards* the reader when the disc is viewed as in Fig. 4.4 (a) while all particles in the lower half have a component acceleration *away from* the reader. The sense and direction of these accelerations are shown in Fig. 4.5, the end view of the disc.

If m is the mass of particle S, and y is perpendicular distance of S from zz, the Force required to give S its Coriolis component of acceleration

$$= \text{Mass} \times \text{Acceleration} = m \cdot 2\omega_p v_s$$
$$= m \cdot 2\omega_p V_s \sin \theta = m \cdot 2\omega_p V_s(y/r_s)$$
$$= m \cdot 2\omega_p \omega r_s y/r_s = 2\omega_p \omega m y$$

It follows that the

Torque which must be exerted about zz to accelerate $S = 2\omega_p\omega my^2$
and, for the whole disc,

$$\text{Total torque} = 2\omega_p\omega\Sigma my^2$$
$$= 2\omega_p\omega I_{zz}$$

But, for a disc

$$I_{zz} = I_{yy}$$

so that

$$2I_{zz} = I_{zz} + I_{yy} = I_{xx}$$

$\therefore$ Total torque required $= I_{xx}\omega\omega_p$

or

$$T = I\omega\omega_p$$

where I is the moment of inertia of the disc about xx.

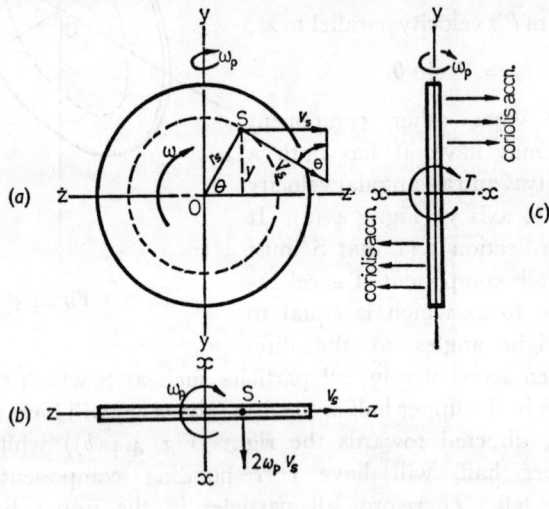

Fig. 4.4

The applied torque T must act in a clockwise sense about zz (from Fig. 4.4 (c)). The gyroscopic torque, will be equal to T in magnitude but opposite in sense.

Example. *In Fig. 4.5 (a) a thin disc complete with shaft has mass 10 kg and radius of gyration 0·14 m. The system is suspended by a cord attached to a bearing at one end of the shaft and caused to rotate at 2700 rev/min anticlockwise as viewed in the direction of arrow X while the shaft is in a horizontal position. Determine the tension in the cord and the magnitude, direction and sense of precession.*

For equilibrium, the

Tension in cord $= mg$

$$= 10 \times 9\cdot81 = 98\cdot1 \text{ N}$$

Angular velocity $(\omega) = 2\pi2700/60$

$$= 90\pi \text{ rad/s}$$

Moment of inertia $(I) = mk^2$

$$= 10 \times 0\cdot14^2 = 0\cdot196 \text{ kg m}^2$$

Applied torque = Clockwise couple caused by tension in cord and the pull of gravity on the system

i.e. $T =$ Force $\times$ Arm of couple

$$= 98\cdot1 \times 0\cdot1 = 9\cdot81 \text{ Nm}$$

From Eq. (4.1)

$$\omega_p = T/I\omega$$
$$= 9\cdot81/0\cdot196 \times 90\pi$$
$$= 0\cdot1771 \text{ rad/s}$$

The applied torque acts horizontally in a clockwise sense. If the system is viewed from above then, using the right hand screw rule and Fig.

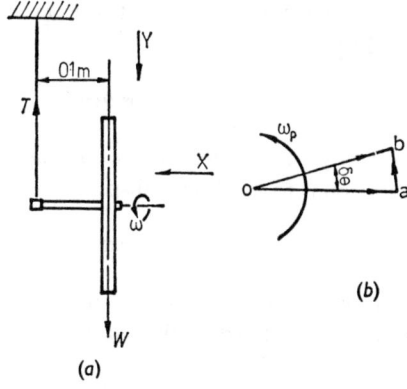

(a)

(b)

Fig. 4.5

4.5 (b), $\overrightarrow{oa}$ represents the angular momentum at the commencement of precession, $\overrightarrow{ab}$ represents the change in angular momentum due to the applied clockwise torque and $\overrightarrow{ob}$ represents the angular momentum after precession has been taking place for some time δt. The precession therefore occurs in a horizontal plane anticlockwise as viewed from above.

4.3 GYROSCOPIC TORQUE AND ANGULAR SIMPLE HARMONIC MOTION

Formulae for angular s.h.m.

Suppose that a thin flexible but inextensible tape is wound once round a cylindrical drum pivoted at O. If a rod AB is attached across one end of the drum, if there is no slip between the tape and drum and if the tape is moved with simple harmonic motion, the motions of the drum

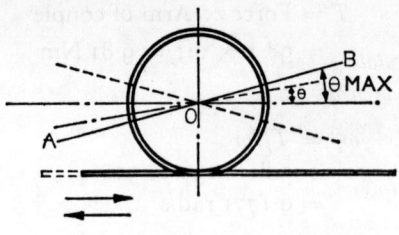

Fig. 4.6

and the attached rod about O will also be simple harmonic. If r is the amplitude of the linear movement of the tape, then the linear displacement from mid-position is

$$x = r \cos \omega t = y \cos 2\pi nt$$

where n is the frequency of the oscillation. Because the angular displacement θ is proportional to the linear displacement x,

$$\theta = \theta_{max} \cos 2\pi nt \qquad (4.2)$$

where θ_{max} is the amplitude of the angular displacement. The angular velocity is then obtained by differentiating and is given by

$$\frac{d\theta}{dt} = 2\pi n\theta_{max} \sin 2\pi nt$$

This is a maximum when $\sin 2\pi nt = 1$, so that

$$\text{Maximum angular velocity} = 2\pi n\theta_{max} \qquad (4.3)$$

Differentiating again

$$\text{Angular acceleration } d^2\theta/dt^2 = -4\pi^2 n^2 \theta_{max} \cos 2\pi nt$$

The maximum occurs when $\cos 2\pi n t = 1$, so that

$$\text{Maximum angular acceleration} = -4\pi^2 n^2 \theta_{max} \qquad (4.4)$$

If T is the periodic time of one complete oscillation, so that $n = 1/T$, then

$$\text{Maximum angular velocity} = (2\pi/T)\theta_{max} \qquad (4.5)$$

$$\text{Maximum angular acceleration} = (2\pi/T)^2\theta_{max} \qquad (4.6)$$

These equations are useful in certain problems involving gyroscopic torque.

Example. *The rotating parts of a ship's turbine have a mass of* 20 000 kg *and a radius of gyration of* 0·6 m. *The speed of the turbine is* 1000 rev/min, *anticlockwise viewed from the stern of the ship. Determine the maximum gyroscopic couple and its effect on the vessel.*

(a) When the ship is moving at 36 km/h *and makes a turn to the right in a curve having a mean radius of* 200 m.

(b) If the ship pitches with simple harmonic motion through an angle of 5° *above and* 5° *below the horizontal with a periodic time of* 15 *seconds and the bows are falling with maximum velocity.*

(a) $\qquad \omega = 2\pi 1000/60 = 100\pi/3$ rad/s

$\qquad \omega_p = $ Linear velocity/radius of turn

$\qquad\qquad = 36$ km/h/(200 m) $= 10$ m/s/(200 m) $= 0\cdot05$ rad/s

$\qquad$ Moment of inertia of turbine parts $= mk^2$

$$= 20\ 000 \times 0\cdot6^2$$

$$= 7200 \text{ kg m}^2$$

Then,

$$\text{Gyroscopic couple } T = I\omega\omega_p$$

$$= 7200(100\pi/3)0\cdot05$$

$$= 37\ 680 \text{ Nm}$$

$$= 37\cdot68 \text{ kNm}$$

Fig. 4.7 (a) $\overrightarrow{ab}$ gives the change in momentum of the turbine rotor etc., and shows that the applied torque must be clockwise looking on the ship in the direction of arrow X. Thus the gyroscopic torque exerted by the rotor will be equal but opposite, or anticlockwise. This tends to lift the bows.

(b) $\theta_{max} = 5(2\pi/360) = \pi/36$ radians.

$$\therefore \text{ Maximum angular velocity} = (2\pi/T)\theta_{max}$$
$$= (2\pi/15)36$$
$$= 0.0366 \text{ rad/s}$$

and this is also the maximum value of ω. It follows that the

$$\text{Maximum gyroscopic couple } T \quad 7200(100\pi/3)0.0366$$
$$= 27\,600 \text{ Nm}$$

From Fig. 4.7 (b), representing the ship from the right side, $\overrightarrow{ab}$ is the change in angular momentum as the bows descend, so that the applied torque

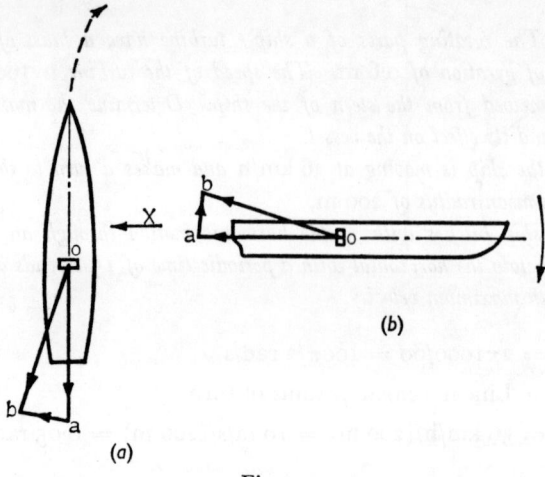

Fig. 4.7

will be anticlockwise looking down on the ship and the gyroscopic reaction will be clockwise, tending to turn the bows to the right.

4.4 DISC INCLINED TO AXIS OF PRECESSION

Fig. 4.8 is an end elevation of a disc with an auxiliary view projected from it to show full face of the disc. The disc is inclined at angle θ to the vertical axis vv, about which it is precessing with total angular velocity ω_p. Resolving this velocity into rectangular components about axes xx and yy, it can easily be shown that the rate of precession about axis yy $= \omega_p \cos \theta$. The gyroscopic torque about axis zz is therefore reduced to

$$T = I\omega\omega_p \cos \theta \tag{4.7}$$

when $\theta = 0°$, $\cos \theta = 1$ and Eq. (4.7) becomes Eq. (4.1). When $\theta = 90°$, $\cos \theta = 0$ and the gyroscopic torque becomes zero, the axis of spin xx coinciding with the axis of precession vv.

Example. *A disc of uniform thickness 50 mm and outside diameter 1 m rolls on its edge round a horizontal circular track of radius 12 m at a speed of 5 m/s. The disc material has a density of 7 Mg/m³. Determine the gyroscopic couple and the angle of inclination to the vertical at which the disc runs uniformly.*

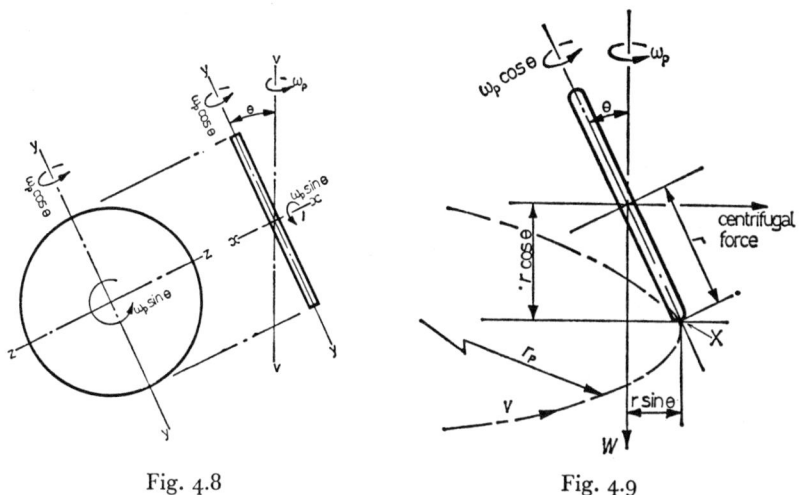

Fig. 4.8   Fig. 4.9

Referring to Fig. 4.9, the disc is in equilibrium about point X under three couples due to gyroscopic torque, centrifugal force and gravity. Equating moments:

$$[(\text{Centrifugal force})r \cos \theta] + [\text{Gyroscopic torque}] = Wr \sin \theta$$

$$\therefore \ [m\omega_p^2 r_p r \cos \theta] + [I\omega\omega_p \cos \theta] = Wr \sin \theta$$

$$\left[m \frac{v^2}{r_p^2} r_p r \cos \theta \right] + \left[mk^2 \frac{v}{r} \cdot \frac{v}{r_p} \cos \theta \right] = mgr \sin \theta$$

Putting $k^2 = r^2/2$ (for a disc),

$$m \frac{v^2}{r_p^2} r_p r \cos \theta + m \frac{r^2}{2} \frac{v}{r} \frac{v}{r_p} \cos \theta = mgr \sin \theta$$

$$\frac{v^2}{r_p} \cos \theta + \frac{v^2}{2r_p} \cos \theta = g \sin \theta$$

$$\frac{3}{2} \frac{v^2}{r_p} \cos \theta = g \sin \theta$$

This means that the inclination of the disc is

$$\tan \theta = \frac{3}{2} \frac{v^2}{r_p g}$$

$$= \frac{3 \times 5^2}{2 \times 12 \times 9 \cdot 81} = 0 \cdot 3185$$

In other words, the angle of inclination is

$$\theta = 17° \; 40'$$

To calculate the gyroscopic couple,

Mass of the disc = Volume × Density

$$= \frac{\pi}{4} \times 1^2 \times \frac{50}{1,000} \times 7000 = 274 \text{ kg}$$

Moment of inertia $I = mk^2 = m(r^2/2)$
$$= 274(0 \cdot 5^2/2) = 34 \cdot 25 \text{ kg m}^2$$
$$\omega = v/r = 5/0 \cdot 5 = 10 \text{ rad/s}$$
$$\omega_p = v/r_p = 5/12 = 0 \cdot 417 \text{ rad/s}$$

Then,

Gyroscopic couple $= I\omega\omega_p$
$$= 34 \cdot 25 \times 10 \times 0 \cdot 417$$
$$= 143 \text{ Nm}$$

4.5 DISC INCLINED TO AXIS OF SPIN

Certain mechanisms, swash pumps for example, contain rotating parts inclined to the axis of spin at some angle other than 90°, see Fig. 4.10 (a). The problem is to find the gyroscopic torque.

Let aa represent the axis of a shaft to which a disc, inclined to the shaft, is keyed. Let the shaft and the disc be rotating with angular velocity ω rad/s clockwise as viewed in the direction of arrow L. Let xx represent the polar axis of the disc and yy and zz represent diametral axes mutually perpendicular to xx. The total angular velocity ω of the disc may be resolved into rectangular components about axes xx and yy so that, from the velocity diagram (Fig. 4.11),

$$\omega \cos \theta = \text{Angular velocity about xx}$$
$$\omega \sin \theta = \text{Angular velocity about yy}$$

(*a*) *Spin about axis* xx. From Fig. 4.12, the axis xx itself rotates about aa with angular velocity ω, forming one mode of precession.

Consider a half length l of axis xx, the free end of which describes a circle of radius r.

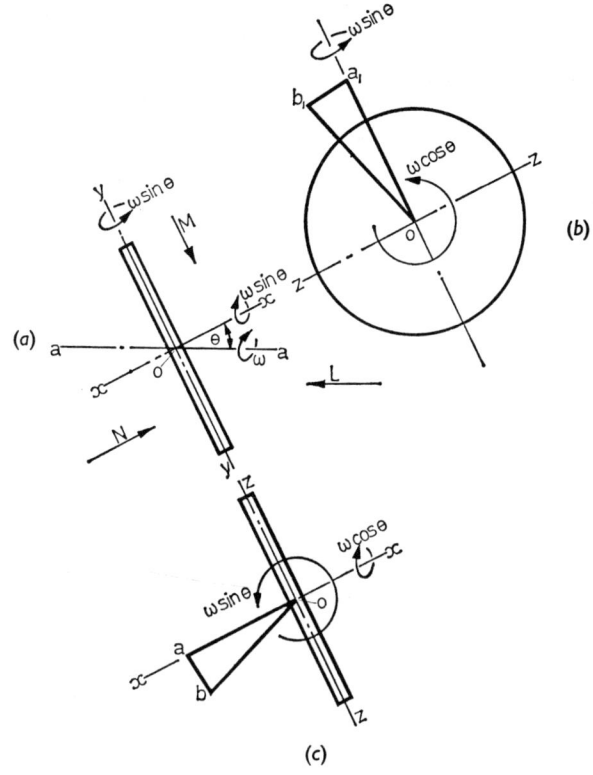

Fig. 4.10

If $\delta\beta$ is the angular displacement of l in time δt, then, for small angles,

$$\omega \cdot \delta t \cdot r \simeq \delta\beta \cdot l$$

When $\delta t \to 0$,

$$d\beta = \omega \, dt \cdot r/l$$

but

$$r/l = \sin\theta$$

$\therefore$ Velocity of precession $d\beta/dt = \omega \sin\theta$

and

Gyroscopic torque due to the precession of axis xx $= I_{xx}\omega \cos\theta\omega \sin\theta$

$$= I_{xx}\omega^2 \sin\theta \cos\theta$$

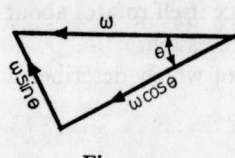

Fig. 4.11

Vewing the disc in the direction of arrow M (Fig. 4.10 (*c*)), $\overrightarrow{ab}$ indicates the magnitude, direction and sense of the change in angular momentum and hence the sense of the externally applied torque. The gyroscopic torque will have opposite sense, that is, clockwise about zz.

(*b*) *Spin about axis* yy. The axis yy also rotates about aa with angular velocity ω (Fig. 4.13) forming a second mode of precession. Suppose that the free end of a half length l_1 of this axe yy, describes a circle of radius r_1 and let $\delta\phi$ be the angular displacement of l_1 in time δt. For small angles

$$\omega \cdot \delta t r_1 \simeq \delta\phi l_1$$

and when $\delta t \to 0$,

$$d\phi = \omega \cdot dt \cdot r_1/l_1$$

But $\qquad\qquad r_1/l_1 = \sin(90 - \theta) = \cos\theta$

$$\therefore \text{Velocity of precession } d\theta/dt = \omega\cos\theta$$

It follows that the

Gyroscopic torque due to the precession of axis yy $= I_{yy}\omega\sin\theta\omega\cos\theta$

$$= I_{yy}\omega^2\sin\theta\cos\theta$$

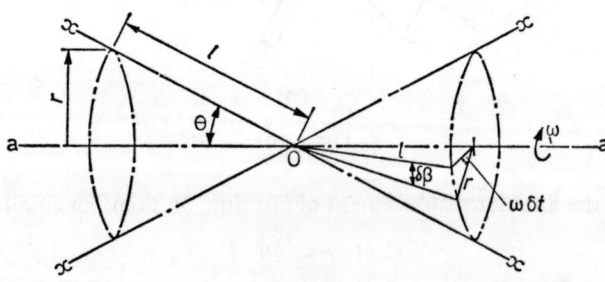

Fig. 4.12

To obtain the direction of the torque, viewing the disc in the direction of arrow N (Fig. 4.10 (*a*)) $\overrightarrow{a_1b_1}$ indicates the magnitude, direction and sense of the change in angular momentum and hence the sense of the externally applied torque. The gyroscopic torque will have opposite sense, and will therefore be anticlockwise about zz as viewed in Fig.

4.10 (*a*) and as indicated in Fig. 4.10 (*c*). The two torques about zz are opposite in sense, so that,

$$\text{Net torque } T \text{ about zz} = I_{xx}\omega^2 \sin\theta \cos\theta - I_{yy}\omega^2 \sin\theta \cos\theta$$
$$= \tfrac{1}{2}[(I_{xx} - I_{yy})\omega^2 \sin 2\theta] \qquad (4.8)$$

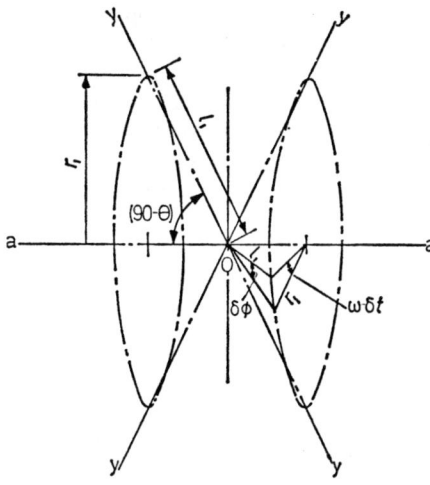

Fig. 4.13

For a thin disc

$$I_{yy} = \tfrac{1}{2}I_{xx}$$

$$\therefore \text{ Net torque } T \text{ for a thin disc} = \tfrac{1}{2}(I_{xx}\omega^2 \sin\theta/\cos\theta)$$
$$= \tfrac{1}{4}(I_{xx}\omega^2 \sin 2\theta) \qquad (4.9)$$

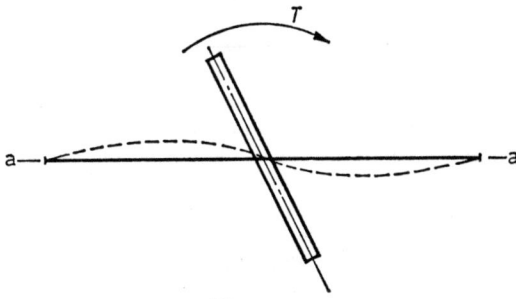

Fig. 4.14

This torque will be clockwise in sense and will be transmitted to the shaft in such a way as to tend to flex it (Fig. 4.14). The plane in which T is applied rotates at the same speed and with the same sense as the shaft, so that the bending moment and flexure on the shaft remain constant for a uniform shaft speed.

PROBLEMS

For tutorials

1. How many degrees of freedom has a gyroscope which is mounted so that its axis of spin can be aligned with any point in space?

2. A boy is bowling a hoop along a level pavement and controlling its movement with a short length of rod. At what point must he touch the hoop, and in what direction must he apply pressure, if he wishes to make the hoop turn to the left?

3. Explain why a cyclist riding without holding the handlebars is able to cause his bicycle to *turn* to the right or left by *inclining* the machine and himself to the right or left respectively.

4. A pair of locomotive wheels rigidly connected by the axle passes over a weak joint in one rail of a level track so that the right hand wheel drops 10 mm and then recovers, the fall and recovery occurring in about $\frac{1}{10}$ of a second. Discuss the gyroscope effects of this on the locomotive frame. Estimate the moment of inertia of the wheel-axle system and so determine the probable magnitude of the maximum gyroscopic torque. (Assume that the fall and recovery both take place with simple harmonic motion.)

General

1. A thin circular hoop of mean diameter 1 m is made from wire weighing 2·5 N/m. It rolls on its edge round a flat horizontal track of 12 m radius at a speed of 6 m/s. Determine the angle of inclination to the vertical at which the hoop runs and the gyroscopic couple.

Ans. 31° 27′, 1·199 Nm

2. A motor vehicle travelling at 48 km/h rounds a curve of 40 m radius. The mass of the vehicle is 1000 kg and its centre of gravity is 0·75 m above the road surface. The effective diameter of the wheels is 0·6 m and the moment of inertia of each road wheel is 1·6 kg m². The rotating parts of the engine and transmission are equivalent to a flywheel of mass 70 kg and radius of gyration 0·1 m. The engine rotates at a speed of 1800 rev/min clockwise when viewed from the front. Find the magnitude of the centrifugal and gyroscopic couples acting on the vehicle when it is turning to the left. If the length of the wheel base is 2·5 m and the width between the wheels is 1·25 m, determine the vertical reaction of the road on each wheel during the turn.

Ans. Centrifugal couple 3330 Nm. Total gyroscopic torque on road wheels 94·8 Nm. Gyroscopic torque on engine, etc. 43·9 Nm. Road reactions: Left hand front wheel 1091·3 N; Left hand rear wheel 1073·7 N; Right hand front wheel 3831·3 N; Right hand rear wheel 3813·7 N.

3. If the vehicle in the previous problem had turned to the right instead of the left, which answers would need modification?

4. Derive from first principles the expression for the gyroscopic torque required to give a precessional velocity ω_p to a body having a moment of inertia I and a velocity of spin ω.

The turbine rotor of a ship has a mass of 6000 kg and a radius of gyration of 0·45 m. It rotates at 2400 rev/min clockwise when viewed from the stern. What will be the magnitude of the gyroscopic torques set up by the rotor in the following circumstances?

(a) When the ship is moving at 30 km/h while making a turn to the left in a curve of 100 m.

(b) When the ship is pitching with simple harmonic motion, the total angular movement being 12° between extreme positions, the periodic time being 24 seconds and the bow of the ship is rising with maximum velocity.

(c) When the ship is rolling and has an angular velocity of 0·08 rad/s at a certain instant.

What is the maximum angular acceleration of the ship in case (b)?

Ans. (a) 26·6 kN m; (b) 8·36 kN m; (c) zero torque; (d) 0·00718 rad/s²

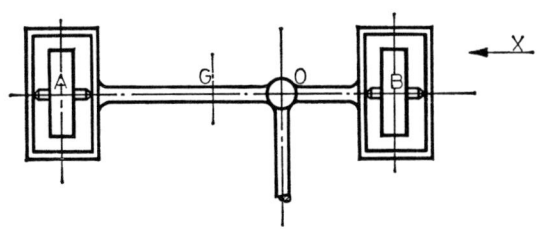

Fig. 4.15

5. A shaft has a wheel keyed to it at the mid-point of its span. The wheel is incorrectly bored so that after being fitted to the shaft, it makes an angle of 0·15° with a plane at right angles to the shaft. If the wheel has a mass of 600 kg and radii of gyration of 0·5 m about a polar axis and 0·35 m about a diametral axis, determine the gyroscopic couple caused by this misalignment when the shaft is running at 1800 rev/min. What will be the maximum bending stress in the shaft due to the gyroscopic couple at the above speed if the shaft diameter is 90 mm and the distance between the bearing centres is 0·9 m. Also find the force exerted on each bearing due to the gyroscopic torque.

Ans. 7,130 Nm, 50 MN/m², 7922 N

6. The combined mass of a motor cycle and rider is 200 kg and its centre of gravity 0·6 m is above ground level when the machine is standing upright. Each of the road wheels has a moment of inertia of 1·4 kg m² and an effective diameter of 0·6 m. The engine speed is six times that of the road wheels and is of the same sense. The moment of inertia of the rotating parts of the engine is 0·25 kg m². Determine the angle to which the rider and machine must lean when travelling in a curve of 30 m at a speed of 64 km/h.

Ans. 50° 17′ to the vertical

7. The frame of a gyroscope shown in Fig. 4.15 is supported at O in such a manner that it can turn freely in all directions. Rotor A is a uniform disc of 0·1 m diameter and mass 2·25 kg with its centre of gravity 0·3 m from O. Rotor B

is a uniform disc of 0·15 m diameter of mass 3·6 kg with its centre of gravity 0·1 m from O. The frame alone has a mass of 1·125 kg with centre of gravity G distant 0·05 m from O on the same side of O as rotor A. If rotor A rotates at 8000 rev/min clockwise and rotor B at 10000 rev/min anticlockwise, both viewed in the direction of the arrow X, with their axles horizontal, find the resultant motion of the system neglecting friction. Determine also the horizontal and vertical components of the *reaction* at the pivot.

Ans. Clockwise precession of 0·447 rad/s when viewed from above; Horizontal reaction 0·074 N towards rotor B; Vertical reaction 68·45 N upwards.

CHAPTER 5

Belt Drives, Pivots and Clutches

5.1 BELT DRIVES

Belt tensions

When a belt is fitted around pulleys so as to transmit power between them, it is given an initial tension to hold it firmly against the pulley surfaces. If a torque is applied, the driving pulley tends to move relative to the belt. This tendency is resisted by the frictional forces between the belt and the pulley, resulting in an increase of tension along one side (the tight side) and a corresponding decrease along the other side (the slack side) of the belt. This difference in tensions exerts a torque on the driven pulley, which will turn when the torque transmitted is great enough to overcome the resistance at the driven pulley.

The change in the tensions between the tight and slack sides takes place over those portions of the belt which, at any instant, are in contact with the pulleys and depends upon the coefficient of friction and the angle of lap of the belt around the pulley. Limiting conditions are reached when the applied torque is great enough to overcome the belt friction, and can occur when the system is at rest as well as in motion. The limiting conditions are also those under which maximum torque can be transmitted, so that the relationships between the belt tensions, friction and angle of lap for these conditions are of great importance.

Fig. 5.1 (a) shows one pulley and a portion of a flat belt. In what follows, T_1 and T_2 are the tensions in the tight and slack sides of the belt respectively, and μ is the limiting coefficient of friction between the belt and the pulley. The angle of lap of the belt is θ.

The belt tension builds up from T_1 to T_2 by the cumulative effect of the friction over the angle of lap between the points of tangency P and Q. An element of the belt with length $r\delta\phi$ at an angle ϕ to OP will be in equilibrium under four forces—the belt tensions T and $T + \delta T$, the pulley reaction R acting normal to the pulley surface and the tangential friction force μR. These forces form a polygon of forces in Fig. 5.1 (b). Because $\delta\phi$ is small,

$$T + \mu R \simeq T + \delta T$$

or $$\mu R \simeq \delta T \qquad\qquad (5.1)$$

But $$R \simeq T\delta\phi$$

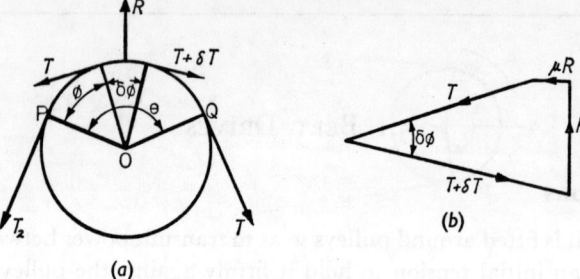

(a)

(b)

Fig. 5.1

Substituting in Eq. (5.1) for R,

$$\mu T\delta\phi \simeq \delta T$$

and, in the limit,

$$\mu T\,\mathrm{d}\phi = \mathrm{d}T$$

or $$\mu\,\mathrm{d}\phi = \mathrm{d}T/T$$

Integrating

$$\int_{\phi=0}^{\phi=\theta} \mu\,\mathrm{d}\phi = \int_{T_2}^{T_1} \mathrm{d}T/T$$

$$\mu\theta = \log_e T_1 - \log_e T_2$$

$$= \log_e T_1/T_2$$

or $$T_1/T_2 = \mathrm{e}^{\mu\theta} \qquad\qquad (5.2)$$

where e, the base of the Naperian logarithms, equals 2·718 approximately. Because μ is the limiting coefficient of friction, Eq. (5.2) gives the ratio of the belt tensions when slip is about to take place.

Fig. 5.2 shows two pulleys of different diameters joined by a belt. The angle of lap is less on the smaller than on the larger pulley so that the maximum permissible ratio of belt tensions is given by $\mathrm{e}^{\mu\theta_1}$ where θ_1 is the angle of lap at the smaller pulley, which is where slip will occur if this ratio is exceeded.

Centrifugal tension. Eq. (5.2) is based on the assumption that the belt speed is sufficiently low for centrifugal effects to be neglected. In its passage around each pulley, the belt follows a curved path and so is subjected to centripetal acceleration. Because of the mass of the belt, a force is required to produce this acceleration and this can only be provided by a pull in the belt itself—a tension within the belt known as the centrifugal tension which is additional to the tensions T_1 and T_2.

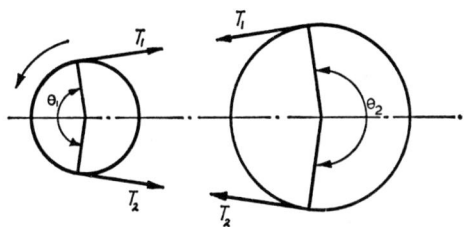

Fig. 5.2

The centrifugal tension can be calculated by considering an element of the belt of length $r\delta\theta$, where r is the radius of the pulley.

$$\text{Mass of the element} = mr\delta\theta$$

where m is the mass of belt per unit length. Then

$$\text{Centripetal acceleration} = \frac{v^2}{r}$$

if v is the belt velocity, and

$$\text{Centripetal force } R = mr\delta\theta v^2/r$$

If T_c is the belt tension necessary to balance the centrifugal reaction equal to R, then Fig. 5.3 (b) shows that if $\delta\theta$ is a very small angle,

$$T_c\delta\theta \simeq R$$

or, substituting for R,

$$T_c\delta\theta \simeq mr\delta\theta v^2/r$$

In the limit,

$$T_c\, d\theta = mr\, d\theta v^2/r$$

or

$$T_c = mv^2 \tag{5.3}$$

The centrifugal tension T_c is thus independent of the angle of lap and so is uniform throughout the belt length. The total tensions in the two sides of the belt are therefore:

$$(T_1 + T_c) \text{ on the tight side}$$

and $\quad\quad\quad\quad (T_2 + T_c) \text{ on the slack side}$

The effective tensions—those causing motion and resulting in power transmission—are still given by T_1 and T_2.

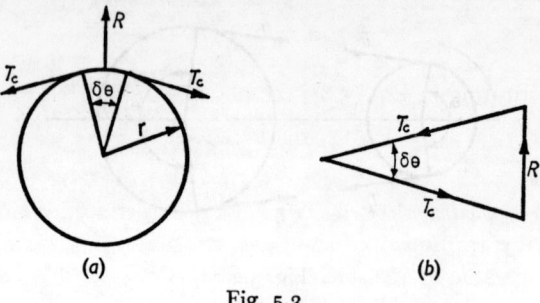

Fig. 5.3

Power transmission by belts

The tensions T_1 and T_2 are effective at the periphery of each pulley (Fig. 5.4) so that the torque exerted can be expressed as,

$$\text{Torque on the smaller pulley} = (T_1 - T_2)r_1 \quad\quad (5.4)$$

$$\text{Torque on the larger pulley} = (T_1 - T_2)r_2 \quad\quad (5.5)$$

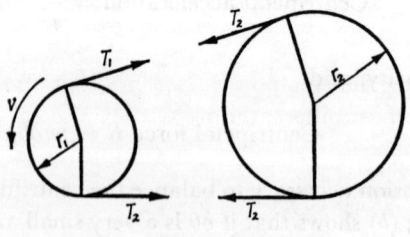

Fig. 5.4

where r_1 and r_2 are the radii of the smaller and larger pulleys respectively. Moreover,

Power transmitted = Torque × Angle turned in unit time

If, therefore, n_1 and n_2 are the speeds in rev/min of the smaller and larger pulleys, and if r_1 and r_2 are measured in metres and T_1 and T_2 in newtons, the power P is

$$P = (T_1 - T_2)2\pi r_1 n_1/60 \text{ Watts} \quad\quad (5.6)$$

Alternatively,

$$P = (T_1 - T_2)2\pi r_2 n_2/60 \text{ Watts} \qquad (5.7)$$

But the belt speed v (in m/s) $= 2\pi r_1 n_1/60 = 2\pi r_2 n_2/60$, so that

$$P = (T_1 - T_2)v \text{ Watts} \qquad (5.8)$$

If it is assumed that the initial tension T_0 is equal to the mean of the tensions on the tight and slack sides (which requires that the belt material follows Hooke's law and that the overall length remains constant),

$$T_2 = 2T_0 - T_1 \qquad (5.9)$$

so that, substituting in Eq. (5.8), the power transmitted

$$P = 2(T_1 - T_0)v \text{ Watts} \qquad (5.10)$$

Transmission of maximum power. It is now possible to determined the conditions for the transmission of maximum power. The first step is to substitute the value of T_2 from Eq. (5.2) in Eq. (5.8), which gives

$$P = (T_1 - T_1/e^{\mu\theta})v$$
$$= T_1(1 - 1/e^{\mu\theta})v = kT_1v$$

where $k = (1 - 1/e^{\mu\theta})$, a constant.

If T is the maximum permissible tension in the belt, centrifugal tension included

$$T = (T_1 + T_c) \quad \text{or} \quad T_1 = (T - T_c)$$

and then
$$P = kv(T - T_c)$$
$$= kv(T - mv^2) \qquad (5.11)$$

Eq. (5.11) indicates that the power transmitted is also dependent on the belt velocity v so that, differentiating with respect to v and equating to zero for a maximum,

$$dP/dv = k(T - 3mv^2) = 0$$
$$\therefore \ T = 3mv^2 \qquad (5.12)$$

or
$$v = \sqrt{(T/3m)} \qquad (5.13)$$

This is the belt velocity for maximum power transmission. Eq. (5.12) shows that $T = 3T_c$. But,

$$T_1 = T - T_c$$

so that
$$T_1 = 3T_c - T_c = 2T_c$$

This means that maximum power will be transmitted when the tension in the right side of the belt is equal to twice the centrifugal tension. From Eq. (5.11), it is also clear that when the belt speed $v = 0$ or $\sqrt{(T/m)}$, the power transmitted will be zero.

Example. *What is the maximum power which a flat belt can transmit to a pulley if the belt is 100 mm wide and 5 mm thick and has a density of 1000 kg/m³? The maximum stress in the belt is not to exceed 1·5 MN/m², the angle of lap is 120° and the coefficient of friction 0·3. Also, what will be the speed of the belt when transmitting maximum power?*

$$\text{Permissible belt tension } T = \text{Stress} \times \text{Area}$$

$$= 1{\cdot}5 \times 10^6 \times (100 \times 5/10^6)$$

$$= 750 \text{ N}$$

$$\text{Centrifugal tension } T_c = T/3$$

$$= 750/3 = 250 \text{ N}$$

$$\text{Mass of unit length of belt} = m = 1000 \times 0{\cdot}1 \times 0{\cdot}005 \times 1$$

$$= 0{\cdot}5 \text{ kg/m}$$

$$\therefore T_c = mv^2 = 0{\cdot}5v^2 = 250 \text{ N}$$

$$\therefore v^2 = 250/0{\cdot}5 = 500$$

or

$$v = 22{\cdot}4 \text{ m/s}$$

This is the speed of the belt when transmitting maximum power. Now $\theta = 120° = 2\pi/3$ radians, $\mu = 0{\cdot}3$ and $T_1/T_2 = e^{\mu\theta}$. Therefore,

$$T_1/T_2 = e^{0{\cdot}2\pi} = 1{\cdot}875$$

$$T_1 = T - T_c = 2T_c = 2 \times 250 = 500 \text{ N}$$

Therefore for maximum power,

$$500/T_2 = 1{\cdot}875$$

or

$$T_2 = 267 \text{ N}$$

$$\therefore \text{ Maximum power} = (T_1 - T_2)v$$

$$= (500 - 267)22{\cdot}4$$

$$= 5219{\cdot}2 \text{ Watts}$$

V-belt drives

V-belt drives. The advantage of the V-belt over the flat belt drive arises from the wedge action between belt and pulley groove, the effect of which is to modify the relationship in Eq. (5.2). In Fig. 5.5 (*a*), let R be the radially outward reaction of the pulley on an element of belt, and P the reaction of the walls of the pulley groove on each side of the elemental length of belt. Then from Fig. 5.5 (*b*),

$$R = 2P \sin \alpha$$

or
$$P = R/2 \sin \alpha \qquad (5.14)$$

where α is half the pulley groove angle.

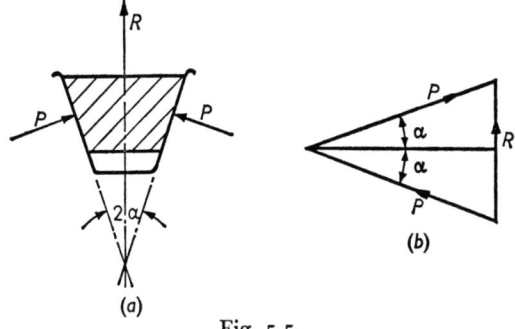

(*a*)

(*b*)

Fig. 5.5

If μ is the limiting coefficient of friction, then the

Force tangential to pulley required to cause the belt to commence slipping

$$= \mu \text{ (total reaction between the belt and pulley)}$$
$$= \mu 2P$$

Substituting for P from Eq. (5.14),

$$\text{Force tangential to pulley} = (\mu/\sin \alpha)R$$
$$= \mu_1 R$$

where μ_1 is referred to as the virtual or false coefficient of friction. This means that the grip of a V-belt is equal to that of a flat belt whose coefficient of friction has been increased to $\mu/\sin \alpha$. If the term μR is replaced by $(\mu/\sin \alpha)R$ in Eq. (5.1) and the subsequent working, Eq. (5.2) will be modified to give

$$T_1/T_2 = e^{\mu\theta/\sin \alpha}$$

or
$$T_1/T_2 = e^{\mu_1\theta} \qquad (5.15)$$

British Standard 1440:1962 *Endless V-belt Drives*, specifies an included angle
of 40° for all V-belts. Pulley diameters, however, must vary to make
possible the necessary speed reductions or increases. The smaller the
pulley, the greater the flexure of the belt as it passes over the pulley and
the greater the tendency for the belt to bulge outwards below the neutral
axis. In order to accommodate this bulge and so prevent too high a
pressure between the belt and the sides of the groove, which would cause
unnecessary wear, the pulley grooves are modified so that the groove
angle decreases as the pulley diameter is reduced. Groove angles of
38°, 36°, 34° and 32° are specified for various pulley diameters and
groove cross sections.

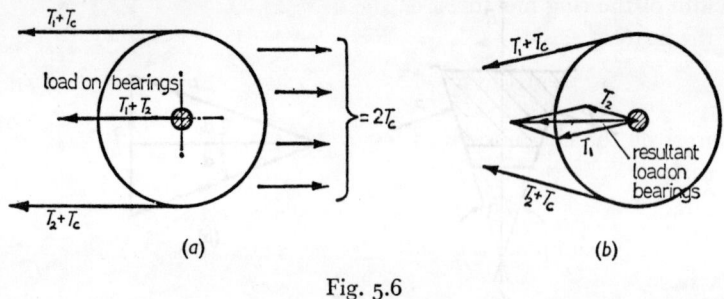

Fig. 5.6

Load on the bearings. When the belt sides are parallel, as when the pulleys
are equal in diameter (Fig. 5.6 (*a*)), the total load on the bearings of
each pulley is given by $T_1 + T_2$. Otherwise, the total load is the vector
sum of the tensions T_1 and T_2 (Fig. 5.6 (*b*)). It is important that the
centripetal tensions T_c are not transmitted to the pulley bearings but
are balanced by the centrifugal reaction of the belt in its passage round
each pulley.

5.2 Pivots

The transmission of power through pivot bearings of various kinds
gives rise to important problems.

Conical pivots

Fig. 5.7 represents a conical pivot with apex angle 2β and with r_1 and r_2
as the maximum and minimum radii of the bearing surface.

If the coefficient of friction μ is constant, the rate of wear of a bearing depends upon the bearing pressure and the rubbing speed. It is usually assumed that either (a) the pressure p remains uniform over the whole bearing surface, in which case the rate of wear will vary with the radius x (since the rubbing speed increases proportionately with radial distance from the pivot axis) or (b) the rate of wear is uniform, in which case p will vary with x and the product of pressure and radius will be constant. (a) *Uniform pressure.* For an elemental ring of the bearing surface of radius x with width (in plan) of δx.

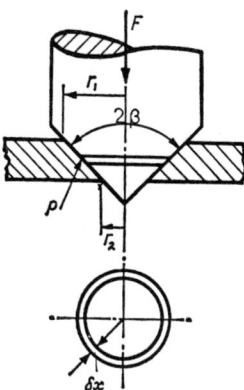

Fig. 5.7

Width of the ring measured on the face on the bearing

$$= \delta x / \sin \beta$$

Thrust on the ring normal to the bearing surface

$$= p \cdot 2\pi x \delta x / \sin \beta$$
$$\text{Axial component of thrust} = (p \cdot 2\pi x \delta x / \sin \beta) \sin \beta$$
$$= p \cdot 2\pi x \delta x$$

The total thrust is the integral of this and must equal the axial load F on the pivot

so that,
$$F = \int_{x=r_2}^{x=r_1} p \cdot 2\pi x \, dx$$

Since p is assumed constant, then

$$F = \pi p (r_1{}^2 - r_2{}^2)$$

or
$$p = F / \pi (r_1{}^2 - r_2{}^2) \qquad (5.16)$$

$$\text{Friction force on the ring} = \mu p 2\pi x \delta x / \sin \beta$$

$$\text{Friction moment on the ring} = \mu p 2\pi x^2 \delta x / \sin \beta$$

$$\text{Total friction torque } T \text{ on the pivot} = (\mu p 2\pi / \sin \beta) \int_{x=r_2}^{x=r_1} x^2 \, dx$$

$$= (\mu p 2\pi / 3 \sin \beta)(r_1{}^3 - r_2{}^3)$$

Substituting for p from Eq. (5.16),

$$T = \frac{2}{3} \frac{\mu F}{\sin \beta} \frac{r_1{}^3 - r_2{}^3}{r_1{}^2 - r_2{}^2} \qquad (5.17)$$

(b) *Uniform wear.* In this case, $px = k$ where k is a constant. Then, the thrust on the elemental ring normal to the bearing surface

$$= p \cdot 2\pi x \delta x / \sin \beta$$

$$= 2\pi k \delta x / \sin \beta$$

$$\text{Axial component of thrust} = (2\pi k \delta x / \sin \beta) \sin \beta$$

$$= 2\pi k \delta x$$

$$\text{Total axial thrust } F = 2\pi k \int_{x=r_2}^{x=r_1} dx$$

$$= 2\pi k (r_1 - r_2)$$

so that
$$k = F / 2\pi (r_1 - r_2) \qquad (5.18)$$

$$\text{Friction force on the ring} = \mu 2\pi k \delta x / \sin \beta$$

$$\text{Friction moment on the ring} = \mu 2\pi k x \delta x / \sin \beta$$

$$\text{Total friction torque } T \text{ on the pivot} = \frac{\mu 2\pi k}{\sin \beta} \int_{x=r_2}^{x=r_1} x \, dx$$

$$= \frac{\mu \pi k}{\sin \beta} (r_1^2 - r_2^2)$$

Substituting for k from Eq. (5.18),

$$T = \frac{\mu \pi F}{2\pi \sin \beta} \frac{(r_1^2 - r_2^2)}{(r_1 - r_2)}$$

$$= \frac{1}{2} \frac{\mu F}{\sin \beta} (r_1 + r_2) \qquad (5.19)$$

Example. *A conical pivot has an apex angle of* 120° *and rests in a bearing surface having outer and inner diameters of* 80 mm *and* 20 mm *respectively. The pivot supports an axial load of* 20 kN *while rotating at* 240 rev/min. *Given a coefficient of friction of* 0·05, *determine the friction torque and the power absorbed in overcoming friction assuming* (a) *uniform pressure* (b) *uniform wear. Which of these criteria should be adopted in the design of the pivot?*

(a) *Uniform pressure.* Substituting the data in Eq. (5.17) gives the friction torque

$$T = \frac{2 \times 0\cdot05 \times 20\,000}{3 \times 0\cdot8660} \left(\frac{0\cdot04^3 - 0\cdot01^3}{0\cdot04^2 - 0\cdot01^2} \right)$$

$$= \frac{2 \times 0\cdot05 \times 20\,000 \times 0\cdot042}{3 \times 0\cdot8660}$$

$$= 32\cdot3 \text{ Nm}$$

Power absorbed in friction $= T \cdot \omega$

$$= 32 \cdot 3 (2\pi 240/60)$$
$$= 812 \text{ W}$$

(b) *Uniform wear.* From Eq. (5.19),

$$T = \frac{0 \cdot 05 \times 20\,000}{2 \times 0 \cdot 8660}\,(0 \cdot 04 + 0 \cdot 01)$$

$$= 0 \cdot 05 \times 20\,000 \times 0 \cdot 05/2 \times 0 \cdot 8660$$

$$= 28 \cdot 9 \text{ Nm}$$

$\therefore$ Power absorbed in friction $= 28 \cdot 9 (2\pi 240/60)$

$$= 726 \text{ W}$$

It is usual to design with full allowance for power absorbed in friction, so that uniform pressure should be assumed.

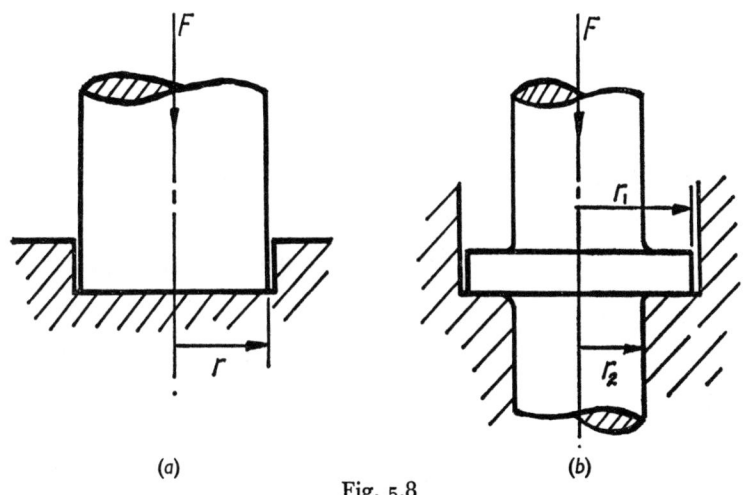

Fig. 5.8

Collar and flat pivot bearings

The conical pivot is converted into a flat pivot or collar bearing of the kinds represented in Fig. 5.8 (a) and (b) respectively by making the apex angle $2\beta = 180°$ ($\sin \beta = 1$). The expressions for torque given in Eqs. (5.17) and (5.19) are therefore modified.

(a) *Uniform pressure.* Using Fig. 5.8 (b), the total friction torque on the bearing,

$$T = \tfrac{2}{3}\,\mu F\,\frac{(r_1{}^3 - r_2{}^3)}{(r_1{}^2 - r_2{}^2)} \tag{5.20}$$

For the bearing in Fig. 5.8 (*a*), where $r_2 = 0$,

$$T = 2/3(\mu F r_1) \qquad\qquad (5.21)$$

(*b*) *Uniform wear.* For the case of Fig. 5.8 (*b*), total friction torque,

$$T = \tfrac{1}{2}\mu F(r_1 + r_2) \qquad\qquad (5.22)$$

For Fig. 5.8 (*a*), where $r_2 = 0$,

$$T = \mu F r_1/2 \qquad\qquad (5.23)$$

Example. *A single collar bearing absorbs* 500 W *in overcoming friction when rotating at* 300 rev/min. *The bearing surface has outside and inside diameters of* 120 mm *and* 60 mm *respectively and the coefficient of friction is* 0·1. *Assuming uniform wear, determine the axial load and the intensity of pressure at the inner and outer radii of the bearing surface.*

$$\text{Power absorbed} = T\omega$$

$$\therefore\ 500 = T . 2\pi 300/60$$

or $\qquad\qquad\qquad T = 50/\pi\ \text{Nm}$

For uniform wear, $\qquad\quad T = (\mu F/2)(r_1 + r_2)$

$$\therefore\ 50/\pi = (0\cdot 1 F/2)(0\cdot 06 + 0\cdot 03)$$

and $\qquad\qquad\qquad F = 50/\pi \times 0\cdot 05 \times 0\cdot 09$

$$= 3550\ \text{N}$$

Also $\qquad\qquad\qquad k = F/2\pi(r_1 - r_2)$

$$= 3550/2\pi \times 0\cdot 03$$

$$= 18\ 870$$

where $\qquad\qquad\qquad k = px$

At the inner radius, $\qquad\quad x = r_2 = 0\cdot 03$

$$\therefore\ \text{Pressure at } r_2 = k/x$$

$$= 18\ 870/0\cdot 03$$

$$= 629\ 000$$

$$= 629\ \text{kN/m}^2$$

At the outer radius, $\qquad\quad x = r_1 = 0\cdot 06$

$$\therefore\ \text{Pressure at } r_1 = 18\ 870/0\cdot 06$$

$$= 314\cdot 5\ \text{kN/m}^2$$

Multicollar thrust bearings

By forming several annular surfaces which share the axial load, larger axial thrusts can be carried with suitably low pressures. Fig. 5.9 represents a multicollar thrust bearing in which the collars are assumed to take equal shares of the load, so that the axial load per collar is F/n where n is the number of collars. Eqs. (5.16) and (5.18) must be modified to give

$$p = F/\pi n(r_1^2 - r_2^2) \text{ for uniform pressure} \tag{5.24}$$

and

$$k = F/2\pi n(r_1 - r_2) \text{ for uniform wear} \tag{5.25}$$

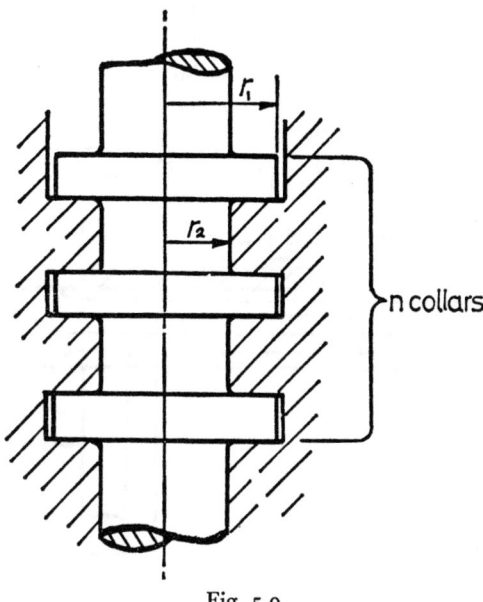

Fig. 5.9

Thus the pressure and hence the torque per collar is reduced n times. The total friction torque for the whole bearing remains unaltered, as will be seen if the modified values of p or k are substituted in the earlier equations for a single collar and then multiplied by n.

5.3 FRICTION CLUTCHES

In power transmission it is often essential to be able to interrupt the drive without reducing the speed of the prime mover, as with a machine tool during examination of the work or when changing the gear of a car.

It may also be desirable to prevent the transmission of undesirable torque fluctuations, and these needs can be met by inserting a friction clutch at a suitable point in the transmission system. Essentially, the clutch is a component with parts fastened to adjacent ends of two co-axial shafts and through which power is transmitted. One part of the clutch is splined, and is moved axially along the shaft when disengaged. Engagement is made by causing the clutch parts to come into contact, when the torque necessary for power transmission is provided by friction at the surfaces.

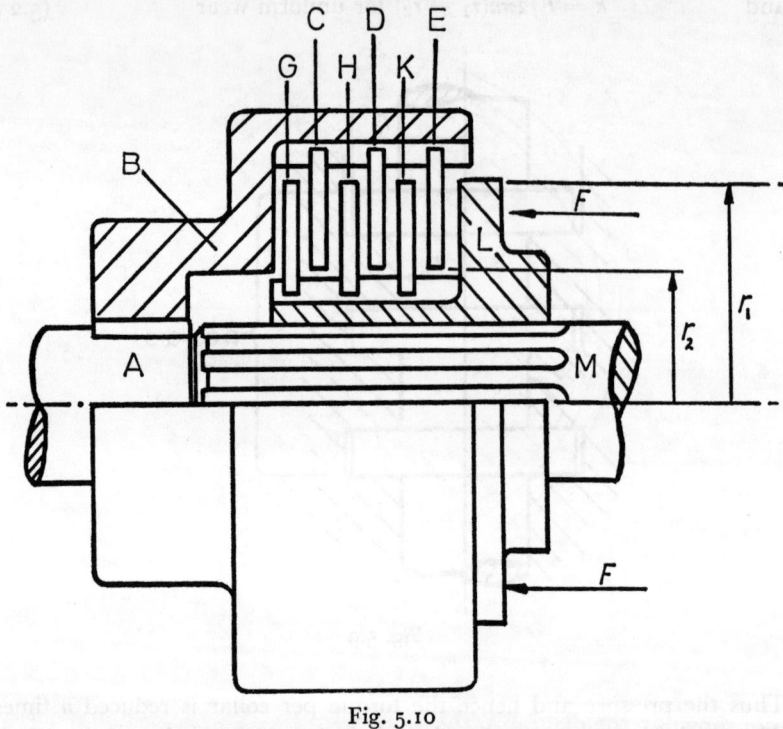

Fig. 5.10

The two main types of friction clutch are the plate clutch and the cone clutch. In the former, the pressure required to hold the plates in contact is usually provided by spring loading, and in certain clutch arrangements, the springs can be adjusted so that the clutch will slip if excessive torque is exerted.

Plate clutches

Fig. 5.10 shows the essential features of a multiplate clutch. The housing B is keyed to and rotates with the driving shaft A. At the periphery, splines are cut internally to engage with slots in the outer circumference of the three annular discs or plates C, D and E. Interleaved with these plates are three other plates G, H and K which engage at their inner circumferences with splines cut externally in the member L, itself carried on the splined shaft M. One set of plates is thus forced to rotate with the member B and the other set with the member L, but all the plates and the component L are free to move axially. In the position shown, the plates are not in contact and torque cannot therefore be transmitted. If, however, a force *F* is applied to L through levers, springs, toggles, etc.

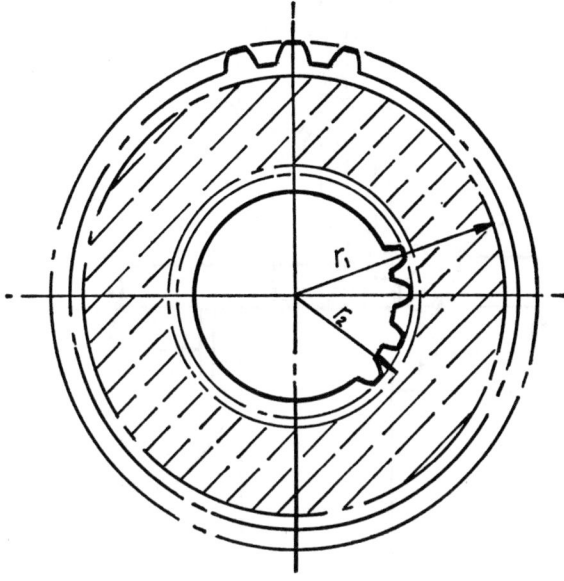

Fig. 5.11

(not shown in the diagram), the plates and the members B and L will be pressed together and torque will be transmitted by means of friction through *each pair* of contacting surfaces to the shaft M. In the clutch shown in Fig. 5.10 the number of pairs of contacting surfaces is 7, but this can vary from one design to another.

The friction area of the contacting plates is annular (Fig. 5.11), the outer and inner radii being r_1 and r_2 respectively. The maximum torque transmitted by a plate clutch is therefore given by the equations for the frictional torque in a collar bearing, provided allowance is made for the number of effective pairs of contacting surfaces.

The axial force *F* holding the plates in contact is transmitted across each pair of surfaces in contact so that, assuming uniform pressure, Eq. (5.16), gives

$$p = F/\pi(r_1^2 - r_2^2)$$

The maximum torque transmitted is

$$T = \tfrac{2}{3}\mu nF(r_1^3 - r_2^3)/(r_1^2 - r_2^2) \qquad (5.26)$$

where n is the number of pairs of contacting surfaces.

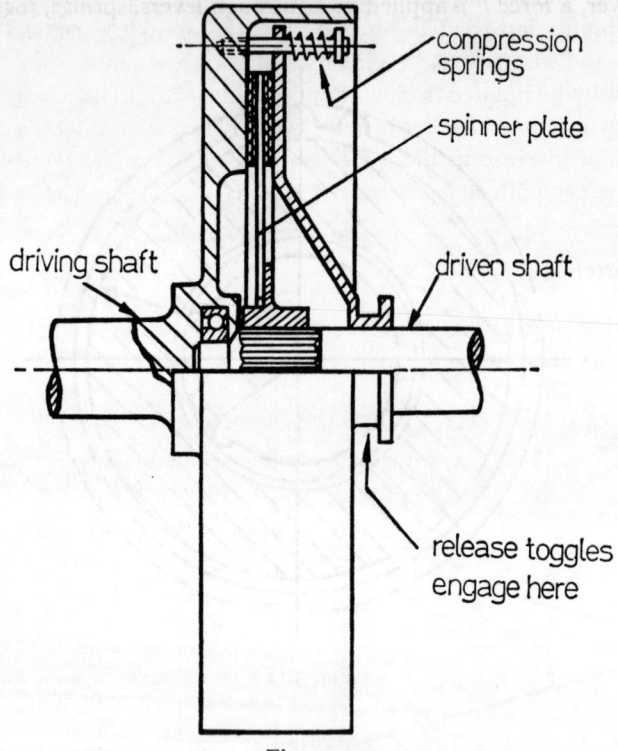

compression springs

spinner plate

driving shaft

driven shaft

release toggles engage here

Fig. 5.12

Assuming uniform wear, the intensity of pressure p at any radius x between r_1 and r_2 is given by:

$$p = k/x$$

k is given by Eq. (5.18), as

$$k = F/2\pi(r_1 - r_2)$$

and

$$T = \tfrac{1}{2}\mu nF(r_1 + r_2) \qquad (5.27)$$

The single-plate clutch has an annular friction surface riveted to each side of the spinner plate (Fig. 5.12), the axial force F being exerted on each side, so that $n = 2$.

Values for p and μ. Permissible values of p vary from about 350 kN/m²
to 2,800 kN/m² depending on the friction material, the environment in
which the clutch operates and the severity of the operation. Recommended values of μ for dry operation vary from 0·20 to 0·27. For oil-
immersed clutches, μ varies from about 0·04 to 0·10 depending upon
pressure and rubbing speeds.

Oil-immersed clutches. Plate clutches may operate dry or immersed in
oil. Oil immersion is advisable in heavy duty units where heat must be
removed and also when the clutch is to be located in a gear box and
cannot easily be isolated from lubricating oil. It brings about a reduction
of the coefficient of friction at the contacting surfaces which must be
offset by an increase in the force holding the plates together. The advantages are smooth engagement and reduced wear of the friction surfaces.

Cone clutches

The driving and driven members of cone clutches have coned surfaces
(Fig. 5.13). The outer female member is usually the driver while the

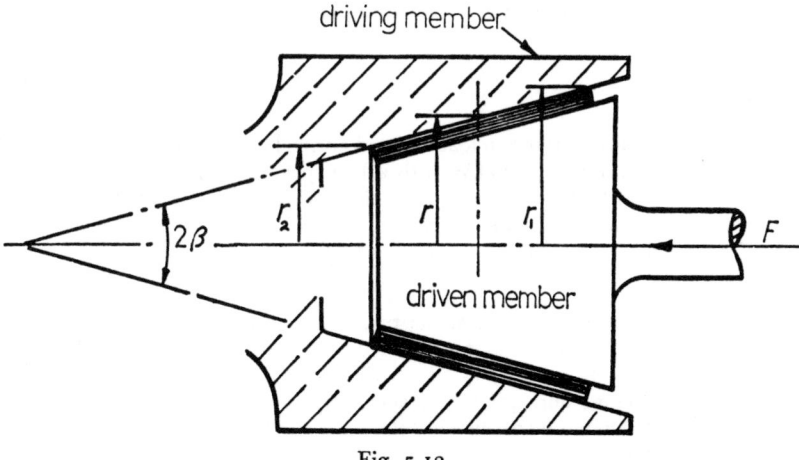

Fig. 5.13

inner male member carries the friction lining riveted to its surface.
The cone clutch is simpler than the plate clutch but engages much more
abruptly, which has restricted its use to relatively small and robust
pieces of machinery.

The Eqs. (5.17) and (5.19) for the frictional torque in a conical pivot
bearing also apply to the cone clutch.

In the design of clutches, the formulae for uniform wear should be used as these give lower, and hence more conservative estimates of the torque which the clutch can transmit.

Problems

For tutorials

1. V-belts must fit the grooves of the pulley so that they do not (a) make contact with the bottom of the grooves, (b) have very small amounts of clearance above the bottom of the grooves. Give one reason in each of the two cases.

2. At a constant speed, the power transmitted by a belt drive is proportional to the *difference* in tensions, while the load transmitted to the pulley bearings is proportional to the *sum* of the tensions. Is it preferable therefore to use a flat or V-belt drive, considering only the implications of the above statement?

3. With reference to pivots, collar bearings and clutches, explain clearly when and why it is preferable to assume (a) uniform wear, (b) uniform pressure.

4. Give reasons why it is usual to make the outer member of a cone clutch the driving member and why the friction lining should be riveted to the inner member. From what materials are the two coned members usually made?

General

1. Determine the difference in load exerted on the bearings of a pulley in a belt drive which is transmitting 230 kW with a belt speed of 25 m/s when a flat belt is exchanged for a V-belt. Assume the ratio is $T_1/T_2 = 3$ for the flat belt and 10 for the V-belt, and that in each case the angle of lap is 180°. Consider only the effective tensions T_1 and T_2, ignoring centrifugal effects.

Ans. Load reduction 7158 N.

2. A belt drive has four V-belts, the pulley groove angles for which are 38°. Each belt has a cross-sectional area of 10^{-4} m² and a density of 1000 kg/m³ and the coefficient of friction is 0·15. The driving and driven pulleys have effective diameters of 150 mm and 450 mm respectively, and their centre distance is 500 mm. If the maximum permissible stress for the belt material, 2 MN/m², is reached when the speed of the driving pulley is 1500 rev/min, determine the effective tensions in each belt and the power transmitted.

Ans. $T_1 = 186·1$ N; $T_2 = 57·8$ N; 6040 W

3. A flat belt drive has pulleys of equal diameter. The belt width and thickness are 100 mm and 7·5 mm respectively, the density of the belt material is 1000 kg/m³ and the coefficient of friction is 0·3. If the maximum permissible stress of the belt material is 1·2 MN/m², determine the maximum power transmitted and the corresponding belt speed.

Ans. 7330 W; 20 m/s

4. A power-operated capstan is used for hauling a load. A rope is attached to the load and wrapped three times around the capstan. The operator exerts a steady pull on the free end. If the load in resisting motion exerts a pull of 5 kN, determine the power supplied by (*a*) the capstan and (*b*) the operator, when hauling at 30 m/min. Assume that the coefficient of friction between the rope and capstan is 0·2.

Ans. (*a*) 2442·3 W; (*b*) 57·7 W

5. What axial load must be applied to a cone clutch which has friction surfaces whose minimum and maximum radii are 80 mm and 100 mm if the coefficient of friction is 0·25 and the maximum pressure normal to the friction surfaces is 350 kN/m²? Assume uniform wear. Also determine the power transmitted by the clutch at a speed of 540 rev/min.

Ans. 3·517 kN; 14·48 kW

6. A shaft of diameter 40 mm is supported vertically by a flat pivot bearing at its lower end and a ball journal bearing at its upper end. The shaft has a flywheel of mass 100 kg and radius of gyration of 0·3 m keyed to it at the mid-point of its length, the plane of the flywheel being horizontal. If the coefficient of friction for the pivot bearing is 0·1 and the friction at the upper bearing can be neglected, determine the number of revolutions the flywheel will make, and the time taken in coming to rest from a speed of 240 rev/min. Neglect the effects of the inertia and weight of the shaft itself. Assume uniform pressure at the pivot.

Ans. 346 revolutions; 2 min 53 seconds

7. A single-plate clutch is used to transmit the drive from a horizontal shaft rotating at a constant speed of 360 rev/min to a second shaft on which is mounted a flywheel and a pulley. The flywheel has a mass of 200 kg and a radius of gyration of 0·4 m. The pulley, the inertial effects of which can be neglected, has an effective diameter of 0·5 m. A rope is wound round the pulley and from its free end is suspended a mass of 100 kg. The radii of the friction surfaces of the clutch are 80 mm and 150 mm and the axial pressure applied at the surfaces when the clutch is engaged is 400 kN/m². If the coefficient of friction for the clutch is 0·25 and constant acceleration can be assumed, determine the angular acceleration of the driven shaft and the time taken for this shaft to attain the same speed as the driving shaft from the moment the clutch is suddenly engaged. Assume that the driven shaft is at rest, that the suspended mass is hanging freely at the moment of engagement and that the pressure at the clutch surfaces is uniform. Also determine the total angle of slip at the clutch surfaces during the period of acceleration.

Ans. 25 rad/s²; 1·51 s; 28·5 radians

CHAPTER 6

Geared Systems

6.1 SIMPLE AND COMPOUND GEAR TRAINS

Ratio of velocities

Geared systems contain two or more gear wheels mounted on shafts and meshed so that angular motion can be transmitted from one shaft to the other. It is theoretically possible to achieve this form of transmission by friction discs held in contact at their peripheries by applying a force R normal to their common tangent. Fig. 6.1 shows the arrangement for spur gearing. In practice, slip will occur whenever the tangential driving force at the point of contact exceeds the frictional force μR.

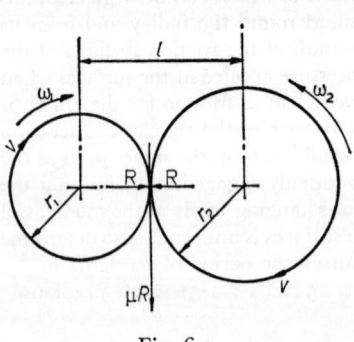

Fig. 6.1

If slip does not occur, the peripheral velocity v of the two discs will be the same, so that

$$v = \omega_1 r_1 = \omega_2 r_2 \qquad (6.1)$$

where r_1 and r_2 are the radii of the smaller and larger discs respectively, and ω_1 and ω_2 are the corresponding angular velocities.

The ratio of the angular velocities of the two wheels is then given by

$$\omega_1/\omega_2 = r_2/r_1 \qquad (6.2)$$

It is worth noting (Fig. 6.1) that

$$r_1 + r_2 = l \qquad (6.3)$$

where l is the distance between the centres of the discs.

Gear wheels retain the kinematic form of friction discs without the possibility of slip by means of teeth spaced or pitched uniformly around the periphery. The teeth are so shaped that when brought into mesh, they interact positively with a continuously constant velocity ratio. The circles on which the teeth are pitched (the pitch circles) represent the circumferences of the kinematically equivalent friction discs and their point of contact with each other is referred to as the *pitch point* (Fig. 6.2). Eqs. (6.1), (6.2) and (6.3) are therefore applicable to gear wheels as well as friction discs.

Involute teeth

Involute gear teeth are in almost cuiversal use because of their strength, the ease with which they can be acnurately formed and their ability to maintain a constant velocity ratio even when the centre distance between two meshing gears is increased slightly, as might occur because of shaft flexure. The geometry of the involute curve is most simply described by considering a cord being unwound from the circumference of a cylinder. If the unwound portion is kept taut, the free end will trace out an involute curve. The end view of the cylinder is represented by a circle called the *base circle* of the involute. Fig. 6.3 shows three successive positions of the free end of the cord q represented by q_1, q_2 and q_3. The corresponding

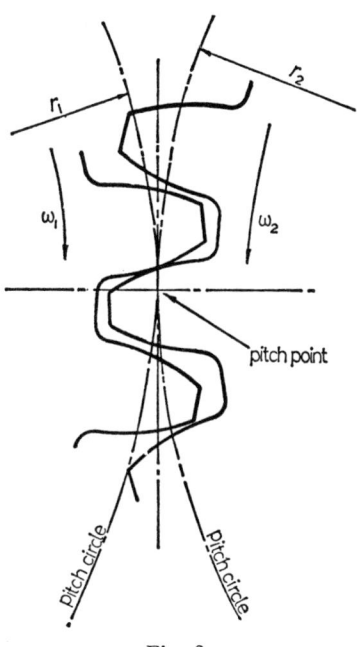

Fig. 6.2

points of contact of the cord with the base circle are I_1, I_2 and I_3 and these must be instantaneous centres for the unwound portions I_1q_1, I_2q_2, etc., so that the directions of motion of q at points q_1, q_2, etc., must coincide with t_1t_1, r_2t_2, etc.—the tangents to the involute at these points. Thus $t_1q_1t_1$, $t_2q_2t_2$, etc., are at right angles to I_1q_1, i_2q_2, etc., and it follows that I_1q_1, I_2q_2, etc. are normals to the involute as well as tangents to the base circle.

Involute teeth have profiles formed by the involutes of circles, and the geometrical arguments above can explain the mode of interaction of gear teeth. In Fig. 6.4 (*a*) and (*b*) represent the centres of two geared wheels. Portions of two involute teeth are shown in mesh, making contact at point Q. The teeth will have a common tangent TT at point Q.

Since I_AQ and I_BQ are normals to the common tangent at point Q, then I_AQI_B is the common normal forming a straight line tangential to both base circles at points I_A and I_B and intersecting the line of centres

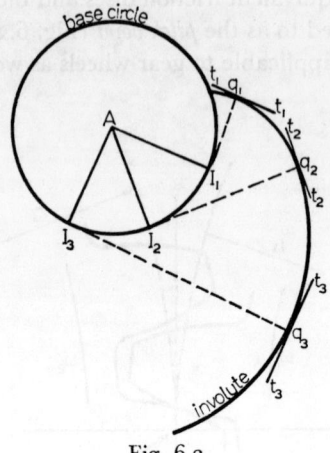

AB at point P. But from the geometry of the involute, the same argument applies to each position of Q during the engagement of the teeth, so that the common normal and its point of intersection P with AB remains fixed. It can be shown that (*a*) for constant velocity ratio, P must be the pitch point of the gears, that is, the common normal must pass through the pitch point, and (*b*) that the involute profile fulfils the conditions for constant velocity ratio stated in (*a*).

Fig. 6.3

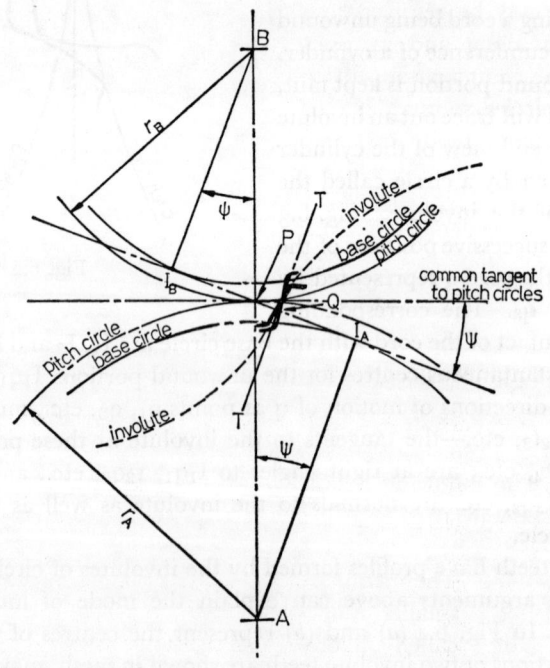

Fig. 6.4

The pressure angle. The triangles AI_AP and BI_BP in Fig. 6.4 are similar (all three angles), so that

$$AI_A/BI_B = AP/BP \qquad (6.4)$$

If ψ is the angle made by the line of centres AB with AI_A and BI_B, then

$$AI_A = AP \cos \psi \qquad (6.5)$$

and

$$BI_B = BP \cos \psi \qquad (6.6)$$

The angle ψ is known as the pressure angle and fixes the sizes of the base circles in relation to the pitch circles and hence the curvature of the tooth profiles. From Fig. 6.4, ψ is also the angle which the common normal I_AI_B makes with the common tangent to the pitch circles. When power is being transmitted, the maximum tooth pressure (neglecting friction at the teeth) is exerted along the common normal through the pitch point. This force may be resolved into components along and at right angles to the common tangent to the pitch circles so that, if F is the maximum tooth pressure (Fig. 6.5),

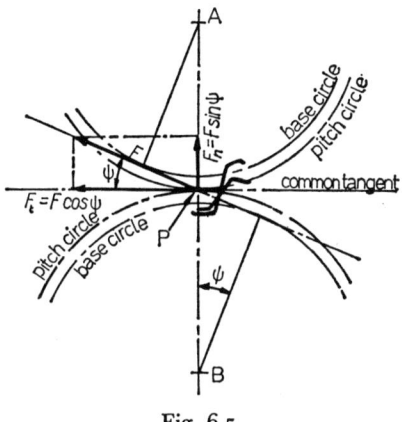

Fig. 6.5

$$\text{Tangential force } F_t = F \cos \psi \qquad (6.7)$$

and
$$\text{Normal force } F_n = F \sin \psi \qquad (6.8)$$

Then

Torque exerted on gear wheel shaft

$$= F_t \times (\text{pitch circle radius of the gear wheel}) \qquad (6.9)$$

Definitions. The following definitions relating chiefly to spur gears are taken from BS 436:1940 and BS 2519:1954 (see also Fig. 6.6).

PINION   The member of an unequal pair of gears that has the smaller number of teeth. (Equal gears are usually called pinions if the diameter is equal to or smaller than the width.)

WHEEL	The member of an unequal pair of gears that has the larger number of teeth. (Equal gears are usually called wheels if the diameter is larger than the width.)
SPUR GEAR	A cylindrical gear with teeth parallel to the axis.
BEVEL GEARS	Gears of conical form designed to operate on intersecting axes.
GEAR RATIO	Of a pair of gears; the ratio of the greater of the two numbers of teeth to the smaller.
PITCH CYLINDERS	Cylinders co-axial with the gears and in peripheral contact, which will roll together without slip.

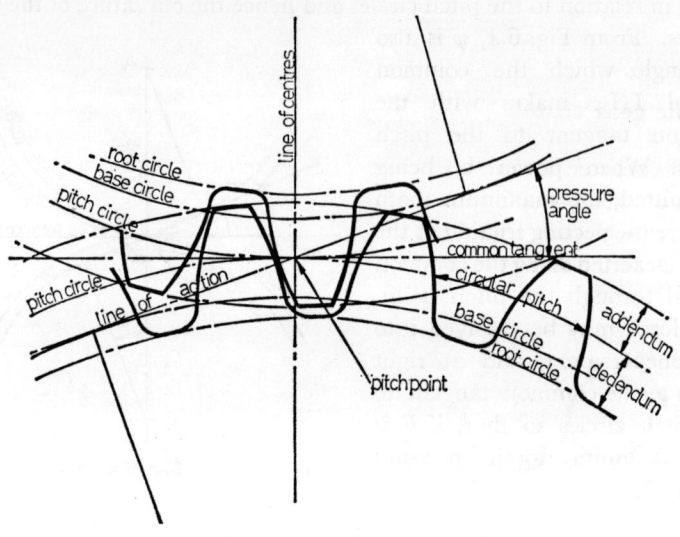

Fig. 6.6

PITCH CIRCLE	Any transverse section of a pitch cylinder normal to the axis.
PITCH POINT	The point of contact of a pair of pitch circles.
PITCH DIAMETER	The diameter of the pitch circle or cylinder.
CIRCULAR PITCH	The length of arc of the pitch circle between similar faces of successive teeth.
DIAMETRAL PITCH	The number of teeth divided by the pitch diameter.
BASE CIRCLE	For involute gears, the circle from which the involutes are derived.
LINE OF ACTION	The common tangent to the two base circles which passes through the pitch point of a pair of mating gears.

MODULE	The pitch diameter divided by the number of teeth; it is the reciprocal of the diametral pitch.
PRESSURE ANGLE	The acute angle between the line of action and the common tangent to the pitch circles at the pitch point.
ADDENDUM	The height from the pitch circle to the tip of the tooth.
DEDENDUM	The depth from the pitch circle to the root of the tooth.
ROOT CYLINDER	The cylinder, co-axial with the gear, that is tangential to the bottom of the tooth spaces.
ROOT CIRCLE	The circle of intersection of the root cylinder by a transverse plane.

Simple gear trains

Systems of three or more wheels in which each wheel is keyed to a separate shaft are called simple gear trains. Figs. 6.7 (*a*) and (*b*) show two simple trains, one of spur and the other of bevel gears. In each case, let gear wheel A be the driver, gear wheel Z the final driven wheel and gear wheel

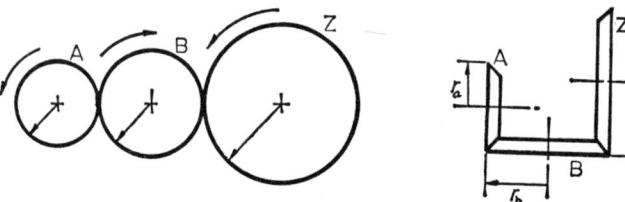

Fig. 6.7

B an intermediate wheel. Also let the radii of the pitch circles for the wheels A, B and Z be r_a, r_b and r_z respectively, let the corresponding angular velocities be ω_a, ω_b and ω_z, and let t_a, t_b and t_z be the numbers of teeth in the three wheels. Because the pitch line velocities of the teeth must be the same for each wheel in the train,

$$\omega_a r_a = \omega_b r_b = \omega_z r_z$$

Picking out the last and first of the terms, which means that the velocity ratio of the train can be expressed in terms of the driving and driven members alone:

$$\omega_a/\omega_z = r_z/r_a = t_z/t_a \tag{6.10}$$

The intermediate wheel of a simple train therefore has no effect on the *magnitude* of the velocity ratio, but does alter the *sense* of the rotation of the driven wheel. The same is true for two or more intermediate wheels.

Compound gear trains

Compound gear trains are systems in which more than one wheel is keyed to each intermediate shaft. An example of compound gearing with spur wheels is shown in Fig. 6.8. Gear A is assumed to be the driver and gear Z the final driven wheel. The intermediate shafts carry gears

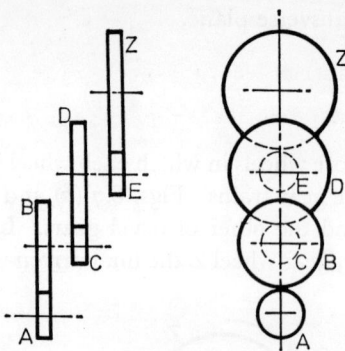

Fig. 6.8

B and C and D and E, of which C and E will be intermediate drivers and gears B and D intermediate driven wheels. Gears A and B alone form a simple train, as do gears C and D and gears E and Z. Therefore

$$\omega_a/\omega_b = t_b/t_a$$

also

$$\omega_c/\omega_d = t_d/t_c$$

and

$$\omega_e/\omega_z = t_z/t_e$$

But ω_c equals ω_b, since gears B and C are keyed to the same shaft and ω_d equals ω_e because gears D and E are keyed to the same shaft. Therefore, since

$$\frac{\omega_a}{\omega_b} \cdot \frac{\omega_c}{\omega_d} \cdot \frac{\omega_e}{\omega_z} = \frac{t_b}{t_a} \cdot \frac{t_d}{t_c} \cdot \frac{t_z}{t_e} = \frac{\omega_a}{\omega_b} \cdot \frac{\omega_b}{\omega_d} \cdot \frac{\omega_d}{\omega_z}$$

In other words,

$$\frac{\omega_a}{\omega_z} = \frac{t_b}{t_a} \cdot \frac{t_d}{t_c} \cdot \frac{t_z}{t_e} = \frac{t_b t_d t_z}{t_a t_c t_e} \tag{6.11}$$

or the ratio of velocities is given by

$$\frac{\omega_a}{\omega_z} = \frac{\text{Product of the teeth on the driven wheels}}{\text{Product of the teeth on the driving wheels}}$$

Example. *A drive is transmitted between two parallel shafts by spur gear wheels. The gear module is 3 mm and the shaft centre distance is 13·5 cm. The velocity ratio of the two wheels is 5 to 1. Find the numbers of teeth in the pinion and wheel. If the teeth are 20° involute, what is the maximum force between the teeth and what is the force tending to separate the gears when the pinion is transmitting 10 kW at 600 rev/min.*

$$\text{Pitch circle radius of pinion} = (1/6)\,13\cdot5 = 2\cdot25 \text{ cm}$$

$$\text{Pitch circle radius of wheel} = (5/6)\,13\cdot5 = 11\cdot25 \text{ cm}$$

$$\text{Number of teeth in pinion} = \text{p.c.d. (mm)/module (mm)}$$

$$= 45/3 = 15$$

$$\text{Number of teeth in wheel} = 225/3 = 75$$

$$\text{Angular velocity of pinion} = \omega = 2\pi 600/60$$

$$= 20\pi \text{ rad/s}$$

$$\text{Power} = \text{Pinion torque} \times \omega$$

$$\therefore \text{ Pinion torque} = \frac{10 \times 1{,}000}{20\pi} = 159 \text{ Nm}$$

$$\text{Force tangential to pitch circle of pinion} = \frac{\text{Torque on pinion}}{\text{p.c. radius of pinion}}$$

$$= (159/2\cdot25)\,100$$

$$= 7070 \text{ N}$$

$$\text{Maximum force between the gear teeth} = 7070/\cos 20°$$

$$= 7070/0\cdot9397$$

$$= 7530 \text{ N}$$

$$\text{Force tending to separate the gears} = 7530 \sin 20°$$

$$= 7530 \times 0\cdot3420$$

$$= 2575 \text{ N}$$

6.2 TORQUE TO ACCELERATE A GEARED SYSTEM

The inertial effects of a geared system can be important. Fig. 6.9 shows a simple geared system with two shafts S_a and S_z to which are keyed wheels A and Z respectively. If ω_a and ω_z are the angular velocities of S_a and S_z, and if α_a and α_z are the corresponding accelerations, I_a and I_z

the moments of inertia of shafts S_a and S_z and their attached masses, and R_{za} the ratio of the velocities and hence the ratio of the accelerations of the shafts S_z and shaft S_a, then

$$R_{za} = \omega_z/\omega_a = \alpha_z/\alpha_a$$

It follows that the

Torque T_a required at S_a to accelerate shaft S_a and its masses $= I_a\alpha_a$

Torque T_z required at S_z to accelerate shaft S_z and its masses $= I_z\alpha_z$

$$= I_z R_{za}\alpha_a$$

since, by definition,

$$R_{za} = \alpha_z/\alpha_a$$

Therefore

$$\text{Torque at } S_a \text{ to accelerate } S_z = R_{za}(I_z R_{za}\alpha_a)$$
$$= I_z R_{za}{}^2\alpha_a$$

It follows that the total Torque T on shaft S_a to accelerate both shafts and their attached masses

$$= T_a + T_z$$
$$= I_a\alpha_a + I_z R_{za}{}^2\alpha_a$$
$$= (I_a + R_{za}{}^2 I_z)\alpha_a \qquad (6.12)$$

The quantity within the brackets is termed the equivalent moment of inertia of the geared system referred to the driving shaft S_a.

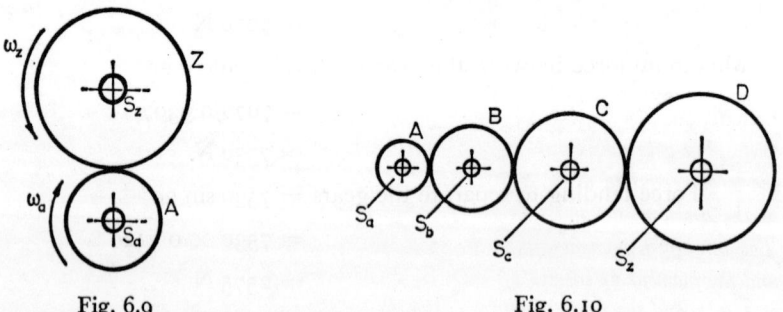

Fig. 6.9 　　　　　　　　　　　Fig. 6.10

If in a train of wheels (Fig. 6.10), intermediate shafts S_b, S_c, etc., and their attached masses, hence moments of inertia I_b, I_c, etc., the total torque T required at shaft S_a to accelerate all these masses is given by:

$$T = T_a + T_b + T_c + \cdots + T_z$$
$$= (\alpha_a I_a) + (\alpha_a R_{ba}{}^2 I_b) + (\alpha_a R_{ca}{}^2 I_c) + \cdots + (\alpha_a R_{za}{}^2 I_z)$$

where R_{ba}, R_{ca}, etc., are the velocity ratios of shafts S_b, S_c, etc., with respect to shaft S_a. This is usually put more compactly in the form:

$$T = [I_a + \Sigma(R^2I)]\alpha_a \qquad (6.13)$$

where I represents the moment of inertia of each shaft and its masses, and R the velocity ratio of each shaft with respect to the driving shaft.

In the compound gear train shown in Fig. 6.11, the shaft S_b carries wheels B and C keyed to it so that the moment of inertia I_b would include the moment of inertia of both wheel B and wheel C, the velocity ratio R of both these wheels being the same with respect to S_a.

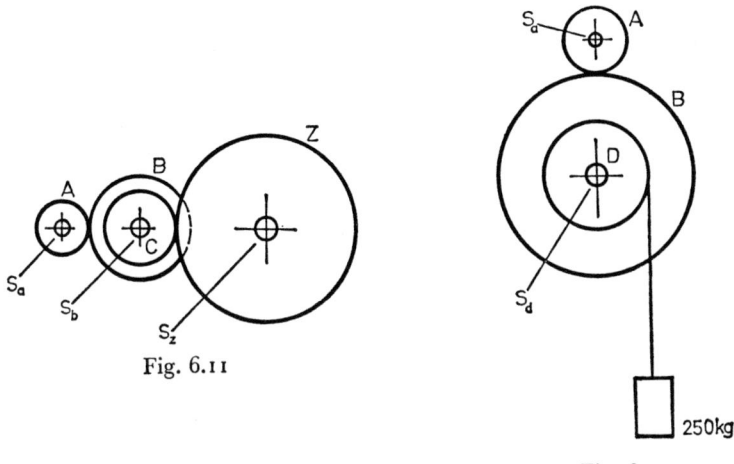

Fig. 6.11

Fig. 6.12

Example. *In Fig. 6.12, shaft S_d has a drum D and a gear wheel B keyed to it. The gear wheel meshes with pinion A keyed to the driving shaft S_a. Determine the torque required at S_a to lift a mass of 250 kg suspended from a cable fastened to the drum if the mass is to be raised with a uniform acceleration of 1 m/s². The combined moment of inertia of shaft S_d, drum and gear wheel is 10 kg m² and the moment of inertia of shaft S_a and the pinion is 1 kg m². The gear ratio is 4 to 1 and the drum diameter is 0·3 m.*

Weight of suspended mass $W = mg$

$$= 250 \times 9\cdot81 = 2452 \text{ N}$$

Total pull in the cable

$F =$ Force to support mass $+$ Force to accelerate mass

$$= W + mf = 2452 + (250 \times 1) = 2702 \text{ N}$$

Torque on S_d due to pull in the cable

$$T_d = F \times \text{Radius of drum}$$

$$= 2702 \times 0.15 = 405.3 \text{ Nm}$$

Also $R_{da} = \omega_d/\omega_a = 1/4 = 0.25$

Torque to support and accelerate the suspended mass alone, referred to

$$S_a = T_d R_{da} = 405.3 \times 0.25 = 101.3 \text{ Nm}$$

Angular acceleration of S_d, $\alpha_d = \dfrac{\text{Peripheral acceleration of drum}}{\text{Radius of drum}}$

$$= 1/0.5 = 6.67 \text{ rad/s}^2$$

$$\therefore \ \alpha_a = \alpha_d/R_{da}$$

$$= 6.67/0.25 = 26.68 \text{ rad/s}^2$$

Equivalent moment of inertia of drum and gears referred to shaft S_a

$$= (I_a + R_{da}{}^2 I_d)$$

$$= 1 + (0.25^2 \times 10) = 1.625 \text{ kg m}^2$$

Therefore total torque required at shaft S_a

$$= (\text{Torque to accelerate geared system})$$
$$+ (\text{Torque to support and accelerate suspended mass})$$

$$= 1.625\alpha_a + 101.3$$

$$= (1.625 \times 26.68) + 101.3$$

$$= 43.4 + 101.3 = 144.7 \text{ Nm}$$

6.3 Epicyclic Gear Trains

Fig. 6.13 shows two gear wheels free to rotate about their axes but kept in mesh by an arm. If wheel A is held at rest and the arm rotated about the axis of A, then wheel B will be forced to rotate *upon* and *around* wheel A. Such motion is described as *epicyclic* (*epi*-upon, and *cyclic*-around) and gear trains arranged so that one or more of their members moves upon and around another member are known as epicyclic trains. They are useful because large reductions of speed can be obtained with compact arrangements.

Epicyclic problems usually involve the determination of the ratio of velocities for the input and output members. There are several methods of solution existing, the most familiar being the tabular method used here, which consists of reducing the relatively complex motion of an epicyclic train to component motions which require a knowledge only of simple or compound trains and which can therefore be analysed by methods already outlined. Vector addition of component angular velocities then gives the required results.

The procedure is illustrated by the problem of Fig. 6.13. Suppose that wheel *A* has 100 teeth and wheel B 50 teeth. If A is held at rest, and the arm rotated once in clockwise direction, how many revolutions does B in fact make? First, the system is assumed to be locked so that its members cannot move relative to each other and, in this condition, the whole system is rotated in the plane of the gear wheels through one complete revolution, say clockwise. In this process wheels A and B and the arm each undergo one clockwise revolution about their own axes. Next, the arm is held at rest while wheel A is rotated once in an anti-clockwise sense. This operation cancels out the revolution already given to A and at the same gives to B a number of further clockwise revolutions equal to $100/50 = 2$. Wheel B therefore makes a total of 3 clockwise revolutions during one clockwise revolution of the arm. The important point is that the second operation involved the motion of a simple train in which the number and the sense of the revolutions made by B can be readily determined. This method of solution is most conveniently shown in tabular form, in which the numbers of revolutions in the above operations are given separately. The sense of each wheel's rotation must be noted, and this has been done by assuming that clockwise rotation is positive and anticlockwise rotation negative. The opposite assumptions would have been equally valid, but whichever convention is used must be strictly adhered to throughout one problem.

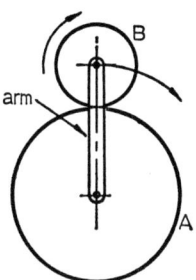

Fig. 6.13

	Operation	A	B	Arm
1	$+1$ revolution to whole system	$+1$	$+1$	$+1$
2	Arm fixed -1 rev to A	-1	$+100/50 = +2$	0
3	Adding individual motions	0	$+3$	$+1$

The final line gives the solution which, in words, indicates that wheel B makes three revolutions and the arm one revolution in the same sense while wheel A remains at rest. It is important that the motion of B as in line 3 is the number of revolutions made by B about its geometrical axis relative to an observer external to the system. Relative to the moving arm or to an axle fixed to the arm, B makes only two revolutions.

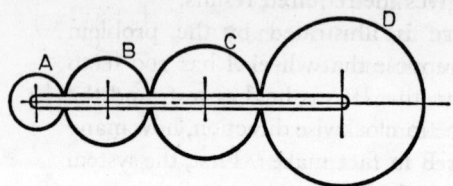

Fig. 6.14

Example. *Wheels* A, B, C *and* D *(Fig. 6.14) have* 50, 75, 100 *and* 150 *teeth respectively. Determine the revolutions made by wheels* B, C *and* D *while the arm rotates five times about the centre of wheel* A, *which is at rest.*

The following table is considered.

	Operation	A	B	C	D	Arm
1	+1 revolution to system	+1	+1	+1	+1	+1
2	−1 revolution to A. Arm at rest	−1	$+\dfrac{50}{75}$ $=+\tfrac{2}{3}$	$-\dfrac{50}{75}\times\dfrac{75}{100}$ $=-\tfrac{1}{2}$	$+\dfrac{50}{75}\times\dfrac{75}{100}\times\dfrac{100}{150}$ $=+\tfrac{1}{3}$	0
3	Adding lines 2 and 3	0	$+1\tfrac{2}{3}$	$+\tfrac{1}{2}$	$+1\tfrac{1}{3}$	+1
4	Multiplying through by +5	0	$+8\tfrac{1}{3}$	$+2\tfrac{1}{2}$	$+6\tfrac{2}{3}$	+5

Thus, during five revolutions of the arm, the wheels B, C and D make $8\tfrac{1}{3}$, $2\tfrac{1}{2}$ and $6\tfrac{2}{3}$ revolutions respectively (relative to an external observer) with the same sense as the arm. As in the previous example, the operation given in line 2 involves calculations for a simple train. The sense of motion of each wheel and hence the sign used in each case is obtained by inspection of the train.

Reverted trains

If a geared system is so arranged that the axes of the driver and the final driven member are co-incident, the train is said to be *reverted*—that is, *turned back*. Various arrangements of the fixed and moving members are possible, but in Fig. 6.15, the driving shaft has keyed to it one end of the arm and carries wheel A relative to which it is free to move. Attached to the other end of the arm is a short axle on which are mounted wheels B and C which are integral with each other or which form a compound wheel and are free to rotate relative to the axle. Wheel B is in mesh with wheel A, and wheel C with wheel D which is keyed to the driven shaft. Wheel A is fixed.

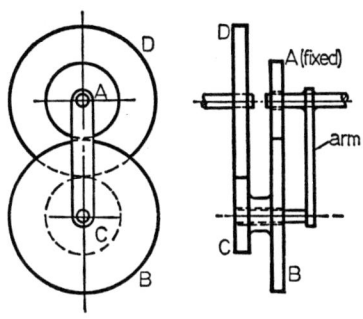

Fig. 6.15

Example. *Determine the velocity ratio of the driving and driven shafts in the reverted train shown in* Fig. 6.15 *and described above. The numbers of teeth in the wheels* A, B, C *and* D *are* 50, 100, 50 *and* 100 *respectively.*

Tabulating the operations:

	Operation	A	B	C	D	Arm
1	$+1$ revolution to system	$+1$	$+1$	$+1$	$+1$	$+1$
2	-1 revolution to A. Arm at rest	-1	$+\dfrac{50}{100} = +\tfrac{1}{2}$	$+\tfrac{1}{2}$	$-\dfrac{50}{100} \times \dfrac{50}{100} = -\tfrac{1}{4}$	0
3	Adding	0	$+1\tfrac{1}{2}$	$+1\tfrac{1}{2}$	$+\tfrac{3}{4}$	$+1$

Thus

$$\text{Velocity ratio} = \frac{\text{Revolutions of driven shaft}}{\text{Revolutions of driving shaft}}$$

$$= (3/4)/1 = +3/4$$

This means that a 3 to 4 reduction is obtained. (Line 2 in the table involves consideration of a compound train.)

Example. *In the epicyclic gear shown in Fig. 6.16, A, B and C are bevel wheels each having 100 teeth. Wheel A is keyed to the driving shaft X and gears with wheel B which is carried on an arm D fixed to the driven shaft Y. Wheel B also gears with a fixed wheel C which is concentric with shaft X. Determine the speed of Y if X rotates at 100 rev/min.*

The speed ratios of bevel gears are determined in the same manner as in spur gearing. The following sequence of operations is tabulated.

	Operation	A	B	C	D
1	$+1$ revolution to the system	$+1$	—	$+1$	$+1$
2	-1 revolution to C. Arm D fixed $\dfrac{100}{100} \times \dfrac{100}{100} = +1$	—	-1	0
3	Adding	$+2$	—	0	$+1$
4	Multiplying by $+50$	$+100$	—	0	$+50$

This means that shaft Y has a speed of 50 rev/min, the sense being the same as that of X.

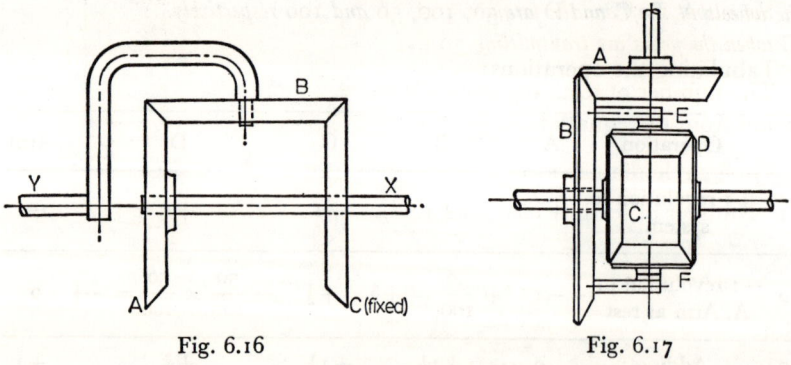

Fig. 6.16   Fig. 6.17

Power transmission

In most epicyclic trains there will be one member through which power is fed into the system and another through which power is transmitted from it. In addition, some other member will usually be fixed. These three members will therefore be subjected to externally applied torques as follows: (*a*) an activating torque T_1 by the drive; (*b*) a reactive torque T_0 by whatever is being driven; and (*c*) a fixing torque T_f by the gear casing, etc. Assuming that power is being transmitted at uniform angular velocity, so that torque is not required to accelerate the gears, then the algebraic sum of the three basic torques must be zero in order that angular equilibrium may be maintained, so that

$$T_i + T_0 + T_f = 0 \qquad (6.14)$$

Furthermore, assuming that losses due to friction can be neglected, the power supplied to the gear systems must equal the power leaving the system. If ω_i and ω_0 are the speeds of the input and output shafts respectively, this means that

$$T_i\omega_i = T_0\omega_0 \qquad (6.15)$$

Example. *In the epicyclic reduction gear shown in Fig. 6.18 the sun wheel D has 20 teeth and is keyed to the input shaft. Two planet wheels B, each having 50 teeth, gear with D and are carried by an arm A fixed to the output shaft. The wheels B also mesh with an internal gear C which is fixed. The input shaft rotates at 2100 rev/min. Determine the speed of the output shaft and the torque required to fix C when the gears are transmitting 30 kW.*

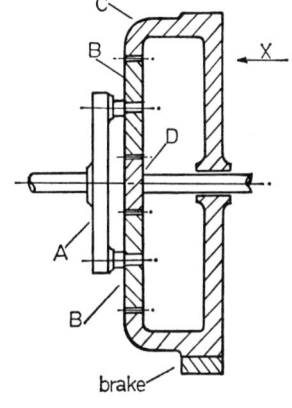

Fig. 6.18

The number of teeth on wheel C can be found from the equation

Pitch circle diameter =

$$\frac{\text{Number of teeth} \times \text{Circular pitch}}{\pi}$$

$$= np/\pi$$

But

Pitch circle diameter of C

$$= \text{Sum of p.c. diameters of sun and planet wheels}$$

Then $\qquad np/\pi = (20p/\pi) + (2 \times 50p/\pi)$

or $\qquad n = 120$

If the sense of rotation of the input shaft (Fig. 6.18) is clockwise when viewed in the direction of arrow X, and considered to be positive, the tabular method will give the solution.

	Operation	Arm A	B	C	D
1	+1 revolution to system	+1	+1	+1	+1
2	−1 revolution to C. Arm fixed	0	$-\dfrac{120}{150}$	−1	$+\dfrac{120}{20} = +6$
3	Adding	+1	—	0	+7
	Dividing by +7 and multiplying by +2100	+300	—	0	+2100

That is, the output shaft rotates at 300 rev/min in the same sense as the input shaft.

$$\text{Power} = T_i \omega_i$$
$$\therefore \ 30 \times 1000 = T_i(2\pi\, 2100/60)$$
$$T_i = 136 \cdot 7 \ \text{Nm (positive)}$$

But

$$\text{Power from the system} = \text{Power put into the system}$$
$$T_0(2\pi\, 300/60) = T_i(2\pi\, 2100/60)$$
$$\therefore \ T_0 = 136 \cdot 7 \times 2100/300$$
$$= 956 \cdot 9 \ \text{Nm}$$

The reaction torque will be opposite in sense to the rotation of the output shaft and will therefore equal −956·9 Nm. For equilibrium

$$T_i + T_0 + T_f = 0$$
$$\therefore \ +136 \cdot 7 - 956 \cdot 9 + T_f = 0$$
$$T_f = +820 \cdot 2 \ \text{Nm}$$

This means that the fixing torque is 820·2 Nm and must be applied in the same sense as the input torque.

PROBLEMS

For tutorials

1. If P in Fig. 6.4 is the pitch point of the gears, show that if the centre distance AB is increased slightly, the common normal will continue to pass through P when the gear teeth are again brought into contact. Will the pressure angle be altered by increasing the centre distance? If so, in what way?

2. Suggest advantages to be gained by using a compound train in place of a simple train.

3. If, in Question 9 (general problems) the power reaches the pinion through a clutch attached to one end of the pinion shaft, to what other form of stress is the pinion shaft subjected? How would the maximum stress in the shaft be determined? Would the presence of the keyway cause the actual maximum stress in the shaft to differ from the theoretical maximum stress? If so, why?

General

1. The desirable centre distance between the pinion and wheel of a spur reduction gear is 320 mm. The pitch of the teeth is 21 mm and the gear ratio 4:1. Determine the number of teeth in each gear wheel and the actual distance between wheel centres to comply as nearly as possible with the desired requirement.

Ans. Number of teeth: Wheel 76, pinion 19; centre distance 317·63 mm

2. Fig. 6.19 shows a train of gear wheels. Shaft S_a carries wheel A and a drum from which is suspended a mass of 100 kg. Wheel A gears with wheel B which, together with wheel C, is mounted on shaft S_{b+c}. Wheel C meshes with D on shaft S_d and D with E on shaft S_e. The drum radius is 100 mm. The total moments of inertia of the rotating masses on each of the four shafts S_a, S_{b+c}, S_d and S_e are respectively $I_a = 1·0$, $I_{b+c} = 1·25$, $I_d = 0·75$ and $I_e = 0·5$ kg m². Determine

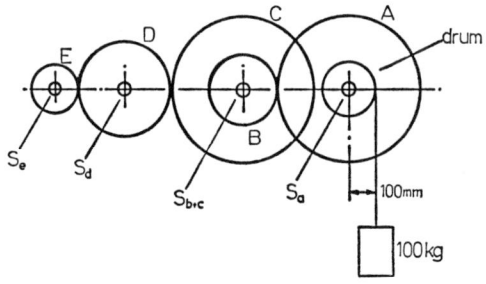

Fig. 6.19

the resultant linear acceleration of the suspended mass if it is permitted to fall freely. The numbers of teeth in the wheels A, B, C, D and E are 150, 75, 150, 100 and 50 respectively.

Ans. 0·31 m/s²

3. An automobile overdrive is shown in Fig. 6.20. The gear box mainshaft D is provided with an arm carrying planet gears C which mesh with an internally toothed ring gear A and a sun wheel B. The internal gear A is connected directly to the propeller shaft which transmits the overdrive to the rear axle. The overdrive is brought into operation by preventing the sun gear from rotating—that is, the sun gear becomes the fixed member. The numbers of teeth on wheels A and B are 75 and 25 respectively. Determine the ratio of velocities for the propeller shaft and the gear box main shaft, and the torque necessary to fix the sun wheel when transmitting 40 kW with a mainshaft speed of 2100 rev/min.

Ans. 1⅓:1, 45·5 Nm

4. In an epicyclic train (Fig. 6.21), the driving shaft X has keyed to it gear wheel A which meshes with wheels B. Wheels B and C are integral with each other and are carried on the shafts P about which they are free to turn. Shafts P are carried on the arm D which is keyed to the driven shaft Y. Wheels C gear

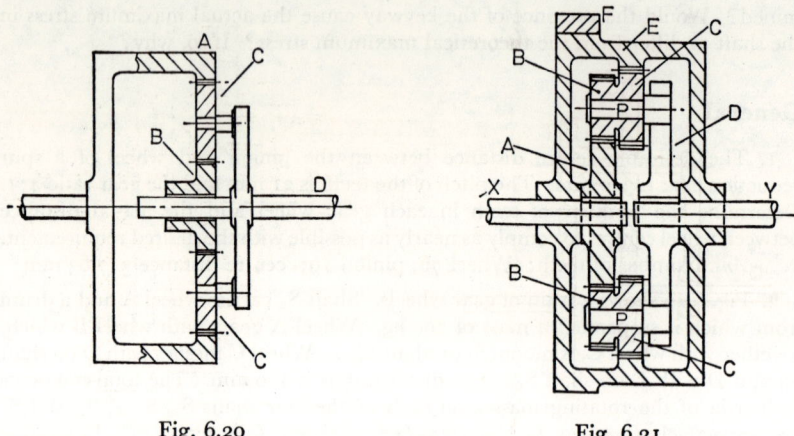

Fig. 6.20   Fig. 6.21

internally with the ring gear attached to the casing E. During operation, the casing E is held at rest by the brake F. The wheels A, B, C and E have 40, 18, 20 and 78 teeth respectively.

(a) If the speed of the driving shaft X is 3000 rev/min, find the speed of the driven shaft Y.

(b) What torque must be applied by the brake to the casing E when 12 kW is being transmitted?

Ans. (a) 1090 rev/min with the same sense as shaft X
          (b) 66 Nm

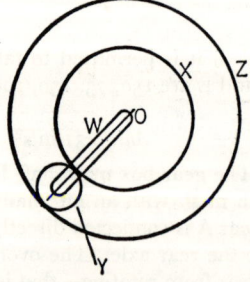

Fig. 6.22

5. In the epicyclic gear shown in Fig. 6.22, the planet wheel Y is free to rotate on a spindle at the outer end of arm W which is pivoted at its opposite end O. The sun wheel X has 30 teeth of pitch 3 cm and is geared with wheel Z through wheel Y. During the time required for X to make 42 revolutions anticlockwise, the arm W makes 6 revolutions anticlockwise and wheel Z is required to make 14 revolutions clockwise. What must be the numbers of teeth in wheels Y and Z and the pitchcircle diameter of Z?

Ans. Y, 12 teeth, Z, 54 teeth; 51·6 cm

6. In the epicyclic gear shown in Fig. 6.23, wheel R is keyed to shaft Q and wheel U is keyed to shaft P. Wheels S and T are integral and are carried on a spindle fixed to arm V. The numbers of teeth in R, S, T and U are 35, 64, 32

and 70 respectively. Find the speed and the direction of rotation of the arm V when P rotates clockwise at 60 rev/min as seen looking in the direction of the arrow A, and when Q rotates at 28 rev/min anticlockwise.

Ans. 89⅓ rev/min clockwise

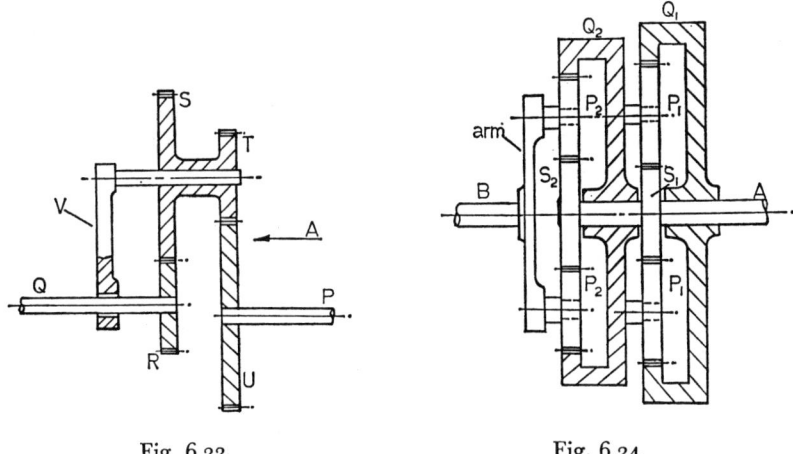

Fig. 6.23   Fig. 6.24

7. A compound epicyclic gear train is shown in Fig. 6.24. The driving shaft A has keyed to it two sun wheels S_1 and S_2 which mesh with the planet wheels P_1 and P_2 respectively. Wheels P_1 and P_2 also mesh with the internal gears Q_1 and Q_2 respectively. Wheels P_1 are free to rotate on axles fixed to Q_2 and wheels P_2 are free to rotate on axles fixed to the arm which drives the output shaft B. The numbers of teeth on the wheels are as follows: $S_1 = 21$, $S_2 = 25$, $Q_1 = 63$ and $Q_2 = 55$.

(*a*) Determine the speed and sense of rotation of Q_1 when the driving shaft A rotates at 1000 rev/min and the output shaft B rotates at 500 rev/min with opposite sense to A.

(*b*) If the power transmitted is 8 kW, what torque will be required to hold the internal gear Q_1 stationary when the driving shaft is rotating at 1000 rev/min? Neglect frictional losses. What is now the speed of the output shaft?

Ans. (*a*) 1909 rev/min with the same sense as shaft B

(*b*) 81·5 Nm, 484·4 rev/min

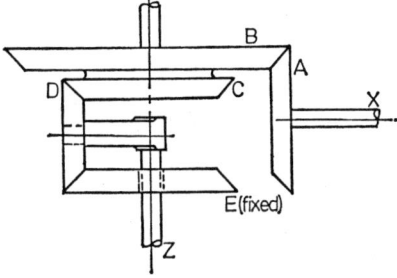

Fig. 6.25

8. In the epicyclic train of bevel wheels shown in Fig. 6.25, wheel A is keyed to the driving shaft X and gears with wheel B. Wheels B and C are integral with each other and keyed to shaft Y. Wheel D gears with wheel C and is carried on a

spindle at the end of an arm attached to shaft Z. Shaft Z passes through the centre of a fixed bevel E. The numbers of teeth on A, B, C, D and E are 80, 160, 100, 60 and 100 respectively. Determine the speed of shaft Z when X is rotating at 160 rev/min.

Ans. 40 rev/min

9. In a certain reduction gear, a pinion having 20° involute teeth is keyed centrally to a shaft of 40 mm diameter and transmits power to a gear wheel. The gear module is 4 mm and the pinion has 25 teeth. If the distance between the pinion shaft bearings is 0·3 m and the shaft can be assumed to be simply supported, what is the maximum bending stress in the shaft when the system is transmitting 30 kW at a pinion speed of 600 rev/min?

Ans. 121·5 MN/m²

CHAPTER 7

Natural Vibrations

7.1 INTRODUCTION

The number of degrees of freedom which a body possesses is the number of co-ordinates necessary to define completely its configuration.

To specify the position of a point on a simple graph, for example, requires two co-ordinates (x and y), which implies that the point can move either in the direction of the x-axis or in the direction of the y-axis, or both. Such a point is said to have two degrees of freedom. If, however, the point were a body with mass distributed somehow about a centroid

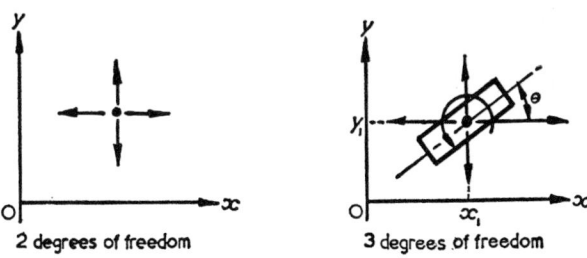

2 degrees of freedom   3 degrees of freedom

Fig. 7.1

the co-ordinates of which are (x_1, y_1), then it is also necessary to specify the orientation of the body relative to the axis Ox (or Oy). The body therefore has three degrees of freedom.

The maximum number of degrees of freedom which a single body can have is six, since it may translate in any of the directions Ox, Oy and Oz and also rotate about each of these three axes.

Natural modes of vibration

To each degree of freedom of a body corresponds a manner, or *mode*, in which it may vibrate. Thus a body with three degrees of freedom will have three natural (or normal) modes of vibration, and may vibrate in any one of these modes, or possibly in a combination of all three. Each natural mode of vibration will have its own natural frequency, and the equation of motion for the body as a whole will have a number of simple harmonic solutions corresponding to these frequencies.

The mathematical analysis of the vibrations of real bodies can be simplified without much error by considering instead equivalent idealised systems of masses without elasticity supported by structures with elasticity but no inertia.

7.2 SIMPLE HARMONIC MOTION

If the motion of a body is such that its acceleration is directly proportional to its distance from some fixed point and directed towards it, the motion is described as simple harmonic motion (S.H.M.). Mathematically,

$$\frac{d^2x}{dt^2} = -\omega_n{}^2 x \tag{7.1}$$

where x is the displacement and $\omega_n{}^2$ is a constant of proportionality which is necessarily greater than zero. If the equation of motion of a body is found (from Newton's second law) to have this form, the motion is S.H.M.

It is often convenient to represent S.H.M. vectorially. Fig. 7.2 (*a*) represents a vector OQ of length a rotating at a constant speed ω_n rad/second and making an angle α with the vertical at time $t = 0$. If P is the projection of the point Q on the vertical diameter the displacement of P from O is given by

$$x = a \cos(\omega_n t + \alpha)$$

By differentiating, the velocity of P is

$$\frac{dx}{dt} = -a\omega_n \sin(\omega_n t + \alpha)$$

and by further differentiation, the acceleration of P is

$$\frac{d^2x}{dt^2} = -a\omega_n{}^2 \cos(\omega_n t + \alpha) = -\omega_n{}^2 x$$

This equation is identical with Eq. (7.1), so that the motion of P is simple harmonic. From Fig. 7.2 it is clear that P oscillates up and down about O as the vector OQ rotates.

Eq. (7.1) is a second order differential equation which has the solution

$$x = a \cos (\omega_n t + \alpha) \tag{7.2}$$

where a and α are constants of the integration. From Fig. 7.2 (c) the constant a is clearly the maximum displacement of P or the amplitude of the vibration. The value of α depends on the position of P when $t = 0$

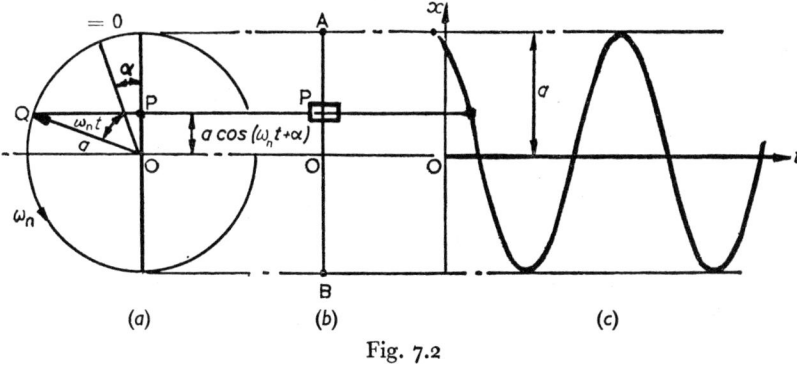

(a) (b) (c)

Fig. 7.2

and is often quite arbitrary. The vector representation shows that the time for one complete oscillation is equal to the time for one complete rotation of the vector, so that

$$\text{Periodic time} = \frac{2\pi}{\omega_n} \text{ seconds} \tag{7.3}$$

$$\text{Frequency} = \frac{\omega_n}{2\pi} \text{ oscillations/second or hertz (Hz)} \tag{7.4}$$

The constant ω_n is known as the natural circular frequency.

Example. *A body of mass* 10 kg *vibrates with simple harmonic motion of frequency* 100 Hz. *The total movement of the body is measured and found to be* 4 mm.

Determine (a) The velocity and acceleration when the body is 0·5 mm *from the extremity of its stroke. (b) The maximum force acting on the body. (c) The maximum kinetic energy of the body.*

The amplitude a is half the total movement or 2 mm. The natural frequency can be calculated from Eq. (7.4) and is

$$\omega_n = 200\pi \text{ rad/s}$$

It is convenient to put $t = 0$ when the displacement is a maximum so that $\alpha = 0$. Then, because the angle QOP is $\omega_n t$,

$$x = a \cos \omega_n t$$

and

$$\frac{dx}{dt} = -a\omega_n \sin \omega_n t$$

In other words, from Fig. 7.3, the velocity

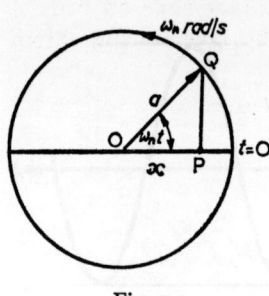

Fig. 7.3

$$V = -a\omega_n(\text{QP}/\text{OQ})$$
$$= -a\omega_n\{\pm\sqrt{(a^2 - x^2)}\}/a$$
$$= \pm\omega_n\sqrt{(a^2 - x^2)}$$
$$= \pm 200\pi\sqrt{(0.002^2 - 0.0015^2)} \text{ m/s}$$
$$= 0.831 \text{ m/s}$$

Acceleration
$$= d^2x/dt^2 = -\omega_n^2 x$$
$$= -40\,000\pi^2\,0.0015 \text{ m/s}^2$$
$$= -592.1 \text{ m/s}^2$$

(*b*) From Newton's second law, the maximum force is mass $\times$ maximum acceleration, or

$$\text{Maximum force} = 10 \times \omega_n^2 a$$
$$= 10 \times 40\,000\pi^2\,0.002$$
$$= 9529 \text{ N}$$

(*c*) Velocity $= dx/dt = -a\omega_n \sin \omega_n t$,

$$\therefore \text{ Maximum velocity} = a\omega_n$$
$$= 0.002 \times 200\pi \text{ m/s}$$
$$= 1.257 \text{ m/s}$$

$$\therefore \text{ Maximum kinetic energy} = \tfrac{1}{2}mv_{max}^2$$
$$= \tfrac{1}{2}10(1.257^2) \text{ joules}$$
$$= 7.9 \text{ joules}$$

Not all vibrations are necessarily *simple* harmonic. The equation of motion for any system is a differential equation and different forms of these are representative of different forms of motion. As a result, it is usually possible to recognise the form of the motion without formal solution of the equations.

7.3 FREE VIBRATIONS

A body of mass m supported by an elastic structure of stiffness S may be represented in the idealised form shown in Fig. 7.4. If the body is constrained to move in vertical guides, there is only one degree of freedom and therefore only one natural mode and frequency. Let the mass be displaced upwards and released and suppose that at time t it is a distance x from the equilibrium position. Then the resultant force on body is $-Sx$. (The negative sign means the the force acts downwards.) From Newton's second law:

$$m(d^2x/dt^2) = -Sx$$

or
$$(d^2xdt^2) + (S/m)x = 0 \qquad (7.5)$$

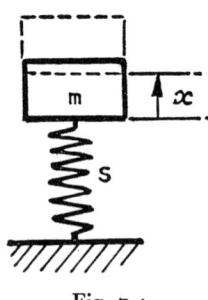

Fig. 7.4

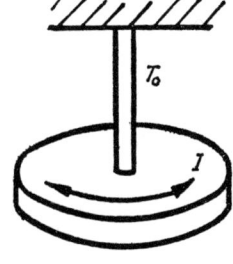

Fig. 7.5

This is a second order differential equation which does not have to be solved formally because a slight re-arrangement shows it to be typical of S.H.M. Putting the constant $S/m = \omega_n^2$, the equation becomes:

$$d^2x/dt^2 = -\omega_n^2 x$$

The solution is

$$x = a \cos(\omega_n t + \alpha) \quad \text{where} \quad \omega_n = \sqrt{(S/m)}$$

This is shown graphically in Fig. 7.2. The natural frequency is given by

$$\frac{\omega_n}{2\pi} = \frac{1}{2\pi}\sqrt{\left(\frac{S}{m}\right)} \qquad (7.6)$$

Angular motions are dealt with similarly. Consider a flywheel of polar moment of inertia I mounted on a shaft of torsional stiffness T_0 (Fig. 7.5). Let the flywheel be given an angular displacement and suppose that at time t it is θ radians from the equilibrium position. The resultant torque on flywheel is $-T_0\theta$ and from Newton's second law,

$$I(\mathrm{d}^2\theta/\mathrm{d}t^2) = -T_0\theta$$

or
$$\mathrm{d}^2\theta/\mathrm{d}t^2 = -(T_0/I)\theta \qquad (7.7)$$

This equation is also similar to the S.H.M. eq. (7.1) if T_0/I is replaced by $\omega_n{}^2$.

$$\therefore \; \theta = \theta_{\max} \cos(\omega_n t + \alpha)$$

where $\omega_n = \sqrt{(T_0/I)}$ and the natural frequency is

$$\frac{\omega_n}{2\pi} = \frac{1}{2\pi}\sqrt{\left(\frac{T_0}{I}\right)} \qquad (7.8)$$

Example. *A solid disc flywheel of mass* 240 kg *and a diameter of* 1 m *is mounted on a mild steel shaft* 40 mm *in diameter and length* 2 m. *Neglecting the inertia of the shaft, calculate the natural frequencies of free torsional, transverse and longitudinal vibrations. For mild steel* $E = 200$ GN/m^2 *and* $G = 82$ GN/m^2.

Moment of intertia of flywheel $I = m\,\dfrac{d^2}{8} = 240 \times \tfrac{1}{8} = 30$ kg m^2

Polar second moment of area of shaft $J = \dfrac{\pi d^4}{32} = \dfrac{\pi(0\cdot040)^4}{32}$

$$= 2\cdot514 \times 10^{-7}\,\mathrm{m}^4$$

Second moment of area of shaft in bending

$$I_{\mathrm{NA}} = \pi d^4/64 = 1\cdot257 \times 10^{-7}\,\mathrm{m}^4$$

Cross sectional area of shaft $A = \pi d^2/4 = 1\cdot257 \times 10^{-3}\,\mathrm{m}^4$

Torsional stiffness $T_0 = GJ/L = (82 \times 10^9 \times 2\cdot514 \times 10^{-7})/2$

$$= 10\,300\ \mathrm{Nm/rad}$$

Transverse stiffness $S_{\mathrm{T}} = 3EI_{\mathrm{NA}}/L^3 = (3 \times 200 \times 10^9 \times 1\cdot257 \times 10^{-7})/8$
$$= 9426\ \mathrm{N/m}$$

Longitudinal stiffness $S_{\mathrm{L}} = EA/L = (200 \times 10^9 \times 1\cdot257 \times 10^{-3})/2$
$$= 125\cdot7 \times 10^6\ \mathrm{N/m}$$

For torsional vibrations, the frequency f is given by

$$f = \frac{1}{2\pi}\sqrt{\left(\frac{T_0}{I}\right)} = \frac{1}{2\pi}\sqrt{\left(\frac{10\,300}{30}\right)} = 2\cdot95\ \mathrm{Hz}$$

For transverse vibrations

$$f = \frac{1}{2\pi} \sqrt{\left(\frac{S_T}{m}\right)} = \frac{1}{2\pi} \sqrt{\left(\frac{9426}{240}\right)} = 1\cdot00 \text{ Hz}$$

For longitudinal vibrations

$$= \frac{1}{2\pi} \sqrt{\left(\frac{S_L}{m}\right)} = \frac{1}{2\pi} \sqrt{\left(\frac{125\cdot7 \times 10^6}{240}\right)} = 115\cdot2 \text{ Hz}$$

7.4 DAMPED VIBRATIONS

In practice, all motion is resisted by frictional forces. In vibrating systems the effect of friction is referred to as damping, and in the idealised system this is represented by means of a dashpot (Fig. 7.6). Many systems have damping devices inserted deliberately, but air resistance and hysteresis in the elastic support will in any case provide a damping force which gradually absorbs the energy of the system. Damping provided by fluid resistance is known as viscous damping, when the force resisting motion is assumed to be directly proportional to the velocity, so that

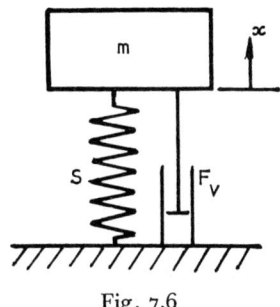

Fig. 7.6

$$\text{damping force} = -F_v(dx/dt) \tag{7.9}$$

where F_v a constant = damping force at unit velocity.

In the idealised system of Fig. 7.6 in which the mass m can move only in the vertical direction; if at time t, the mass is displaced upwards a distance x from its equilibrium position, the

$$\text{resultant force on mass} = -Sx - F_v(dx/dt)$$

From Newton's second law:

$$m(d^2x/dt^2) = -Sx - F_v(dx/dt)$$

or $$m(d^2x/dt^2) + F_v(dx/dt) + Sx = 0 \tag{7.10}$$

This is the second order differential equation typical of a natural damped vibration. Its solution is easier by writing it as follows:

$$d^2x/dt^2 + (F_v/m)(dx/dt) + (S/m)x = 0$$

or $$(d^2x/dt^2) + 2k(dx/dt) + \omega_n^2 x = 0 \tag{7.11}$$

where $2k = F_v/m$ and $\omega_n^2 = S/m$. The constant ω_n is the undamped natural circular frequency.

The next step is to look for solutions of Eq. (7.11) in the form $x = A\,e^{pt}$ where p is a constant. Substituting for x,

$$Ap^2\,e^{pt} + 2kAp\,e^{pt} + \omega_n^2 A\,e^{pt} = 0$$

or

$$p^2 + 2kp + \omega_n^2 = 0 \qquad (7.12)$$

This is the auxiliary equation which gives the values of the constant p as

$$p = \{-2k \pm \surd(4k^2 - 4\omega_n^2)\}/2$$
$$= -k \pm \surd(k^2 - \omega_n^2) \qquad (7.13)$$

The solution of Eq. (7.11) is evidently

$$x = A\,e^{\{-k+\surd(k^2-\omega_n^2)\}t} + B\,e^{\{-k-\surd(k^2-\omega_n^2)\}t} \qquad (7.14)$$

where A and B are constants.

The motion described by Eq. (7.14) will depend on the value of the damping constant k. Three cases must be considered.

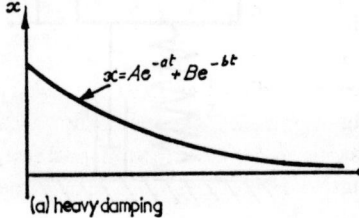

(a) heavy damping

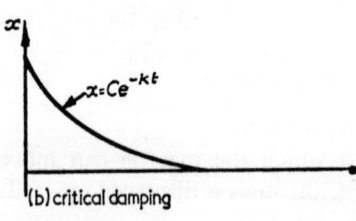

(b) critical damping

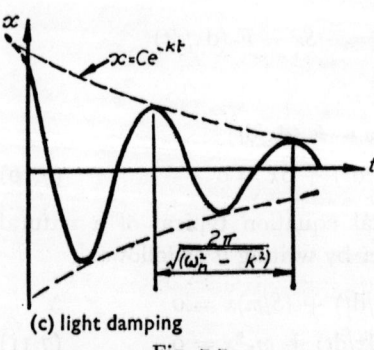

(c) light damping

Fig. 7.7

If $k > \omega_n$ so that $(k^2 - \omega_n^2)$ is positive and both values of p are real, the damping is said to be *heavy*. In this case Eq. (7.14) takes the form

$$x = A\,e^{-at} + B\,e^{-bt}$$

and this solution is shown graphically in Fig. 7.7 (a), which makes it clear that no oscillations occur and the mass simply returns to its equilibrium position exponentially. The motion is said to be aperiodic.

Critical damping occurs when $k = \omega_n$ so that $\surd(k^2 - \omega_n^2)$ is zero and Eq. (7.14) reduces to

$$x = A\,e^{-kt} + B\,e^{-kt} = C\,e^{-kt} \qquad (7.15)$$

where $C = A + B$. This solution (Fig. 7.7 (b)) is again aperiodic but the mass returns without overshooting to its equilibrium position in the shortest possible time.

Light damping occurs when $k < \omega_n$ so that $(k^2 - \omega_n^2)$ is negative and the values of p have imaginary parts, i.e.

$$p = -k \pm j\sqrt{(\omega_n{}^2 - k^2)} \quad \text{where} \quad j = \sqrt{(-1)}$$

Eq. (7.14) may now be written

$$x = A\,e^{\{-k+j\sqrt{(\omega_n{}^2-k^2)}\}t} + B\,e^{\{-k-j\sqrt{(\omega_n{}^2-k^2)}\}t}$$

$$= e^{-kt}[A\,e^{+j\{\sqrt{(\omega_n{}^2-k^2)}\}t} + B\,e^{-j\{\sqrt{(\omega_n{}^2-k^2)}\}t}]$$

Since $e^{j\theta} = \cos\theta + j\sin\theta$, this may be written

$$x = e^{-kt}[A\cos\theta + Aj\sin\theta + B\cos\theta - Bj\sin\theta]$$

where $\theta = \{\sqrt{(\omega_n{}^2 - k^2)}\}t$. This means that

$$x = e^{-kt}[C_1\cos\theta + jC_2\sin\theta]$$

$$= e^{-kt}\sqrt{(C_1{}^2 - C_2{}^2)}[\cos\theta\cos\alpha - \sin\theta\sin\alpha]$$

$$= C\,e^{-kt}\cos[\{\sqrt{(\omega_n{}^2 - k^2)}\}t - \alpha] \qquad (7.16)$$

where $\tan\alpha = -jC_2/C_1$ and $\sqrt{(C_1{}^2 - C_2{}^2)} = C$.

In this case (Fig. 7.7 (c)) the motion is periodic of circular frequency $\sqrt{(\omega_n{}^2 - k^2)}$ and amplitude $C\,e^{-kt}$ which reduces exponentially with time. Although the mass eventually returns to its equilibrium position, it overshoots because of its inertia, and the oscillations may take some considerable time to die away. Then

$$\text{Natural damped frequency} = \{\sqrt{(\omega_n{}^2 - k^2)}\}/2\pi \qquad (7.17)$$

With motion of this type, the ratio of successive amplitudes on the side of the mean after one complete period is called the *amplitude reduction factor*. At time t,

$$\text{Amplitude} = C\,e^{-kt}$$

and one period later,

$$\text{Amplitude} = C\,e^{-k[t+2\pi/\sqrt{(\omega_n{}^2-k^2)}]}$$

$$\therefore \text{Amplitude reduction factor} = \frac{C\,e^{-kt}}{Ce^{-k[t+2\pi/\sqrt{(\omega_n{}^2-k^2)}]}}$$

$$= e^{2\pi k/\sqrt{(\omega_n{}^2-k^2)}} \qquad (7.18)$$

It is found convenient in practice to relate the degree of damping to the critical value of damping. The damping ratio c is defined as

$$c = \frac{\text{Actual damping}}{\text{Critical damping}} \qquad (7.19)$$

and provides a dimensionless measure of the degree of damping. For critical damping, $k = \omega_n$, so that

$$c = k/\omega_n \tag{7.20}$$

The expression for the natural damped frequency in Eq. (7.17), may thus be written

$$\frac{\omega_n}{2\pi} \sqrt{(1 - c^2)} \tag{7.21}$$

Example. *A mass of 50 kg is supported by an elastic structure of stiffness 10 000 N/m. The mass is set vibrating and it is observed that after 2 complete oscillations the amplitude is reduced to one tenth of its initial value.*

Determine (a) the damping ratio, (b) the damping force at 1 m/s, and (c) the natural frequency of the oscillations.

Natural undamped circular frequency $= \omega_n{}^2 = S/m = 10\,000/50$

$$= 200/s^2$$

Let initial amplitude be $C\,e^{-kt}$.

Amplitude after 2 complete oscillations $= C\,e^{-k(t+2T)}$, where

$$T = \text{periodic time} = 2\pi/\sqrt{(\omega_n{}^2 - k^2)}$$

$$\therefore C\,e^{-kt}/C\,e^{-k(t+T)} = 10$$

$$\therefore e^{2kT} = 10$$

$$\therefore 2kT = \log_e 10 = 2\cdot303$$

$$\therefore 2k \cdot 2\pi/\sqrt{(200 - k^2)} = 2\cdot303$$

From which $\qquad k^2 = 6\cdot495$

$$\therefore k^2/\omega_n{}^2 = 6\cdot495/200 = 0\cdot03248$$

$$\therefore \text{Damping ratio } c = \sqrt{(0\cdot03248)} = 0\cdot18$$

Damping force at 1 m/s, Fv $= m(2k) = 50 \times 2\sqrt{(6\cdot495)} = 255\,\text{N}$

Natural frequency $= \{\sqrt{(200 - 6\cdot495)}\}/2\pi = 2\cdot215\,\text{Hz}$

Similar considerations apply to damped angular oscillations. Consider the idealised system of Fig. 7.8 consisting of a flywheel of polar moment of inertia I mounted on a shaft of stiffness T_0 and fitted with a damping device providing a frictional torque T_F at unit angular velocity.

If at time t the flywheel has an angular displacement θ from its equilibrium position, then

$$\text{resultant torque on flywheel} = -T_0\theta - T_F(d\theta/dt)$$

From Newton's second law

$$I(d^2\theta/dt^2) = -T_0\theta - T_F(d\theta/dt)$$

or $\quad I(d^2\theta/dt^2) + T_F(d\theta/dt) + T_0\theta = 0 \quad (7.22)$

This may be written in the form

$$(d^2\theta/dt^2) + 2k(d\theta/dt) + \omega_n^2\theta = 0 \quad (7.23)$$

where $2k = T_F/I$ and $\omega_n^2 = T_0/I$. This is identical with eq. (7.11), so that the analysis need be taken no further.

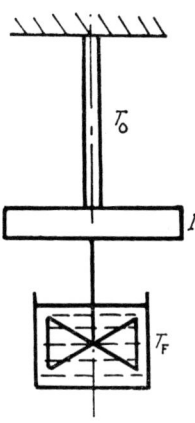

FIG. 7.8

Example. *A flywheel of moment of inertia 50 kg m²
is mounted on a shaft 2 m long and 40 mm in diameter.
Calculate the damping torque required at 1 rad/s
to give critical damping. If a torsional damping device is fitted to give a damping
ratio of 0·125 determine the resulting natural frequency of torsional oscillations
and the amplitude reduction after one complete oscillation. Assume $G = 80$ GN/m².*
Torsional stiffness of shaft

$$= GJ/L = 80 \times 10^9 \times \pi(0.04)^4/64 = 10\,050 \text{ N m/rad}$$
$$\therefore \; \omega_n^2 = T_0/I = 10\,050/50 = 201/\text{s}^2$$

For critical damping,

$$k = \omega_n = \sqrt{201} = 14.18 \text{ rad/s}$$
$$\therefore \text{ Damping torque at 1 rad/s} = T_F = 2kI$$
$$= 2 \times 14.18 \times 50 = 1418 \text{ N m}$$

With damping ratio of 0·125,

$$k = 0.125\omega_n = 1.772 \text{ rad/s}$$
$$\text{Natural frequency} = \{\omega_n\sqrt{(1 - c^2)}\}/2\pi$$

or $\qquad\qquad f = 14.18\sqrt{\{1 - (0.125)^2\}}/2\pi$
$$= 2.24 \text{ Hz}$$
$$\text{Periodic time} = 1/f = 1/2.24 = 0.4464 \text{ s}$$
$$\text{Amplitude reduction factor} = C\,e^{-kt}/C\,e^{-k(t+T)} = e^{kT}$$
$$= e^{1.772 \,.\, 0.4464} = e^{0.79} = 2.206$$

i.e. after one complete oscillation, the amplitude is 45·6 per cent of its initial value or has been reduced by 54·4 per cent.

PROBLEMS

For tutorials

1. Why do systems vibrate freely when disturbed?

2. Three concentrated masses are all free to slide on a smooth horizontal surface. If the masses are connected together by three light springs so as to form a triangle, how many degrees of freedom would the system have?

3. In each case discussed in this chapter the characteristic of the elastic support has been assumed to be linear, whereas in practice they are frequently non-linear. Bearing in mind that the study of vibrations is aimed at their reduction or prevention, does the presence of non-linearities seriously undermine the validity of the theory?

4. The response of a single-mass damped system to a disturbance is shown in Fig. 7.7 (c). What would the response have been if the damping ratio had been negative? Discuss the possibility of negative damping occurring in practice, and consider the implications as regards the total energy of the system.

5. The following equations of motion describe the motion of various systems. By inspection of these equations deduce the type of system to which they refer, sketch the graphical solutions, and calculate the natural frequencies:

(a) $50(d^2x/dt^2) + 10^6x = 0$

(b) $10(d^2/dt^2) + 10^5(dx/dt) + 10^7x = 0$

(c) $20(d^2x/dt^2) + 400(dx/dt) + 10\ 000x = 0$

(d) $5(d^2x/dt^2) + 100(dx/dt) + 500x = 0$

(e) $(d^2x/dt^2) - 20(dx/dt) + 400x = 0$

General

1. A mass of 5 kg is suspended from a spring of stiffness 180 N/m. The mass is displaced 40 mm from it equilibrium position and then released. Determine the natural frequency of the resulting oscillations. Find also,

(a) the velocity and acceleration of the mass when 20 mm from its equilibrium position,

(b) the total energy of the system,

(c) the time taken for the mass to move from a position 30 mm to a position 10 mm from the equilibrium position.

Ans. 0·955 Hz (a) 0·208 m/s, 0·72 m/s²

(b) 0·144 joules

(c) 0·099 s

2. A disc is mounted on a mild steel shaft 5 mm in diameter and 1 m long. The shaft is mounted vertically with its upper end fixed and the disc oscillates about its polar axis. If the periodic time of the torsional oscillations is 0·4 s calculate the moment of inertia of the disc.

The wheel of a motor vehicle is now attached symmetrically to the disc and the system again oscillates. The time for 20 complete torsional oscillations is found to be 1·75 minutes. If the wheel has a mass of 40 kg calculate its radius of gyration. For mild steel G = 82 GN/m². Neglect the inertia of the shaft.

Ans. 0·0204 kg m²; 296 mm

3. A mass of 10 kg is supported by a structure having a stiffness of 1000 N/m. Movement of the mass is controlled by a dashpot, giving a damping ratio of 0·1. If the mass is displaced 40 mm from its equilibrium position and then released, determine:

(*a*) natural frequency of damped vibrations,
(*b*) the displacement, velocity and acceleration of the mass after 1 sec,
(*c*) the amplitude after 5 complete oscillations.

Ans. (*a*) 1·58 Hz; (*b*) −14 mm, +0·075 m/s, +1·2 m/s²; (*c*) 1·8 mm

4. A rectangular door of mass 40 kg is hinged about one vertical edge, about which it has a radius of gyration of 0·6 m. It is to be fitted with a spring and dashpot which will automatically close the door after opening. The specifications for the system are that it should be critically damped and that, after being opened to the 90° position, it should return to within 1 degree of the fully closed position within 1 second from release.

Determine the required torsional stiffness about the hinge and the damping torque at unit velocity.

Ans. 291·6 Nm/radian; 129·6 Nm at 1 rad/s.

5. A mass of 5 kg suspended from a spring performs vertical damped oscillations. The amplitude of the vibrations is found to reduce to 50 per cent of the original value after 10 complete oscillations, and 100 oscillations are completed in a time of 32 seconds.

Calculate the damping ratio and the stiffness of the spring.

Ans. 0·011; 1·93 kN/m.

CHAPTER 8

Forced Vibrations

8.1 FORCES AND RESPONSES

Forcing functions

Forced vibrations are sustained by some external source of energy. Unlike natural vibrations, which eventually die out as a result of damping, forced vibrations continue so long as the external source continues to supply energy to the system. Clearly the manner in which this energy is supplied determines the motion or *response* of the system. The time dependent force by means of which energy is supplied to the system is called the *forcing function* $F(t)$.

It is helpful to regard the forcing function as the input to the system and the response as the output (Fig. 8.1). System dynamics is mainly

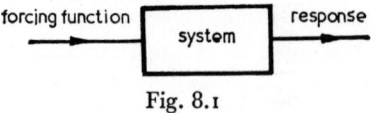

Fig. 8.1

concerned with the relationship between output and input, and the ratio system response/forcing function is known as the *transfer function* of the system (see Chapter 10).

There are three distinct categories of forcing functions:

Impulsive. A series of impulses at regular intervals; inbetween impulses, the system vibrates naturally.

Continuous harmonic. The disturbing force acts continuously but varies harmonically with time—for example,

$$F(t) = F_0 \cos \omega t \tag{8.1}$$

This is a force of maximum value $\pm F_0$ varying sinusoidally at a frequency $\omega/2\pi$ known as the *forcing frequency*.

Self-induced. This is also a harmonic disturbing force caused by the oscillation itself. The external source of energy is some body or substance with which the system is in contact. The classical example is that of a bow passing over a violin string, and an idealised version of this system is shown in Fig. 8.2. The force due to dry friction or Coulomb friction as it is called, decreases as the relative velocity increases until a limiting force, the kinetic friction, is reached. (If the system is lubricated, viscous drag leads to an increase of frictional force at large velocities as shown in Fig. 8.3, but the argument which follows is still valid.) To begin with, the static friction causes the spring to deflect. The force in the spring increases with the displacement, however, so that the point is reached at

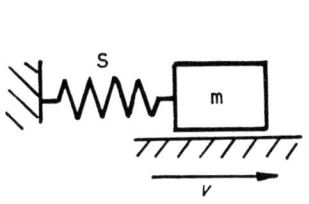

Fig. 8.2

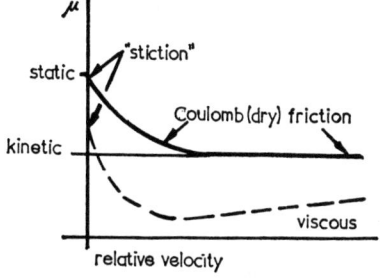

Fig. 8.3

which the spring force exceeds static friction, and the mass starts to accelerate towards its undisplaced position. But this means that the relative velocity increases so that the frictional force reduces to the kinetic value. The motion continues until the spring force is less than kinetic friction, when relative motion ceases and the whole cycle of events is repeated. This type of "stick-slip" vibration can be dangerous because the falling friction characteristic may result in "negative damping." The equation of motion reduces to one similar to Eq. (7.11) but with a negative value of k. The result is not just a sustained vibration but a vibration with amplitude growing exponentially with time. Excitation of this type is responsible for the squeal of brakes, shaft whipping due to the action of oil in journal bearings, "flutter" of aircraft wings and tool chatter in machining operations.

Transient and steady-state responses

When a spring-mass system is excited the equation of motion which expresses Newton's second law is usually a linear second order differential equation. The solution of such an equation consists of two parts known as the "complementary function" and the "particular integral." These would be better named "transient solution" and "steady-state solution" respectively.

The complementary function or transient solution describes the natural response of the system to the initial disturbance. It is given by Eq. (7.16) and is independent of the forcing function. The natural response is eventually damped out, or is "transient." In the study of vibrations, the transient response is of little importance because it is only evident during the starting-up period, although large amplitudes are possible during this period by the superposition of the transient and steady-state responses.

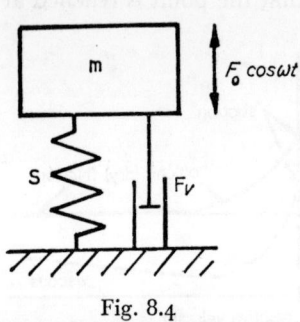

Fig. 8.4

The particular integral or steady-state solution is the response of the system imposed by the forcing function. Although the frequency of the response is always the same as the forcing frequency, the amplitude depends not only on the magnitude of the forcing function but also on the physical properties of the system. If the forcing frequency differs appreciably from the natural frequency of the system, the system responds less readily to the forcing function, with the result that more energy is dissipated and the amplitude is small. If, on the other hand, the forcing frequency is similar to the natural frequency of the system, the system offers less resistance to the imposed motion, less energy is dissipated, and the amplitude is correspondingly greater. In this respect the system is said to have "mechanical impedance."

8.2 RESPONSE TO HARMONIC EXCITATION

Force of constant amplitude

In the idealised spring-mass system with viscous damping shown in Fig. 8.4, the mass is acted on by the simple harmonic disturbing force $F_0 \cos \omega t$. If the mass m is constrained to move in vertical guides, the system has only one degree of freedom. If at some time t, the mass is displaced upwards a distance x from its equilibrium position

resultant force on the mass $= -Sx - F_v \, (dx/dt) + F_0 \cos \omega t$

From Newton's second law,

$$m(d^2x/dt^2) = -Sx - F_v(dx/dt) + F_0 \cos \omega t$$

or $\qquad m(d^2x/dt^2) + F_v(dx/dt) + Sx = F_0 \cos \omega t \qquad (8.2)$

This is a second order differential equation typical of a forced vibration with simple harmonic excitation which can conveniently be written

$$(d^2x/dt^2) + (F_v/m)(dx/dt) + (S/m)x = (F_0/m) \cos \omega t \qquad (8.3)$$

or
$$(d^2x/dt^2) + 2k(dx/dt) + \omega_n{}^2 x = (F_0/m) \cos \omega t \qquad (8.4)$$

where, as before, $2k = F_v/m$, and $\omega_n{}^2 = S/m$. The constant ω_n will be recognised as the undamped natural circular frequency of the system.

The transient solution is independent of the forcing function (see previous section) and is therefore obtained by equating the left-hand side of Eq. (8.4) to zero. This has been done in Section 7.4, and assuming light damping, the transient response is given by Eq. (7.16)

$$x = C\,e^{-kt} \cos [\{\sqrt{(\omega_n{}^2 - k^2)}\}t - \alpha_0]$$

where α_0 is a constant.

The steady-state solution is known from practical experience to be simple harmonic with frequency equal to the forcing frequency, so that it is reasonable to assume a solution of the form:

$$x = A \cos \omega t \qquad (8.5)$$

where A is a constant equal to the amplitude of the steady-state response. If this is to be the correct solution, substitution for x in Eq. (8.4) gives

$$-A\omega^2 \cos \omega t - 2kA\omega \sin \omega t + \omega_n{}^2 A \cos \omega t = (F_0/m) \cos \omega t$$

i.e. $\qquad A(\omega_n{}^2 - \omega^2) \cos \omega t - 2kA\omega \sin \omega t = (F_0/m) \cos \omega t$

i.e. $\quad A\sqrt{\{(\omega_n{}^2 - \omega^2)^2 + 4k^2\omega^2\}} \cos (\omega t + \alpha) = (F_0/m) \cos \omega t \quad (8.6)$

where $\qquad\qquad\qquad \tan \alpha = 2k\omega/(\omega_n{}^2 - \omega^2) \qquad (8.7)$

Examination of Eq. (8.6) leads to the conclusion that the assumed solution is only correct if

$$\alpha = 0$$

and also

$$A\sqrt{\{(\omega_n{}^2 - \omega^2)^2 + 4k^2\omega^2\}} = F_0/m$$

From Eq. (8.7), if $\alpha = 0$, $k = 0$ (i.e. zero damping). Thus, Eq. (8.5) is a solution only if the system is undamped, so that, putting $k = 0$,

$$A = \frac{F_0/m}{(\omega_n{}^2 - \omega^2)} \qquad (8.8)$$

In practice, there is always, some damping, so that the solution represented by Eq. (8.5) must therefore be incorrect and must be modified to

$$x = A \cos (\omega t - \alpha) \qquad (8.9)$$

Eq. (8.6) now becomes:

$$A\sqrt{\{(\omega_n{}^2 - \omega^2)^2 + 4k^2\omega^2\}} \cos\{(\omega t - \alpha) + \alpha\} = (F_0/m)\cos\omega t \quad (8.10)$$

Examination of this equation leads to the conclusion that this modified solution is correct only if

$$A\sqrt{\{(\omega_n{}^2 - \omega^2)^2 + 4k^2\omega^2\}} = F_0/m$$

i.e.
$$A = \frac{F_0/m}{\sqrt{\{(\omega_n{}^2 - \omega^2)^2 + 4k^2\omega^2\}}} \quad (8.11)$$

Thus, the steady-state response (Fig. 8.5) of a spring-mass system with viscous damping to simple harmonic excitation is a simple harmonic

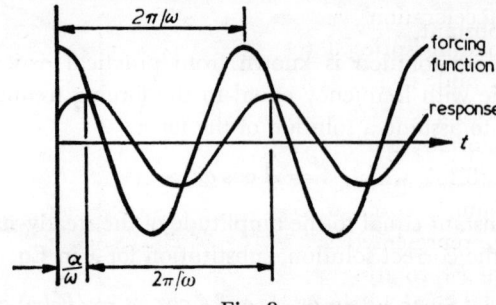

Fig. 8.5

motion of frequency equal to the forcing frequency $\omega/2\pi$ and of amplitude A (Eq. (8.11)). The response is out of phase with the forcing function, lagging by an amount represented by the angle α (Eq. (8.7)).

Vector solution for the steady-state response. The equation of motion in the form of Eq. (8.2) may be written in words as

inertia force + damping force + spring force
$$= \text{disturbing force} \quad (8.12)$$

Each of these forces acts in the direction of motion of the vibrating mass and varies simple harmonically, and as in Section 7.2, in the same way, each of these simple harmonic forces may be represented by a rotating vector.

If, as before, the steady-state solution is:

$$x = A\cos\omega t$$

then
$$(dx/dt) = -A\omega\sin\omega t$$

and
$$(d^2x/dt^2) = -A\omega^2\cos\omega t$$

$$\text{Spring force, } Sx = SA\cos\omega t$$

and is a vector of length $+SA$ rotating at constant speed ω in phase with the displacement vector.

$$\text{Damping force, } F_v(dx/dt) = -F_vA\omega \sin \omega t$$

and is a vector of length $-F_vA\omega$ rotating at constant speed ω, but lagging the displacement vector by 90°.

$$\text{Inertia force, } m(d^2x/dt^2) = -mA\omega^2 \cos \omega t$$

and is a vector of length $-mA\omega^2$ rotating at constant speed ω, in phase with the displacement vector.

(The negative length of the last two vectors effectively reverses their direction.)

Fig. 8.6 shows the displacement, velocity, and acceleration vectors and the corresponding spring force, damping force, and inertia force vectors. The three force vectors may now be added vectorially as in the left-hand side of Eq. (8.12), which shows that the vector sum is equal to the disturbing force, represented by a vector of length F_0 rotating at a constant speed ω. Since all vectors rotate at the same speed, they are fixed relative to each other.

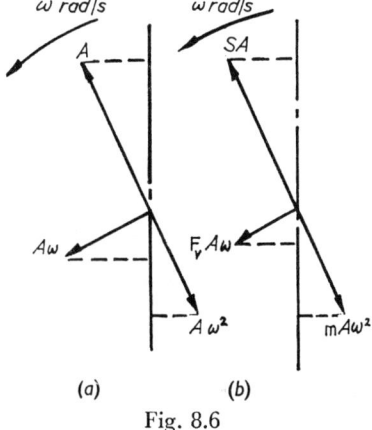

(a) (b)

Fig. 8.6

The summation of the vectors is shown in Fig. 8.7, two alternative arrangements being given. The vector diagram immediately shows that, provided the damping vector $F_vA\omega$ is not zero, the forcing vector F_0 must lead the spring force vector, and therefore the displacement, by some angle α. The assumed solution is thus shown to be slightly incorrect and requires modifying to

$$x = A \cos (\omega t - \alpha)$$

This modification has been incorporated in Fig. 8.7, from which the amplitude and phase-shift of the steady-state response can readily be extracted. Thus

$$(SA - mA\omega^2)^2 + (F_vA\omega)^2 = F_0^2 \text{ (Pythagoras)}$$

so that
$$A = F_0/\sqrt{\{(S - m\omega^2)^2 + (F_v\omega)^2\}}$$

$$= \frac{F_0/m}{\sqrt{\{(\omega_n^2 - \omega^2)^2 + (2k\omega)^2\}}} \tag{8.13}$$

where $\omega_n{}^2 = S/m$ and $2k = F_v/m$. Further,

$$\tan \alpha = F_v A \omega / (SA - mA\omega^2)$$

or $$\tan \alpha = \frac{(F_v/m)\omega}{(S/m) - \omega^2} = \frac{2k\omega}{\omega_n{}^2 - \omega^2} \quad (8.14)$$

Example. *A mass of 50 kg is supported by an elastic structure of total stiffness 20 kN/m. The damping ratio of the system is 0·2. A simple harmonic disturbing force acts on the mass, and at any time t seconds the force in newtons is 60 cos 10t. Determine the amplitude of the steady-state vibrations and the phase-shift caused by the damping.*

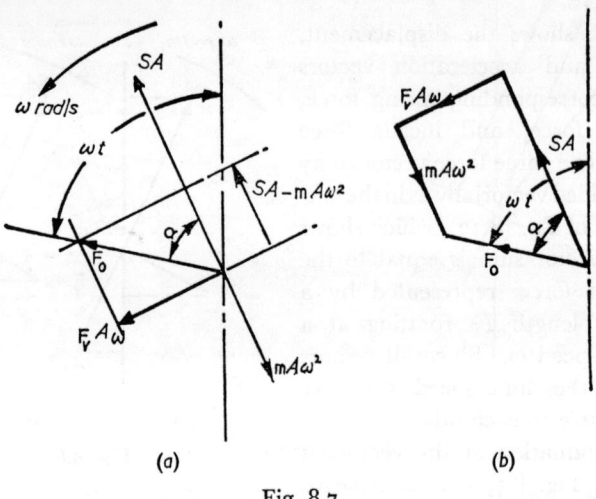

Fig. 8.7

Damping ratio $c = k/\omega_n$ and $\omega_n{}^2 = S/m = 20\,000/50 = 400/\text{s}^2$

$$\therefore k = c\omega_n = 0\cdot2\sqrt{(400)} = 4 \text{ rad/s}$$

$\therefore$ Damping force at unit velocity $= F_v = 2km = 2 \times 4 \times 50 = 400$ N

Equation of motion is

$$50(\mathrm{d}^2x/\mathrm{d}t^2) + 400(\mathrm{d}x/\mathrm{d}t) + 20\,000x = 50 \cos 10t$$

If A is the amplitude of the steady-state response,

Maximum spring force $= SA = 20\,000A$ newtons

Maximum damping force $= F_v A\omega = 400A10 = 4000A$ newtons

Maximum inertial force $= mA\omega^2 = 50A10^2 = 5000A$ newtons

From the vector diagram shown in Fig. 8.8,

$$(15\,000A)^2 + (4000A)^2 = 60^2 \text{ (Pythagoras)}$$
$$\therefore A^2 = 3600/(225 + 16)10^6$$
$$\therefore A = 3 \cdot 865 \times 10^{-3} \text{ m} = 3 \cdot 865 \text{ mm}$$

The phase shift is given by:

$$\tan \alpha = 4000A/15\,000A$$

or
$$\alpha = \tan^{-1}(4/15) = 14° \, 56'$$

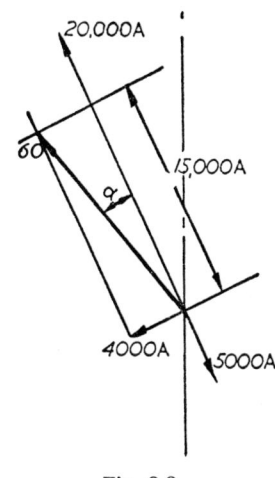

Fig. 8.8

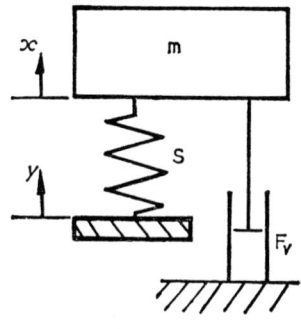

Fig. 8.9

Periodic displacement of the support

Dashpot not connected to the support. Suppose (Fig. 8.9) a periodic disturbing force is not applied directly to the mass, but is transmitted to the mass through the spring from a support which is moving with S.H.M. If the displacement of the support from its mean position at time t is y, where

$$y = a \cos \omega t$$

the output, or response of the system, is the displacement of the mass from its mean position. Let this be x at the time t and measured in the same sense as y. The extension of the spring $= x - y$, so that

$$\text{spring force} = S(x - y)$$

The spring force is now proportional to the *relative* displacement between the mass and the support and not, as in previous cases, to the absolute displacement of the mass. Viscous damping is provided by a dashpot unconnected to the moving support but which acts between the mass and the ground (assumed rigid). As in previous cases, damping force $= F_v(dx/dt)$, so that the resultant force on the mass is $-S(x-y) - F_v(dx/dt)$. It follows from Newton's second law that

$$m(d^2x/dt^2) = -S(x-y) - F_v(dx/dt)$$

Thus

$$m(d^2x/dt^2) + F_v(dx/dt) + Sx = Sy = Sa\cos\omega t \qquad (8.15)$$

This equation is identical with Eq. (8.2) with F_0 equal to Sa. Excitation of this type is therefore identical with the simple harmonic disturbing force of constant amplitude already dealt with. The amplitude A of the steady-state vibration is given by Eq. (8.13), where

$$F_0 = Sa = \text{amplitude of the equivalent disturbing force} \qquad (8.16)$$

The simple harmonic motion of the support may be represented by a rotating vector of length a and the equivalent disturbing force by a rotating

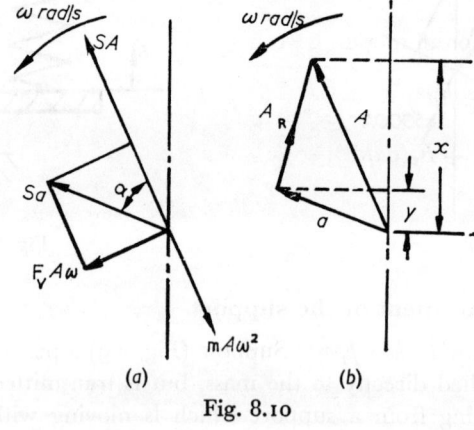

(a)   (b)

Fig. 8.10

vector of length Sa. Because the disturbing force is proportional to y, these vectors are in phase.

The force vector diagram corresponding to Eq. (8.15) is shown in Fig. 8.10 (*a*). Fig. 8.10 (*b*) shows the vector generating the motion of the support (input) and the vector generating the motion of the mass (output) which, because of damping, lags by an angle α, given by Eq. (8.14). Fig. 8.10 (*b*) also enables the relative displacement of load and support to be calculated. At any instant the

$$\text{relative displacement } x - y = A\cos(\omega t - \alpha) - a\cos\omega t$$

$$= A_R\cos(\omega t - \beta)$$

where A_R is the vector difference $(A - a)$ as indicated in Fig. 8.10 (b), and is called the relative amplitude.

Dashpot between mass and support. If the dashpot acts between the mass and support (Fig. 8.11), the damping force will be proportional to the *relative* velocity between them. The shock absorbers fitted to a motor car are an example. If when the support moves a distance $y(= a \cos \omega t)$, the mass moves a distance x in the same direction,

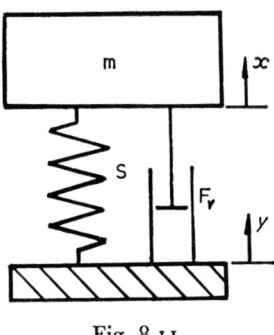

$$\text{spring force} = S(x - y)$$

and

Fig. 8.11

$$\text{damping force} = F_v\{(dx/dt) - (dy/dt)\}$$

Thus the equation of motion becomes

$$m(d^2x/dt^2) = -S(x - y) - F_v\{(dx/dt) - (dy/dt)\}$$

i.e. $m(d^2x/dt^2) + F_v(dx/dt) + Sx$

$$= Sy + F_v(dy/dt) \tag{8.17}$$

$$= Sa \cos \omega t - F_v a\omega \sin \omega t$$

$$= [\sqrt{\{(Sa)^2 + (F_v a\omega)^2\}}] \cos (\omega t + \varepsilon)$$

$$= F_0 \cos (\omega t + \varepsilon) \tag{8.18}$$

where $\quad F_0 = \sqrt{\{(Sa)^2 + (F_v a\omega)^2\}} \tag{8.19}$

Eq. (8.18) is, again, very similar to Eq. (8.2). The equivalent disturbing force is $F_0 \cos (\omega t + \varepsilon)$. The vector generating the equivalent disturbing force now leads the vector generating the motion of the support by the angle ε, where

$$\tan \varepsilon = F_v a\omega/Sa = 2k\omega/\omega_n^2 = 2c(\omega/\omega_n) \tag{8.20}$$

The steady-state solution to Eq. (8.17) is

$$x = A \cos \{(\omega t + \varepsilon) - \alpha\} \tag{8.21}$$

where A and α are given by Eqs. (8.13) and (8.14) respectively. Fig. 8.12 (a), (b) and (c) show, respectively, the vectors generating the equivalent disturbing force F_0; the vector solution of Eq. (8.18), and the vectors generating the input motion of the support $y = a \cos \omega t$ and the output motion of the mass $x = A \cos \{(\omega t + \varepsilon) - \alpha\}$.

The relative amplitude may be obtained, if required, from Fig. 8.12 (c). Notice that the output A now lags the input a by the angle $(\alpha - \varepsilon)$.

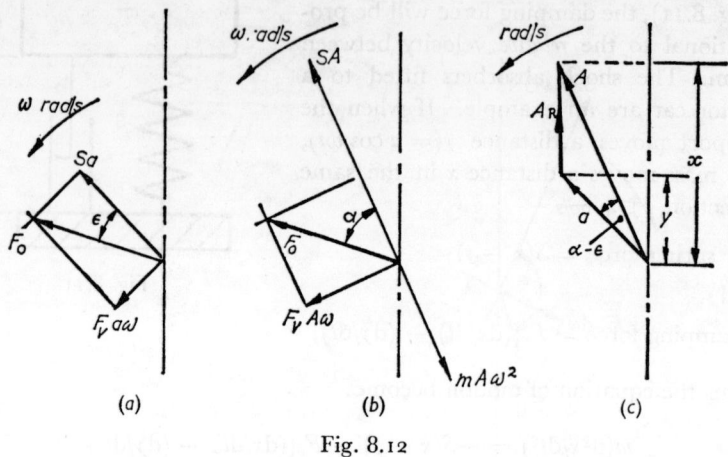

Fig. 8.12

Example. *A mass of 5 kg is suspended from a support by springs, which deflect 49·05 mm under its static weight. Both mass and support are constrained to move vertically, and the support is moved by a tappet rod activated by a simple harmonic cam. The throw of the cam is 2 mm, and its speed of rotation is 147 rev/min. Movement of the mass is controlled by a dashpot connected between the mass and the ground, which may be assumed rigid. Under these conditions the amplitude of the mass is measured and found to be 4 mm. Determine (a) the maximum damping force, (b) the damping ratio, (c) the maximum spring force.*

The system described is similar to that shown in Fig. 8.9

$$\text{Total stiffness of springs} = mg/\text{static deflection}$$

i.e.
$$S = 5 \times 9\cdot81/0\cdot04905$$
$$= 1000 \text{ N/m}$$

$$\text{Natural circular frequency } \omega_n = \sqrt{(S/m)} = \sqrt{(1000/5)}$$
$$= 14\cdot14 \text{ rad/s}$$

$$\text{Forcing frequency (circular) } \omega = 147 \times 2\pi/60$$
$$= 15\cdot4 \text{ rad/s}$$

$$\text{Amplitude of support } a = 1 \text{ mm} = 0\cdot001 \text{ m}$$

$$\text{Maximum equivalent disturbing force } F_0 = Sa = 1000 \times 0\cdot001$$
$$= 1\cdot0 \text{ N}$$

Maximum equivalent spring force $= SA = 1000 \times 0.004$
$$= 4.0 \text{ N}$$

Maximum inertia force $= mA\omega^2$
$$= 5 \times 0.004 \times (15.4)^2$$
$$= 4.743 \text{ N}$$

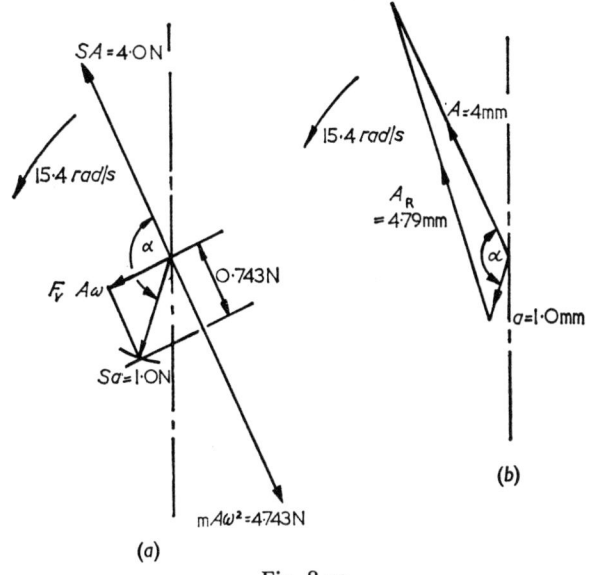

Fig. 8.13

The force vector diagram may now be drawn, to scale if preferred, as shown in Fig. 8.13 (*a*), and from this diagram:

Maximum damping force $= F_v A\omega = 0.669 \text{ N}$

Phase shift angle $= \alpha = 138°$

Damping ratio, $c = k/\omega_n$ where $2k = F_v/m$

Thus $$2kmA\omega = 0.669 \text{ N}$$

so that

$$k = 0.669/(2 \times 5 \times 0.004 \times 15.4) = 1.086 \text{ rad/s}$$

and

Damping ratio, $c = 1.086/14.14 = 0.0768$

The displacement vector diagram is shown in Fig. 8.13 (*b*), from which

relative amplitude $= A_R = 4.79 \text{ mm}$

$\therefore$ Maximum spring force $= SA_R = 1000 \times 0.00479 = 4.79 \text{ N}$

Example. *A two-wheeled trailer, whose tyres may be assumed rigid, is drawn at 15 m/s over an undulating road surface. The maximum vertical height between hollow and crest is 100 mm, and the length between adjacent crests is 5 m. Springs and shock absorbers are fitted between the axle and the trailer body, whose mass is 500 kg. The total stiffness of the springs is 50 kN/m and the shock absorbers provide a viscous resistance giving a damping ratio of 0·25.*

Assuming the surface undulations to be sinusoidal, and that the wheels never leave the surface at any time, determine: (a) the amplitude of vertical vibrations of the trailer body, (b) the maximum spring force, (c) the maximum damping force, (d) the maximum force transmitted to the road.

The system described is essentially that of Fig. 8.11, the support in this case being the axle which will be constrained to move in accordance with the road surface undulations.

$$\text{Amplitude of support } a = 50 \text{ mm} = 0·05 \text{ m}$$

$$\text{Forcing frequency } \omega = 2\pi/T$$

where T is the time taken from crest to crest $= (5/15)$ seconds.

$$\therefore \ \omega = 2\pi/(1/3) = 6\pi \text{ rad/s}$$

Natural circular frequency

$$\omega_n = \sqrt{(S/m)} = \sqrt{(50\,000/500)} = 10 \text{ rad/s}$$

Damping constant,

$$k = 0·25 \times \omega_n = 2·5 \text{ rad/s}$$

$$\therefore \ F_v = 2km = 2 \times 2·5 \times 500 = 2500 \text{ N s/m}$$

From Eq. (8.19),
  Equivalent disturbing force,

$$F_0 = \sqrt{\{(Sa)^2 + (F_v a\omega)^2\}}$$

$$= \sqrt{\{(50\,000 \times 0·05)^2 + (2500 \times 0·05 \times 6\pi)^2\}}$$

$$= 3435 \text{ N}$$

The generation of this disturbing force by vectors is shown in Fig. 8.14 (a), from which $\varepsilon = 43° 18'$. Alternatively, from Eq. (8.20),

$$\tan \varepsilon = 2 \times 0·25(6\pi/10) = 0·9426$$

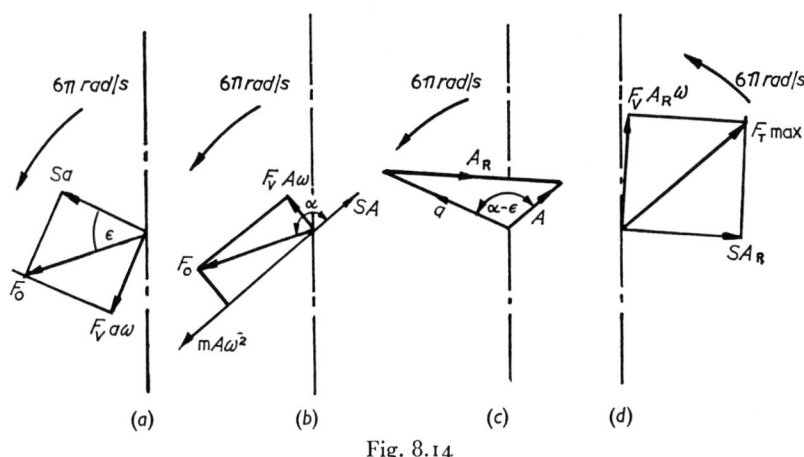

Fig. 8.14

The force vector diagram representing the solution of the equation of motion (8.18) is shown in Fig. 8.14 (b), and referring to this diagram

$$\text{Vector } SA = 50000A$$
$$\text{Vector } F_v A\omega = 2500 \times A \times 6\pi = 47130A$$
$$\text{Vector } mA\omega^2 = 500 \times A \times (6\pi)^2 = 177700A$$

where A is the amplitude of the trailer body.

$$SA - mA\omega^2 = (50000 - 177700)A = -127700A$$

By Pythagoras

$$F_0 = \sqrt{\{(127700A)^2 + (47130A)^2\}} = 3435 \text{ N}$$

so that $\qquad A = 0.02524 \text{ m} = 25.24 \text{ mm}$

Also, $\qquad \tan \alpha = F_v A\omega/(SA - mA\omega^2) = -47130/127700$

$$\therefore \alpha = 180° - 20° 15' = 159° 45'$$

The vector generating the input motion is of length a ($= 50$ mm) in phase with the vector Sa (Fig. 8.14 (a)). The vector generating the output motion is of length A ($= 25.24$ mm) in phase with the vector SA (Fig. 8.14 (b)). Fig. 8.14 (c) shows the vectors generating input and output motions and gives

$$\alpha - \epsilon = 159° 45' - 43° 18' = 116° 27'$$

Hence, by the cosine rule (or scale drawing)

$$\text{Relative amplitude } A_R = 65.28 \text{ mm}$$
$$\therefore \text{ Maximum spring force} = SA_R = 50000 \times 0.06528 \text{ N}$$
$$= 3.264 \text{ kN}$$

At any time, t

    Relative displacement $= A_R \cos(\omega t - \beta)$ where β is a constant

    $\therefore$ Relative velocity $= -A_R \omega \sin(\omega t - \beta)$

$\therefore$ Maximum relative velocity $= A_R \omega$

    $\therefore$ Maximum damping force $= F_v A_R \omega = 2500 \times 0.06528 \times 6\pi$ N

                                  $= 3.075$ kN

The force transmitted from the trailer body to the ground via the spring and the dashpot is the resultant of the spring force and the damping force. Since spring force is proportional to relative displacement and damping force to relative velocity, these forces are 90° out of phase (Fig. 8.14 (d)).

    $\therefore$ Force transmitted to the ground $= \sqrt{(3.264^2 + 3.075^2)}$

                                          $= 4.485$ kN

Dynamic magnification and phase shift

S.H. disturbing force of constant amptitude. The magnification factor (M.F.) or dynamic magnifier is the ratio between the output response and the corresponding input. This ratio is non-dimensional and is expressed either as the ratio of amplitude or as the ratio of forces, i.e.

$$\text{M.F.} = \frac{\text{Actual amplitude of forced vibration}}{\text{Amplitude at zero forcing frequency (static)}} \quad (8.22)$$

$$= \frac{\text{Maximum force in spring}}{\text{Maximum value of disturbing force}} \quad (8.22)$$

The equality of these two ratios is demonstrated in Eq. (8.23). In the system of Fig. 8.5, the

$$\text{Forcing function} = F_0 \cos \omega t$$

so that at zero forcing frequency, $\omega = 0$, it is just F_0, a constant.

    The static deflection of the spring due to the application of a constant force F_0 is F_0/S. When the forcing frequency is not zero, the amplitude is given by Eq. (8.13), so that

$$\text{M.F.} = \frac{A}{(F_0/S)} = \frac{SA}{F_0} \quad (8.23)$$

$$= \frac{S}{F_0} \times \frac{F_0/m}{\sqrt{\{(\omega_n^2 - \omega^2)^2 + (2k\omega)^2\}}}$$

$$= \frac{S/m}{\sqrt{\{(\omega_n^2 - \omega^2)^2 + (2k\omega)^2\}}}$$

But $S/m = \omega_n{}^2$

$$\therefore \text{M.F.} = \omega_n{}^2/\sqrt{\{(\omega_n{}^2 - \omega^2)^2 + (2k\omega)^2\}}$$

$$= 1/\sqrt{[\{1 - (\omega/\omega_n)^2\}^2 + \{2k\omega/\omega_n{}^2\}^2]}$$

$$= \frac{1}{\sqrt{[\{1 - (\omega/\omega_n)^2\}^2 + \{2c(\omega/\omega_n)\}^2]}} \qquad (8.24)$$

Eq. (8.24) expresses the M.F. conveniently in terms of the damping ratio c and the frequency ratio ω/ω_n. The variation of M.F. with ω/ω_n for various values of c is shown in Fig. 8.15 (a). As the ω/ω_n is increased from zero, the amplitude grows until a maximum value is reached. This critical condition is known as *resonance*, and the forcing frequency at resonance is called the resonant frequency. For zero damping, resonance occurs at a frequency equal to the natural undamped frequency (i.e. $\omega/\omega_n = 1 \cdot 0$), and the amplitude becomes infinite. However, in practice there is always some damping, and this limits the amplitude. For lightly damped systems, the resonant condition can be extremely dangerous and failure of the elastic structure can result.

As the frequency ratio is increased beyond resonance, the amplitude decreases, and, even for zero damping, frequency ratios greater than $\sqrt{2}$ produce amplitudes which are less than the static deflection ($\omega = 0$). For large values of frequency ratio, the amplitude tends to zero, the "vibrating" mass becoming almost motionless, its inertia being such that it is unable to follow the constantly changing input commands.

The effect of damping is important only in the resonant region, where two effects are noteworthy:

1. The resonant amplitude is considerably reduced if the damping ratio is increased.

2. The resonant frequency is reduced by increasing the damping, but the effect is slight for light damping and is then sometimes ignored.

Resonance occurs at a value of ω/ω_n for which the M.F. Eq. (8.24) is a maximum, i.e. when $[\{1 - (\omega/\omega_n)^2\}^2 + \{2c(\omega/\omega_n)\}^2]$ is a minimum. Putting $r = (\omega/\omega_n)$, this condition holds if

$$\frac{\mathrm{d}}{\mathrm{d}r}[(1 - r^2)^2 + 4c^2r^2] = 0$$

i.e. $$2(1 - r^2)(-2r) + 8c^2r = 0$$

From which

$$r = \sqrt{(1 - 2c^2)} \qquad (8.25)$$

This equation for resonant frequency ratio is only valid for values of

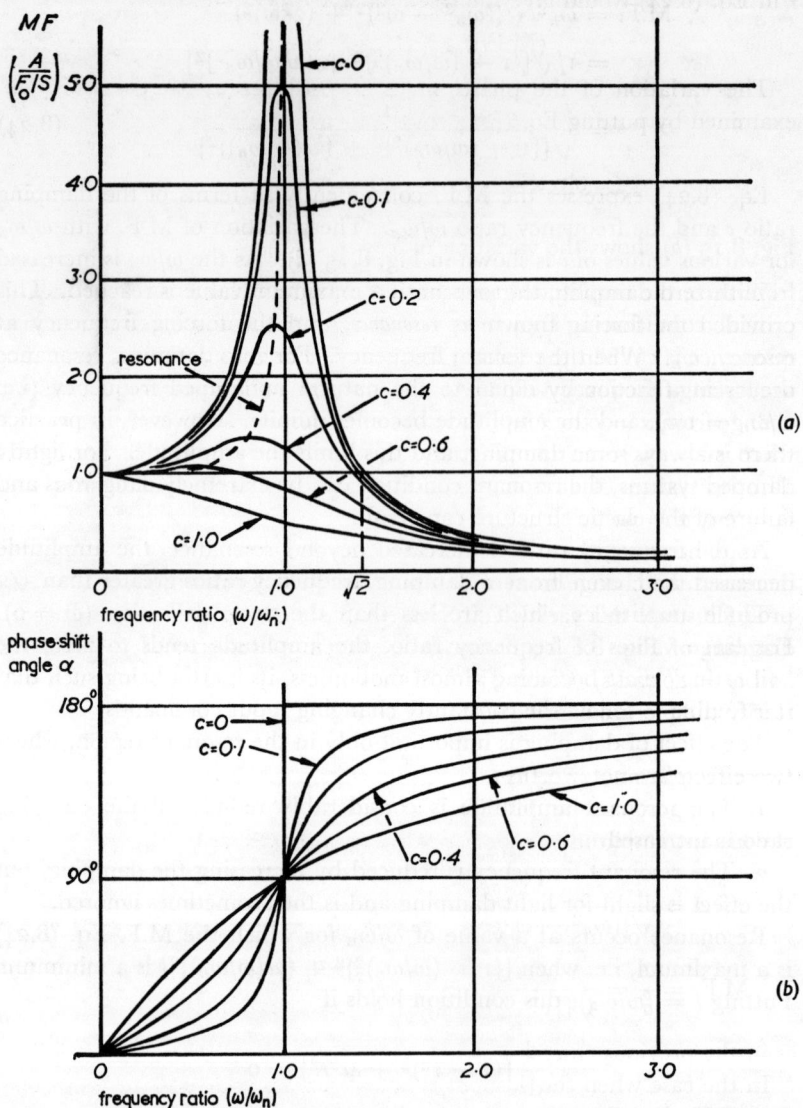

Fig. 8.15

$c^2 \leqslant 0.5$, otherwise r becomes imaginary. Substitution of this value of r in Eq. (8.24) would give the resonant magnification factor

$$(\text{M.F.})_{\text{res}} = 1/2c\sqrt{(1 - c^2)} \qquad (8.26)$$

The variation of the phase angle, α, with frequency ratio may be examined by putting Eq. (8.14) in the form

$$\tan \alpha = \frac{2k\omega}{\omega_n^2 - \omega^2} = \frac{2c(\omega/\omega_n)}{1 - (\omega/\omega_n)^2} \qquad (8.27)$$

Fig. 8.15 (*b*) shows the variation of α with (ω/ω_n).

With zero damping, the response is in phase with the forcing function provided the forcing frequency is less than the natural undamped frequency, ω_n. When the forcing frequency exceeds ω_n, the response lags the forcing function by $180°$. At resonance there is, apparently, a sudden change from $\alpha = 0$ to $\alpha = 180°$. It may be shown however that $\alpha = 90°$, when $\omega = \omega_n$, even though this is not obvious from the diagram. At all other values of damping ratio the diagram clearly shows that when $\omega = \omega_n$, $\alpha = 90°$.

Note, however, that this is not the resonant condition. As damping is increased the change from $\alpha = 0$ to $\alpha = 180°$ becomes more gradual.

To illustrate this variation in phase angle it is suggested that the vector diagram of Fig. 8.8 be re-drawn for forcing frequencies $\omega = 20$ rad/s and $\omega = 40$ rad/s. Comparison of the three diagrams thus obtained will clarify the picture enormously.

Periodic displacement of support. The case when the dashpot is not connected to the support can be described by substituting a S.H. disturbing force of constant amplitude Sa for F_0. Thus from Eq. (8.13)

$$A = \frac{Sa/m}{\sqrt{\{(\omega_n^2 - \omega^2)^2 + (2k\omega)^2\}}} = a\,\frac{\omega_n^2}{\sqrt{\{(\omega_n^2 - \omega^2)^2 + (2k\omega)^2\}}}$$

$$\therefore \ \text{M.F.} = \frac{A}{a} = \frac{1}{\sqrt{[\{1 - (\omega/\omega_n)^2\}^2 + \}2c(\omega/\omega_n)\}^2]}}$$

as in the previous case.

In the case when the dashpot is between mass and support, the equivalent S.H. disturbing force is given by

$$F_0 = \sqrt{\{(Sa)^2 + (F_v a\omega)^2\}} = a\sqrt{(S^2 + F_v^2\omega^2)}$$

Substituting in Eq. (8.13)

$$A = \frac{\{a\sqrt{(S^2 + F_v^2\omega^2)}\}/m}{\sqrt{\{(\omega_n^2 - \omega^2)^2 + (2k\omega)^2\}}} = a\,\frac{\sqrt{\{\omega_n^4 + (2k\omega)^2\}}}{\sqrt{\{(\omega_n^2 - \omega^2)^2 + (2k\omega)^2\}}}$$

From which

$$\text{M.F.} = \frac{A}{a} = \frac{\sqrt{[1 + \{2c(\omega/\omega_n)\}^2]}}{\sqrt{[\{1 - (\omega/\omega_n)^2\}^2 + \{2c(\omega/\omega_n)\}^2]}} \qquad (8.28)$$

The variation of this quantity with frequency ratio is shown in Fig. 9.2.

Harmonic disturbing force with variable amplitude. An important case occurs in practice where the disturbing force is generated by the lack of balance

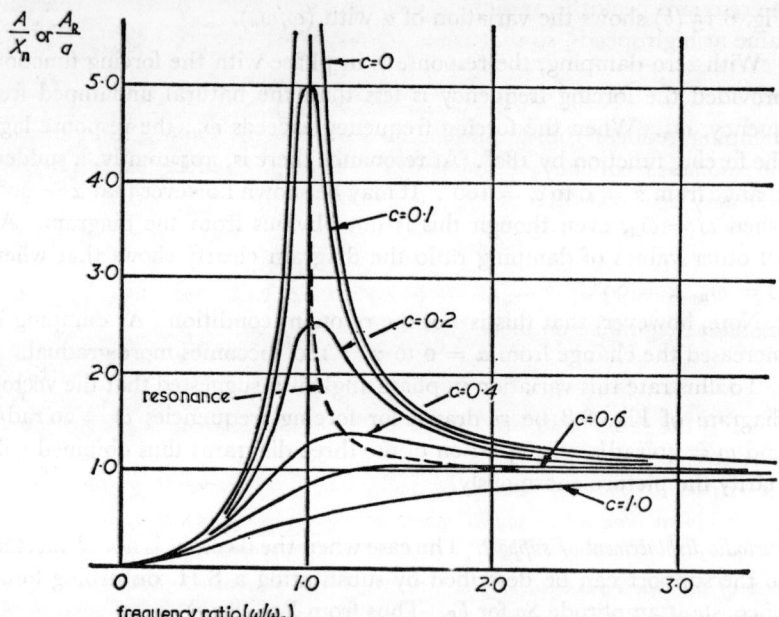

Fig. 8.16

of the rotating part of a machine. The maximum disturbing force F_0 is $m_1 r_1 \omega^2$, the centrifugal force due to an unbalanced mass m_1 at radius r_1. Substituting for F_0 in Eq. (8.13)

$$A = \frac{m_1 r_1 \omega^2/m}{\sqrt{\{(\omega_n^2 - \omega^2)^2 + (2k\omega)^2\}}} = \frac{m_1}{m} r_1 \frac{\omega^2}{\sqrt{\{(\omega_n^2 - \omega^2)^2 + (2k\omega)^2\}}}$$

But

$$m_1 r_1/m = m_1 r_1 \omega_n^2/S = X_n \qquad (8.29)$$

where X_n is the static deflection of the elastic support when subjected to a force equal to the unbalanced centrifugal force at a speed ω_n rad/s.

The magnification factor in this case may be defined in terms of this parameter as

$$\text{M.F.} = \frac{A}{X_n} = \frac{(\omega/\omega_n)^2}{\sqrt{[\{1 - (\omega/\omega_n)^2\}^2 + \{2c(\omega/\omega_n)\}^2]}} \qquad (8.30)$$

The variation of this quantity with frequency ratio is shown in Fig. 8.16. At zero frequency ratio, there is no disturbing force and therefore no displacement. The amplitude gradually builds up to a resonant peak as the amplitude of the disturbing force $m_1 r_1 \omega^2 \cos \omega t$ increases Beyond the resonant peak the amplitude decreases and tends towards a constant value at high speeds, so that when (ω/ω_n) is very large,

$$A = X_n \qquad (8.31)$$

Damping reduces the amplitude in the resonant region but increases the resonant frequency. Elsewhere, the effects of damping are negligible.

It may be shown that resonance occurs when:

$$(\omega/\omega_n) = 1/\sqrt{(1 - 2c^2)} \qquad (8.32)$$

Substitution of this value of the frequency ratio in Eq. (8.30) gives the resonant magnification factor, as

$$(\text{M.F.})_{\text{res}} = 1/2c\sqrt{(1 - c^2)} \qquad (8.33)$$

which is the same as in Eq. (8.26).

Example. *A mass of 500 kg is mounted on supports having a total stiffness of 100 kN/m, and which provide viscous damping, the damping ratio being 0.4. The mass is constrained to move vertically, and is subjected to a vertical disturbing force of the type $F_0 \cos \omega t$. F_0 is constant, but ω may vary. Determine the frequency at which resonance will occur, and the maximum allowable value of F_0 if the amplitude at resonance is to be restricted to 5 mm.*

Natural circular frequency $\omega_n = \sqrt{(S/m)} = \sqrt{(100 \times 10^3/500)}$

$$= \sqrt{200} = 14 \cdot 14 \text{ rad/s}$$

From Eq. (8.25), the frequency ratio at resonance is given by:

$$(\omega/\omega_n)_{\text{res}} = \sqrt{(1 - 2c^2)} = \sqrt{(1 - 2 \times 0 \cdot 4^2)} = \sqrt{(0 \cdot 68)}$$

$$\therefore \quad \omega_{\text{res}} = 14 \cdot 14 \cdot \sqrt{(0 \cdot 68)}$$

$\therefore$ Resonant frequency,

$$\omega_{\text{res}}/2\pi = 14 \cdot 14 \sqrt{(0 \cdot 68)}/2\pi = 1 \cdot 856 \text{ Hz}$$

From Eq. (8.24)

$$(M.F.)_{res} = 1/\sqrt{[(1 - 0.68)^2 + 4 . (0.4)^2 0.68)]} = 1.364$$

From Eq. (8.23), M.F. $= SA/F_0$, so that if the amplitude at resonance is 5 mm

$$1.364 = 100000 \times 0.005/F_0$$

and

$$F_0 = 500/1.364 = 366.6 \text{ N}$$

Example. *A machine has a mass of 1000 kg and is mounted on supports providing a total stiffness of 1 MN/m. When in operation the machine vibrates vertically due to the unbalance of the rotating parts of the machine.*

As the speed is increased from zero, the machine is observed to pass through the resonant condition, and a maximum amplitude of 0.25 mm is recorded. At very high speeds of operation the amplitude of vibration is found to be 0.05 mm and is apparently unaffected by changes of speed.

Estimate the amount of rotating unbalance in the machine, and also the damping ratio.

$$\omega_n{}^2 = S/m = 10^6/1000 = 1000 \text{ (rad/s)}^2$$

The disturbing force is proportional to the square of the speed.

From Eqs. (8.29) and (8.31):

Amplitude at high speeds, $A = X_n = m_1 r_1 \omega_n{}^2/S$

$$\therefore 0.05 \times 10^{-3} = m_1 r_1 . (1000/10^6)$$

$$\therefore m_1 r_1 = 0.05 \text{ kg m}$$

This means that the rotating unbalance is equivalent to a mass of 0.05 kg at 1 m radius, or 5 kg at 10 mm radius.

From Eq. (8.33)

$$(M.F.)_{res} = 1/2c\sqrt{(1 - c^2)} = A_{max}/X_n = 0.25/0.05 = 5$$

i.e.

$$1/4c^2(1 - c^2) = 25$$

This reduces to a quadratic in c^2 whose roots are

$$c^2 = 0.9899 \quad \text{or} \quad c^2 = 0.0101$$

The first of these roots gives a damping ratio close to the critical, which is clearly not the case.

$$\therefore \text{Damping ratio, } c = \sqrt{0.0101} = 0.1005$$

PROBLEMS

For tutorials

1. What may be deduced regarding the frequency ratio of a forced oscillation from the fact that the phase-shift angle is (*a*) between O and 90°, (*b*) equal to 90°, (*c*) between 90° and 180°?

2. Define the term "resonance." Why is it incorrect to say that resonance occurs when the forcing frequency is equal to the undamped natural frequency? Under what circumstances is the statement approximately true?

3. Eq. (8.25) gives the frequency ratio at which resonance occurs. This equation is only valid for values of c^2 less than 0·5. What happens to the resonant peak when c^2 is greater than 0·5?

4. A flywheel of moment of inertia I is mounted on the end of a shaft of torsional stiffness T_0. A simple harmonic disturbing torque $T \cos \omega t$ acts on the flywheel while the other end of the shaft is fixed. Motion of the flywheel is restricted by a viscous damping of magnitude T_F at unit angular velocity. Write down the equation of motion for the system, and hence obtain the response.

5. A cylinder, of mass m, is supported by an elastic structure, of stiffness S. The cylinder is fitted with a piston which is made to perform S.H.M. described by the equation: $y = a \cos \omega t$. Assuming that the resistance to relative motion between piston and cylinder is proportional to velocity, write down the equation of motion for the cylinder.

6. Using the amplitude response vector A and the phase-shift angle α as polar co-ordinates, plot the locus of A (as shown in Fig. 8.17) as the frequency ratio varies from o to infinity. Assume a forcing function $F_0 \cos \omega t$ and for convenience let the static deflection due to F_0 be unity.

Compare the diagram thus obtained with that of Fig. 8.15, and discuss their relative merits.

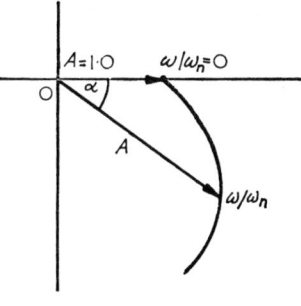

Fig. 8.17

7. An alternating electric current may be regarded as an example of a forced vibration, the current $I \cos (\omega t - \alpha)$ being the response to the forcing potential $V \cos \omega t$. In a circuit possessing resistance R, inductance L and capacitance C, the relationship between peak values of current and voltage is:

$$I = V/\sqrt{\{R^2 + (L\omega - 1/C\omega)^2\}} = V/Z$$

where Z is defined as the impedance of the circuit.

Defining "mechanical impedance" as the ratio of the maximum value of the disturbing force to the maximum velocity of the response show that R is analogous to the viscous resistance F_v, L to the mass m, and $1/C$ to the elastic stiffness S.

Discuss the similarities between vibration theory and alternating current theory, and comment on the practical importance of this similarity.

General

1. A mass of 1 kg is supported by an elastic structure having a stiffness of 244 N/m. The mass is acted upon by a simple harmonic disturbing force of magnitude 13 cos 10t newtons, where t is the time in seconds. The system is subjected to viscous damping and at a velocity of 1 m/s the damping force is 6 N.

Write down the equation of motion for the system and sketch the vector diagram representing it. Hence determine the amplitude of the motion and the phase angle between the disturbing force and the motion.

What is the magnification factor?

Ans. 83·3 mm; 22° 37'; 1·563

2. An electric motor, of total mass 30 kg, rests on elastic supports having a total vertical stiffness of 200 kN/m. The armature of the motor has a mass of 10 kg, and due to imperfections in manufacture its centre of gravity is 1·0 mm from the axis of rotation. When the motor runs at 1050 rev/min the amplitude of vertical vibrations is found to be 0·5 mm.

Assuming the system is subject to viscous damping determine,

   (*a*) the frequency ratio,
   (*b*) the damping force at 1 m/s,
   (*c*) the damping ratio,
   (*d*) the phase difference between the vertical deflection and the rotating disturbance force,
   (*e*) the amplitude when the motor speed is increased to a very large value.

Ans. (*a*) 1·347,   (*b*) 1·627 kN,   (*c*) 0·332
(*d*) 132° 21', (*e*) 0·333 mm

3. Define the terms "frequency ratio," "damping ratio," and "magnification factor."

A mass of 5 kg is supported by an elastic structure of stiffness 600 N/m. The mass vibrates, and the vibration is sustained by a simple harmonic disturbing force having a constant peak value of 20 N. If the damping ratio is 0·5, determine the resonant frequency and the amplitude at this frequency.

Ans. 1·233 Hz; 38·5 mm

4. A single-cylinder vertical reciprocating engine is supported on four springs, each having a stiffness of 250 kN/m. The total mass of the engine is 500 kg, and the mass of the reciprocating parts is 5 kg. The stroke of the piston is 160 mm, and the engine speed is 300 rev/min.

Assuming that the motion of the piston in the cylinder is simple harmonic, and that all rotating masses are balanced, find the amplitude of steady-state vertical vibrations of the engine frame (*a*) if the damping is negligible, (*b*) if the vibrations are critically damped.

Ans. (*a*) 0·780 mm, (*b*) 0·264 mm

5. (*a*) A mass is connected by means of a spring to an inelastic support, and when the system is allowed to hang freely from the support the static deflection is 50 mm. The support is given a vertical S.H.M. of frequency 2·0 Hz and amplitude 10 mm. If the damping is negligible determine the amplitude of vibration of the mass.

(*b*) The mass is now lowered into a bath of oil, so that it is totally immersed at all times, and the support given the same motion as before. The effect of the oil bath is to reduce the amplitude of vibration of the mass to 20 per cent of its previous value. Find the damping ratio, the phase shift between the motions of the support and the mass, and the relative amplitude between the mass and the support.

(Assume $g = 9.81$ m/s^2.)

Ans. (*a*) 51·36 mm; (*b*) 0·532; 78° 28′; 12·83 mm

6. The body of a vehicle has a mass of 1 tonne when fully loaded and 250 kg when empty. The suspension is such that the total stiffness of the springs is 400 kN/m, and the shock absorbers provide a damping ratio of 0·5 *when the vehicle is fully loaded*. The vehicle is travelling along a sinusoidal track of 20 mm amplitude and 5 m between crests at a speed of 25 m/s. Assuming the effective stiffness of the tyres to be very large and that the wheels do not leave the ground, find the amplitude of vertical vibrations of the vehicle body (*a*) when fully loaded (*b*) when empty.

(Note that since the damping force, F_v, at unit velocity remains constant and $F_v = 2c\omega_n m$, the damping ratio c will vary with the mass m.)

Ans. (*a*) 17·32 mm; (*b*) 23·0 mm

CHAPTER 9

Vibration Isolation and Seismic Instruments

9.1 TRANSMISSIBILITY

When a simple harmonic disturbing force $F_0 \cos \omega t$ is applied to a mass m supported by a spring of stiffness S (Fig. 8.4), force is transmitted by means of the spring and dashpot to the fixed support. This force is equal to the vector sum of the spring force and the damping force, and from the vector diagram, Fig. 9.1, the force transmitted = $F_T \cos (\omega t - \phi,)$ where

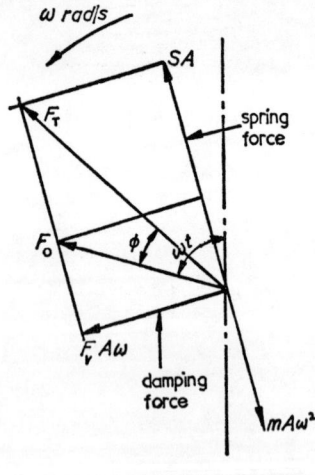

$$F_T = \sqrt{\{(SA)^2 + (F_v A\omega)^2\}} \quad (9.1)$$

The ratio F_T/F_0 is known as the *isolation factor* or *transmissibility ratio* (T.R.) and works out as

$$F_T/F_0 = \sqrt{\{(SA)^2 + (F_v A\omega)^2\}}/F_0$$
$$= Am\sqrt{\{(S/m)^2 + (F_v/m)^2\omega^2\}}/F_0$$

Substituting for A from Eq. (8.13),

$$\text{T.R.} = \frac{\sqrt{\{\omega_n^4 + (2k\omega)^2\}}}{\sqrt{\{(\omega_n^2 - \omega^2)^2 + (2k\omega)^2\}}}$$

$$= \frac{\sqrt{[1 + \{2c(\omega/\omega_n)\}^2]}}{\sqrt{[\{1 - (\omega/\omega_n)^2\}^2 + \{2c(\omega/\omega_n)\}^2]}}$$

$$(9.2)$$

Fig. 9.1

This equation is identical with Eq. (8.28), so that when the excitation is due to a periodic displacement of the support (as in Fig. 8.11), rather than to a disturbing force applied to the mass, the proportion of *motion* transmitted to the mass (A/a) is also given by Eq. (9.2).

The variation of T.R. with frequency ratio is shown in Fig. 9.2, which shows that the frequency ratio must be greater than $\sqrt{2}$ if the force or motion transmitted is to be less than the impressed force or motion. Furthermore, when the frequency ratio is greater than $\sqrt{2}$ the effect of damping is to increase the transmissibility.

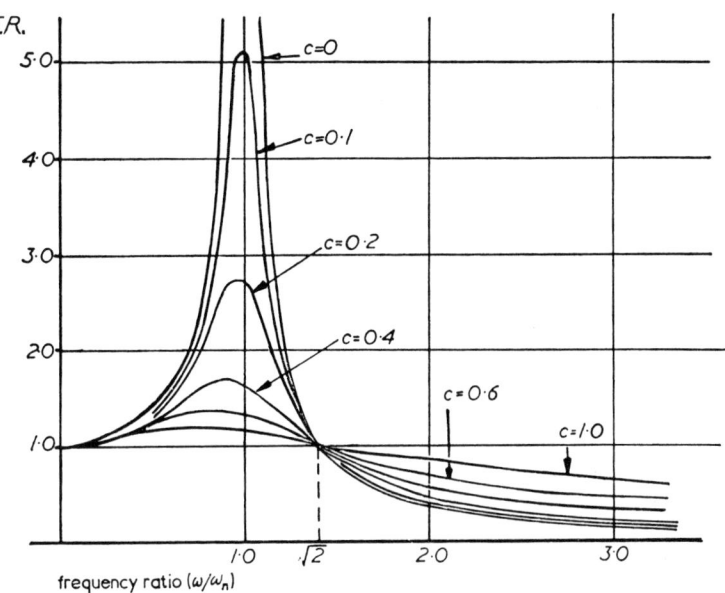

Fig. 9.2

Example. *The instrument panel in the cockpit of an aircraft has a mass of 25 kg and is mounted on isolators having a stiffness of 50 kN/m. Assuming that damping is negligible, estimate the percentage of motion transmitted to the instruments when the airframe vibrates at a frequency of 2000 vibrations/min.*

$$\text{Natural circular frequency } \omega_n = \sqrt{(S/m)}$$
$$= \sqrt{(50\,000/25)} = 44\cdot72 \text{ s}^{-1}$$
$$\text{Forcing frequency } \omega/2\pi = 2000/60$$
$$\therefore \omega = 4000\pi/60 = 209\cdot4 \text{ s}^{-1}$$
$$\therefore \text{ Frequency ratio } \omega/\omega_n = 209\cdot4/44\cdot72 = 4\cdot68$$

Since the damping is negligible, $c = 0$, Eq. (9.2) gives T.R. $= 0\cdot0479$, so that only $4\cdot8$ per cent of the vibratory motion of the airframe is transmitted to the instruments.

Example. *A centrifugal fan has a mass of 50 kg and runs at a constant speed of 1200 rev/min. The fan has a rotating unbalance equivalent to a mass of 10 kg at a radius of 10 mm. Assuming a damping ratio of 0·25, determine the total stiffness of springs suitable for the mounting of the fan if only 10 per cent of the unbalanced force is to be transmitted to the floor. Calculate also the maximum value of the force transmitted under these conditions.*

For convenience, let the frequency ratio, $\omega/\omega_n = r$. The damping ratio $c = 0·25$ and the transmissibility ratio $= 0·10$. Eq. (9.2) gives

$$0·10 = \sqrt{\{1 + (2 \times 0·25r)^2\}}/\sqrt{\{(1 - r^2)^2 + (2 \times 0·25r)^2\}}$$

Squaring both sides and clearing fractions

$$0·01\{(1 - r^2)^2 + 0·25r^2\} = 1 + 0·25r^2$$

which simplifies to $r^4 - 26·75r^2 - 99 = 0$.

The only possible solution of this quadratic in r^2 is

$$r^2 = 30·05$$
$$\therefore r = \omega/\omega_n = 5·48$$
$$\therefore \omega_n = \omega/5·48$$

But
$$\omega = 1200 \times 2\pi/60 = 40\pi \text{ rad/s}$$
$$\therefore \omega_n = 40\pi/5·48 = 22·93 \text{ rad/s}$$
$$S/m = \omega_n{}^2 = 22·93^2$$
$$\therefore S = 22·93^2 \times 50 = 26\ 290 \text{ N/m}$$

i.e.

Total stiffness of support springs $= 26·29$ kN/m

Maximum force transmitted $= 0·10 \times F_0$
$$= 0·10 \times m_1 r_1 \omega^2$$
$$= 0·10 \times 10 \times 0·010 \times (40\pi)^2$$
$$= 158 \text{ N}$$

If the fan had been rigidly mounted, the floor would have been subjected to a fluctuating force of magnitude ± 1580 N.

9.2 Vibration Isolation

Active isolation

All machines with moving parts must generate periodic disturbing forces which cause the machine itself to vibrate and which are also transmitted to its foundations, causing the floor to vibrate. These ground vibrations can crack concrete floors, damage under-floor services and are transmitted over surprisingly great distances to other structures and machines, impairing their efficiency and inflicting damage over long periods. Obviously these effects must be avoided if possible. Unfortunately, it is often impossible to eliminate the disturbing force by balancing or by fitting of vibration absorbers. By mounting a machine on suitably designed flexible supports, however, the transmission of undesirable vibrations to its foundations can be reduced. This way is called active isolation (Fig. 8.4).

Passive isolation

When the operation of a precision machine or instrument is adversely effected by a vibration transmitted to it through its foundations, it can be protected by mounting it on flexible supports in such a way that the system has a low transmissibility ratio. This is called passive isolation (Fig. 8.11). One difficulty is that passive isolation must shield against vibrations of the wide range of frequencies emanating from the surroundings. By contrast, active isolation must be designed for the well-defined frequencies generated by the machine itself.

Anti-vibration mountings

An anti-vibration mounting, active or passive, should have a transmissibility ratio as small as possible. Fig. 9.2 shows that for frequency ratios between 0 and $\sqrt{2}$, rigid mountings are best, for flexible mountings result in a T.R. greater than unity. It follows that for anti-vibration mountings to be effective, the frequency ratio must be greater than $\sqrt{2}$ and, ideally, as large as possible. At large frequency ratios, however, damping has an adverse effect on the isolation properties of the mountings. The ideal mounting should therefore provide a low natural frequency and a low damping ratio. Unfortunately, both objectives encounter difficulties, so that practical solutions must inevitably be compromises.

A low natural frequency can be provided only by reducing the stiffness of the mountings ($\omega_n^2 = S/m$), which implies a large static deflection under the weight of the machine. For a large machine, springs of suitably low stiffness and sufficient strength would assume massive proportions which are too bulky to be housed.

Low damping ratios are perfectly satisfactory provided the frequency ratio is greater than $\sqrt{2}$ (Fig. 9.2). In order to reach this frequency ratio, however, it is necessary to increase the speed of the machine from zero, and in so doing to pass through the resonant zone. Although resonant speeds are passed through as quickly as possible, in starting-up and shutting-down procedures, there has to be some damping to reduce amplitudes of vibration during these periods to safe proportions. Damping is also desirable as protection against shock loads.

Steel springs are probably the most suitable anti-vibration mountings, for they can withstand static and dynamic loads and have relatively constant static and dynamic properties. Damping action may be provided separately by the action of viscous fluids such as oil, grease or silicones, or by the action of air contained in rubber bellows with a small orifice.

Rubber and cork are frequently used isolation materials, but have the disadvantage that their moduli of elasticity are different under dynamic and static loading. The dynamic modulus of cork is between 2 and 5 times greater than the static modulus, while the ratio is between 1·5 and 3 for rubber. These materials are used in compression or, in the case of rubber, in shear. Generally damping action relies upon the elastic hysteresis of these materials. Although rubber can be very flexible, its bulk modulus is surprisingly high (about the same as water), so that space must be allowed for it to bulge considerably when compressed.

For the sake of stability, mountings should ideally be arranged so that the centroid of the machine being isolated lies in the plane for the mountings. For this reason, large machines and, in particular, tall machines are sometimes rigidly fastened to a concrete inertia block which is supported by anti-vibration mountings in a well. The inertia block effectively lowers the centroid of the supported system as well as increasing its effective mass. Some typical arrangements of anti-vibration mountings are given in Fig. 9.3. For a mounting to be effective there must be no contact between the machine and the ground except through the elastic medium.

9.3 SEISMIC PICK-UPS

The system shown in Fig. 8.11, in which the forcing function consists of a periodic displacement of the support, forms the basis for a series of primary elements in measuring systems collectively referred to as "seismic pick-ups" or "vibrometers." By suitable selection of the system parameters, they may be used to measure amplitude, frequency or acceleration. Moreover, suitable calibration of an acceleration measuring pick-up, or accelerometer, enables quantities such as force or pressure to be measured.

The support of the spring-mass system is attached to the object under investigation, the motion of which provides the input to the system. The output is the motion of the seismic mass. Since no other datum is available, the instrument is capable of measuring only the *relative* motion between output and input.

Let y be the displacement of the support (input) and x the displacement of the seismic mass (output). Then the relative displacement, x_R is

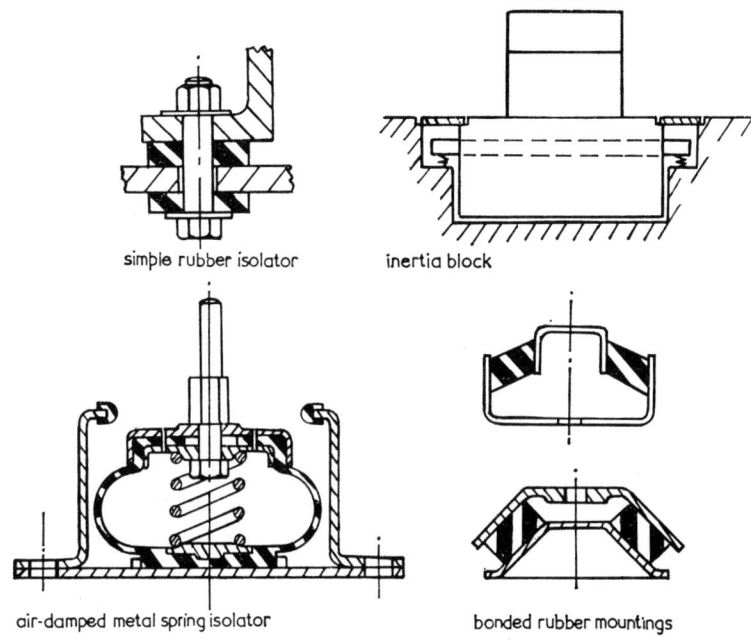

simple rubber isolator · · · inertia block

air-damped metal spring isolator · · · bonded rubber mountings

Fig. 9.3

$x - y$. The equation of motion for the system is given by Eq. (8.17) which may be re-written as

$$m(\mathrm{d}^2x/\mathrm{d}t^2) + F_v\{(\mathrm{d}x/\mathrm{d}t) - (\mathrm{d}y/\mathrm{d}t)\} + S(x - y) = 0$$

i.e.

$$m\{(\mathrm{d}^2x/\mathrm{d}t^2) - (\mathrm{d}^2y/\mathrm{d}t^2)\} + F_v\{(\mathrm{d}x/\mathrm{d}t) - (\mathrm{d}y/\mathrm{d}t)\} + S(x - y)$$
$$= -m(\mathrm{d}^2y/\mathrm{d}t^2)$$

But $y = a \cos \omega t$, so that

$$-m(\mathrm{d}^2y/\mathrm{d}t^2) = ma\omega^2 \cos \omega t$$

$$\therefore\ m(\mathrm{d}^2x_R/\mathrm{d}t^2) + F_v(\mathrm{d}x_R/\mathrm{d}t) + Sx_R = ma\omega^2 \cos \omega t \qquad (9.3)$$

This equation is the same as Eq. (8.2), but the disturbing force F_0 is now proportional to the square of the forcing frequency. The response of the system is therefore giving by Eq. (8.30) and is depicted in Fig. 8.16.

However, in this case

$$X_n = ma/m = a \qquad (9.4)$$

Eq. (8.30) may thus be re-written

$$A_R/a = r^2/\sqrt{\{(1 - r^2)^2 + (2cr)^2\}} \qquad (9.5)$$

where A_R is the relative amplitude and r is the frequency ratio (ω/ω_n). Eq. (9.5) therefore represents the response of a seismic pick-up, and this is shown graphically in Fig. 8.16.

Amplitude measurements

If a seismic pick-up is to be used to measure the amplitude a of the input motion, the frequency ratio should be very large. The response curves of Fig. 8.16 show that for large values of ω/ω_n, A_R/a tends to unity.

For zero damping Eq. (9.5) reduces to

$$A_R/a = r^2/(1 - r^2)$$

i.e.

$$a = A_R \left(\frac{1}{r^2} - 1\right) \qquad (9.6)$$

When $r = \infty$, $a = -A_R$ and the amplitude recorded by the pick-up is a measure of the amplitude of the support. The negative sign indicates that the relative amplitude is 180° out of phase with the input amplitude.

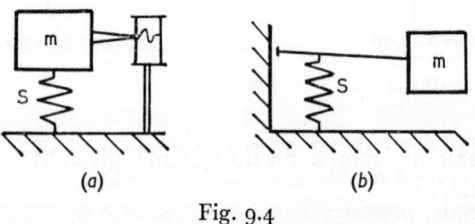

(a)   (b)

Fig. 9.4

Fig. 9.4 (a) shows the principle of a seismograph used to record the amplitude of earthquake vibrations. Ideally, as the support vibrates, the mass m remains stationary in space so providing a datum from which to measure the amplitude of the support. To provide a large frequency ratio, ω_n should be small and this means making m large and S small. The result can be a rather bulky system, but this problem can be simplified by attaching the mass to the free end of a pivoted arm (Fig. 9.4 (b)). An equivalent torsional system is obtained by attaching a flywheel of relatively high inertia to an oscillating shaft by means of a flexible torsional spring.

Example. *A seismic pick-up is used to record the amplitude of vibrations of frequency* 1200 vib/min. *If the seismic mass is* 4 kg *and the spring stiffness is* 1 kN/m, *estimate the percentage error of the amplitude recorded.*

Natural circular frequency $\omega_n{}^2 = S/m = 1000/4 = 250/\mathrm{s}^2$

$\therefore$ Natural frequency $= \sqrt{250}/2\pi = 2\cdot517\,\mathrm{Hz}$

Impressed frequency $= 1200$ vibrations/min $= 20\,\mathrm{Hz}$

$\therefore$ Frequency ratio $r = 20/2\cdot517 = 7\cdot947$

If A_R is the recorded (i.e. relative) amplitude and a is the actual amplitude

$$A_R/a = r^2/(1 - r^2) = 7\cdot947^2/(1 - 7\cdot947^2) = -1\cdot016$$

This means that $A_R = 1\cdot016a$, so that the recorded amplitude is $1\cdot6$ per cent larger than the actual amplitude.

This example suggests the question of how far the frequency ratio may be lowered before (say) a 5 per cent error occurs in the amplitude measurement. For zero damping, the phase-shift for frequency ratios greater than unity is 180°, so that when the error is 5 per cent, $A_R/a = -1\cdot05$ and if r is the frequency ratio,

$$-1\cdot05 = r^2/(1 - r^2)$$

from which

$$r^2 = 21 \quad \text{and} \quad r = 4\cdot583$$

The calculation shows that, with an undamped vibrometer, the frequency ratio must be greater than $4\cdot6$ to avoid an error of 5 per cent, which may be unacceptable.

To avoid rapid increase of error at lower frequencies (see Fig. 8.16), damping may be used to "flatten" the response curve. A damping ratio of $0\cdot6$, for example, provides a response which is almost constant from a frequency ratio of about $1\cdot25$ upwards (see Fig. 8.16), the maximum error in this range being $4\cdot2$ per cent at the resonant frequency ratio of $1\cdot89$. All vibrometers for amplitude measurement should obviously have this degree of damping in order to make their range of operation as wide as possible.

Example. *A vibrometer is to be used for amplitude measurements. Determine the damping ratio required to ensure that the error never exceeds* $+3$ *per cent. Find also the minimum frequency ratio at which the instrument may operate if the reading is to be guaranteed to within* ±3 *per cent.*

When the error is +3 per cent, the instrument will record an amplitude 1·03 times the actual amplitude, and this must occur at the resonant frequency ratio. From Eq. (8.33):

$$1·03 = 1/2c\sqrt{(1 - c^2)}$$

i.e. $$4c^2(1 - c^2) = 1/1·03^2 = 0·9425$$

which gives $c^2 = 0·62$ or $0·38$. Although, this apparently gives two possible damping ratios, substitution of $c^2 = 0·62$ in Eq. (8.32) gives an imaginary value for the resonant frequency ratio. The correct value of damping ratio is therefore given by $c^2 = 0·38$ or $c = 0·6165$.

When the error is −3 per cent the instrument will record an amplitude 0·97 times the actual amplitude. If $r = \omega/\omega_n$,

$$0·97 = r^2/\sqrt{\{(1 - r^2)^2 + 4c^2r^2\}}$$

This reduces to the following quadratic in r^2:

$$r^4 + 7·612r^2 - 15·87 = 0$$

from which

$$r^2 = +1·704 \quad \text{or} \quad -9·316$$

The negative root may, of course, be ignored, so that

$$r = \sqrt{1·704} = 1·306$$

This is the minimum frequency ratio at which the instrument should be used.

Because a vibrometer of this type cannot operate below a minimum frequency ratio, it is important that the natural frequency of the seismic mass-spring arrangement should be *as small as possible* if low frequency vibrations are to be measured.

Vibration detection and frequency measurement

If the natural frequency of an undamped vibrometer is equal to the frequency of the vibration being measured, the relative amplitude becomes very large. Such an instrument could be used to detect small vibrations, although the natural frequency would have to be variable. This is the principle of the reed vibrometer, which consists of a thin strip of spring steel clamped at one end to form a cantilever. The fixed end is firmly attached to the vibrating surface, or body, and its free length is adjusted until resonance occurs. The reed may be calibrated so that the natural frequency is indicated by a scale marked off along its length, when the frequency of the vibration may be read off directly.

Accelerometers

If the natural frequency of a seismic system is very high, the frequency ratio is likely to be very low. For zero damping, if $r = \omega/\omega_n$,

$$A_R/a = r^2/(1 - r^2) = r^2(1 - r^2)^{-1}$$
$$= r^2(1 + r^2 + r^4 + r^6 + \ldots)$$

i.e.
$$A_R/a = r^2 + r^4 + r^6 + \ldots$$

Since r is low (less than 0·5, say), r^2 will be small and r^4 and all higher powers of r will be negligible.

$$\therefore \quad A_R/a = r^2 = \omega^2/\omega_n{}^2 \tag{9.7}$$
$$\therefore \quad a\omega^2 = A_R\omega_n{}^2$$

But if a is the amplitude of the vibration under investigation,

$$\text{Displacement} = y = a \cos \omega t$$
$$\text{Velocity} = dy/dt = -a\omega \sin \omega t$$
$$\text{Acceleration} = d^2y/dt^2 = -a\omega^2 \cos \omega t$$

It follows that the

$$\text{Maximum acceleration of the vibration} = \pm a\omega^2$$

or

$$\text{Maximum acceleration} = a\omega^2 = A_R\omega_n{}^2 \tag{9.8}$$

Provided the natural frequency of the instrument is high, the relative amplitude recorded (A_R) is proportional to, and is therefore a measure of, the acceleration of the vibration. The factor $\omega_n{}^2$ is the constant of proportionality, or calibration factor, of the instrument.

A seismic pick-up for acceleration measurements, or an accelerometer, should therefore have *a very high* natural frequency. The frequency ratios of undamped accelerometers should be less than 0·5, but this figure can be increased by the introduction of damping, when the response curve (Fig. 8.16) more closely resembles the parabola of Eq. (9.7) up to frequency ratios of about 0·8.

Another advantage of damping in accelerometers is that the vibration being measured may not be pure simple harmonic, or that it may contain harmonics of higher frequencies. If these high frequencies should happen to coincide with the natural frequency of the accelerometer, their amplitudes might be magnified to the same extent as the fundamental, which would mean considerable distortion of the signal. Damping prevents this *amplitude distortion* by reducing the magnification in the resonant zone.

It is usually most practicable and desirable to convert the signal from an accelerometer into an electrical signal, which lends itself to the amplification which is frequently necessary. A further advantage is that integrating circuits may be used to give the corresponding velocity and displacement. Because of the high natural frequency, accelerometers can also be made quite small compared with vibrometers, which may mean that an accelerometer with integrated output is preferable to a vibrometer of low natural frequency for the measurement of displacement.

Because, force is proportional to acceleration, accelerometers may be calibrated to measure force. These devices, especially with electrical output, are therefore most versatile measuring instruments.

Example. *An accelerometer has a natural frequency of* 500 Hz *and a damping ratio of* 0·6. *When attached to the surface of a body vibrating at* 100 Hz, *a relative amplitude of* 0·5 mm *is recorded. Determine* (a) *the maximum acceleration of the vibrating surface as measured by the accelerometer;* (b) *the probable error in this reading.*

$$\text{Measured acceleration} = A_R \omega_n{}^2$$
$$= 0 \cdot 0005 (2\pi \times 500)^2$$
$$= 4935 \text{ m/s}^2 = 503g$$

($g = 9 \cdot 81$ m/s^2 = the acceleration due to gravity.)

$$\text{Frequency ratio } r = 100/500 = 0 \cdot 2$$
$$\text{Damping ratio } c = 0 \cdot 6$$
$$\text{Measured acceleration} = A_R \omega_n{}^2 = a\omega^2 (A_R \omega_n{}^2 / a\omega^2)$$
$$= \text{Actual acceleration} \times (A_R/ar^2)$$

But
$$A_R/a = r^2 / \sqrt{\{(1 - r^2)^2 + 4c^2 r^2\}}$$
$$\therefore A_R/ar^2 = 1/\sqrt{\{(1 - 0 \cdot 04)^2 + (4 \times 0 \cdot 36 \times 0 \cdot 04)\}}$$
$$= 1/0 \cdot 9895 = 1 \cdot 0106$$

$\therefore$ Measured acceleration = $1 \cdot 0106 \times$ Actual acceleration

Probable error = $+1 \cdot 06$ per cent

Transducers

A transducer is a device which changes a signal which is in one physical form to a corresponding signal in another physical form.

A bourdon tube, for example, is a transducer because it converts a pressure signal into a displacement, thereby facilitating the indication of the pressure on a calibrated scale. A loud-speaker is a transducer, for it converts an electrical signal into sound. A photo-electric cell is a transducer because it converts a light signal into an electric signal. Similarly, the primary elements of all the many different forms of thermometers are transducers. Transducers are often connected in series, so that, for example, a pressure signal from a flow measuring device may operate a bellows, converting the signal into a displacement; the displacement may then be converted into an electrical signal by connecting the bellows to a rheostat or potentiometer; finally, after transmission to some recording device, the electrical signal may be converted back into a displacement.

Now, however, the word transducer is associated with devices which convert mechanical quantities such as force, displacement, velocity, acceleration, pressure and temperature into electrical signals and vice

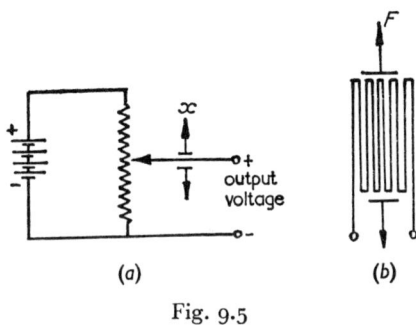

(a) (b)

Fig. 9.5

versa. Many electronic devices now exist for the amplification, integration, differentiation and display of such signals. In vibration measurements a number of different types of transducer are concerned and may be classified by knowing whether their operation depends on changes of resistance, inductance or capacitance.

Resistive types include potentiometers and strain gauges. The potentiometer consists of an electric contact arranged to move over a slide-wire, either linearly or angularly (Fig. 9.5 (a)). In the strain gauge (Fig. 9.5 (b)) the change in resistance is due to the change in the physical dimensions of the wire when subjected to elastic strain. Strain gauges may be used for vibration measurements, for example by fixing them to the upper or lower surfaces of a vibrating beam.

Inductive transducers (Fig. 9.6) are widely used and are based on the movement of an iron core in the magnetic field of an electric coil energised by an alternating current. The seismic mass of a vibrometer or accelerometer controls the movement of the core in the inductance coil, so that the output voltage is proportional to the relative displacement of the seismic mass. Amplification of the electrical signal enables the transducer to measure relative displacements of the order of microns.

Capacitive transducers depend on the variation in distance between two elements of a condenser (Fig. 9.7). Yet another type of transducer, depends on the fact that certain substances (in particular, quartz crystals) generate electrical potentials when mechanically stressed. This is known as the "piezo-electric effect."

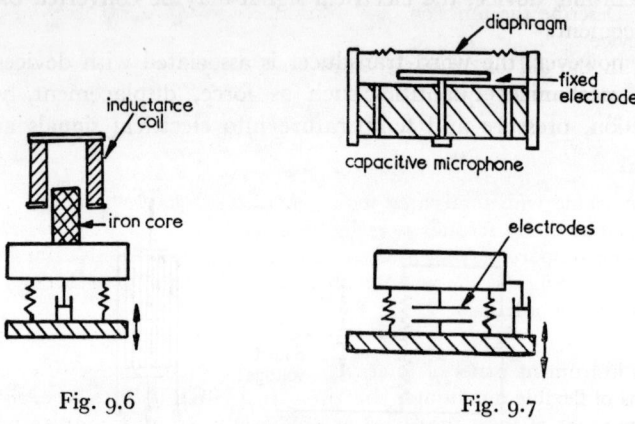

Fig. 9.6   Fig. 9.7

PROBLEMS

For tutorials

1. (*a*) Explain the difference between "active" and "passive" isolation.

(*b*) Define "transmissibility ratio" and show that the ratio is the same for both active and passive isolation systems.

2. (*a*) Describe the effects of viscous damping on a vibration isolation system for frequency ratios in excess of $\sqrt{2}$.

(*b*) Give two reasons to explain why damping is desirable in an anti-vibration mounting.

(*c*) Under what circumstances would a rigid mounting be preferable to a flexible mounting?

3. What is an "inertia block"? Explain the purpose of inertia blocks as used in anti-vibration systems. Illustrate your answer with suitable sketches.

4. A machine of mass 100 tonne generates a simple harmonic disturbing force when operating at a speed of 200 rev/min. To protect the floor and surrounding machinery, it is proposed to mount the machine on helical springs so that the transmissibility ratio is reduced to 0·1. Assuming the damping to be negligible determine the total stiffness of the springs required, and show that the static deflection of the springs, due to the weight of the machine, would be about 250 mm. Design a suitable springing arrangement, and comment on the practical difficulties of such a system.

5. Seismic pick-ups may be used to measure (a) amplitude, (b) frequency and (c) acceleration. What are the essential differences between these three types of instruments?

6. (a) Explain why damping is introduced into vibrometers and accelerometers.

(b) On the same axes, plot a graph of A_R/a against r for the functions $A_R/a = r^2$ and $A_R/a = r^2/(1 - r^2)$ if r lies between 0 and 1·0. Discuss the significance of these curves in relation to accelerometers.

7. Explain what is meant by "amplitude distortion" in an accelerometer. What do you think is meant by "phase distortion"?

8. (a) Describe the function of a transducer, and give six different examples of the device.

(b) With the aid of diagrams, show how transducers may be employed in vibration measurements.

General

1. A machine with total mass 500 kg generates a simple harmonic disturbing force which at time t seconds is equal to 500 cos 30t newtons. The machine is mounted on supports having a total stiffness of 100 kN/m and a damping ratio of 0·25. Determine (a) the transmissibility ratio; (b) the maximum force transmitted to the foundations.

Ans. (a) 0·3986; (b) 199·3 N

2. An instrument panel of total mass 400 kg is to be supported at six points by means of flexible mountings. Isolation from vibration is to be effective for a frequency range of 10 to 100 Hz, over which range no more than 10 per cent of any impressed vibration is to be transmitted. To avoid excessive amplitudes during resonance, viscous damping is to be provided and the damping ratio is 0·25. Determine the maximum stiffness of each mounting if the specified conditions are to be achieved.

Ans. 8·758 kN/m

3. An undamped vibrometer has a seismic mass of 5 kg and is to be used to measure the amplitude of vibrations of frequency 2000 vib/min. Specify the maximum permissible spring stiffness if the error is not to exceed 1 per cent.

Ans. 2·172 kN/m

4. A vibrometer has a seismic mass of 2 kg which is supported by a spring of stiffness 0·5 kN/m. Viscous damping is provided to give a damping ratio of 0·6. Estimate the errors in the amplitudes recorded by the instrument at frequencies of 2·517 Hz, 4·755 Hz and 10 Hz. What is the significance of the first two frequencies?

Ans. −16·67 per cent; +4·2 per cent; +1·6 per cent

5. An accelerometer has a seismic mass of 0·05 kg and a damping ratio of 0·6. The elastic support of the seismic mass has a stiffness of 200 kN/m.

(a) This accelerometer is attached to a part of a machine which operates at 3000 rev/min, and the maximum acceleration recorded is 100 m/s². What is the relative amplitude of the seismic mass under these conditions?

(b) Determine the ranges of frequency for which this accelerometer may be used without incurring errors of more than ±2 per cent.

Ans. (a) 0·025 mm; (b) 0 to 90·2 Hz, also between 220·5 Hz and 251·4 Hz. (*Note:* Between 90·2 Hz and 220·5 Hz the maximum error is +4·2 per cent.)

CHAPTER 10

Automatic Control

10.1 INTRODUCTION

A control system is a series of physical components connected together in such a way that the system is able to control or regulate either itself or some other quantity to which it is applied. In some systems, one of the components is a human being who opens or closes a valve to achieve a desired flow rate, turns a handle to obtain a correct tool position or adjusts the position of the steering wheel of a car. Such control systems may be described as "manually operated," although the operation may be power-assisted. An *automatic* control system, by contrast, is one in which the human operator is replaced by an inanimate device.

Because the technologies of process control and of servo-mechanisms have developed independently, the terminology of automatic control is extensive and there may be some confusion because several terms all have the same meaning. The British Standard terminology is set out in B.S. 1523, but a great deal of American terminology is in use. The best known of these is the term "cybernetics," which is derived from a Greek word meaning "steersman," and describes the entire field of control and communication theory.

Automatic control systems are as old as life itself. Every life-form is a complex of control systems, and the human body contains many examples of what are called feed-back control mechanisms. The mechanism of temperature control in the human body, the automatic operation of the human eye subjected to continual changes of light intensity and focal length, the maintenance of balance and the simple act of picking up some object with the hand are all examples of biological automatic control systems.

The first essential in a control system is a means of detecting or measuring the variable. The degree of control possible will be governed by the accuracy of the measuring system. It will frequently be necessary to transmit signals from primary elements or "pick-ups," or information recorded on the instrument itself, to other parts of the control system, sometimes over quite long distances. Thus in the design of a control system, measurement systems and data transmission systems are inevitably involved. Both of these are in themselves control systems, the objective of which is that the output should be a faithful reproduction of the input. Unfortunately, in practice the output can never be completely faithful to the input. As a signal is transmitted through the system, spurious information will be added and some of the original information lost. This is inescapable and a consequence of the general principle, described in Chapter 27 as "the second law of thermodynamics," that in all natural processes disorder tends to increase. Time delays and inertia in the system also tend to cause instability.

Control systems may be classified by the fields in which they operate. A process control system is designed to control a process—a physical or chemical change in the property of matter or a conversion of energy. The control of pressure, temperature, electrical potential, level, acidity, rate of flow and speed, are all examples of process control. A control process which maintains a quantity at a constant value is called a regulator. A kinetic control system is one which controls the displacement, velocity or acceleration of a member, and is sometimes referred to as a remote position control (r.p.c.). When an automatic kinetic control system contains a power amplifier, it is called a servo-mechanism.

Basically all these systems are the same, and are governed by the same theory, the chief difference between them being in the magnitude of the time values involved. A far more important method of classification is the division of control systems into open-loop and closed-loop systems.

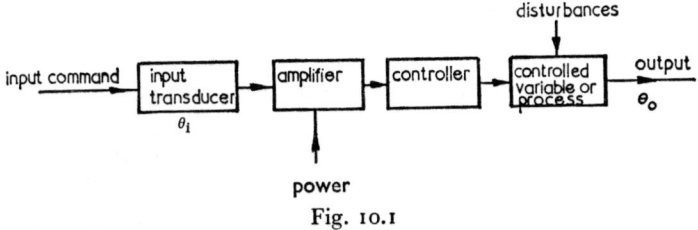

Fig. 10.1

Open-loop systems

In an open-loop or calibrated system, the action taken by the controller is independent of the output from the system. The general arrangement of an open-loop system is shown in Fig. 10.1, from which it is clear that the amount of control action taken is dependent on the input command

θ_i (also referred to as the desired value, or set point). For a given value of θ_i, the controller will always take the same action in spite of disturbances or changes outside the system. The result is that the output θ_0 may vary even though θ_i is unchanged.

Most measuring instruments are open-loop control systems, and for the same input signal the readings will depend on such things as ambient temperature and pressure. The simple Bourdon tube pressure gauge is an example. Another example of open-loop control might be the regulation of the speed of an engine by the adjustment of the throttle valve. The position of the throttle control could be calibrated to indicate on a scale the corresponding engine speed. However, a change in the load torque on the engine, or of the calorific value of the fuel, would render the calibration useless.

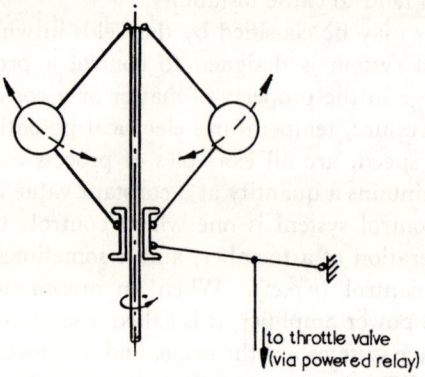

to throttle valve
(via powered relay)

Fig. 10.2

Closed-loop systems

In the last example, a better arrangement would be to measure the engine speed at the crankshaft and to provide a human operator to note the speed indicated and make any necessary adjustments to the throttle valve. With this arrangement, the control action is no longer independent of the output—the crankshaft speed. The human operator provides a *feed-back link* by monitoring the output and using the information to make a suitable adjustment to the throttle to achieve the desired value of the output.

If the human operator in this system were replaced by a mechanical device, then the system would be automatic. Such is the case with a fly-ball governor. This device relies on the centrifugal force exerted by rotating masses to raise or lower a sleeve which is free to slide on the governor spindle (Fig. 10.2). The position of this sleeve is a measure of the spindle speed, and therefore of the crankshaft speed, since the governor spindle is driven through gearing by the crankshaft. In the mechanical tachometer this principle is used for speed measurement, the movement of the sleeve being transmitted to the instrument pointer via a quadrant. In the engine governor, however, the sleeve movement is used to alter the throttle valve position so as to keep the engine speed constant. If the governor action, by itself is too weak to operate the throttle valve, amplification is provided by inserting a powered relay between the governor sleeve and the throttle valve.

Systems such as these, which use feed-back, are known as closed loop systems.

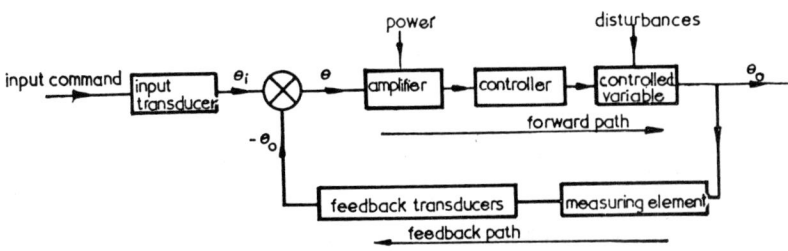

Fig. 10.3

10.2 Block Diagrams and Transfer Functions

Control systems can be represented by block diagrams as in Fig. 10.3. All closed-loop systems consist basically of:

(*a*) A system, or process, a characteristic of which is being controlled.

(*b*) A measuring element to monitor the output value θ_0 of this characteristic.

(*c*) A feed-back link containing the measuring element and, possibly, tranducers and transmission elements.

(*d*) A summation device to add the desired value, or input command θ_i to the *negative* feed-back—θ_0 of the output signal, thereby comparing θ_0 with θ_i and producing the error signal θ, where $\theta = \theta_i - \theta_0$.

(*e*) A control element capable of converting the error signal θ into a suitable control action which will adjust the output to the desired value.

The reduction of a control system to a block diagram greatly facilitates the analysis of the system performance, or response. Each block in the diagram (Fig. 10.3) represents an element, either a forward path element or a feed-back path element, for each of which there is an input signal and an output signal. In some cases, the output is simply an amplified or attenuated form of the input signal as, for example, in a simple gearbox. In general, the output signal will be of a different form from the input signal (as in a transducer or an electric motor). It is convenient to define the *transfer function** of an element as the ratio of its output to its input signal. Thus

$$\text{Transfer function} = \theta_0/\theta_i \qquad (10.1)$$

The output from an element may thus be obtained by multiplying the input signal by the transfer function. For example, the transfer function of a 4 to 1 reduction gear box is $\frac{1}{4}$. In many instances, the law relating the output and input of an element is a differential equation, and in such cases the transfer function will be some function of the operator* D, where $D = d/dt$.

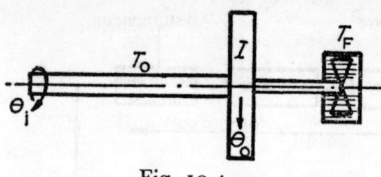

Fig. 10.4

For example, Fig. 10.4 shows a shaft of stiffness T_0 which is used to position a load of moment of inertia I. Movement of the load is resisted by a viscous frictional torque T_F per unit angular velocity. If the input signal is θ_i, then at some time t

$$\text{Twist in the shaft} = \theta_i - \theta_0$$
$$\text{Torque transmitted to load} = T_0(\theta_i - \theta_0)$$
$$\text{Frictional torque} = -T_F(d\theta_0/dt)$$

From Newton's second law,

$$I(d^2\theta_0/dt^2) = T_0(\theta_i - \theta_0) - T_F(d\theta_0/dt)$$
$$I(D^2\theta_0) + T_F(D\theta_0) + T_0\theta_0 = T_0\theta_i \qquad (10.2)$$
$$D^2\theta_0 + 2kD\theta_0 + \omega_n^2\theta_0 = \omega_n^2\theta_i$$

where $2k = T_F/I$, and $\omega_n^2 = T_0/I$. This may be written

$$(D^2 + 2c\omega_nD + \omega_n^2)\theta_0 = \omega_n^2\theta_i$$

where $c = k/\omega_n$ and is the damping ratio.

$$\therefore \text{Transfer function } \theta_0/\theta_i = \omega_n^2/(D^2 + 2c\omega_nD + \omega_n^2)$$
$$= 1/(T^2D^2 + 2cTD + 1) \qquad (10.3)$$

---

* The term "function" is used here in its general sense. Strictly speaking the term "transfer function" is reserved for the case when the ratio is expressed in terms of Laplace transforms. When expressed in terms of the operator D, the mathematically correct term for θ_0/θ_i is the *transfer operator*. However, the term "transfer function" expresses a greater physical meaning and is preferred in this chapter.

where $T = 1/\omega_n$ and is the periodic time of the undamped natural oscillations of the system divided by 2π.

The response of the load, as described by θ_0, will depend on the nature of the input θ_1. If, for example, θ_1 were sinusoidal, then comparison of Eq. (10.2) with Eq. (8.15) would give the corresponding response. Examples of the responses due to various inputs will be dealt with later in this chapter.

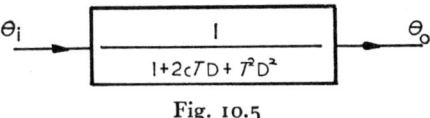

Fig. 10.5

If the system of Fig. 10.4 were itself an element in a control system, it could be represented by a block on its own, with a transfer function of its own, as shown in Fig. 10.5. If several such elements are connected in series, the overall transfer function of the series is the product of the individual transfer functions. Thus if $\theta_1/\theta = F_1(D)$, $\theta_2/\theta_1 = F_2(D)$ and $\theta_3/\theta_2 = F_3(D)$ are the individual transfer functions of three elements in series, then

$$\frac{\theta_3}{\theta} = \frac{\theta_1}{\theta} \times \frac{\theta_2}{\theta_1} \times \frac{\theta_3}{\theta_2} = F_1(D)F_2(D)F_3(D) = KG(D) \qquad (10.4)$$

where K is a constant representing the overall amplification or gain, and $G(D)$ is some function of the operator D. Eq. (10.4) is true only if there is no interaction between the elements, that is, the output from one element is not affected by its connection to the subsequent element.

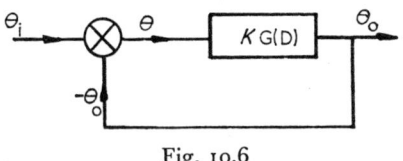

Fig. 10.6

Open-loop transfer function

Fig. 10.6 represents a closed-loop control system. All the elements in the forward path have been combined into a single block in accordance with Eq. (10.4). It has also been assumed that the output θ_0 is measured in the same units as the input θ_i, so that $-\theta_0$ can be fed back directly to the summation device to give the deviation θ. Such a system is referred to as a unity feed-back system.

The open-loop transfer function is defined as the overall transfer function of the forward path elements, i.e.

$$\text{Open-loop transfer function} = \frac{\theta_0}{\theta} = \frac{\theta_0}{\theta_i - \theta_0} = KG(\text{D}) \qquad (10.5)$$

Closed-loop transfer function

This is the overall transfer function of the entire control system.

$$\text{Closed loop transfer function} = \frac{\text{Output}}{\text{Input}} = \frac{\theta_0}{\theta_i} \qquad (10.6)$$

From Eq. (10.5),

$$\theta_0 = \theta_i KG(\text{D}) - \theta_0 KG(\text{D})$$

$$\therefore \ \theta_0(1 + KG(\text{D})) = \theta_i KG(\text{D})$$

$$\therefore \ \frac{\theta_0}{\theta_i} = \frac{KG(\text{D})}{1 + KG(\text{D})} \qquad (10.7)$$

The block diagram may be further simplified as in Fig. 10.7 where the entire system is represented by a single block.

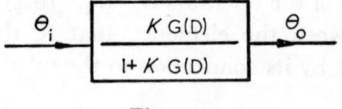

$$\theta_i \quad \boxed{\begin{array}{c} K\,G(\text{D}) \\ \hline 1 + K\,G(\text{D}) \end{array}} \quad \theta_0$$

Fig. 10.7

Example. *The motion of a pointer over a scale is resisted by a viscous damping torque of magnitude 0·6 N m at an angular velocity of 1 rad/s. The pointer, of negligible inertia, is mounted on the end of a relatively flexible shaft of stiffness 1·2 N m/radian, and this shaft is driven through a 4 to 1 reduction gear box (Fig. 10.8 (a)).*

Sketch the block diagram for the system and determine its overall transfer function. If the input shaft to the gear box is suddenly rotated through 1 complete revolution, determine the time taken for the pointer to reach a position within 1 per cent of its final value.

For the gear box, transfer function,

$$\theta/\theta_1 = \tfrac{1}{4}$$

Consider the shaft and pointer. Since the inertia of the pointer is negligible, the torque generated by the twisting of the shaft has merely to overcome the damping torque. Then

$$T_0(\theta - \theta_0) = T_F(d\theta_0/dt)$$

$$\therefore \ T_0\theta - T_0\theta_0 = T_F D\theta_0$$

from which

$$\frac{\theta_0}{\theta} = \frac{1}{1 + TD} \tag{10.8}$$

where

$$T = T_F/T_0 = \frac{0\cdot6 \text{ N m s/rad}}{1\cdot2 \text{ N m/rad}} = 0\cdot5 \text{ s}$$

and is known as the *time constant* of the system.

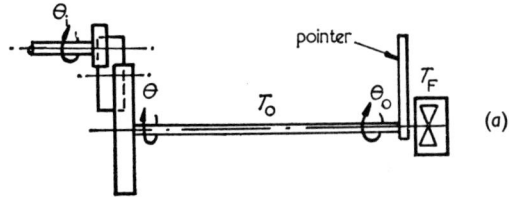

(a)

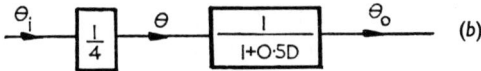

(b)

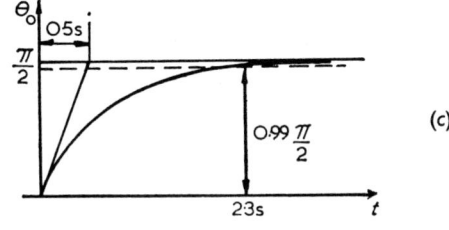

(c)

Fig. 10.8

The block diagram for the system is shown in Fig. 10.8 (b). Overall transfer function,

$$\frac{\theta_0}{\theta_i} = \frac{\theta}{\theta_i}\frac{\theta_0}{\theta} = \frac{1}{4}\frac{1}{(1 + 0\cdot5D)}$$

This may be written

$$(1 + 0.5D)\theta_0 = \tfrac{1}{4}\theta_1$$

i.e. $$0.5(d\theta_0/dt) + \theta_0 = \tfrac{1}{4}\theta_1$$

In this case $\theta_1 = 2\pi$, a constant.

$$\therefore\ 0.5(d\theta_0/dt) + \theta_0 = \tfrac{1}{4} \times 2\pi = \pi/2$$

i.e. $$0.5(d\theta_0/dt) = (\pi/2) - \theta_0$$

Separating the variables and integrating, this becomes

$$\int \frac{d\theta_0}{(\pi/2 - \theta_0)} = 2\int dt$$

i.e. $$-\log_e (\pi/2 - \theta_0) = 2t + \text{constant}$$

When $t = 0$, $\theta_0 = 0$.

$$\therefore\ \text{constant} = -\log_e (\pi/2)$$

$$\therefore\ \log_e (\pi/2 - \theta_0) = -2t + \log_e (\pi/2)$$

from which

$$(\pi/2 - \theta_0)/(\pi/2) = e^{-2t}$$

i.e. $$\theta_0 = \frac{\pi}{2}(1 - e^{-2t}) \qquad (10.9)$$

The response curve represented by Eq. (10.9) is shown in Fig. 10.8 (c).

The output θ_0 will be within 1 per cent of its final value when $\theta_0 = 0.99\,(\pi/2)$ i.e. when

$$0.99 = 1 - e^{-2t}$$

$$e^{-2t} = 0.01$$

$$\therefore\ 2t = \log_e 100 = 4.6$$

$$\therefore\ t = 2.3\ \text{s}$$

The type of response depicted in Fig. 10.8 (c) is referred to as a simple exponential time delay, and Eq. (10.8) is typical of such a response.

Example. *A position control servo-mechanism controls the angular displacement of a load of moment of inertia I by applying to it a torque proportional to its deviation from the desired position. Rotation of the load is resisted by a viscous damping torque, T_F. Draw a block diagram representing the servo-mechanism, and determine the open-loop and closed-loop transfer functions.*

If $I = 800\ \text{kg m}^2$, and $T_F = 20\ \text{kN m s/rad}$, and the servomotor delivers 50 kN m/radian of misalignment, investigate the behaviour of the load when the reference input is rotated at a constant speed of $0.1\ \text{rad/s}$.

Since the servo-amplifier and motor (see Fig. 10.9) deliver a torque proportional to the deviation θ, the transfer function of this element is a constant, K_1 (say).

$$\therefore \text{ Torque applied to load} = K_1\theta + \text{Damping torque}$$
$$= K_1\theta - T_F(d\theta_0/dt)$$

From Newton's second law

$$I(d^2\theta_0/dt^2) = K_1\theta - T_F(d\theta_0/dt)$$
$$\therefore ID^2\theta_0 + T_F D\theta_0 = K_1\theta$$
$$\therefore (ID^2 + T_F D)\theta_0 = K_1\theta$$

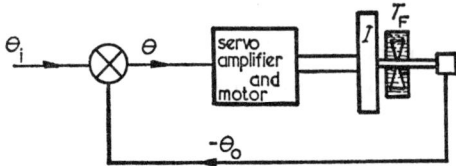

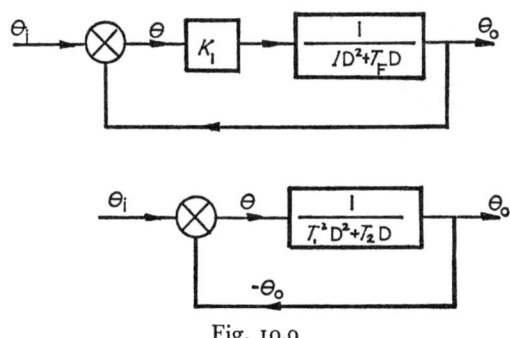

Fig. 10.9

Open-loop transfer function,

$$\theta_0/\theta = K_1/(ID^2) + T_F D) = 1/(T_1^2 D^2 + T_2 D) = KG(D)$$

where $T_1 = \sqrt{(I/K_1)}$, and $T_2 = T_F/K_1$, are time constants.

(In fact it may be shown that $T_2 = 2cT_1$, where c is the damping ratio.)

Closed-loop transfer function,

$$\theta_0/\theta_i = KG(D)/\{1 + KG(D)\}$$
$$= (T_1^2 D^2 + T_2 D)^{-1}/\{1 + (T_1^2 D^2 + T_2 D)^{-1}\}$$
$$= 1/(T_1^2 D^2 + T_2 D + 1)$$

For the values given

$$T_1{}^2 = I/K_1 = 800/(50 \times 10^3) = 0\cdot016 \text{ s}^2$$
$$T_2 = T_F/K_1 = (20 \times 10^3)/(50 \times 10^3) = 0\cdot4 \text{ s}$$
$$\therefore \; \theta_0/\theta_i = 1/(0\cdot016D^2 + 0\cdot4D + 1)$$

If the input is rotated at a constant speed $0\cdot1$ rad/s, then after t seconds, $\theta_i = 0\cdot1t$ radian.

$$\therefore \; (0\cdot016D^2 + 0\cdot4D + 1)\theta_0 = \theta_i = 0\cdot1t$$

i.e. $$(d^2\theta_0/dt^2) + 25(d\theta_0/dt) + 62\cdot5\theta_0 = 6\cdot25t \qquad (10.10)$$

The transient solution is given by

$$(d^2\theta_0/dt^2) + 25(d\theta_0/dt) + 62\cdot5\theta_0 = 0$$

This is typical of a damped vibration (see Chapter 7, Section 7.4), and comparison with Eq. (7.11) shows that $k = 12\cdot5$, and $\omega_n{}^2 = 62\cdot5$ so that the damping ratio $c = k/\omega_n = 12\cdot5/\sqrt{62\cdot5} = 1\cdot58$. Since the damping ratio is greater than unity, no oscillations will occur and the load will approach its equilibrium position, or θ_0 will approach its steady-state value, exponentially.

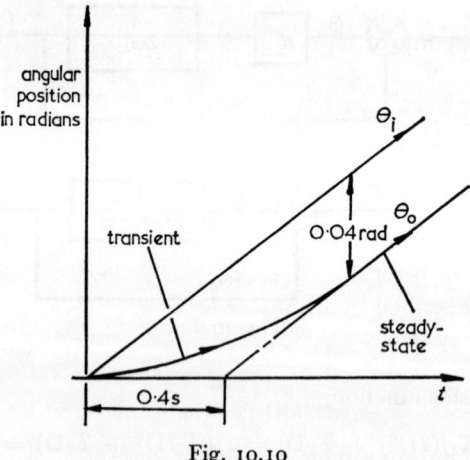

Fig. 10.10

When the transient motion dies away, $(d^2\theta_0/dt^2)$ will tend to zero, but since θ_i increases with time at a constant rate, the output velocity $(d\theta_0/dt)$ will tend towards a constant. The steady-state solution will be of the form

$$\theta_0 = At + B$$

where A and B are constants.

Substituting in Eq. (10.10),

$$0 + 25A + 62 \cdot 5(At + B) = 6 \cdot 25t$$

This is true if

$$62 \cdot 5A = 6 \cdot 25$$

and $25A + 62 \cdot 5B = 0$, from which $A = 0 \cdot 1$ rad/s and $B = -0 \cdot 4A = -0 \cdot 04$ radian.

The steady-state solution is

$$\theta_0 = 0 \cdot 1t - 0 \cdot 04 \text{ radian} \qquad (10.11)$$

Comparison of this with the input $\theta_i = 0 \cdot 1t$ shows that, after the transient motion has died out, the load rotates at the same speed as the reference input (i.e. $0 \cdot 1$ rad/s) but lags behind it by $0 \cdot 04$ radian. This is called a steady-state error.

The transient and steady-state response of the system is shown graphically in Fig. 10.10.

10.3 System Order

In general the response of any system is described by a differential equation of the form

$$a_n(d^n\theta_0/dt^n) + a_{n-1}(d^{n-1}\theta_0/dt^{n-1}) + \ldots + a_1(d\theta_0/dt)$$
$$+ a_0\theta_0 = a_i\theta_i \quad (10.12)$$

The word "system" is used here in its general sense, and may refer to an entire control system or merely to one component element. Such an equation is said to be of the nth order, and this classification is also used for the system described by the equation.

A first order system would therefore be represented mathematically by the equation

$$a_1(d\theta_0/dt) + a_0\theta_0 = a_i\theta_i \qquad (10.13)$$

If the input and output signals are of the same form—if both are angular positions, temperatures, etc.—then the constants a_0 and a_1 will be dimensionally (and frequently numerically) equal. Eq. (10.13) will therefore reduce to

$$TD\theta_0 + \theta_0 = K\theta_i$$

where $\qquad T = (a_1/a_0)$ and has the dimensions of time

and $\qquad K = (a_1/a_0)$ and is dimensionless

Thus, for a first order system

Transfer function, $\dfrac{\theta_0}{\theta_i} = \dfrac{K}{1 + T\mathrm{D}}$ (10.14)

A similar transfer function has already been considered Eq. (10.8) and its response, is shown in Fig. 10.8. The response of any first order system to a suddenly applied (step) input signal is a simple exponential curve similar to those of Fig. 10.8 and Fig. 10.18.

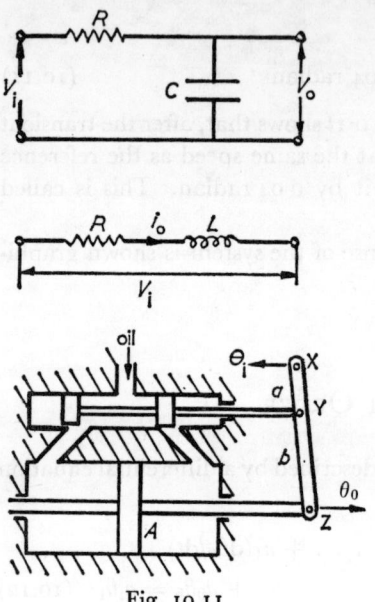

Other first order systems (see Fig. 10.11) will be found to have a *single* form of energy storage, and this is the physical characteristic by which a first order system may be recognised.

Example. *Show that the resistance-capacitor network of Fig. 10.11 is a first order system.*

Suppose that t seconds after V_i is applied, the output voltage is V_0. Then the charge on the condenser is given by $q = CV_0$.

Current flowing to condenser,

Fig. 10.11

$$i = \mathrm{d}q/\mathrm{d}t = C(\mathrm{d}V_0/\mathrm{d}t)$$

The voltage drop across resistance is $(V_i - V_0)$ so that from Ohm's law

$$i = (V_i - V_0)/R$$
$$\therefore\ C(\mathrm{d}V_0/\mathrm{d}t) = (V_i - V_0)/R$$
$$\therefore\ RC\mathrm{D}V_0 = V_i - V_0$$
$$\therefore\ V_0/V_i = 1/(1 + RC\mathrm{D}) = 1/(1 + T\mathrm{D})$$

where $T = RC$ and is the time constant for the network.

This shows that the circuit possesses a simple exponential time lag, i.e. it is a first order system.

Example. *Show that the hydraulic relay of Fig. 10.11 is a first order system.*

A displacement θ_i given to X causes the relay valve to move to the left, thus admitting oil to the left hand side of the piston in the power cylinder which will be displaced to the right by an amount θ_0.

The flow of oil per unit time into the power cylinder must equal the volume swept by the piston per unit time. The rate of flow of the oil will be proportional to the valve displacement. Thus

$$A(d\theta_0/dt) = K\theta$$

where K is a valve port constant.

The displacement θ of point Y may be found by supposing the link XYZ to pivot first about Z, then about X and then superimposing.

Displacement of Y with Z fixed $= \theta_i b/(a+b)$

Displacement of Y with X fixed $= -\theta_0 a/(a+b)$

$\therefore$ Resultant displacement of Y, $\theta = (\theta_i b - \theta_0 a)/(a+b)$

$\therefore A(d\theta_0/dt) = K(\theta_i b - \theta_0 a)/(a+b)$

$\therefore A(a+b)D\theta_0/K = b\theta_i - a\theta_0$

$T . D\theta_0 = (b/a)\theta_i - \theta_0$

where
$T = A(a+b)/Ka$ is a time constant

$\therefore \theta_0/\theta_i = (b/a)/(1 + TD)$

This is of the same form as Eq. (10.14), so that the system is of the first order.

The general equation for a second order system is of the form

$$a_2(d^2\theta_0/dt^2) + a_1(d\theta_0/dt) + a_0\theta_0 = a_i\theta_i \qquad (10.15)$$

If θ_i and θ_0 are of the same form, constants a_0 and a_i will again be dimensionally (and frequently numerically) equal, so that the equation may be written

$$T_1^2 D^2\theta_0 + T_2 D\theta_0 + \theta_0 = K\theta_i$$

where $T_1^2 = (a_2/a_0)$ with the dimensions of $(\text{time})^2$;

$T_2 = (a_1/a_0)$ with the dimensions of time;

and
$K = (a_i/a_0)$ and is dimensionless.

Thus for a second order system

$$\text{Transfer function } \frac{\theta_0}{\theta_i} = \frac{K}{T_1^2 D^2 + T_2 D + 1} \qquad (10.16)$$

Physically, a second order system will be found to contain *two* elements capable of storing energy, together with an element which dissipates energy during energy transfer.

Eq. (10.15) may be rearranged to give

$$(d^2\theta_0/dt^2) + 2c\omega_n(d\theta_0/dt) + \omega_n^2\theta_0 = K_1\omega_n^2\theta_i \qquad (10.17)$$

where $2c\omega_n = a_1/a_2$, $\omega_n^2 = a_0/a_2$ and $K_1 = a_i/a_0$.

The significance of the coefficients of the left hand side of Eq. (10.17) will be understood by referring to Section 7.4 and Chapter 8. The constant ω_n corresponds to the natural circular frequency of the system and c to the damping ratio. If c is less than unity, the transient solution of Eq. (10.17) will be oscillatory. The response of a second order system with

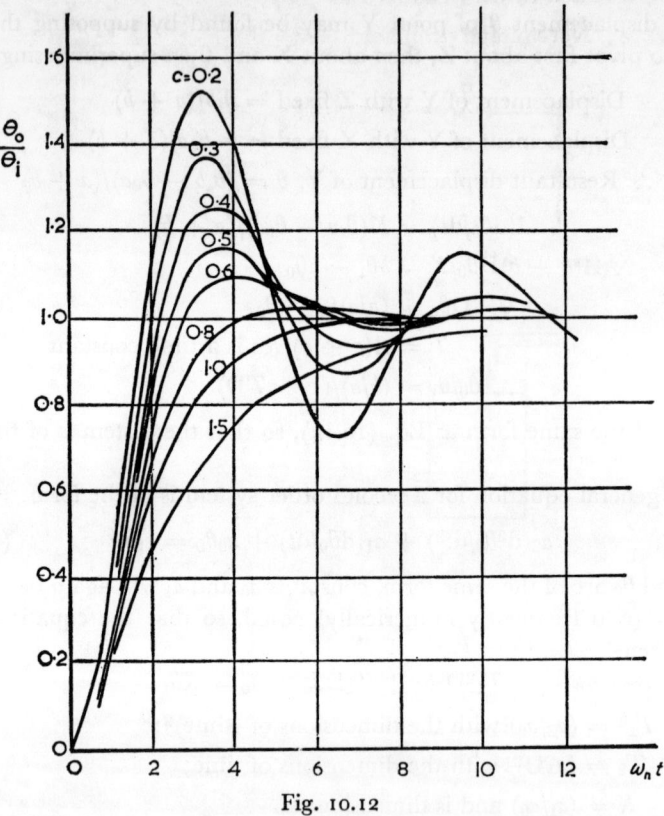

Fig. 10.12

light damping to a suddenly applied input signal will be a damped oscillation about the new equilibrium position. Typical response curves for a system in which $K_1 = 1 \cdot 0$ (usually the case) are shown in Fig. 10.12.

Most measuring instruments constitute second order systems and exhibit response characteristics as in Fig. 10.12. A damping ratio of about $0 \cdot 6$ is usually chosen to give the best compromise between a fast response rate and minimum overshoot. Examples of second order systems are given in Figs. 10.4, 10.9 and 10.13.

Example. *Show that the electrical network shown in* Fig. 10.13 *is a second order system. Find the natural undamped frequency of the system and its damping ratio.*

Suppose that t seconds after V_1 is applied, the output voltage is V_0, and the current flowing through R is i.

$$i = \mathrm{d}q/\mathrm{d}t = C(\mathrm{d}V_0/\mathrm{d}t)$$

Voltage drop across R and $L = (V_1 - V_0)$

Back E.M.F. Induced in $L = -L(\mathrm{d}i/\mathrm{d}t)$

$\therefore$ Net voltage drop across $R = (V_1 - V_0) - L(\mathrm{d}i/\mathrm{d}t)$

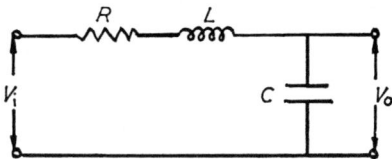

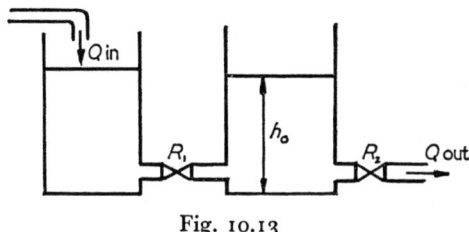

Fig. 10.13

From Ohm's law

$$(V_1 - V_0) - L(\mathrm{d}i/\mathrm{d}t) = Ri$$

Substituting for i

$$V_1 - V_0 - LC(\mathrm{d}^2V_0/\mathrm{d}t^2) = RC(\mathrm{d}V_0/\mathrm{d}t)$$

$$\therefore LCD^2V_0 + RCDV_0 + V_0 = V_1$$

$$\therefore V_0/V_1 = 1/(T_1{}^2D^2 + T_2D + 1)$$

where
$$T_1 = \sqrt{(LC)} \quad \text{and} \quad T_2 = RC$$

This equation is of the same form as Eq. (10.16) and the system is therefore of the second order. Alternatively,

$$(\mathrm{d}^2V_0/\mathrm{d}t^2) + (R/L)(\mathrm{d}V_0/\mathrm{d}t) + (1/LC)V_0 = (1/LC)V_1$$

Comparing this with Eq. (10.17),

$$\omega_n{}^2 = 1/LC$$

$\therefore$ Natural undamped frequency, $\omega_n/2\pi = 1/2\pi\sqrt{(LC)}$

$$2c\omega_n = R/L$$

$\therefore$ Damping ratio $= R/2L\omega_n = \tfrac{1}{2}R\sqrt{(C/L)}$

Combination of time delays

If two non-interacting first order systems are connected in series, as shown in Fig. 10.14, the overall transfer function is given by Eq. (10.4)

$$\frac{\theta_0}{\theta_i} = \frac{1}{(1 + T_1 D)(1 + T_2 D)} = \frac{1}{1 + (T_1 + T_2)D + T_1 T_2 D^2} \quad (10.18)$$

This is in the same form as Eq. (10.16) and so the series combination results in a second order system. Such a second order system will, however, always be overdamped. Eq. (10.18) may be written

$$D^2\theta_0 + \{(T_1 + T_2)D\theta_0/T_1 T_2\} + (\theta_0/T_1 T_2) = (\theta_i/T_1 T_2)$$

From which

$$\omega_n^2 = 1/T_1 T_2$$

and

$$2c\omega_n = (T_1 + T_2)/T_1 T_2$$

$$\therefore \text{ Damping ratio, } c = (T_1 + T_2)/2\sqrt{(T_1 T_2)}$$

The minimum value of this expression is 1·0 and occurs when $T_1 = T_2$.

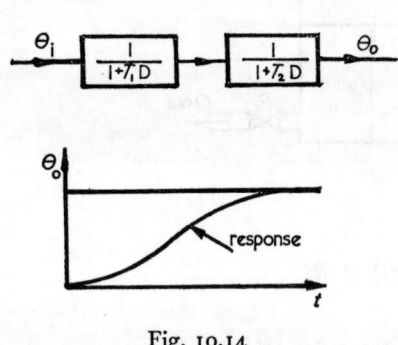

Thus two non-interacting identical first order systems connected in series form a critically damped second order system. If T_1 and T_2 differ, the resulting second order system will have a damping ratio greater than the critical. Conversely, provided a second order system is over-damped, it may be split into two first order systems in series or cascade.

Fig. 10.14

10.4 CONTROL ACTION

In all error-actuated automatic control systems, the difference between output and input signals is measured and the resulting error signal is fed into an error modifying device, or controller, the output of which is some function of the error.

The output from the controller, called the control action, is usually not powerful enough to control the process directly and must be amplified before use. The way in which the controller modifies the error signal will depend not only on the characteristics of the system to be controlled but also on the degree of accuracy desired. In the design of control systems, it is important not to aim at unnecessary accuracy, because this usually involves disproportionate costs. If possible, the plant should be designed with characteristics which facilitate its control.

Four common types of control action, sometimes used in conjunction with each other, are described below.

Two-step action or on-off control

The control action in which there are only two positions for the regulating device is called two-step action or on-off control. In a flow process, for example, the valve controlling the rate of flow may be either at maximum or at minimum opening—either fully open or fully closed, for example, but at no other position. Control action is initiated simply by using limit switches which may, for example, switch on electrical power to a solenoid to open the valve when the flow rate is too low and switch it off when the flow rate is too high. (A spring returns the valve to its minimum opening position.) The limit switches are frequently located in the measurement recorder, and may be simple electrical contacts, readily adjustable and operated by the movement of the instrument monitoring the output signal. Another form of switching device uses the principle of interrupted air jets to make or break a pneumatic circuit.

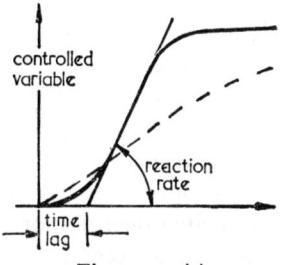

Fig. 10.15 (a)

Because of its simplicity, two-step control is by far the cheapest form of control, but its usefulness depends on the plant or system to which it is applied. The response of a plant to a variation of input energy is usually of the form shown in Fig. 10.15 (a). The curve may be approximated to by a straight line, the slope of which represents the *reaction rate*. This straight line does not start at zero time; instead, there is a time delay

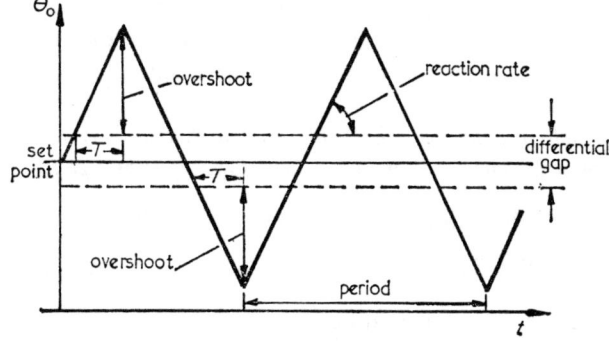

Fig. 10.15 (b)

after control action is applied before a detectable response occurs. In practice, it is necessary to leave a *differential gap* between the limit switches so that both switching operations do not occur simultaneously, thus completely baffling the regulating device. In Fig. 10.15 (*b*) the controlled variable rises and falls continuously and overshoots the limits imposed by the switches in a manner dependent on the plant reaction rate and dead time.

The following points should be noted:

1. A slower reaction rate would increase the period of cycling but decrease the amount of overshoot.

2. For a given reaction rate, an increase in the dead time causes an increase in both the period of cycling and the overshoot.

The last point means that for a plant with a fast reaction rate and an appreciable dead time, the controlled variable will oscillate wildly about the desired value. In such cases on-off control is invariably unsatisfactory. It may be concluded that a plant having a characteristic similar to that shown dotted in Fig. 10.15 (*a*) is much more easily controlled.

Two-step control can cater for changes of load only up to a certain limit. Normally, the system is arranged so that each limit switch operates for 50 per cent of the time. An increase of load, which requires an increase of output power from the regulating device, means that the "on" switch has to operate longer than the "off" switch. A limit is reached when one switch operates for the whole time, and no further increase in load can then be accommodated.

On-off control probably accounts for a very large proportion of automatic control systems, especially in temperature control.

Example. *The head of liquid in a process tank is to be controlled at 2 m by an on-off control system. The outflow from the tank is constant and equal to 0·010 m³/s, and control is to be effected by varying the inlet valve from its fully open position to its fully closed position. In the fully open position the inlet valve permits 0·015 m³/s of liquid to flow into the tank. The tank has a cross-sectional area of 0·4 m². The limit switches are set with a differential gap of 100 mm, and the dead time is 5 s.*

Plot a graph showing how the head varies with time, and hence deduce the maximum and minimum heads and the period of cycling to be expected.

Suppose the inlet valve is open. Then

$$\text{Net flow into tank} = 0\cdot015 - 0\cdot010 = 0\cdot005 \text{ m}^3/\text{s}$$

∴ Reaction rate (i.e. rise in head)

$$= 0\cdot005/0\cdot4 = 0\cdot0125 \text{ m/s} = 12\cdot5 \text{ mm/s}$$

∴ Time taken to reach upper limit switch from set point
$$= 50/12 \cdot 5 = 4 \text{ s}$$
Overshoot due to dead time $= 5 \times 12 \cdot 5 = 62 \cdot 5 \text{ mm}$

When the inlet valve is closed,

Net flow from tank $= 0 \cdot 010 \text{ m}^3/\text{s}$

∴ Reaction rate (i.e. fall in head)
$$= 0 \cdot 010/0 \cdot 4 = 0 \cdot 025 \text{ m/s} = 25 \text{ mm/s}$$

∴ Time taken to reach lower limit switch after leaving maximum head
$$= (100 + 62 \cdot 5)/25 = 6 \cdot 5 \text{ s}$$
Overshoot due to dead time $= 5 \times 25 = 125 \text{ mm}$

Time taken to reach set point from minimum load
$$= (125 + 50)/12 \cdot 5 = 14 \text{ s}$$

The graph of head against time is shown in Fig. 10.16.

Maximum head $= 2 + 0 \cdot 050 + 0 \cdot 0625 = 2 \cdot 1125 \text{ m}$

Minimum head $= 2 - 0 \cdot 050 - 0 \cdot 125 = 1 \cdot 825 \text{ m}$

Period of cycling $= 4 + 5 + 6 \cdot 5 + 5 + 14 = 34 \cdot 5 \text{ s}$

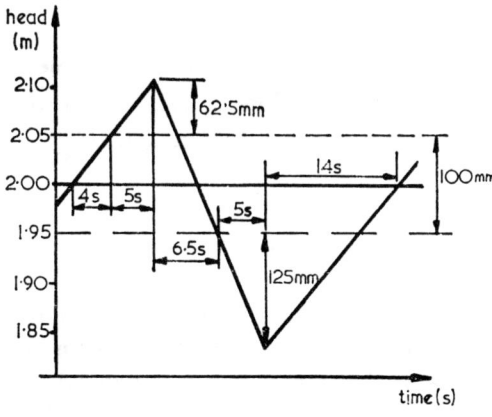

Fig. 10.16

Proportional control action

The natural development of two-step control action is that in which the regulating device can take up *any* position within its range. The logical position is one which is directly proportional to the error signal, so that when the error is large, a large amount of action is taken, and when there is only a small error, only a small amount of action is taken. The aim is to avoid the inevitable overshoot and consequent cycling produced by on-off control.

Proportional control action may be expressed mathematically as follows:

$$\text{Proportional control action} = V = K_1\theta \qquad (10.19)$$

where K_1 = sensitivity constant, or controller gain and $\theta = \theta_i - \theta_0 =$ deviation. If K_1 is large, a small error will produce a large control action, and it is clear that an increase of K_1 will eventually create a situation in which the slightest change in θ will cause the regulating device to move rapidly from one extreme to the other, much as in on-off control. An

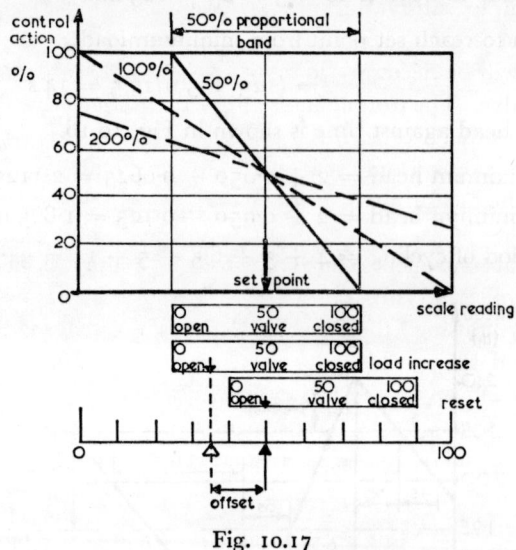

Fig. 10.17

over-sensitive control action will therefore tend to produce hunting, or cycling, of the variable—the very thing the design is intended to avoid. On the other hand, a high degree of accuracy requires that the controller should be sensitive to small changes of the variable. As always, the best compromise must be sought and, of course, an intelligent use of damping is of considerable assistance.

The sensitivity of a controller is often indicated by the term *proportional band*, which is the range of values of the controlled variable which causes the regulating unit to move from one end of its travel to the other. The narrower the proportional band, the smaller the change in variable to produce a given control action. Zero proportional band is equivalent to on-off control. Proportional band is usually expressed as a percentage of the range of the scale of the measuring instrument. In reality, it is the reciprocal of the sensitivity K_1 (see Fig. 10.17). All the control systems so far considered, with the exception of the example of two-step control action, have used proportional control action, so that the derivation of typical transfer functions does not need to be repeated.

One disadvantage of proportional control is its response to changes of load. This can be seen from the example of a control system for maintaining the water in a storage tank at some desired temperature. Heat is supplied by means of a steam heating coil immersed in the tank. Suppose that to begin with, water at the desired temperature is being drawn off from the tank at a certain rate and that the control valve is supplying enough steam to satisfy this process demand. In this condition, the deviation from set point will be zero. If the demand for hot water is now increased, it will be necessary to increase the supply of steam by opening the control valve. The trouble is that this can be done only if there is some deviation. In other words, if the increased load is sustained, additional steam can be supplied only if there is a decrease of the temperature as recorded by the measuring instrument. The resulting sustained deviation is called "offset."

Proportional controllers are usually provided with a manual adjustment to eliminate offset. With the water tank, for example, offset could be eliminated by "raising" the set point so that the actual temperature (still lagging behind the set point) is increased to the value really desired. More usually, however, the proportional band is shifted relative to the set point (see Fig. 10.17). This is called manual reset. For the case quoted, the control action could probably be expressed mathematically as

$$\text{Control action} = P + K_1\theta$$

where P represents the control action necessary to satisfy process demand when $\theta = 0$. Manual reset is equivalent to increasing P so that increased control action may be obtained while the deviation θ remains at zero. If the plant is liable to frequent changes of load, a pure proportional controller is unlikely to be satisfactory, for if the controller requires resetting manually at frequent intervals, then the process may as well be controlled entirely manually.

An example of offset has been met in the example of the position control servo-mechanism of Fig. 10.9, where there is a steady-state error Eq. (10.11) due to the viscous torque applied to the load and caused by its constant velocity. This viscous torque represents an additional sustained load, so that offset is unavoidable with proportional control.

Integral control action

Integral control action is that in which the control action is proportional to the time integral of the deviation. In other words,

$$\text{Integral control action} = V = K_2\!\int\!\theta \,.\, \mathrm{d}t \qquad (10.20)$$

where $K_2 = $ Integral action sensitivity constant.

Eq. (10.20) may also be written:

$$\mathrm{d}V/\mathrm{d}t = K_2\theta \qquad (10.21)$$

This implies that the rate at which control action is taken is proportional to the deviation. If the regulating unit is a valve, for example, the valve will keep moving at a speed proportional to the deviation so long as deviation exists. Permanent offset due to a sustained change of load is, therefore, impossible.

Because integral control eliminates offset, it is sometimes called "automatic reset." Another name is "floating control," for the valve position is not fixed by the magnitude of the deviation (as in proportional control) but will adjust itself to cater for load changes.

Integral control is usually used not by itself but in conjunction with proportional control. Thus, if the controller of the position control servo-mechanism of Fig. 10.9 delivered an output proportional to the integral of the deviation as well as to the deviation itself, the equation of motion would be:

$$I(\mathrm{d}^2\theta_0/\mathrm{d}t^2) + T_F(\mathrm{d}\theta_0/\mathrm{d}t) = K_1\theta + K_2\!\int\!\theta \,\mathrm{d}t \qquad (10.22)$$

Differentiating, this becomes:

$$I(\mathrm{d}^3\theta_0/\mathrm{d}t^3) + T_F(\mathrm{d}^2\theta_0/\mathrm{d}t^2) = K_1(\mathrm{d}\theta/\mathrm{d}t) + K_2\theta$$

Since $\theta = (\theta_i - \theta_0)$, this may be written:

$$ID^3\theta_0 + T_FD^2\theta_0 + K_1D\theta_0 + K_2\theta_0 = K_1D\theta_i + K_2\theta_i \qquad (10.23)$$

i.e. $$[T_1{}^2D^3 + T_2D^2 + D + (1/T_I)]\theta_0 = [D + (1/T_I)]\theta_i$$

where $T_1 = \sqrt{(I/K_1)}$, $T_2 = T_F/K_1$ and $T_I = K_1/K_2$. These ratios all have the dimensions of time. T_I is known as the *integral action time*, defined (see B.S. 1523) as the time taken for the integral action to produce the same control effect as the proportional action when the deviation is constant. The closed-loop transfer function is given by

$$\frac{\theta_0}{\theta_i} = \frac{(T_ID + 1)}{(T_IT_1{}^2D^3 + T_IT_2D^2 + T_ID + 1)} \qquad (10.24)$$

which is indicative of a third order system. But third order systems have a tendency towards instability. An unfortunate combination of time constants in the system described above could, for example, lead to sustained hunting, possibly with increasing amplitude, when the system is disturbed. If integral control is to be added to an existing proportional control system to eliminate offset, it is advisable to investigate the possible instability of the resulting system.

In short, integral control improves the steady-state response of a control system but causes a deterioration of the transient response.

Derivative control action

The position of a mass supported by a spring is controlled by the force in the spring. When the mass is displaced, the spring force takes action proportional to the deviation to return the mass to its equilibrium position. On reaching the equilibrium position, however, the momentum of the mass carries it beyond this position, oscillations occur and frictional damping has to restore the original situation. This is a simple example of a proportional position control. If an intelligent human being was restoring the mass to its original position, he would decelerate the mass so that its velocity on arrival at the equilibrium position was zero, thereby avoiding overshoot. In doing so, the human being would be taking action dependent on *the rate of change of the deviation*. A spring is not equipped to do this, and overshoot of a spring-controlled mass can only be avoided by the use of excessive damping (greater than critical), which involves long time delays and the wasteful dissipation of energy.

The conclusion is that a proportional control system can be stabilised (i.e. overshoot and consequent oscillation reduced) by the addition of a control action proportional to the first derivative of the deviation

$$\text{Derivative control action} = V = K_3(d\theta/dt) \qquad (10.25)$$

where $K_3 =$ Derivative action sensitivity constant. Derivative control action is only used in conjunction with proportional action, so that:

$$\text{Total control action} = K_1\theta + K_3(d\theta/dt)$$

Applying this to the position control servo-mechanism previously considered

$$I(d^2\theta_0/dt^2) + T_F(d\theta_0/dt) = K_1\theta + K_3(d\theta/dt) \qquad (10.26)$$

Since $\theta = (\theta_i - \theta_0)$, this may be written:

$$I(d^2\theta_0/dt^2) + (T_F + K_3) \cdot (d\theta_0/dt) + K_1\theta_0 = K_1\theta_i + K_3(d\theta_i/dt) \qquad (10.27)$$

The transient response is obtained by equating the left-hand side of this equation to zero. This equation is that of a damped oscillation and differs from the pure proportional controller only in the increase of the effective damping torque from T_F to $(T_F + K_3)$. Thus, even if T_F is zero, derivative control action may be used to damp out the transient oscillation, and to do so without wasteful dissipation of energy. Derivative control action thus enables frictional torque to be reduced to a minimum without unstable transient response—indeed, the transient response may be improved by increasing the value of K_3.

The steady-state response is also improved by derivative control action. Suppose, for example, that the input is rotated at a constant velocity ω so that at time t, $\theta_i = \omega t$. When the transient motion has died away (damped out by the constant K_3), $d^2\theta_0/dt$ will be zero and $d\theta_0/dt$ will be equal to ω, so that Eq. (10.27) becomes:

$$(T_F + K_3)\omega + K_1\theta_0 = K_1\omega t + K_3\omega$$

The steady-state response is then given by

$$\theta_0 = \omega t + (K_3/K_1)\omega - (T_F + K_3)\omega/K_1$$
$$= \omega t - (T_F/K_1)\omega \qquad (10.28)$$

Eq. (10.28) shows that the steady-state response is quite independent of K_3, so that if T_F has been reduced to a minimum, the steady-state error $-(T_F/K_1)\omega$ has also been reduced.

This reduction in the amount by which the output lags the input also applies when the input is sinusoidal, which is why derivative action is described as "phase advance" (see Figs. 8.11 and 8.12).

The general expression for the transfer function of controllers with proportional as well as derivative action may be deduced by re-writing Eq. (10.27) in the operator form

$$ID^2\theta_0 + (T_F + K_3)D\theta_0 + K_1\theta_0 = K_1\theta_i + K_3D\theta_i$$

i.e. $\qquad [T_1{}^2D^2 + (T_2 + T_D)D + 1]\theta_0 = (1 + T_DD)\theta_i$

where $T_1 = \sqrt{(I/K_1)}$, $T_2 = T_F/K_1$ and $T_D = K_3/K_1$. These ratios all have the dimensions of time. T_D is known as the *derivative action time*, defined (see B.S. 1523) as the time taken for the proportional action to produce the same control effect as the derivative action when the deviation is changing at a constant rate.

The closed-loop transfer function is given by

$$\frac{\theta_0}{\theta_i} = \frac{(1 + T_DD)}{T_1{}^2D^2 + (T_2 + T_D)D + 1} \qquad (10.29)$$

This is indicative of a second order system.

Three-term control action

Control systems liable to frequent load changes may require the addition of integral action to avoid steady-state errors, and this will have a detrimental effect on the transient response of the system, possibly rendering it unstable. In such circumstances the addition of derivative action is the antidote required to correct the situation. The outcome is a controller with proportional, integral and derivative action which is called a three-term controller.

$$\text{Control action } V = K_1\theta + K_2\int \theta \,.\, \mathrm{d}t + K_3(\mathrm{d}\theta/\mathrm{d}t) \qquad (10.30)$$

$$= K_1\left[\theta + \frac{1}{T_\mathrm{I}}\cdot\frac{\theta}{D} + T_\mathrm{D}D\theta\right]$$

$$= K_1[1 + (1/T_\mathrm{I}D) + T_\mathrm{D}D)]\theta \qquad (10.31)$$

where $D = \mathrm{d}/\mathrm{d}t$,
$T_\mathrm{I} = K_1/K_2 =$ Integral action time, and
$T_\mathrm{D} = K_3/K_1 =$ Derivative action time.

Applying this to the position control servo-mechanism, gives the following equation of motion

$$ID^2\theta_0 + T_\mathrm{F}D\theta_0 = K_1[1 + (1/T_I D) + T_\mathrm{D}D]\theta$$

or $$T_1{}^2D^2\theta_0 + T_2D\theta_0 = [1 + (1/T_I D) + T_\mathrm{D}D]\theta$$

Writing $\theta = \theta_i - \theta_0$, the closed-loop transfer function may be derived as

$$\frac{\theta_0}{\theta_i} = \frac{(1 + T_I D + T_I T_\mathrm{D}D^2)}{T_I T_1{}^2D^3 + (T_I T_2 + T_I T D)D^2 + T_I D + 1} \qquad (10.32)$$

This is indicative of a third order system.

10.5 SYSTEM PERFORMANCE

The ideal control system would be one in which output is, at every instant, identical with the input command. This performance is unattainable, since friction will slow down the response rate, making output lag behind input. Inertial and time lags will also tend to make the system overshoot and oscillate about its steady-state value.

The performance of control system may be regarded as consisting of two distinct parts:

(a) The transient response, during which the system requires time to respond and may oscillate or hunt.

(b) The steady-state response, which becomes apparent once the transient has died away.

By making simplifying assumptions (usually designed to reduce the order of the system), a differential equation describing the system performance may be written down. The complementary function gives the transient response and the particular integral gives the steady-state response. Of the two, the transient response will differ most from the ideal and the control system designer is more preoccupied with this. Except in the most simple cases, the analytical approach can only give an *estimate* of system performance and an experimental investigation is frequently necessary if an accurate assessment is required, possibly by means of an electrical analogue.

Although the ideal cannot be attained, the following are the properties of a control system to aim at:

1. The attainment of the desired value under all conditions of loading within specified limits.

2. The attainment of the desired value within a reasonable time.

3. Stability.

In assessing performance, analytically or experimentally, it is helpful to consider the response to relatively simple input signals—a suddenly applied constant signal or step input; a signal which increases at a constant rate, usually known as a ramp input; and a sinusoidal input.

Response to step input

A step function is defined thus: $\theta_i = 0$, for $t < 0$; $\theta_i = \Theta_i$, for $t \geqslant 0$; where Θ_i is the magnitude of the step input.

(a) For a first order system with a transfer function

$$\theta_0/\theta_i = 1/(1 + TD)$$

the response to the step input is given by

$$T(d\theta_0/dt) + \theta_0 = \Theta_i$$

The steady-state solution (particular integral) is $\theta_0 = \Theta_i$. The transient solution (complementary function) is

$$T(d\theta_0/dt) + \theta_0 = 0$$

or
$$\theta_0 = A\,e^{-t/T}$$

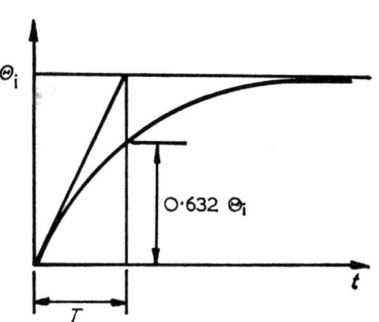

where A is a constant. The complete solution is $\theta_0 = \Theta_i + A\,e^{-t/T}$. But when $t = 0$, $\theta_0 = 0$, so that $A = -\Theta_i$. Therefore the response is given by

$$\theta_0 = \Theta_i(1 - e^{-t/T}) \quad (10.33)$$

This is shown graphically in Fig. 10.18. There is no overshoot, T is the time for the output to reach the steady-state value if the initial

Fig. 10.18

response rate is maintained and in time T, the output actually reaches only 63·2 per cent of its final value.

(*b*) For a second order system, a typical transfer function would be

$$\theta_0/\theta_i = 1/(1 + 2cTD + T^2D^2)$$

If $\theta_i = \Theta_i$, the equation describing the response is

$$T^2(d^2\theta_0/dt^2) + 2cT(d\theta_0/dt) + \theta_0 = \Theta_i$$

or
$$(d^2\theta_0/dt^2) + 2c\omega_n(d\theta_0/dt) + \omega_n^2\theta_0 = \omega_n^2\Theta_i$$

where $\omega_n = 1/T$. The steady-state solution (particular integral) is $\theta_0 = \Theta_i$. The transient solution (complementary function) is given by equating the left-hand side of this equation to zero, which gives an equation similar to Eq. (7.11). For light damping ($C < 1\cdot0$), the transient solution is thus given by Eq. (7.16), or by

$$\theta_0 = C\,e^{-c\omega_n t}\cos\{\omega_n t\sqrt{(1 - c^2)} - \alpha\}$$

The complete solution is

$$\theta_0 = \Theta_i + C\,e^{-c\omega_n t}\cos\{\omega_n t\sqrt{(1 - c^2)} - \alpha\}$$

When $t = 0$, $\theta_0 = 0$ and $d\theta_0/dt = 0$. Substitution of these initial conditions gives

$$\tan\alpha = c/\sqrt{(1 - c^2)}$$

and
$$C = -\Theta_i\sec\alpha = -\Theta_i/\sqrt{(1 - c^2)}$$

The response is therefore given by

$$\theta_0 = \Theta_i[1 - \{e^{-c\omega_n t}/\sqrt{(1 - c^2)}\} \cos \{\omega_n t \sqrt{(1 - c^2)} - \alpha\}] \quad (10.34)$$

This response is plotted non-dimensionally in Fig. 10.12 for $\Theta_i = 1\cdot0$. For a given value of ω_n, the transient response may be altered by varying the damping ratio c. The following quantities are frequently used to describe the transient:

(i) The magnitude of the first overshoot.
(ii) The settling time or the time taken to reach and remain within specified limits of the final value.
(iii) The rise time, or the time elapsed between 10 per cent and 90 per cent of the final value in the initial swing.

Response to ramp input

A typical ramp input is $\theta_i = \omega t$ where ω is a constant. For example, the input member of a remote position controller might be rotated at a constant velocity ω rad/s.

(*a*) For a first order system, the equation describing the response would be $T(d\theta_0/dt) + \theta_0 = \omega t$. The steady-state solution (particular integral) is of the form $\theta_0 = \omega t + A$, where A is a constant, which means that $d\theta_0/dt = \omega$. Thus $T\omega + (\omega t + A) = \omega t$, from which $A = -\omega T$. The transient solution (complementary function) is of the form $\theta_0 = B\,e^{-t/T}$, so that the complete solution is

$$\theta_0 = \omega t - \omega T + B\,e^{-t/T}$$

When $t = 0$, $\theta_0 = 0$, so that $B = \omega T$, and the response is given by

$$\theta_0 = \omega[t - T(1 - e^{-t/T})] \quad (10.35)$$

This is shown graphically in Fig. 10.19.

(*b*) For a second order system, the equation describing the response would be

$$T^2(d^2\theta_0/dt^2) + 2cT(d\theta_0/dt) + \theta_0 = \omega t$$

or $\qquad (d^2\theta_0/dt^2) + 2c\omega_n(d\theta_0/dt) + \omega_n^2\theta_0 = \omega_n^2\omega t$

where $\omega_n = 1/T$.

The steady-state solution (particular integral) is again of the form $\theta_0 = \omega t + A$, where A is a constant. Then $d\theta_0/dt = \omega$ and $(d^2\theta_0/dt^2) = 0$. Substituting in the response equation, $A = -2cT\omega$.

Equating the left-hand side of the equation to zero to obtain the transient response (complementary function) again gives the equation of a damped oscillation. If the damping is less than critical ($c < 1 \cdot 0$), the transient will be a decaying oscillation about the steady-state value.

The complete response (Fig. 10.20) is given by:

$$\theta_0 = \omega(t - 2cT) + \{\omega t \, e^{-c\omega_n t}/\sqrt{(1 - c^2)}\} \cos \{\omega_n t \sqrt{(1 - c^2)} - \beta\}$$
(10.36)

where $\omega_n = 1/T$ and $\tan \beta = (2c^2 - 1)/\{2c\sqrt{(1 - c^2)}\}$.

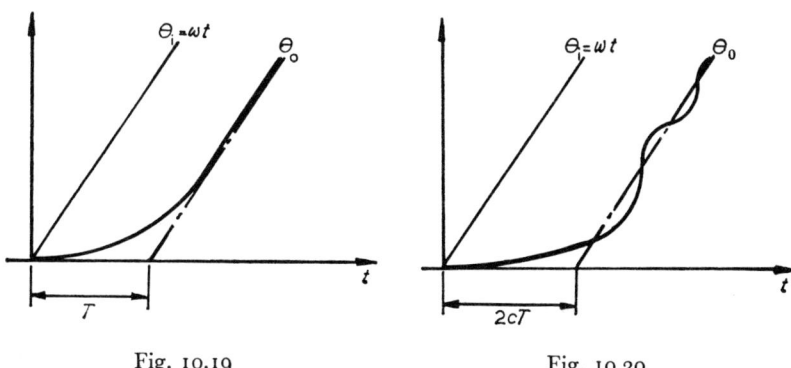

Fig. 10.19   Fig. 10.20

Response to sinusoidal input

A typical sinusoidal input is $\theta_i = \Theta_i \cos \omega t$ and may be generated by a vector of length Θ_i rotating at a constant speed ω rad/s. Transients quickly decay in first and second order systems and so are usually unimportant. They will not be considered here.

(*a*) For a first order system, the equation describing the response is

$$T(d\theta_0/dt) + \theta_0 = \Theta_i \cos \omega t$$

The steady-state solution is of the form

$$\theta_0 = \Theta_0 \cos (\omega t - \alpha)$$

Thus

$$-T\Theta_0\omega \sin (\omega t - \alpha) + \Theta_0 \cos (\omega t - \alpha) = \Theta_i \cos \omega t$$

from which

$$\Theta_0/\Theta_i = 1/\sqrt{(1 + \omega^2 T^2)}$$
(10.37)

and

$$\tan \alpha = \omega T$$
(10.38)

The vector solution of this equation is shown in Fig. 10.21.

(b) For a second order system, the equation describing the response is

$$T^2(d^2\theta_0/dt^2) + 2cT(d\theta_0/dt) + \theta_0 = \Theta_i \cos \omega t$$

or $\qquad (d^2\theta_0/dt^2) + 2c\omega_n(d\theta_0/dt) + \omega_n^2\theta_0 = \Theta_i\omega_n^2 \cos \omega t$

where $\omega_n = 1/T$. This equation is identical in form to Eq. (8.4), and its steady-state solution is given by Eqs. (8.9) and (8.11).

Polar plots

Sinusoidal inputs and outputs have the same frequency, and are related by an amplitude ratio (i.e. gain or attenuation) and a phase-shift angle (see Chapter 8). At any frequency, the input and output signals may be

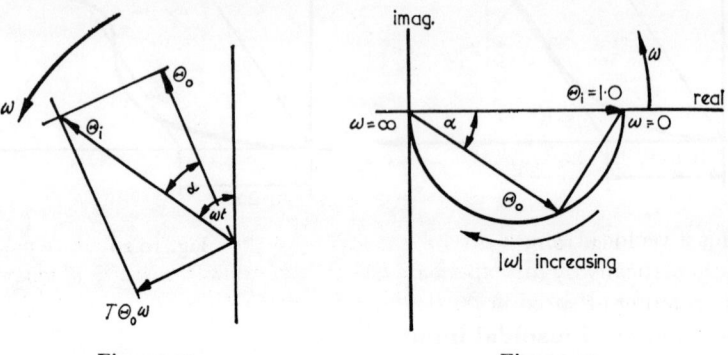

Fig. 10.21   Fig. 10.22

generated by rotating vectors, as for example in Fig. 10.21. This vector diagram may be re-drawn with the input vector Θ_i horizontal and of unit length, as shown in Fig. 10.22. Keeping the input vector in this position, the length and relative position of the output vector may be calculated from Eqs. (10.37) and (10.38) for a whole range of values of the forcing frequency. The locus of the output vector thus obtained provides a useful two-dimensional display of the frequency response spectrum. The use of a polar plot in this way makes it possible to replace two diagrams such as those in Fig. 8.15 by a single diagram. A polar plot corresponding to the frequency response curves of Fig. 8.15 is shown in Fig. 10.23.

The magnitude and relative angular position of a vector may be conveniently represented by means of complex numbers and the Argand diagram. For example, a vector of length A rotating at ω rad/s is shown in Fig. 10.24. If the vector initially coincides with the real axis, it will make an angle ωt radians with this axis after t seconds, when the vector then has a real component $A \cos \omega t$ and an imaginary component $jA \sin \omega t$, so that at time t the rotating vector is given by the vector sum

$$A \cos \omega t + jA \sin \omega t = A(\cos \omega t + j \sin \omega t)$$

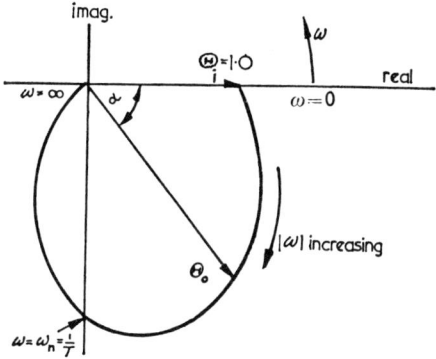

Fig. 10.23

But it may be shown that

$$e^{j\omega t} = \cos \omega t + j \sin \omega t$$

Thus a vector of length A which rotates at a speed ω may be represented mathematically by the expression: $A e^{j\omega t}$. Similarly, a vector of length B, also rotating at speed ω but leading the first vector by an angle ϕ, may

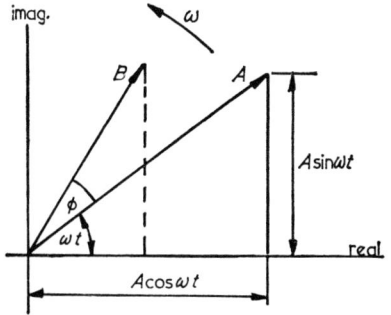

Fig. 10.24

be represented by the expression $B e^{j(\omega t + \phi)}$. This method of representation is particularly useful for vectors generating sinusoidal signals.

Suppose $$\theta = A e^{j\omega t}$$

Then $$d\theta/dt = j\omega A e^{j\omega t} \quad \text{or} \quad D\theta = j\omega\theta$$

so that the operator "D" is evidently equivalent to "$j\omega$."

A further point to note (see Fig. 10.25) is that in general the complex number

$$a + jb = A\,e^{j\alpha} \tag{10.39}$$

where $A = \sqrt{(a^2 + b^2)}$, and

$$\alpha = \tan^{-1}(b/a)$$

Consider, for example, a first order system with a transfer function

$$\theta_0/\theta_i = 1/(1 + TD)$$

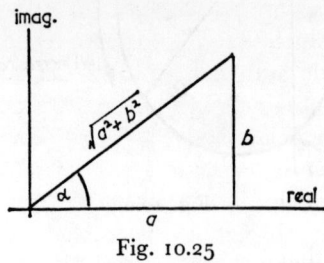

Fig. 10.25

For a sinusoidal input, this may be written

$$\theta_0/\theta_i = 1/(1 + T\,.\,j\omega)$$

If θ_i is of unit amplitude, i.e. $\Theta_i = 1\cdot0$ as in Fig. 10.22, then

$$\theta_i = 1 + j\,.\,0 = 1\cdot0\,e^{j0} \quad \text{and} \quad \theta_0 = 1/(1 + j\omega T)$$

i.e.
$$\theta_0 = 1/[\sqrt{(1 + \omega^2 T^2)}\,e^{j\,.\,\tan^{-1}\omega T}]$$

in accordance with the Eqs. (10.39)

$$\therefore\; \theta_0 = (1 + \omega^2 T^2)^{-\frac12}\,e^{-j(\tan^{-1}\omega T)}$$

Thus, the output is given by

$$\theta_0 = \Theta_0\,e^{-j\alpha} \tag{10.40}$$

where $\Theta_0 = 1/\sqrt{(1 + \omega^2 T^2)}$ and $\tan\alpha = \omega T$.

Eq. (10.40) represents a vector of magnitude Θ_0 lagging the input vector $1\cdot0\,e^{j0}$ by the angle α (Fig. 10.22). The above values of Θ_0 and α are seen to agree with Eqs. (10.37) and (10.38).

The second order transfer function may similarly be written

$$\theta_0/\theta_i = 1/(1 + 2cTj\omega + T^2 j^2 \omega^2)$$

provided the input is sinusoidal. Remembering that $j^2 = -1$ and that

$T = 1/\omega_n$, this expression may be rationalised and reduced to the same form as Eq. (10.40). It will, however, be found that in this case:

$$\Theta_0 = 1/\sqrt{[\{1 - (\omega/\omega_n)^2\}^2 + \{2c(\omega/\omega_n)\}^2]}$$

an expression already encountered in Chapter 8.

Frequency response

The practical measurement of transients is difficult, which is why harmonic or frequency response tests are usually preferred. Sinusoidal input signals are used and only the steady-state output signals need be measured. The relative amplitude and phase-shift of the output over a range of frequency values are obtained and plotted either separately (Fig. 8.15) or, more usually, in a polar plot (Figs. 10.22 and 10.23). One advantage of frequency response testing is that individual elements of a system may be tested separately before deducing the overall response for a composite system in which the elements are connected in series. If, for example, three elements which are to be connected in series are each subjected to a unit sinusoidal input $\theta_i = e^{j\alpha t}$, and the resulting outputs are, respectively, $A_1 e^{j\alpha_1}$, $A_2 e^{j\alpha_2}$ and $A_3 e^{j\alpha_3}$. Then from Eq. (10.4), the overall transfer function is

$$(A_1 e^{j\alpha_1})(A_2 e^{j\alpha_2})(A_3 e^{j\alpha_3}) = A_1 A_2 A_3 e^{j(\alpha_1 + \alpha_2 + \alpha_3)} \qquad (10.41)$$

All that is required is to multiply the individual gain factors and add the phase-shift angles. In practice, of course, α_1, α_2 and α_3 are usually negative.

Example. *Three non-interacting elements, an integrator and two simple exponential delays, are connected in series as shown in Fig. 10.26 (a). Sketch the polar plot of the overall frequency response of the system.*

From Eq. (10.4), the overall transfer function is given by

$$\theta_0/\theta = 1/D(1 + D)(1 + 2D)$$

For a unit *sinusoidal input*, this may be written:

$$\theta_0 = 1/[j\omega(1 + j\omega)(1 + 2j\omega)]$$

$$j\omega = 0 + j\omega = \omega\, e^{j\, \tan^{-1} \infty} = \omega\, e^{j\pi/2} = A_1 e^{j\alpha_1}$$

$$1 + j\omega = \sqrt{(1 + \omega^2)}\, e^{j\, \tan^{-1} \omega} = A_2 e^{j\alpha_2}$$

$$1 + j2\omega = \sqrt{(1 + 4\omega^2)}\, e^{j\, \tan^{-1} 2\omega} = A_3 e^{j\alpha_3}$$

Thus, using Eq. (10.41)

$$\theta_0 = 1/(A_1\,e^{j\alpha_1}A_2\,e^{j\alpha_2}A_3\,e^{j\alpha_3}) = 1/A_1A_2A_3\,e^{j(\alpha_1+\alpha_2+\alpha_3)}$$

$$\therefore\ \theta_0 = \Theta_0\,e^{-j\alpha}$$

where $\Theta_0 = 1/A_1A_2A_3 = 1/\omega\sqrt{(1+\omega^2)}\,.\,\sqrt{(1+4\omega^2)}$

$$\alpha = \alpha_1 + \alpha_2 + \alpha_3$$

$$= 90° + \tan^{-1}(\omega) + \tan^{-1}(2\omega)$$

(Note that α is an angle of *lag*, as indicated by the negative sign, and may be measured clockwise on the Argand diagram.)

(a)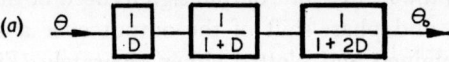

(b)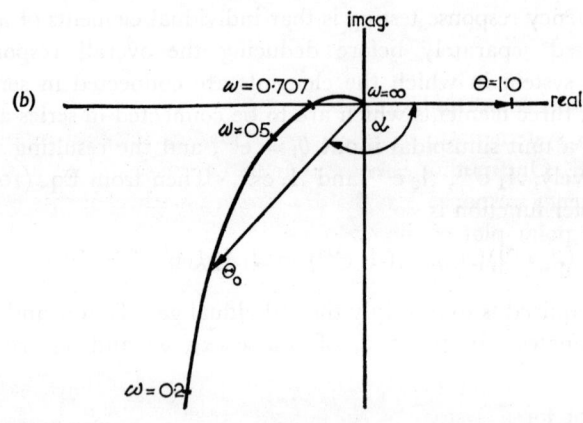

Fig. 10.26

When $\omega = 0$,

$$\Theta_0 = \infty \quad \text{and} \quad \alpha = 90°$$

When $\omega = \infty$,

$$\Theta_0 = 0 \quad \text{and} \quad \alpha = 90° + 90° + 90° = 270°$$

Also, $\alpha = 180°$, when

$$\tan^{-1}(\omega) + \tan^{-1}(2\omega) = 90°$$

i.e. when

$$(\omega + 2\omega)/(1 - \omega\,.\,2\omega) = \infty$$

i.e. when

$$1 - 2\omega^2 = 0, \quad \text{or} \quad \omega = 0.707 \text{ rad/s}$$

$\therefore$ When $\omega = 0.707$,

$$\Theta_0 = 1/0.707\sqrt{1.5}\sqrt{3} = 0.67$$

and $\qquad\qquad \alpha = 180°$

When $\omega = 1.0$,

$$\Theta_0 = 1/1\sqrt{2}\sqrt{5} = 0.316$$

and $\qquad\qquad \alpha = 90° + 45° + 63° \ 26' = 198° \ 26'$

When $\omega = 0.5$,

$$\Theta_0 = 1/0.5\sqrt{1.25}\sqrt{2} = 1.265$$

and $\qquad\qquad \alpha = 90° + 26° \ 34' + 45° = 161° \ 34'$

Other values may similarly be calculated. The polar frequency response is shown in Fig. 10.26 (*b*).

Stability

One of the more important aims of a frequency response test, when applied to a control system, is to establish the degree of stability or instability of the system. One method is to disconnect the feed-back link and to carry out a frequency response test on the forward path elements only, thus obtaining a polar plot of the open-loop frequency response. For a unit sinusoidal input, the open-loop transfer function Eq. (10.5) may be written:

$$\theta_0/\theta = KG(\mathrm{D}) = KG(j\omega) \qquad\qquad (10.41)$$

The polar plot of this function is known as the *Nyquist diagram*. A typical Nyquist plot for a system of the fourth order is shown in Fig. 10.27. Also shown on this diagram is the vector addition: $\theta + \theta_0 = \theta_1$, which must given the input signal required by the closed-loop to generate a unit deviation (since $\theta = \theta_1 - \theta_0$).

It is evident from this diagram that if the open-loop frequency response is such that the output θ_0 at a particular frequency is numerically equal to θ (i.e. unity) and is 180° out-of-phase, which would be the case if the locus passed through the point $(-1, 0)$, then the input signal θ_i to the corresponding closed-loop system which is required to create the unit error signal will be zero. This means that any random disturbance of the input command will cause the output to oscillate, and this oscillation will be sustained because the negative feed-back of a signal of unit gain with 180° phase-shift is, in effect, 100 per cent *positive* feed-back. If, under these conditions, the open-loop gain is greater than unity, the amplitude of the oscillations will grow. Thus, the open-loop frequency response can give an indication of the stability of the closed-loop system. Fig. 10.27 shows Nyquist diagrams for systems which are stable, on the verge of instability and unstable. It is clear that a stable fourth order system can be made unstable by increasing the open-loop gain factor K (see Eq. (10.41)). This also applies to the third order system in Fig. 10.26, but the first and second order systems of Fig. 10.22 and Fig. 10.23 respectively obviously cannot be rendered unstable by an increase in the gain factor.

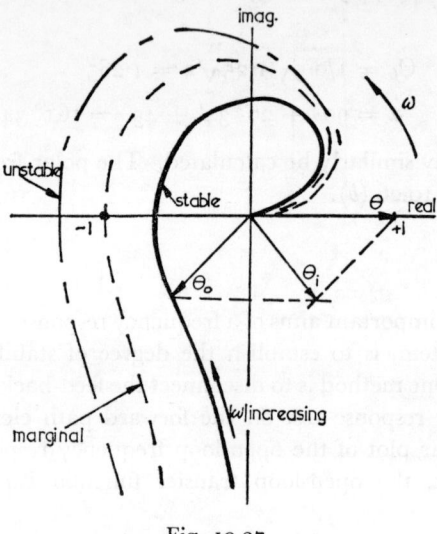

Fig. 10.27

It was Nyquist who first demonstrated the use of the harmonic response locus in the investigation of stability. The *Nyquist stability criterion* may be stated as follows:

A closed-loop system will be stable, provided the open-loop frequency response locus does not encircle the point $(-1, 0)$ *as it is traced from* $\omega = 0$ *to* $\omega = \infty$.

An important advantage of the Nyquist stability criterion is that it also indicates the degree of stability of a system, or the closeness of the response locus to the point $(-1, 0)$.

The margin of stability can be expressed in two ways.

The phase margin. The difference between the phase angle at unit gain and 180°, indicated as ϕ in Fig. 10.28, is known as the phase margin and is quickly obtained by drawing a circle of unit radius to intersect the locus at the *gain cross-over* point.

The gain margin. There are two definitions in current use. That adopted in the following examples is that the gain margin is the distance of the *phase cross-over* point from $(-1, 0)$ in the diagram (Fig. 10.28). Alternatively, gain margin is defined as the reciprocal of the open-loop gain when $\alpha = 180°$ (i.e. $1/x_{180}$). In this form, the gain margin is frequently expressed in decibels, $20 \log (1/x_{180})$.

Experience indicates that the phase margin should exceed $30°$ and that the gain margin (as given by the first definition) should exceed 0.6.

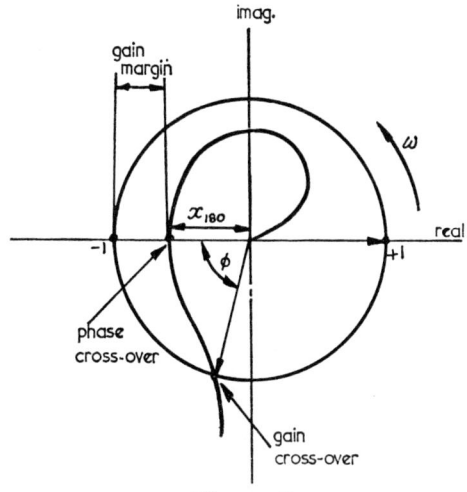

Fig. 10.28

Example. *The open-loop transfer function of a controller is:*

$$\theta_0/\theta = K/D(1 + D)(1 + D + D^2)$$

Determine a suitable value for K so that the gain margin will be 0.6 and sketch the Nyquist diagram. Find also the phase margin corresponding to this value of K.

For unit sinusoidal input

$$\theta_0 = K/j\omega(1 + j\omega)(1 + j\omega + j^2\omega^2)$$
$$= K/[(0 + j\omega)(1 + j\omega)(1 - \omega^2 + j\omega)]$$
$$= K/\{A_1 e^{j\alpha_1} A_2 e^{j\alpha_2} A_3 e^{j\alpha_3}\}$$

where $A_1 = \sqrt{(0^2 + \omega^2)} = \omega$ and $\alpha_1 = \tan^{-1}(\omega/0) = 90°$

$A_2 = \sqrt{(1 + \omega^2)}$ and $\alpha_2 = \tan^{-1}(\omega/1)$

$A_3 = \sqrt{\{(1 - \omega^2)^2 + \omega^2\}}$ and $\alpha_3 = \tan^{-1}\{\omega/(1 - \omega^2)\}$

Thus $\theta_0 = \Theta_0 e^{-j\alpha}$

where $\Theta_0 = K/\omega\sqrt{\{(1 + \omega^2)[(1 - \omega^2)^2 + \omega^2]\}}$

and $\alpha = 90° + \tan^{-1}(\omega) + \tan^{-1}\{\omega/(1 - \omega^2)\}$

The phase cross-over point occurs when $\alpha = 180°$ or when

$$\tan^{-1}(\omega) + \tan^{-1}\{\omega/(1 - \omega^2)\} = 90°$$

i.e. when

$$\frac{\omega + \omega/(1 - \omega^2)}{1 - \omega \cdot \omega/(1 - \omega^2)} = \tan 90° = \infty$$

or

$$1 - \{\omega^2/(1 - \omega^2)\} = 0$$

from which

$$\omega^2 = \tfrac{1}{2} \quad \text{and} \quad \omega = 0.707 \text{ rad/s}$$

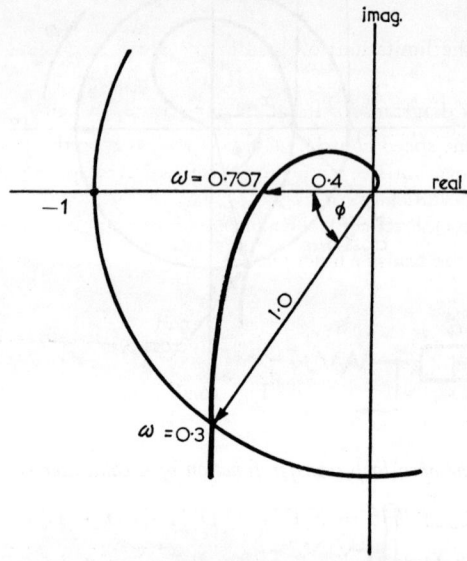

Fig. 10.29

Thus, when

$$\omega = 0.707 \text{ rad/s}, \quad \alpha = 180°$$

and

$$\Theta_0 = K/0.707\sqrt{(1.5 \times 0.75)} = 4K/3$$

$$\therefore \text{ Gain margin} = 1 - 4K/3 = 0.6$$

$$\therefore K = 0.3$$

Corresponding values of ω, Θ_0 and α are given below, and the Nyquist diagram is shown in Fig. 10.29.

ω rad/s	0	0·1	0·3	0·4	0·5	0·707	1·0	1·414	2·0	∞
Θ_0	∞	3·0	1·0	0·75	0·60	0·40	0·21	0·07	0·02	0
α	90°	101·5°	125°	137·3°	150·3°	180°	225°	270°	300°	360°

The gain cross-over point occurs when $\Theta_0 = 1\cdot0$. This is obtained by drawing a circle of unit radius on the Nyquist diagram. From the diagram

$$\text{Phase margin} = \phi = 55°$$

Problems

For tutorials

1. Discuss the importance of the measuring element in an automatic control system.

2. What are the limitations of open-loop systems and what facts justify their continued use?

3. Draw block diagrams for the following control systems:

(*a*) An engine speed control, using a fly-ball governor with a powered relay between the governor sleeve and the throttle valve.

(*b*) A thermostatically controlled electric furnace.

(*c*) A ball-cock level control as employed in domestic water supply tanks.

4. Determine the transfer functions for the systems shown in Fig. 10.30.

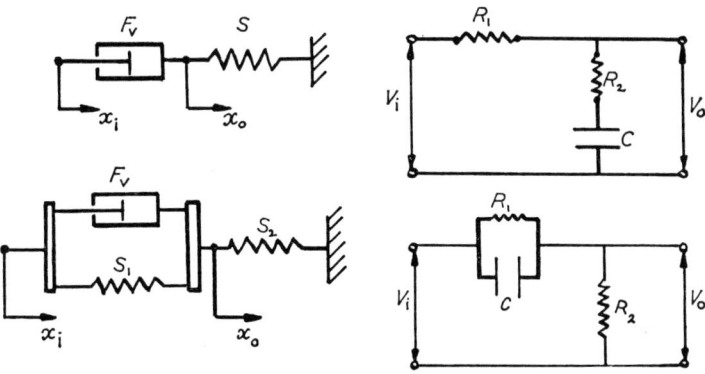

Fig. 10.30

5. The rate at which the temperature of a fluid rises is directly proportional to the rate at which it receives heat. Show that the transfer function of a liquid-in-bulb thermometer is therefore of the form:

$$\theta_0/\theta_1 = 1/(1 + TD)$$

where θ_0 is the temperature indicated by the thermometer and θ_1 is the actual temperature surrounding the bulb. Upon what will the value of T depend?

6. Fig. 10.16 shows the variation with time of the output of an on-off level control system. Devise a suitable arrangement for controlling level in this way.

7. Draw the vector solution for Eq. (10.27), assuming a sinusoidal input, and hence illustrate the meaning of the term "phase advance."

8. The advantages of the use of control action proportional to the derivative of the error θ have been discussed in Section 10.4. Discuss the possibility of employing control action proportional to the derivative of the output θ_0. What effect would the use of this type of control action have on the steady-state response described by Eq. (10.28)?

9. Explain why it would be undesirable to remove all frictional and damping effects from a remote position control employing pure proportional action. What adverse effects can friction or damping have on the transient and steady-state responses of such a controller? Why does the addition of derivative of error action make the reduction of friction and damping forces a feasible proposition?

10. The additional of integral control action increases the system order to that of the third degree, and thus renders the system liable to instability. Explain this statement by referring to typical Nyquist diagrams for first, second and third order systems, and to the Nyquist stability criterion.

11. Discuss what equipment would be necessary to carry out an open-loop frequency response test on a practical control system.

12. Discuss how the concepts of gain margin and phase margin can be of assistance in the design of a control system.

General

1. A thermocouple has a simple exponential time delay, its time constant being 5 seconds. Calculate the attenuation of an oscillatory input signal of frequency $1 \cdot 5/2\pi$ Hz.

Ans. $0 \cdot 133$

2. A thermometer having a steady reading of $10°C$ is placed in a liquid, the temperature of which is $60°C$. After 4 seconds, the reading is $40°C$. Assuming the response of the thermometer is a simple exponential delay, calculate its time constant.

The same thermometer is used to record a temperature which is rising steadily at the rate of $2°C$ per second. What is the steady-state error of its reading?

Ans. $4 \cdot 37$ s; $8 \cdot 74°C$

3. The temperature of 20 kg of water in a tank is thermostatically controlled. The thermostat controls the supply of heat to the water, the output of the controller being 100 watts/°C deviation from the thermostat setting. The water loses heat at the rate of $3\theta_0$ watts, where θ_0 is the actual temperature of the water in °C. Initially the water is at $10°C$ when the thermostat setting is moved to $50°C$.

Determine:

    (a) the temperature of the water ultimately reached,

    (b) the time taken for the temperature to rise to within $1°C$ of this final value.

Ans. (a) $48 \cdot 5°C$; (b) $49 \cdot 4$ min

4. A servo-mechanism controls the position of a load of moment of inertia 500 kg m². Motion of the load is resisted by a viscous damping torque of 10

kNms/radian. The servo-motor delivers a torque of 100 kNm/radian of mis-alignment. If the input member of the controller is suddenly rotated through one complete revolution, find:

    (*a*) the damping ratio;
    (*b*) the periodic time of the subsequent oscillations;
    (*c*) the magnitude of the first overshoot, and the time when this occurs;
    (*d*) the time which elapses before the error is always less than 1°.

        Ans. (*a*) 0·707; (*b*) 0·6284 s; (*c*) 15·6°, 0·3142 s; (*d*) 0·405 s

    5. (*a*) A simple position control system is used to move a load of moment of inertia 50 kg m². The torque delivered by the servo-motor is proportional to the misalignment, and the damping ratio is 0·6.

When the input member is rotated at a constant speed of 100 rev/min, the load is found to lag by 10°. Determine the magnitude of the torque delivered by the servo-motor per radian of misalignment, and the damping torque at 1 radian/s.

    (*b*) To reduce the steady-state velocity error to 2°, the viscous damping torque is reduced and the effective damping ratio is maintained at 0·6 by the introduction of derivative of error control action. Determine the new value of the damping torque at unit velocity, and the required value of the derivative action sensitivity constant.

        Ans. (*a*) 259·2 kNm; 4·32 kNm
                  (*b*) 0·864 kNm; 3·456 kNms/rad

    6. A control system has an open-loop transfer function:

$$\theta_0/\theta = 2(D + 1)/D(D + 5)$$

    (*a*) What parts of this transfer function are associated with (i) the controller, and (ii) the load?

    (*b*) What type of control action is being employed?

    (*c*) Plot the Nyquist diagram, and determine the forcing frequency necessary to give the function unit value.

        Ans. (*c*) 0·436/2π Hz

    7. A controller has an open-loop transfer function:

$$\theta_0/\theta = 3/D(1 + 0·5D)(1 + 0·25D)$$

Plot the Nyquist diagram, and determine the gain margin and phase margin.

        Ans. 0·5; 20°

CHAPTER 11

Control System Elements

11.1 BLOCK DIAGRAMS COMPONENTS

It is not possible to deal with all the components used in automatic control systems in a small space, so this discussion is mainly limited to control action generators of the pneumatic and hydraulic types.

In a block diagram distinct groups of elements are referred to. For example, there is the measuring group, which may consist of a primary detector, transducing elements, transmitting elements, indicating and recording elements. The controller group usually consists of error modifying elements, an amplifier and the final regulating element such as a valve motor, power cylinder or electric motor. A crucial element is the differential device, which compares the measured output with the desired input and produces the all-important error signal. These groups are defined in the functional sense. In practice physical boundaries tend to become blurred in some systems. For example, in pneumatic control systems the control action generator is almost always to be found in the recorder or indicator which, of course, belongs to the measuring group.

11.2 THE MEASURING GROUP

The measuring group is of vital importance, for no variable can be controlled until it is measured, and the limits within which it is controlled cannot be finer than the limits of accuracy of the measuring instrument. Measurement systems consist in general of a primary detection element, which is in effect a transducer as well as an indicating and/or recording device. Sometimes there may also be intermediate elements to modify, amplify and transmit the primary signal to the recording device. A recording device is not always essential, but it is advisable to have some means of assessing the quality of the control. The table below shows a list of variables which may be the subject of control and of the types of primary detection elements or transducers usually employed.

Variable	Transducer or detecting element
Position	Linear wound potentiometer (Fig. 9.5 (*a*)), rotary or linear. Inductive and capacitive elements (compare Figs. 9.6 and 9.7). The gyroscope (see Chapter 4), employing either its "spatial memory" property to provide a fixed reference datum or the reaction torque generated when restrained by a spring, as in the "rate gyroscope." The gyroscope is much used in inertial guidance systems.
Speed	Mechanical tachometer (see fly-ball mechanism, Fig. 10.2). Tachogenerator in which a voltage proportional to the speed is generated.
Pressure	Bourdon tube, diaphragm, and bellows—sometimes combined with resistive, inductive, or capacitive transducers. Manometers.
Flow	Orifice plate, venturi tube, flumes, notches, and weirs. Pitot-static tubes. (All develop pressure signals.)
Temperature	Liquid-in-bulb expansion thermometers, developing pressure signals. Resistance thermometers and thermocouples.
Acidity	The glass electrode, in which liquids of different acidity generate an electric potential across a thin glass membrane. This potential is proportional to the difference of pH value between the liquids. (*N.B.* the pH of a liquid is the negative logarithm to the base 10 of the hydrogen ion concentration.)
CO_2 content (Boiler control)	The CO_2 content of a flue gas affects its thermal conductivity and hence the rate of cooling of a heated platinum wire surrounded by the gas; consequent resistance changes are measured, as in a resistance thermometer.
O_2 content (Boiler control)	Lehrer's apparatus for the measurement of oxygen (Fig. 11.1). Oxygen is distinguished from the other gases by being strongly paramagnetic, and is therefore drawn into the horizontal tube at one end by the asymmetrically positioned magnetic field. The heated platinum windings are consequently cooled at the inlet end by the uni-directional flow of oxygen, thus unbalancing the bridge.

11.3 Differential Devices

In a feed-back control system, some device is required to compare the
measured output with the desired input and to produce a signal which is
proportional to the deviation. The output of the differential device is

$$\theta = k(\theta_i - \theta_0) \qquad (11.1)$$

where k is a constant, not necessarily unity, which forms part of the loop gain.

The error signal is often obtained by means of a simple mechanical
linkage (Fig. 11.2). The link ABC is fixed by the desired value setting
θ_i while the link CDE pivots about C as the
measured output θ_0 varies. The distance BD
is proportional to the deviation θ. A pneumatic force balance method of achieving
the same result is shown in Fig. 11.3, where
the input and output signals are pressures
and where relative variations cause the balance
arm to pivot about its fulcrum, thus producing
a deflection proportional to the difference of
pressure.

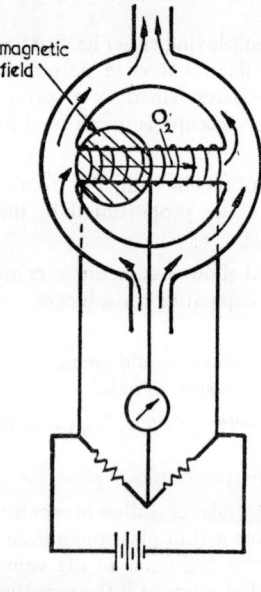

magnetic field

Where the signals are electric potentials, a
self-balancing potentiometer may be used.
The voltage signal from the primary measuring
element is compared with the output from
a slide wire supplied with a standard voltage. The difference of voltage is fed to an
amplifier, the output of which drives a
servo-motor which adjusts the position of
the tapping point on the slide wire until the
voltage difference is zero (Fig. 11.4).
At the same time, the motor also adjusts one
of two tapping points on a second slide wire

Fig. 11.1

while the other tapping point represents the desired value setting, so
that the voltage across the two tapping points is proportional to the
deviation.

11.4 Generation of Control Action

This section deals with methods of generating proportional, derivative,
and integral control actions of the types described in Chapter 10. There
are pneumatic, hydraulic and electrical methods.

Pneumatic

The error signal leaving the differential device may often be in the form of a mechanical displacement, in which case a proportional air pressure can be generated by the device shown in Fig. 11.5. Air,

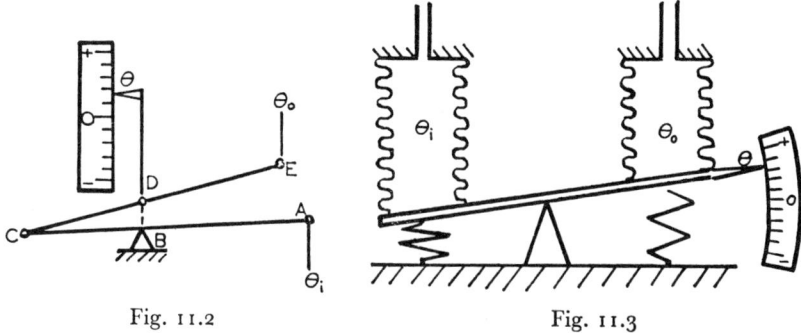

Fig. 11.2   Fig. 11.3

at approximately 140 kN/m² (1·4 bar), is supplied to a nozzle N through a restrictor R. A pivoted flapper-bar may be held against the nozzle aperture, thus sealing it off and causing a pressure build-up between R and N. Alternatively, the flapper may be moved away, allowing air to escape to atmosphere and the gauge pressure between R and N to fall

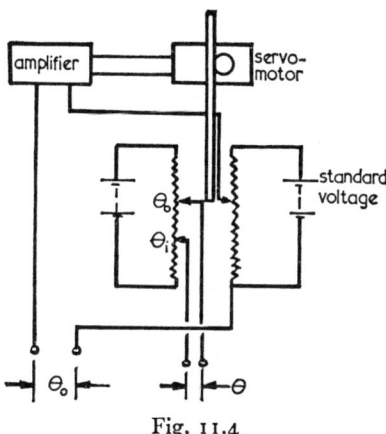

Fig. 11.4

to zero. At some intermediate position of the flapper (very close to the nozzle aperture but not sealing it off completely), the pressure between R and N will be maintained at some value between 140 kN/m² and zero. In fact, variation of the flapper position (input) produces an output pressure between R and N which varies as shown in Fig. 11.6.

In practice, the movement of the flapper to change the output pressure from its maximum to zero is of the order of only 0·025 mm and the flapper-nozzle characteristic is non-linear. Over a limited range, however, the characteristic is approximately linear and within this range the device

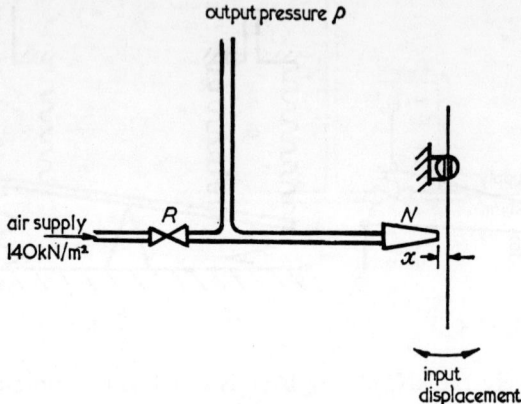

Fig. 11.5

behaves as a power-amplifying transducer capable of converting a small displacement signal into a proportional output pressure.

Even so, the device of Fig. 11.5 suffers from the faults of (1) over-sensitivity, and (2) non-linearity. These may be corrected by fitting a *proportional feed-back bellows* (Fig. 11.7). Movement of the flapper towards the nozzle causes the output pressure, and thus the pressure in the bellows, to rise. The resultant expansion of the bellows, against the action of the spring, moves the flapper pivot away from the nozzle, thereby reducing the effective movement of the flapper towards the nozzle. This reduces the sensitivity of the device and renders negligible the effects of the non-linearity of the flapper-nozzle characteristic.

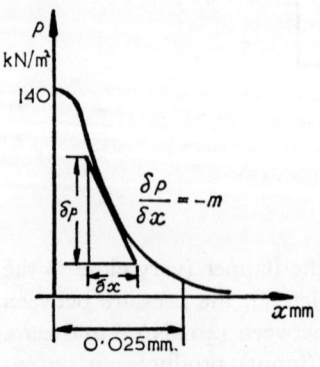

Fig. 11.6

If the output pressure is p when the flapper is a distance x from the nozzle suppose that an input displacement θ *towards* the nozzle causes an increase in the output pressure of δp. Then, if A is the cross-sectional area of the bellows

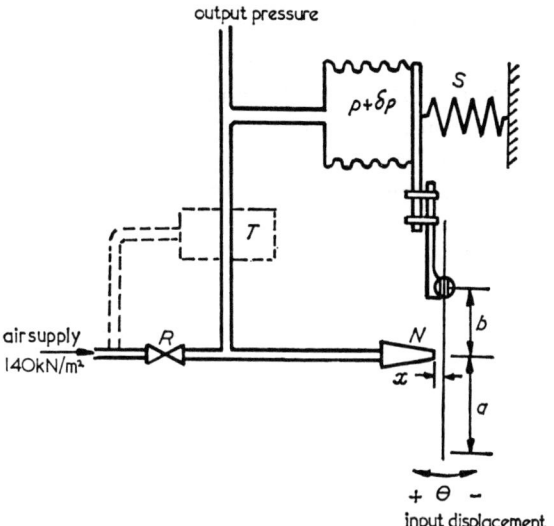

output pressure

Fig. 11.7

and S is the combined stiffness of bellows and spring, the

$$\text{Bellows displacement} = A\delta p/S = k_s\, \delta p$$

where $k_s = A/S$. The

Flapper movement at the nozzle due to bellows displacement

$$= k_s\delta pa/(a + b) = k_a k_s\, \delta p$$

where $k_a = a/(a + b)$. The

Flapper movement at the nozzle due to input displacement θ

$$= -\theta\, b/(a + b) = -k_b\theta$$

where $k_b = b/(a + b)$. The

Total flapper movement at the nozzle

$$= \delta x = k_a k_s\, \delta p - k_b\, \theta \qquad (11.2)$$

But from the flapper-nozzle characteristic (Fig. 11.6)

$$\delta p = -m\, \delta x$$
$$= -m\, (k_a k_s\delta p - k_b\theta)$$

where m is the slope of this characteristic. Solving for δp,

$$\delta p = \frac{mk_b\theta}{1 + mk_a k_s} = \frac{k_b\theta}{(1/m) + k_a k_s} \qquad (11.3)$$

But m is very large, so that $1/m$ is negligible.

$$\therefore\ \delta p = (k_b/k_a k_s)\theta = K_1\theta \qquad (11.4)$$

Eqs. (11.3) and (11.4) show how the addition of the feed-back bellows makes the output pressure independent of m. This cures the over-sensitivity of the device and also means that variation of m due to the non-linearity of the characteristic is unimportant. The output pressure is strictly proportional to θ, and the constant of proportionality (Eq. (11.4)) or the controller gain K_1, is given by

$$K_1 = k_b/k_a k_s = bS/aA \qquad (11.5)$$

The controller gain or proportional band can therefore be varied by altering the spring rate S or, more usually, by adjustment of the feed-back link b.

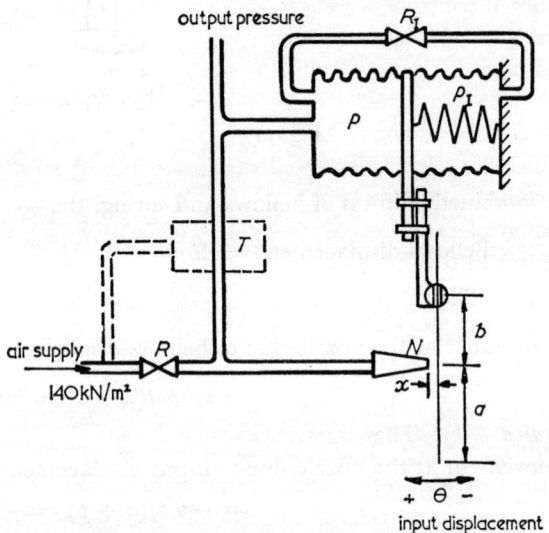

output pressure

air supply
140kN/m²

+ θ −
input displacement

Fig. 11.8

This analysis assumes that the feed-back bellows responds instantly to changes of nozzle pressure. In practice, the relatively large air space of the bellows (and of course the valve motor or pneumatic jack to which the output is connected) requires time to be changed, but this time delay is reduced to negligible proportions by fitting an air relay between the nozzle and the bellows (at point T). This relay acts as a 1:1 transmitter, but can provide large quantities of air very rapidly.

Integral control action may be added to the proportional controller of Fig. 11.7 by incorporating a second bellows, identical with the proportional bellows but acting in opposition to it (Fig. 11.8). Air is bled from the proportional bellows into the *integral bellows* through the *integral restrictor* R_I. Under steady conditions with zero deviation, the pressures in both bellows are equal to the output pressure. A deviation θ will move the flapper towards the nozzle, the output pressure and the pressure in the feed-back bellows will build up rapidly as before, but the integral restrictor prevents a rapid build-up in the integral bellows. The resulting pressure difference across R_I causes air to flow into the integral chamber, so that the pressures gradually equalise. The flapper position will then correspond to zero deviation again, but *the output pressure will be greater than before*, thus making it possible for the regulator to cater for a sustained load increase without offset.

The same effect can also be achieved by increasing the force in the spring S, and this is the method usually employed for manually resetting pure proportional controllers subjected to sustained changes of load.

The integral restrictor is a needle valve, the thimble of which may be calibrated with an integral action time scale. (If the integral restrictor is fully closed, the integral bellows becomes inoperative and a pure proportional control action results, while if fully opened the integral bellows nullifies the effect of the feed-back bellows and an over-sensitive or "on-off" control action results.)

The variation of pressure output is obtained as follows. If p and p_I are the pressures in the feed-back and integral bellows respectively, then initially

$$p = p_I = p_0 \text{ say}$$

For a deviation θ, the net flapper movement at the nozzle is given by Eq. (11.2) as

$$\delta x = k_a k_s \delta p - k_b \theta$$

where δp is the increase in the output pressure p. Assuming this increase occurs rapidly, the pressure in the integral chamber will remain p_0, so that

$$p = p_0 + \delta p \qquad \text{with } p_I = p_0$$
$$\therefore \ \delta p = p - p_0 = p - p_I$$
$$\therefore \ \delta x = k_a k_s (p - p_I) - k_b \theta$$

Dividing by $k_a k_s$, this becomes

$$(\delta x / k_a k_s) = (p - p_I) - K_1 \theta$$

where $K_1 = k_b / k_a k_s$ the controller gain, as before. The term $(\delta x / k_a k_s)$ is negligible, since δx is originally of the order of microns (see Fig. 11.6) and is reduced still further by the feed-back action.

$$\therefore \ p - p_I = K_1 \theta \qquad\qquad (11.6)$$

Under laminar flow conditions, the mass flow through the integral restrictor will be proportional to the pressure difference across the restrictor, or, using the electrical analogy, equal to $(p - p_I)/R_I$. But if M is the mass of air, in the integral chamber, then from the gas laws,

$$M = p_I V/RT = p_I C_I/RT$$

where C_I is the capacity of the integral chamber, and which will vary by only a very small amount. Neglecting any change in C_I and assuming isothermal conditions, the mass flow into the integral chamber is:

$$dM/dt = (C_I/RT) \cdot (dp_I/dt)$$

Equating expressions for mass flow:

$$(C_I/RT)Dp_I = (p - p_I)/R_I \qquad \text{(where D} = d/dt)$$

i.e. $\qquad (R_I C_I/RT)Dp_I = p - p_I$

i.e. $\qquad T_I Dp_I = p - p_I \hfill (11.7)$

where $T_I = R_I C_I/RT =$ Integral action time constant.

It follows from Eqs. (11.6) and (11.7), that $T_I Dp_I = K_1\theta$, so that

$$p_I = K_1\theta/T_I D = \frac{K_1}{T_I}\int \theta \, dt + p_0 \hfill (11.8)$$

where p_0 is the constant of integration.

Substitution for p_I in Eq. (11.6) gives the output pressure p, i.e.

$$p - (K_1\theta/T_I D) = K_1\theta$$

i.e. $\qquad p = K_1[1 + (1/T_I D)]\theta = K_1\theta + K_2\int \theta \, dt + p_0$

where $K_2 = K_1/T_I =$ Integral action sensitivity constant. Thus the

Variation in output pressure $= \delta p = p - p_0 = K_1\theta + K_2\int \theta \, dt \quad (11.9)$

Both K_2 and T_I are defined in Section 10.4.

Derivative action can be generated by inserting a variable restrictor R_D between the output and the proportional feed-back bellows (Fig. 11.9). This *derivative restrictor* will also take the form of a needle valve, calibrated to give the derivative action time. The variation of output pressure is obtained as follows.

Let p and p_F be the output pressure and the pressure in the feed-back bellows and suppose that initially $p = p_F = p_0$. If there is a disturbance causing a deviation θ, the amount of feed-back is governed by the feed-back pressure p_F so that the net flapper movement at the nozzle is given by a modification of Eq. (11.2), i.e.

$$\delta x = k_a k_s \, \delta p_F - k_b \theta$$

In this case,

$$\delta p_F = p_F - p_0$$
$$\therefore \quad \delta x = k_a k_s (p_F - p_0) - k_b \theta$$

i.e.
$$(\delta x / k_a k_s) = (p_F - p_0) - K_1 \theta$$

where, as before, $K_1 = k_b / k_a k_s$ = the controller gain. Again $(\delta x / k_a k_s)$ is negligible, since δx is of the order of microns (Fig. 11.6).

$$\therefore \quad p_F - p_0 = K_1 \theta$$

or
$$p_F = p_0 + K_1 \theta \qquad (11.10)$$

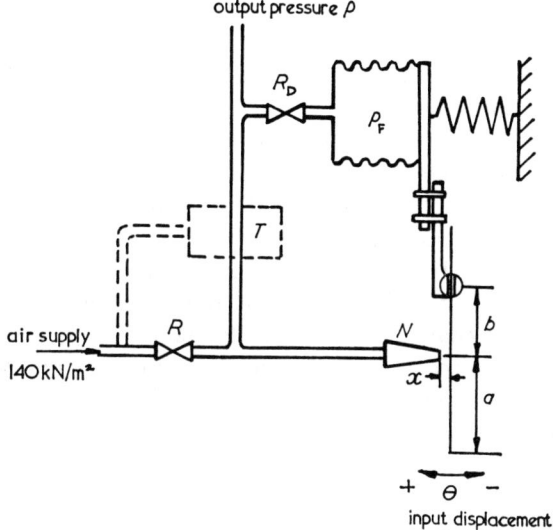

Fig. 11.9

Under the laminar flow conditions, the mass flow through the derivative restrictor is $(p - p_F)/R_D$. If M is the mass of air in the proportional feedback bellows,

$$M = p_F V / RT = p_F C_F / RT$$

where C_F is the capacity of the feed-back bellows, the percentage change of which will be small. Neglecting any change in C_F and assuming isothermal conditions, the mass flow into the feed-back bellows is:

$$dM/dt = (C_F / RT)(dp_F / dt)$$

Equating expressions for mass flow:

$$(C_F/RT)Dp_F = (p - p_F)/R_D$$

i.e.
$$(R_DC_F/RT)Dp_F = p - p_F$$

i.e.
$$T_DDp_F = p - p_F \tag{11.11}$$

where $T_D = (R_DC_F/RT) =$ Derivative action time constant.

From Eq. (11.11), $p_F = p/(1 + T_DD)$, and substitution for p_F in Eq. (11.10) gives:

$$p/(1 + T_DD) = p_0 + K_1\theta$$

i.e.
$$p = (p_0 + K_1\theta)(1 + T_DD)$$
$$= p_0 + K_1\theta + T_DDp_0 + K_1T_DD\theta$$

But $Dp_0 = 0$, since p_0 is a constant so that the

Variation in output pressure $\delta p = p - p_0 = K_1\theta + K_3(d\theta/dt)$ (11.12) where $K_3 = K_1T_D =$ Derivative action sensitivity constant. (Both K_3 and T_D are defined Section in 10.4.)

A full three-term controller may be obtained by combining the derivative restrictor and the integral restrictor and bellows, or by super-imposing Figs. 11.8 and 11.9. In practice, however, the three terms will interact, for an alteration of either integral action time or derivative action time will bring about a change in the controller gain K_1. Eq. (10.31) which describes three term control action indicates that the integral and derivative terms may be altered without effecting the proportional term, but analysis of the controller described here would lead to the following expression for the

Variation in output pressure $\delta p = K_1\left[\left(1 + \dfrac{2T_D}{T_I}\right) + (1/T_ID) + T_DD)\right]\theta$

$$\tag{11.13}$$

Comparison with Eq. (10.31) shows that the proportional term is multiplied by $(1 + 2T_D/T_I)$, which is called the *interaction factor*.

Hydraulic

The basic hydraulic relay mechanism is shown in Fig. 11.10. Movement of the piston valve regulates the supply of high pressure oil to each side of a power piston so that an input displacement or deviation θ causes an output displacement at the power piston. The system is a power amplifier because only a small force is required to move the valve even though the force exerted by the power piston may be considerable.

Initially, the valve ports will be closed by the piston valve but with the displacement θ, oil will be admitted to one side of the power piston at a rate proportional to the valve port opening or the displacement. Since the oil, for all practical purposes, is incompressible, the rate at which it enters the power cylinder is equal to the volume swept by the power piston in unit time.

$$\therefore \; A(d\theta/dt) = k\theta \qquad (11.14)$$

where A is the cross-sectional area of the power cylinder and k is a valve port constant. This may be written

$$\theta_0 = \frac{1}{T_1} \int \theta \, dt = \frac{1}{T_1 D} \theta \qquad (11.15)$$

where $T_1 = A/k =$ Relay time constant.

This hydraulic device therefore produces an output proportional to the time integral of the error. Even if pure integral control action were required, however, it would be of little practical use because T_1 would

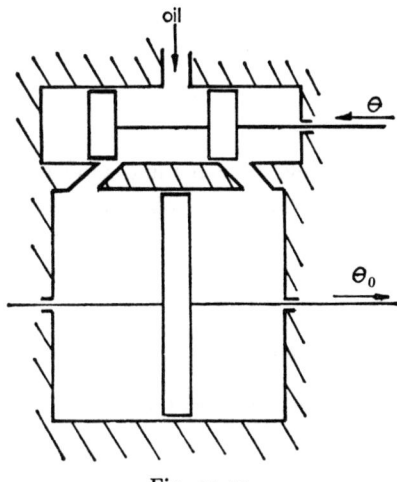

Fig. 11.10

be far too small, resulting in large output velocities for small input signals. Variation of T_1 would also be difficult. In any case, integral action is usually needed in combination with proportional action.

The arrangement of Fig. 11.10 may be modified to generate an output proportional to the input by the addition of a mechanical feed-back link (Fig. 11.11). When the piston valve is moved to the left, the resulting motion of the power piston is fed back through the pivoted lever to a sliding sleeve in which the valve moves. Since the motion of the sleeve is in the same direction as the valve, the valve port opening is thus reduced and the sensitivity of the system is decreased, just as the sensitivity of the pneumatic flapper-nozzle device is decreased by the proportional feed-back bellows.

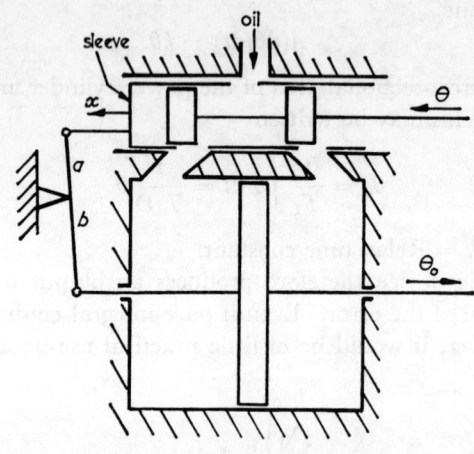

Fig. 11.11

From Fig. 11.11, the sleeve movement x is given by $x/a = \theta_0/b$ so that $x = (a/b)\theta_0$. The relative valve port opening is $\theta - x$ and the oil flow into power cylinder $k(\theta - x)$ where k is the valve port constant. Assuming the oil is incompressible,

$$A\,D\theta_0 = k(\theta - x)$$

$$(A/k)D\theta_0 = \theta - (a/b)\theta_0$$

$$(Ab/ka)D\theta_0 = (b/a)\theta - \theta_0$$

$$TD\theta_0 = K_1\theta - \theta_0$$

where $T = Ab/ka$ is a time constant, and $K_1 = (b/a)$

$$\therefore (1 + TD)\theta_0 = K_1\theta \tag{11.16}$$

Thus after a simple exponential delay, the output θ_0 is now proportional to the error θ. This time delay is very small, so that the steady-state output is given by:

$$\theta_0 = K_1\theta \tag{11.17}$$

where $K_1 = (b/a)$ is the controller gain.

If integral action is required in addition to the proportional action, this may be achieved by the addition of the spring and dashpot arrangement shown in Fig. 11.12. If x is the sleeve movement (as before), and y the displacement of the upper end of the feed-back link, then

$$y = (a/b)\theta_0$$

If the inertia of the sleeve is neglected, the spring force and dash-pot forces are equal, so that

$$Sx = F_v[Dy - Dx]$$
$$x = T_I D(a/b)\theta_0 - T_I Dx$$

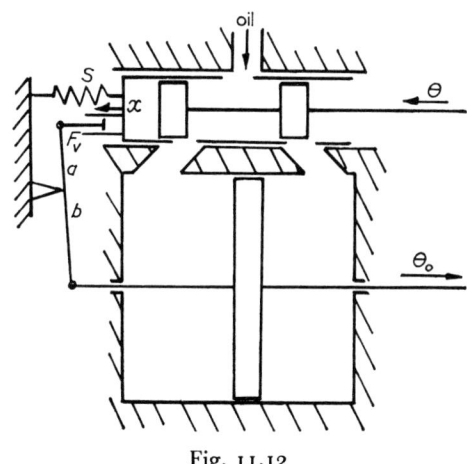

Fig. 11.12

where $T_I = F_v/S = $ Integral action time constant, i.e.

$$K_1(1 + T_I D)x = T_I D\theta_0 \quad \text{where} \quad K_1 = (b/a)$$
$$\therefore x = T_I D\theta_0/K_1(1 + T_I D)$$

But from continuity of flow,

$$AD\theta_0 = k(\theta - x), \text{ as before}$$
$$T_1 D\theta_0 = \theta - \{T_I D\theta_0/K_1(1 + T_I D)\}$$

where $T_1 = A/k$, and is normally small enough to be neglected. Thus, neglecting T_1,

$$T_I D\theta_0/K_1(1 + T_I D) = \theta$$
$$\therefore \theta_0 = K_1(1 + T_I D) \theta/T_I D$$

i.e. $$\theta_0 = K_1[1 + (1/T_I D)]\theta = K_1\theta + K_2\int\theta \, dt \quad (11.18)$$

where $K_2 = K_1/T_I$ is the integral action sensitivity constant. This equation is characteristic of proportional plus integral control action (see Eqs. (10.20), (10.22), (10.31) and (11.9)). Both T_I and K_2 are defined in Section 10.4.

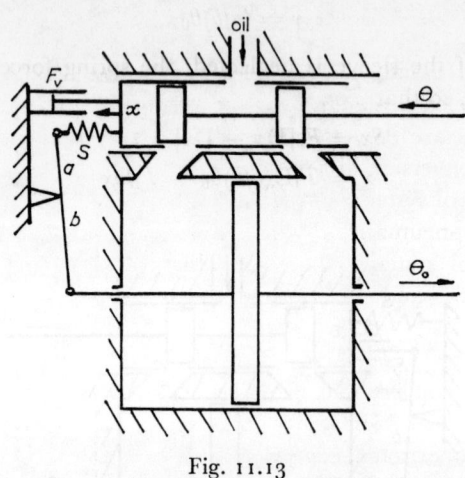

Fig. 11.13

Proportional plus derivative control action may be obtained by interchanging the spring and dashpot as shown in Fig. 11.13.

$$\text{Dash-pot force} = F_v(\mathrm{d}x/\mathrm{d}t) = F_v\mathrm{D}x$$
$$\text{Spring force} = S[(a/b)\theta_0 - x]$$

∴ Neglecting sleeve inertia:

$$F_v\mathrm{D}x = S[(a/b)\theta_0 - x]$$
or
$$T_\mathrm{D}\mathrm{D}x = (\theta_0/K_1) - x$$

where $T_\mathrm{D} = F_v/S$ = Derivative action time constant, and
$K_1 = (b/a)$ = Controller gain.

Thus
$$x = \theta_0/K_1(1 + T_\mathrm{D}\mathrm{D})$$

and by continuity

$$A\mathrm{D}\theta_0 = k(\theta - x)$$
or
$$T_1\mathrm{D}\theta_0 = \theta - \{\theta_0/K_1(1 + T_\mathrm{D}\mathrm{D})\}$$

where $T_1 = A/k$ and which may be neglected if T_1 is small compared with T_D. Thus:

$$\theta_0 = K_1(1 + T_\mathrm{D}\mathrm{D})\theta = K_1\theta + K_3(\mathrm{d}\theta/\mathrm{d}t) \qquad (11.19)$$

where $K_3 = K_1 T_D$ is the derivative action sensitivity constant. This equation is indicative of proportional plus derivative of error control action (see Eqs. (10.25), (10.26), (10.31) and (11.12)).

Electrical

Any mechanical signal, whether displacement or pressure, can be converted into an electrical signal by an appropriate transducer. Because electrical signals are easy to manipulate, it is sometimes convenient to carry out this conversion in a hydraulic or pneumatic control system, so that as well as completely pneumatic, hydraulic or electrical control systems, there are also electro-pneumatic and electro-hydraulic control systems.

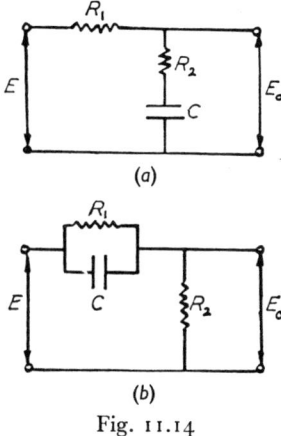

Fig. 11.14

In an electrical control system, the transducer will convert the error signal into a proportional voltage, so that to obtain pure proportional control action, it is only necessary to amplify the signal by an electrical or electronic amplifier. If integral or derivative actions are needed the voltage signal must be passed through a suitable network.

Consider, for example, the network shown in Fig. 11.14 (a). The

$$\text{Input voltage } E = R_1 i + R_2 i + \frac{1}{C}\int i\, dt$$

$$= [R_1 + R_2 + (1/CD)]i$$

$$\text{Output voltage} = E_0 = R_2 i + \frac{1}{C}\int i\, dt = [R_2 + (1/CD)]i$$

$$\therefore\ E_0/E = [R_2 + (1/CD)]/[R_1 + R_2 + (1/CD)]$$
$$= (R_2 CD + 1)/[(R_1 + R_2)CD + 1]$$
$$= a(T_1 D + 1)/(T_1 D + a)$$

where $T_1 = R_2 C$, and $a = R_2/(R_1 + R_2)$. If the values of R_1 and R_2 are chosen so that a is very small, then

$$E_0/E = a(T_1 D + 1)/T_1 D = a[1 + (1/T_1 D)]$$

or $\qquad E_0 = a[1 + (1/T_1 D)]E \qquad\qquad\qquad (11.20)$

This is similar to Eq. (11.18), and is typical of proportional plus integral control action and T_I is the integral action time constant.

In the network of Fig. 11.14 (b), the potential difference across R_1 and $C = E - E_0$,

$$\therefore i_1 = (E - E_0)/R_1 \quad \text{and} \quad i_2 = CD(E - E_0)$$

where $D = d/dt$. But,

$$i_1 + i_2 = i = E_0/R_2$$
$$\therefore \{(E - E_0)/R_1\} + CD(E - E_0) = E_0/R_2$$

It follows that

$$E_0/E = a(1 + T_DD)/(1 + aT_DD)$$

where $T_D = R_1C$, and $a = R_2/(R_1 + R_2)$. Again, if R_2 is very small compared with $(R_1 + R_2)$, then a is very small and the above equation reduces to

$$E_0/E = a(1 + T_DD)$$

or

$$E_0 = a[1 + T_DD]E \tag{11.21}$$

which is similar to Eq. (11.19) and is typical of proportional plus derivative control action, with T_D as the derivative action time constant.

CHAPTER 12

Stress and Strain—1

12.1 COMPOUND BARS

Compound bars are members subjected to axial loading (either tensile or compressive) and consisting of elements of two or more different materials arranged in parallel and rigidly connected at their ends. The arrangement shown in Fig. 12.1 is symmetrical, as will be all those dealt with in this chapter. The reason for this is that the members of a loaded compound bar have different stresses, so that if the arrangement is not symmetrical, the line of action of the resisting force will not pass through the centroid of the cross-section of the compound bar. If, as is usual, the load is applied centrally, the outcome will be "eccentric load-ing" (see Chapter 15) and stress cal-culations will be difficult.

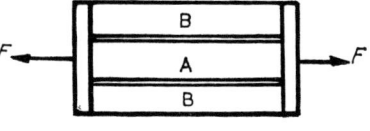

Fig. 12.1

Stresses in a compound bar

The compound bar (Fig. 12.1) is subjected to a tensile load F, A_A is the cross-section of the central bar, A_B the combined cross-section of the two outer bars, E_A and E_B are moduli of elasticity, σ_A and σ_B the stresses and ε_A and ε_B the strains of the central and outer bars respectively.

The stresses in the compound bar are calculated from the following facts:

1. The applied load F must be shared between the members, so that the sum of the individual loads must be equal to F, or

$$\sigma_A A_A + \sigma_B A_B = F \tag{12.1}$$

2. Since the end connections are rigid, both materials must have the

same extensions and since they had the same length originally they must have equal strains. In other words,

$$\varepsilon_A = \varepsilon_B$$

$$\therefore \quad \sigma_A/E_A = \sigma_B/E_B$$

or $\qquad\qquad\qquad \sigma_A/\sigma_B = E_A/E_B \qquad\qquad\qquad (12.2)$

From Eqs. (12.1) and (12.2), stresses σ_A and σ_B may be found and from these the individual loadings and the extension of a compound bar may be calculated. The equations apply equally well to compressive loadings.

Example. *A steel rod* 2 m *long and* 20 mm *in diameter is placed inside a copper tube* 20 mm *internal diameter and* 30 mm *external diameter. The tube and rod are rigidly joined at their ends and a tensile force of* 50 kN *is applied. Find (a) the stresses in the steel and the copper, (b) the loads carried by each component, and (c) the extension of the compound bar under load. Take E for steel* = 200 GN/m² *and E for copper* = 110 GN/m².

The cross-sectional areas of the two materials are

$$\text{Steel, } A_S = \tfrac{1}{4}\pi \times 0.02^2 = 3.142 \times 10^{-4} \text{ m}^2$$

$$\text{Copper, } A_C = \tfrac{1}{4}\pi(0.03^2 - 0.02^2) = 3.927 \times 10^{-4} \text{ m}^2$$

The subscripts S and C refer to steel and copper respectively.

(*a*) Because the strains in the two materials are equal, Eq. (12.2) gives

$$\sigma_S = \sigma_C E_S/E_C = \sigma_C \times 200/110 = 1.818\sigma_C$$

The total load carried = 50 kN, so that

$$\sigma_S A_S + \sigma_C A_C = 50 \times 10^3$$

Thus $\qquad 3.142 \times 10^{-4}\sigma_C + 3.927 \times 10^{-4}\sigma_C = 50 \times 10^3$

or $\qquad\qquad\qquad 3.142\sigma_C + 3.927\sigma_C = 50 \times 10^7$

Substituting $\sigma_S = 1.818\sigma_C$

$$5.712\sigma_C + 3.927\sigma_C = 50 \times 10^7$$

$\therefore$ Stress in copper $\sigma_C = 51.88 \times 10^6$ N/m² $= 51.88$ MN/m²

Stress in steel $\sigma_S = 1.818\sigma_C = 94.38 \times 10^6$ N/m²

$$= 94.38 \text{ MN}/m^2$$

(*b*) Load carried by steel $= \sigma_S A_S = 94.38 \times 10^6 \times 3.142 \times 10^{-4}$

$$= 29.63 \times 10^3 \text{ N} = 29.63 \text{ kN}$$

Load carried by copper $= \sigma_C A_C = 51.88 \times 10^6 \times 3.927 \times 10^{-4}$

$$= 20.37 \times 10^3 \text{ N} = 20.37 \text{ kN}$$

(c) Both rod and tube have same extension, so that either can be considered. Considering the steel rod,

$$\text{Extension } x_S = l_S \varepsilon_S = l_S \sigma_S / E_S$$
$$= (2 \times 94 \cdot 38 \times 10^6)/(200 \times 10^9)$$
$$= 9 \cdot 438 \times 10^{-4} \text{m} = 0 \cdot 9438 \text{ mm}$$

12.2 TEMPERATURE STRESSES

Coefficient of linear expansion

All engineering materials expand when heated and this expansion is usually equal in all directions. If a bar of material of length l has its temperature increased by t degrees, the increase of length x is (1) directly

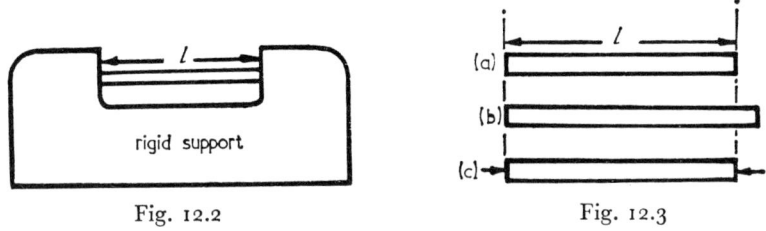

Fig. 12.2   Fig. 12.3

proportional to the original length l and also (2) directly proportional to the temperature change t. Hence

$$x \propto lt = \text{Constant} \times lt$$

The constant is known as the coefficient of linear expansion of the material α, so that

$$x = \alpha lt \tag{12.4}$$

Induced stress in a constrained bar

When a material is heated and not allowed to expand freely, stresses are induced which are known as "temperature stresses." Their values depend on the temperature change and the restraint. The extreme case is when the expansion is completely prevented. Fig. 12.2 shows a bar of length l placed between supports which can be considered perfectly rigid. If the bar is heated so that its temperature increases by t degrees, the resulting stress may be found by considering the process to take place in two stages (see Fig. 12.3).

1. The bar is allowed to expand freely (Fig. 12.3 (*b*)) so that

$$\text{Expansion } x = \alpha l t$$

2. A compressive force is then applied so as to restore the bar to its original length (Fig. 12.3 (*c*)). The previous expansion *x* now becomes the compression of the bar under load. Thus

$$\text{Compressive strain } \varepsilon = x/l = \alpha t$$

and

$$\text{Compressive stress} = E\varepsilon = E\alpha t \qquad (12.5)$$

Example. *A brass bar* 600 mm *long is being turned between centres in a lathe. The tailstock centre is adjusted to take up all clearance, and a finishing cut brings the diameter of the bar to* 100 mm. *The heat generated during this operation raises the temperature of the bar to a uniform* 95°C. *Assuming that the bar was originally at the room temperature of* 20°C *and under no axial load, find the thrust exerted by the bar,* (*a*) *if the lathe is considered rigid,* (*b*) *if the lathe deflects, resulting in* 0·6 mm *relative movement of the centres.* (*For brass E* = 90 GN/m² *and the coefficient of linear expansion is* 18 × 10⁻⁶ *per* °C.)

(*a*)   Stress in bar = $E\alpha t$ = 90 × 10⁹ × 18 × 10⁻⁶ × (95 − 20)

$$= 121·5 \times 10^6 \text{ N/m}^2$$

Thrust = Stress × Area = 121·5 × 10⁶ × $\tfrac{1}{4}\pi(0·1)^2$

$$= 954·3 \times 10^3 \text{ N} = 954·3 \text{ kN}$$

(*b*) Free expansion of bar

$$= l\alpha t = 0·6 \times 18 \times 10^{-6} \times (95 - 20)$$

$$= 8·1 \times 10^{-4} \text{ m} = 0·81 \text{ mm}$$

Movement of supports = 0·6 mm

∴ Elastic compression of bar

$$x = 0·81 - 0·6 = 0·21 \text{ mm}$$

Stress = $E\varepsilon = Ex/l$

$$= 90 \times 10^9 \times 0·21 \times 10^{-3}/0·6 = 315 \times 10^6 \text{ N/m}^2$$

Thrust = Stress × Area = 315 × 10⁶ × $\tfrac{1}{4}\pi(0·1)^2$

$$= 228·4 \times 10^3 \text{ N} = 228·4 \text{ kN}$$

Induced stresses in a compound bar

If a compound bar consisting of a central bar of material A and two outer bars of material B is subjected to an increase in temperature, the outer material will tend to expand more than the other if α_B is greater than α_A.

Since the bars are rigidly joined at their ends, however, the expansion of B is restricted and that of A increased. Thus compressive stress is induced in B and tensile stress in A.

The process may be considered to take place in two stages (Fig. 12.4):

1. The bars are allowed to expand freely (Fig. 12.4 (*b*)).

2. B is compressed and A extended, by the application of equal and opposite forces, until their ends are in line. The end connection is then re-applied (Fig. 12.4 (*c*)). As shown,

$$l\alpha_B t - l\alpha_A t = x_A + x_B$$

or, dividing by *l*,

$$(\alpha_B - \alpha_A)t = \varepsilon_A + \varepsilon_B \qquad (12.6)$$

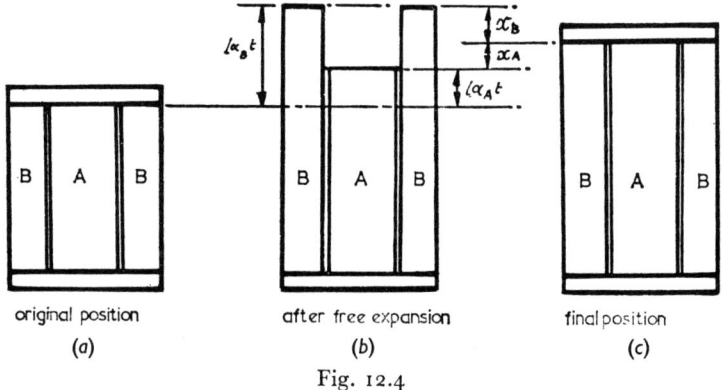

original position · after free expansion · final position

(a) · (b) · (c)

Fig. 12.4

This means that the excess of free expansion per unit length is the sum of the strains. Further, since the elements of the compound bar exert equal and opposite forces on each other,

$$\sigma_A A_A = \sigma_B A_B \qquad (12.7)$$

Example. *A compound bar* 1·5 m *long is made by placing a steel bar between two copper bars and riveting the three together at each end. All of them are* 50 mm *wide by* 10 mm *thick. What will be the stress in each material and the extension of the bar when its temperature is raised by* 60°C? *For steel E* = 200 GN/m², *α* = 12 × 10⁻⁶ *per* °C; *for copper E* = 110 GN/m², *α* = 17 × 10⁻⁶ *per* °C.

The excess of free expansion per unit length is the sum of the strains, or

$$(\alpha_C - \alpha_S)t = \varepsilon_S + \varepsilon_C$$
$$= \sigma_S/E_S + \sigma_C/E_C$$

That is,

$$(17 - 12) + 10^{-6} \times 60 = \sigma_S/(200 \times 10^9) + \sigma_C/(110 \times 10^9)$$
$$6 \times 10^7 = \sigma_S + 1\cdot818\sigma_C$$

The load on the steel equals the load on the copper, or

$$\sigma_S A_S = \sigma_C A_C$$

whence
$$\sigma_S = \sigma_C A_C/A_S = 2\sigma_C$$

If this is substituted in the above equation,

$$6 \times 10^7 = 2\sigma_C + 1\cdot818\sigma_C$$
$$\sigma_C = 1\cdot571 \times 10^7 \text{ N/m}^2 = 15\cdot71 \text{ MN/m}^2 \text{ (compressive)}$$
$$\therefore \ \sigma_S = 2\sigma_C = 31\cdot42 \text{ MN/m}^2 \text{ (tensile)}$$

For the steel bar,

$$\text{Final extension} = \text{Free expansion} + \text{Extension due to load}$$
$$= l\alpha_S t + l\sigma_S/E_S$$
$$= (1\cdot5 \times (12 \times 10^{-6}) \times 60)$$
$$+ (1\cdot5 \times 31\cdot42 \times 10^6/200 \times 10^9)$$
$$= 13\cdot16 \times 10^{-4} \text{ m} = 1\cdot316 \text{ mm}$$

(The same result is obtained by considering the copper bar alone, in which case the final extension is equal to the free expansion − compression due to load.)

Combined loading and temperature change

If a loaded compound bar is subjected to a temperature change, the final stress may be found from the principle of superposition (see Chapter 15), which states that the final stress in any element of the compound bar is the algebraic sum of the stress due to the temperature change alone, and the stress due to the load alone.

Example. *The compound bar of the previous example has applied to it a tensile load of 40 kN and its temperature is raised by 60°C. Find the final stresses in the steel and in the copper.*

For the temperature change alone, the stresses are those already calculated; $\sigma_S = 31\cdot42$ MN/m^2 (tensile) and $\sigma_C = 15\cdot71$ MN/m^2 (compressive). For loading without temperature change,

$$\text{Strain in steel} = \text{Strain in copper}$$

or
$$\sigma_S/E_S = \sigma_C/E_C$$
$$\therefore \ \sigma_S = \sigma_C \times E_S/E_C = \sigma_C \times 200/110 = 1\cdot818\sigma_C$$

Load on steel + Load on copper = 40 kN,

$$\sigma_s A_s + \sigma_c A_c = 40 \times 10^3$$
$$5 \times 10^{-4} \times \sigma_s + 10^{-3} \times A_c = 40 \times 10^3$$

Substituting $\sigma_s = 1 \cdot 818 \sigma_c$,

$$9 \cdot 09 \times 10^{-4} \sigma_c + 10^{-3} \sigma_c = 40 \times 10^3$$
$$\sigma_c = 20 \cdot 95 \times 10^6 \text{ N/m}^2 = 20 \cdot 95 \text{ MN/m}^2$$
$$\therefore \ \sigma_s = 1 \cdot 818 \sigma_c = 47 \cdot 96 \text{ MN/m}^2 \quad \text{(Both stresses are tensile.)}$$

By the principle of superposition,

Final stress in steel

$$= 31 \cdot 42 \text{ MN/m}^2 \text{ (tensile)} + 47 \cdot 96 \text{ MN/m}^2 \text{ (tensile)}$$
$$= 79 \cdot 38 \text{ MN/m}^2 \text{ (tensile)}$$

Final stress in copper

$$= 15 \cdot 71 \text{ MN/m}^2 \text{ (compressive)} + 20 \cdot 95 \text{ MN/m}^2 \text{ (tensile)}$$
$$= 5 \cdot 24 \text{ MN/m}^2 \text{ (tensile)}$$

12.3 Stress Analysis

The loads applied to a material are balanced by internal forces and stress is defined as the resisting force per unit area in some plane chosen within the material. There are two kinds of stresses; direct stress (tensile or compressive) in which the force is normal to the plane considered and shear stress, in which the force is tangential to the plane considered. Forces which are oblique to this plane may be resolved into normal and tangential components, which gives a combination of direct and shear stresses.

For simple loading, stress is calculated as the applied load divided by the cross-sectional area. This, however, is true only if the plane considered is at right angles to the applied force for direct loading, or parallel to it for shear loading. On other planes, combinations of direct and shear stresses are found.

Stresses on oblique planes

Fig. 12.5 (*a*) shows a bar of material of cross-sectional area A under axial tensile loading F. XX is any plane making an angle θ with the normal plane, and the objective is to find the stresses acting on this plane.

Let the direct stress be σ_θ and the shear stress be τ_θ and consider the equilibrium of the right-hand half of the bar (Fig. 12.5 (*b*)). Resolving forces perpendicular to the plane XX, $F_n = F\cos\theta$. The area of section $XX = A/\cos\theta$, so that the

$$\text{Normal or direct stress } \sigma_\theta = (F/A)\cos^2\theta = \sigma\cos^2\theta \qquad (12.8)$$

where σ is the stress on the normal cross-section.

Resolving forces parallel to the plane XX,

$$F_t = F\sin\theta$$

and   Tangential or shear stress $\tau_\theta = (F/A)\sin\theta\cos\theta$

$$= \sigma\sin\theta\cos\theta$$

$$= \tfrac{1}{2}\sigma\sin 2\theta \qquad (12.9)$$

Thus on planes other than the normal cross-section, there are varying combinations of direct tensile stress and shear stress. Maximum shear stress will be when $2\theta = 90°$, that is when $\theta = 45°$, and its value will be $\tfrac{1}{2}\sigma$. Shear stresses on oblique planes are sometimes responsible for failure

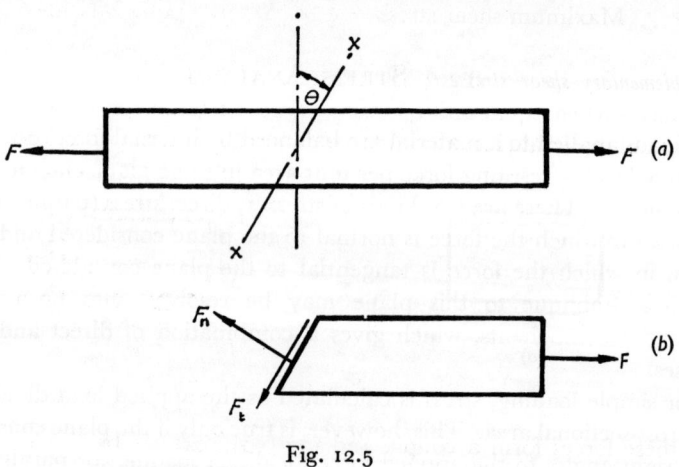

Fig. 12.5

of a material. Brittle materials, for example, are usually much weaker in shear than in direct compression, so that under simple compression they fail by shearing on planes making approximately 45° with the direction of the applied compressive force.

Example. *A wire 2 mm diameter hangs vertically from a support and a mass of 10 kg is attached to its lower end. Find the stress on a normal cross-section,* *and the stresses on a plane making an angle of* 55° *with the axis of the wire. What is the maximum shear stress in the material and on what planes does it act?*

Stress on normal cross-section $F/A = mg/\frac{1}{4}\pi d^2$

$$= 10 \times 9\cdot81/\tfrac{1}{4}\pi \times 0\cdot002^2$$

$$= 31\cdot23 \times 10^6\,\text{N/m}^2$$

$$= 31\cdot23\,\text{MN/m}^2\ \text{(tensile)}$$

On a plane at 55° to the axis,

Direct stress $\sigma_\theta = \sigma \cos^2 \theta$

and $\qquad\qquad\theta = (90° - 55°) = 35°$

$$\therefore\ \sigma_\theta = 31\cdot23 \cos^2 35° = 20\cdot95\,\text{MN/m}^2\ \text{(tensile)}$$

Shear stress $\tau_\theta = \tfrac{1}{2}\sigma \sin 2\theta$

$$= \tfrac{1}{2} \times 31\cdot23 \sin 70°$$

$$= 14\cdot67\,\text{MN/m}^2$$

Maximum shear stress occurs when $\theta = 45°$, or on planes making angles of 45° with the axis of the wire. For $\theta = 45°$, $\tau_\theta = \tfrac{1}{2}\sigma$

$$\therefore\ \text{Maximum shear stress} = \tfrac{1}{2} \times 31\cdot23 = 15\cdot62\,\text{MN/m}^2$$

Complementary shear stresses. Shear stress is often described as the stress produced when equal and opposite forces act along the parallel faces of a block of material (Fig. 12.6 (*a*)). This arrangement is in fact unrealistic,

(a)

(b)

Fig. 12.6

for these forces form a couple and the block cannot be in equilibrium. For equilibrium an opposing couple is required and this can only be supplied by forces acting on the remaining faces of the block (Fig. 12.6 (*b*)). Thus there must be two shear stresses in the material:

$$\tau\ \text{due to } F \quad \text{and} \quad \tau^1\ \text{due to } F^1$$

Equating the couples acting on the block,

$$F \times \text{QR} = F^1 \times \text{PQ}$$

Hence $\qquad\qquad F/\text{PQ} = F^1/\text{QR}$

If the thickness of the block is t,

$$\tau = F/(PQ \times t) \quad \text{and} \quad \tau^1 = F^1/(QR \times t)$$

Hence $$\tau = \tau^1$$

Thus a shear stress is always accompanied by another shear stress of equal magnitude acting on a perpendicular plane and known as the complementary shear stress.

Fig. 12.7 (a) shows a rectangular block of material of unit thickness subjected to complementary shear stresses τ. The objective is to find the direct stress σ_θ and shear stress τ_θ on a plane XX making an angle θ with the upper face of the block.

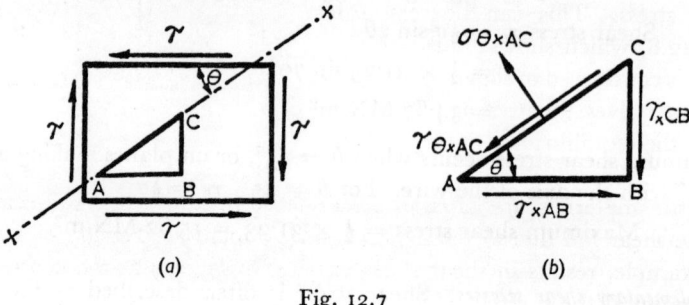

Fig. 12.7

The triangular prism of material ABC having face AC in the plane XX, and faces AB and BC parallel to the faces of the block is in equilibrium. Since the block is of unit thickness, the areas of the faces of the prism will be equal to the lengths of the sides of the triangle ABC and the forces acting on the prism will be as shown in Fig. 12.7 (b). Resolving forces perpendicular to the face AC,

$$\sigma_\theta . AC = \tau . CB . \cos \theta + \tau . AB . \sin \theta$$

$$\therefore \; \sigma_\theta = \tau . (CB/AC) . \cos \theta + \tau . (AB/AC) . \sin \theta$$

$$= \tau \sin \theta \cos \theta + \tau \cos \theta \sin \theta$$

$$= 2\tau \sin \theta \cos \theta$$

$$= \tau \sin 2\theta \text{ (tensile)} \qquad (12.10)$$

Resolving forces parallel to the face AC,

$$\tau_\theta . AC = \tau . AB . \cos \theta - \tau . CB . \sin \theta$$

$$\therefore \; \tau_\theta = \tau . (AB/AC) \cos \theta - \tau . (CB/AC) \sin \theta$$

$$= \tau \cos^2 \theta - \tau \sin^2 \theta$$

$$= \tau \cos 2\theta \qquad (12.11)$$

If θ is between 90° and 180°, sin 2θ is negative and σ_θ is therefore negative, which means that the direction of σ_θ is the opposite of that shown. In other words, σ_θ becomes compressive. In the same way, if θ is between 45° and 135°, τ_θ will be negative, meaning that the direction of τ_θ is reversed. The stress is still a shear stress, however, and the direction of a shear stress is a matter only of convention.

The maximum and minimum values of direct stress will occur when $\theta = 45°$ and $\theta = 135°$, giving $\sigma_\theta = +\tau$ and $\sigma_\theta = -\tau$ respectively. For these values of θ, $\tau_\theta = 0$, so that on the two planes making angles of 45° with the planes of a simple shear stress, there are pure tensile and pure compressive stresses, both numerically equal to the simple shear stress. This can be seen from Fig. 12.8, which shows that if the 45° plane is considered to separate the block into two halves, the stress on it is obvious from the equilibrium of one of the halves.

Fig. 12.8

Brittle materials are often weak in tension, and when subjected to simple shear fail due to tensile stress on 45° planes. Torsion of a shaft, for example, results in shear stress varying from zero at the centre to a maximum at the surface, which means that tensile stress will be a maximum at the surface and on planes at 45° to the axis. The failure of brittle materials in torsion is initiated by such tensile stresses, leading to the characteristic helical fracture.

PROBLEMS

For tutorials

1. A straight steel steam pipe 5 m long connects a boiler to a turbine. If the pipe was unstressed when installed at 20°C, calculate the stress in the pipe when at a temperature of 150°C, assuming (a) both boiler and turbine to be immovable, (b) that the thrust of the pipe causes a total displacement of 6 mm.

The elastic limit stress for steel is 250 MN/m². In view of this, can answer (a) be correct? State which of the basic assumptions made in arriving at it is at fault and give a reasoned estimate of the true stress (assuming the pipe is constrained so as to prevent bending). What are the practical implications of this problem? For steel, $E = 200$ GN/m² and $\alpha = 12 \times 10^{-6}$ per °C.

Ans. (a) 312 MN/m²; (b) 72 MN/m² (both compressive)

2. Two similar copper pipes are joined by a brass sleeve fitted with small clearance and attached to the pipes by soft soldering. If the solder solidifies at 183°C and can be assumed to connect the two metals rigidly together at this and all lower temperatures, what will be the stresses in the two metals after cooling

to room temperature (20°C)? The cross-sectional areas of pipe and sleeve are in the ratio 1:2·5. For copper, $E = 110$ GN/m² and $\alpha = 17 \times 10^{-6}$ per °C and for brass $E = 90$ GN/m² and $\alpha = 18 \times 10^{-6}$ per °C.

Many components are assembled by soldering or brazing. Will temperature stresses be present in all cases? Could failure occur because of such stresses?

Ans. Copper, 12·04 MN/m² (tensile): brass, 4·82 MN/m² (compressive)

General

1. A short column is made by filling a cast-iron tube with concrete. The internal and external diameters of the tube are respectively 200 mm and 300 mm and its length is 500 mm. Find the stress in each material and the shortening of the column under a compressive load of 400 kN. Take E for cast-iron as 110 GN/m² and E for concrete as 20 GN/m².

Ans. Cast-iron, 8·894 MN/m²; concrete, 1·617 MN/m² (both compressive); 0·0404 mm.

2. A steel rod 30 mm diameter has a brass sleeve 40 mm external diameter shrunk on to it. Assuming the sleeve and rod to be rigidly connected, what percentage of the total axial loading will be carried by the steel rod? For steel, $E = 200$ GN/m² and for brass $E = 90$ GN/m².

Ans. 74·1%

3. A compound bar 300 mm long consists of an aluminium tube 25 mm internal and 30 mm external diameter enclosing a steel rod 20 mm diameter, the tube and rod being rigidly connected at their ends. Find the stress in each material and the extension of the compound bar when its temperature is raised by 100°C. For steel, $E = 200$ GN/m² and $\alpha = 12 \times 10^{-6}$ per °C: for aluminium, $E = 70$ GN/m² and $\alpha = 23 \times 10^{-6}$ per °C.

Ans. Steel, 42·67 MN/m² (tensile); aluminium 62·07 MN/m² (compressive); 0·424 mm.

4. A steel rod 25 mm diameter passes centrally through a copper tube 30 mm internal and 40 mm external diameter. The ends of the bar are threaded, and it is fitted with nuts and rigid washers. The nuts are tightened until the rod is under a tensile stress of 20 MN/m², and the temperature is then raised by 50°C. Find the final stresses in the steel and in the copper. For steel, $E = 200$ GN/m² and $\alpha = 12 \times 10^{-6}$ per °C: for copper, $E = 110$ GN/m² and $\alpha = 17 \times 10^{-6}$ per °C.

Ans. Steel 39·06 MN/m² (tensile); copper 34·88 MN/m² (compressive)

5. A tie bar is to carry a tensile load of 150 kN. Calculate the required diameter of the bar (a) if the tensile stress is not to exceed 70 MN/m², (b) if the shear stress is not to exceed 50 MN/m².

Ans. (a) 52·2 mm; (b) 43·7 mm

6. At a point on the neutral axis of a beam, there is a pure shear stress of 40 MN/m². Show by a diagram of a rectangular element the complementary shear stresses, and find the stresses on a plane making 25° with the axis of the beam. What are the maximum direct stresses in the material? Indicate on your diagram the planes on which these stresses act.

Ans. (1) shear, 25·71 MN/m²; direct, 30·64 MN/m² (tensile or compressive depending on the direction in which the 25° angle is measured); (2) 40 MN/m² (tensile) and 40 MN/m² (compressive).

CHAPTER 13

Stress and Strain—2

13.1 COMPLEX STRESSES

In practice components are often acted on simultaneously by two or more stresses so that calculations based on simple stress do not apply. Rarely, for example, is a material subjected to simple shear. Even with a circular shaft in torsion, there is usually some lateral loading as well, either as a result of the weight of the shaft or from forces associated with the torque, so that bending stresses are also present. Moreover (see Chapter 12), simple stresses are simple only when perpendicular planes are considered. In what follows, the analysis of simple stresses will be extended to the more general case.

Stress analysis will be restricted to two-dimensional stress systems, which means that an elementary cube of material has forces of various kinds on four of its faces but no force of any kind acting on the remaining two (opposite) faces. Real materials are, of course, three-dimensional, and a complete analysis would consider stresses acting in all three dimensions. For practical purposes, however, many stress systems are two-dimensional. A common example is that of a circular shaft in torsion which is often subject to bending forces, resulting in a combination of direct and shear stresses. At any given point in the shaft, both stresses will act in the same plane, and at right angles to this plane (i.e. radially) there will be no stress.

Analysis of two-dimensional systems

Fig. 13.1 (*a*) shows rectangular block of material forming part of a component in which stresses are two-dimensional. The block is considered to be infinitely small, so that the stresses in it may be referred to as those "at a point within the material." The face ABCD lies in the plane of the stress system, so that there are no forces acting at right angles to the plane of the paper. Forces will act on faces AB, BC, CD and DA, and each may be resolved into components normal and tangential to the face concerned (Fig. 13.1 (b)). For equilibrium these forces must comprise

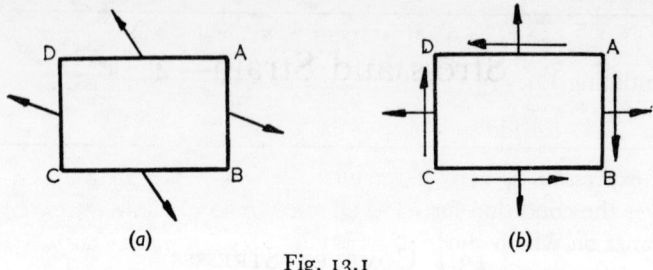

(a)   (b)

Fig. 13.1

two pairs of equal and opposite forces and two equal and opposing couples, so that the block ABCD is acted on by two direct stresses and, in addition, by complementary shear stresses.

The direct stresses acting on the horizontal and vertical faces are designated σ_x and σ_y respectively, the shear stress as τ (Fig. 13.2 (a)), and the stresses on a plane such as CP, inclined at an angle θ to the horizontal face CB, as σ_θ (direct) and τ_θ (shear).

The equilibrium of the prism PBC may now be considered. If the prism is of unit thickness, the areas of its faces will be equal to the lengths

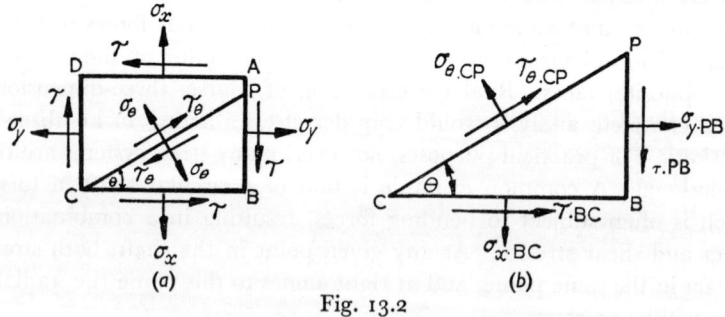

(a)   (b)

Fig. 13.2

of the sides of the triangle PBC and the forces acting on the prism will be those shown in Fig. 13.2 (b). Resolving forces perpendicular to the face PB,

$$\sigma_\theta \, \mathrm{CP} = \sigma_x \, \mathrm{BC} \cos \theta + \sigma_y \, \mathrm{PB} \sin \theta + \tau \, \mathrm{BC} \sin \theta + \tau \, \mathrm{PB} \cos \theta$$

$$\therefore \ \sigma_\theta = \sigma_x \cos^2 \theta + \sigma_y \sin^2 \theta + \tau \cos \theta \sin \theta + \tau \sin \theta \cos \theta$$

$$= \tfrac{1}{2}\sigma_x(1 + \cos 2\theta) + \tfrac{1}{2}\sigma_y(1 - \cos 2\theta) + \tau \sin 2\theta$$

$$= \tfrac{1}{2}(\sigma_x + \sigma_y) + \tfrac{1}{2}(\sigma_x - \sigma_y) \cos 2\theta + \tau \sin 2\theta \qquad (13.1)$$

In the same way, resolving forces parallel to the face PB,

$$\tau_\theta = \tfrac{1}{2}(\sigma_x - \sigma_y)\sin 2\theta - \tau \cos 2\theta \qquad (13\cdot2)$$

From Eq. (13.2), it follows that

$$\tau_\theta = 0 \quad \text{if} \quad \tan 2\theta = 2\tau/(\sigma_x - \sigma_y) \qquad (13\cdot3)$$

Differentiating Eq. (13.1) with respect to θ,

$$d\sigma_\theta/d\theta = -(\sigma_x - \sigma_y)\sin 2\theta + 2\tau \cos 2\theta$$

This expression is zero if $\tan 2\theta = 2\tau/(\sigma_x - \sigma_y)$, so that Eq. (13.3) also gives the condition for σ_θ to be a maximum or minimum. It means that planes on which shear stress is zero are also planes on which direct stress is either a maximum or a minimum. The stresses on these planes may be determined if $\cos 2\theta$ and $\sin 2\theta$ are known, which may be done graphically (i.e., by drawing right-angled triangles).

Maximum and minimum direct stress. Eq. (13.3) is satisfied by two values of 2θ (Fig. 13.3), but the values of 2θ differ by 180°, and so represent values of θ differing by 90° or two planes at right angles to each other. For the angle in the first quadrant,

$$\sin 2\theta = \frac{2\tau}{\sqrt{(\sigma_x - \sigma_y)^2 + 4\tau^2}} \qquad (13\cdot4)$$

and

$$\cos 2\theta = \frac{\sigma_x - \sigma_y}{\sqrt{(\sigma_x - \sigma_y)^2 + 4\tau^2}} \qquad (13\cdot5)$$

Substituting Eqs. (13.4) and (13.5) in Eq. (13.1), the value of σ_θ obtained represents the maximum value of the direct stress.

$$\therefore \ \sigma_{\theta\,max} = \tfrac{1}{2}(\sigma_x + \sigma_y) + \frac{\tfrac{1}{2}(\sigma_x - \sigma_y)(\sigma_x - \sigma_y)}{\sqrt{(\sigma_x - \sigma_y)^2 + 4\tau^2}} + \frac{2\tau^2}{\sqrt{(\sigma_x - \sigma_y)^2 + 4\tau^2}}$$

$$= \tfrac{1}{2}(\sigma_x + \sigma_y) + \frac{(\sigma_x - \sigma_y)^2 + 4\tau^2}{2\sqrt{(\sigma_x - \sigma_y)^2 + 4\tau^2}}$$

$$= \tfrac{1}{2}(\sigma_x + \sigma_y) + \tfrac{1}{2}\sqrt{(\sigma_x - \sigma_y)^2 + 4\tau^2} \qquad (13\cdot6)$$

For the angle 2θ in the third quandrant, $\sin 2\theta$ and $\cos 2\theta$ will be as in Eqs. (13.4) and (13.5) but negative in sign. The result will therefore correspond to a minimum of direct stress,

$$\sigma_{\theta\,min} = \tfrac{1}{2}(\sigma_x + \sigma_y) - \tfrac{1}{2}\sqrt{(\sigma_x - \sigma_y)^2 + 4\tau^2} \qquad (13\cdot7)$$

Maximum shear stress. Differentiating Eq. (13.2) with respect to θ,

$$d\tau_\theta/d\theta = (\sigma_x - \sigma_y)\cos 2\theta + 2\tau \sin 2\theta$$

For a maximum, this must be zero, or

$$\tan 2\theta' = -(\sigma_x - \sigma_y)/2\tau \qquad (13.8)$$

From Eq. (13.3),

$$\tan 2\theta' = -\cot 2\theta$$

This relationship indicates (Fig. 13.4) that the angles 2θ and $2\theta'$ differ by 90°, or θ and θ' differ by 45°. Planes of maximum shear stress are inclined at 45° to planes of maximum (or minimum) direct stress.

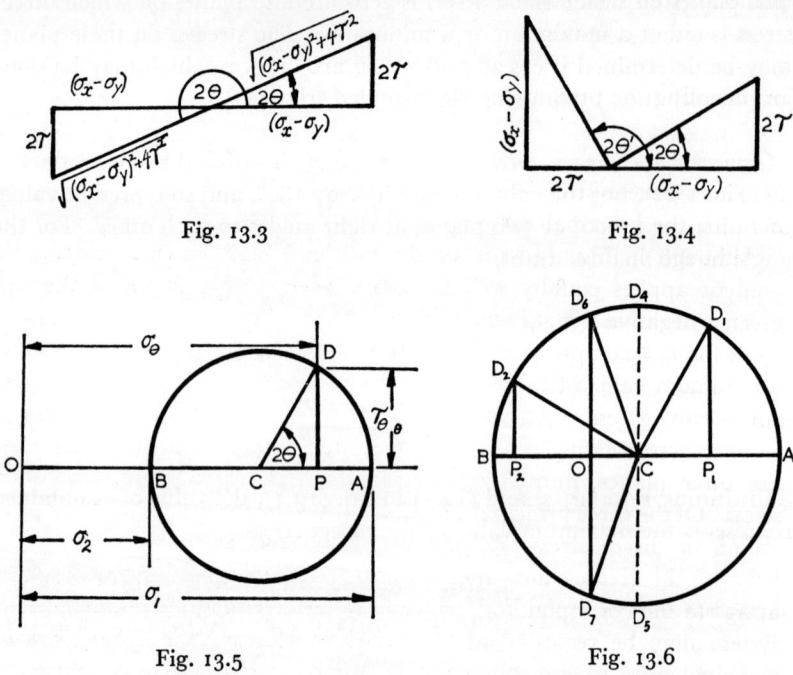

Fig. 13.3   Fig. 13.4

Fig. 13.5   Fig. 13.6

Eq. (13.8) is satisfied by two values of $2\theta'$ (in the second and fourth quadrants respectively). If values of $\sin 2\theta'$ and $\cos 2\theta'$ are found, by constructing triangles, and substituted in Eq. (13.2), it is found that

$$\tau_{\theta\,(\max)} = \tfrac{1}{2}\sqrt{(\sigma_x - \sigma_y)^2 + 4\tau^2} \qquad (13.9)$$

and

$$\tau_{\theta\,(\min)} = -\tfrac{1}{2}\sqrt{(\sigma_x - \sigma_y)^2 + 4\tau^2} \qquad (13.10)$$

Eqs. (13.9) and (13.10) are identical except for sign. Since the corresponding values of $2\theta'$ differ by 180°, they represent equal and opposite shear stresses on two planes at right angles to each other, or two complementary shear stresses.

Principal planes and stresses

This analysis shows that in any two-dimensional stress system, there exist two mutually perpendicular planes on which there is only direct stress (i.e., no shear stress) and that the stresses on these planes are the maximum and minimum direct stresses in the material. These planes are termed the *principal planes* and the stresses on them the *principal stresses.*

Moreover, maximum shear stress occurs on planes at 45° to the principal planes. From Eqs. (13.6), (13.7) and (13.9),

$$\tau_{\theta(max)} = \tfrac{1}{2}[\sigma_{\theta(max)} - \sigma_{\theta(min)}]$$

or, denoting the principal stresses by σ_1 and σ_2,

$$\tau_{\theta(max)} = \tfrac{1}{2}(\sigma_1 - \sigma_2) \tag{13.11}$$

Maximum shear stress is therefore numerically equal to *half the difference* between the principal stresses.

Although in Fig. 13.2 (*a*) all direct stresses are shown as tensile, the analysis applies equally well to compressive stresses provided they are given a negative sign. Thus, if in any system, Eq. (13.1) gives a negative value for σ_θ, a compressive stress is indicated. $\sigma_{\theta(min)}$ may represent either a minimum value of tensile stress or a (numerically) maximum value of compressive stress.

The corresponding result for a three-dimensional system is that there are three planes, mutually perpendicular, on which there is no shear stress. Of the three principal stresses, two are the maximum and minimum values of direct stress. Maximum shear stress is half the difference between the greatest and the least principal stresses and occurs on planes at 45° to the corresponding principal planes. Thus the two-dimensional system may be regarded as a three-dimensional system in which one principal stress is zero. Similarly, a "simple direct stress" is a system in which two of the three principal stresses are zero.

Mohr's stress circle

If, in any system, it is desired to find the principal stresses or to find the stresses on a particular plane, this may be done either by considering a prism of material or by using the results of a general analysis (Eqs. (13.1) to (13.10)). An alternative is to use a graphical construction which may take the form of an ellipse, or, as shown by Otto Mohr, a circle (Fig. 13.5).

The case of two tensile principal stresses σ_1 and σ_2 is dealt with as follows:

(i) On a horizontal axis distances OA and OB from a point O, represent stresses σ_1 and σ_2 to a suitable scale.

(ii) AB is bisected in C and, with C as centre, a circle is drawn which passes through A and B.

(iii) The stresses σ_θ and τ_θ on a plane making an angle θ (measured anti-clockwise) with the plane of σ_1 may be found by constructing a radius CD at an angle 2θ to CA (also measured anti-clockwise) and a perpendicular DP from the point D to the horizontal axis.

The direct stress σ_θ is now represented (to scale) by OP and the shear stress τ_θ (to the same scale) by PD.

This can be shown as follows. In Fig. 13.5,

$$OP = OC + CP$$
$$= \tfrac{1}{2}(OA + OB) + CD \cos 2\theta$$

CD is a radius of the circle and BA a diameter, so that

$$CD = \tfrac{1}{2}BA = \tfrac{1}{2}(OA - OB)$$
$$OP = \tfrac{1}{2}(OA + OB) + \tfrac{1}{2}(OA - OB) \cos 2\theta$$

or $\qquad \sigma_\theta = \tfrac{1}{2}(\sigma_1 + \sigma_2) + \tfrac{1}{2}(\sigma_1 - \sigma_2) \cos 2\theta$

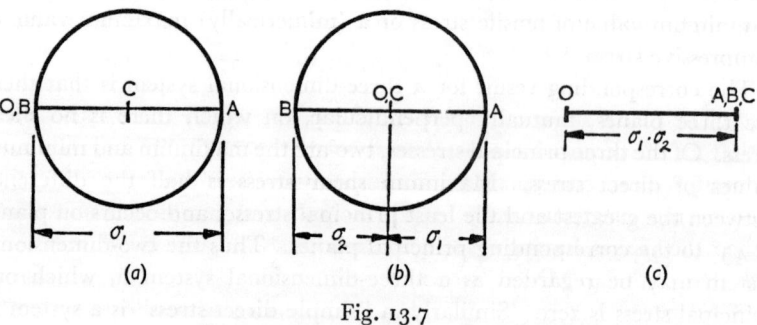

(a)   (b)   (c)

Fig. 13.7

This can be obtained from Eq. (13.1) by putting

$$\sigma_x = \sigma_1, \ \sigma_y = \sigma_2 \quad \text{and} \quad \tau = 0$$

so that Mohr's construction is valid.

Further $\qquad PD = CD \sin 2\theta = \tfrac{1}{2}(OA - OB) \sin 2\theta$

or $\qquad \tau_\theta = \tfrac{1}{2}(\sigma_1 - \sigma_2) \sin 2\theta$

This is equivalent to (13.2) with

$$\sigma_x = \sigma_1, \sigma_y = \sigma_2 \quad \text{and} \quad \tau = 0$$

so that the expressions are seen to be identical.

Principal stresses of opposite sign. If one of the principal stresses is compressive (and regarded as negative), the construction of the circle diagram is that previously described, with the difference that OB is measured in the opposite direction to OA so that the point O is now within the circle.

Example. *At a point in a material, the principal stresses are 40 MN/m²* *(tensile) and 20 MN/m² (compressive). Construct the Mohr stress circle and find* *the stresses on planes making angles of (a) 30°, (b) 75° with the plane of the* *40 MN/m² stress. Find also (c) the maximum shear stress, and (d) the positions* *of the planes on which there is no direct stress (or on which the stress is pure shear).*

The stress circle (Fig. 13.6) is approximately to scale and planes (*a*) and (*b*) are represented by points D_1 and D_2. Values of stress are:

(*a*) $\sigma_\theta = 25 \cdot 0$ MN/m² (tensile)
 $\tau_\theta = 26 \cdot 0$ MN/m².
(*b*) $\sigma_\theta = 16 \cdot 0$ MN/m² (compressive)
 $\tau_\theta = 15 \cdot 0$ MN/m².

(*c*) Maximum distances measured vertically from the axis of the diagram are those to D_4 and D_5, which represent the planes of maximum shear stress, and $\tau_{\theta \, max} = 30$ MN/m².

(*d*) The direct stress on any plane is represented by the distance OP, so that for the required planes, P must coincide with O. Two such planes are seen to exist, corresponding to the points D_6 and D_7.

From the diagram, angle $ACD_6 = 110°$ and angle $ACD_7 = 250°$. Hence the planes having no direct stress make angles of 55° and 125° with that of the 40 MN/m² stress.

Special cases. An advantage of the stress circle is that it displays *all* the available information relating to the stresses in a system, so that types of stress systems may be recognised from their stress circles. Three cases produce distinctive diagrams.

(i) *Simple direct stress* corresponds to a system with only one principal stress, the other being zero, when points O and B coincide (Fig. 13.7 (*a*)).
(ii) *Simple shear stress* is equivalent to two equal principal stresses of opposite sign, so that C coincides with O (Fig. 13.7 (*b*)).

(iii) *Equal and similar principal stresses,* $\sigma_1 = \sigma_2$, *when points A, B and C coincide and the circle is reduced to a point.*

In such a system, all planes have the same direct stress which is unaccompanied by shear stress.* Equal and similar principal stresses occur, for example, in thin spherical shells under internal pressure and in shafts on which collars have been shrunk.

When the principal planes are not known the constuction of the circle can be carried out by the method of the following example.

Example. *At a point in a material, the stresses on two mutually perpendicular planes are* 50 MN/m² *and* 20 MN/m², *both tensile, accompanied by complementary shear stresses of* 10 MN/m². *Find the positions of the principal planes and the values of the principal stresses.*

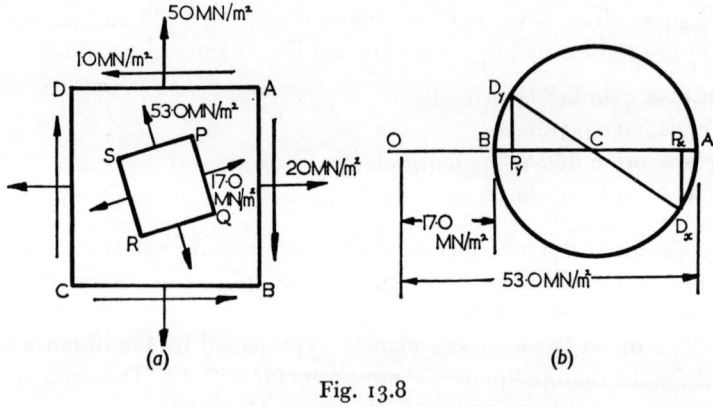

(a)   (b)

Fig. 13.8

Fig. 13.8 (*a*) shows the stresses acting on a small cube ABCD. The given information means that planes x and y are separated by an angle of 90° and that

$$\sigma_x = +50 \text{ MN/m}^2, \quad \tau_x = -10 \text{ MN/m}^2, \quad \sigma_y = +20 \text{ MN/m}^2$$

and $$\tau_y = +10 \text{ MN/m}^2$$

Thus points D_x and D_y may be constructed and, since these must be diametrally opposite, the diagram may be completed (Fig. 13.8 (*b*)).

---

* This does not mean that there is no shear stress in the material but only that there is no shear stress on planes perpendicular to that of the stress system. In three-dimensional systems, this is a case in which the third principal stress σ_3 is zero and hence $\tau_{\theta \text{ (max)}} = \frac{1}{2}(\sigma_1 - \sigma_3) = \frac{1}{2}\sigma_1$. Only a system with all three principal stresses equal and similar would have no shear stress on any plane.

Angles D_xCA and D_xCB are $34°$ and $146°$ respectively and so the angles made by the principal planes with plane x (i.e., that of the 50 MN/m² stress) are $17°$ (measured anticlockwise) and $73°$ (measured clockwise). The principal stresses, represented by OA and OB, are $53\cdot0$ MN/m² and $17\cdot0$ MN/m² respectively (both tensile). These are shown acting on the cube of material PQRS in Fig. 13.8 (*a*).

In this example, positive shear stress has by convention been taken as that which, acting alone, would tend to rotate an element clockwise, and is represented by the upper half of the circle diagram.

13.2 COMBINED STRESSES

Thin cylinders

Cylindrical shells are usually considered *thin* if their thickness is not more than 1/20 their diameter, in which case (i) radial stresses may be neglected and (ii) circumferential stresses may be considered uniform.

In the equilibrium of a longitudinal section of a cylinder (Fig. 13.9 (*a*)), if the ends have no effect, the forces acting are those due to the internal pressure *p* and the circumferential stress σ_C. Equating these for a section of axial length *l*,

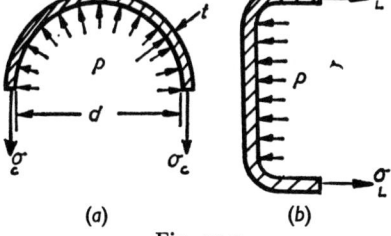

$$p \, dl = \sigma_C \times 2lt$$
$$\therefore \ \sigma_C = pd/2t \qquad (13.12)$$

(*a*)   (*b*)

Fig. 13.9

In the equilibrium of that part of the cylinder on one side of a transverse plane (Fig. 13.9 (*b*)), equating forces gives

$$p \times \tfrac{1}{4}\pi d^2 = \sigma_L \times \pi \, dt$$
$$\therefore \ \text{Longitudinal stress } \sigma_L = pd/4t \qquad (13.13)$$

Example. *A compressed air tank has a diameter of 300 mm and is 10 mm thick. Find, for the normal working pressure of 4 MN/m² (40 bar), (a) the maximum shear stress in its cylindrical wall, (b) the direct stress which accompanies this shear stress.*

From Eq. (13.12),

$$\sigma_C = pd/2t$$
$$= 4 \times 10^6 \times 0\cdot3/2 \times 0\cdot01$$
$$= 60 \times 10^6 \text{ N/m}^2 = 60 \text{ MN/m}^2 \text{ (tensile)}$$

Similarly, from Eq. (13.13),

$$\sigma_L = pd/4t = 30 \text{ MN/m}^2 \text{ (tensile)}$$

Since no shear stresses act on the planes of σ_C and σ_L, it follows that they are the principal stresses in the cylindrical wall. The Mohr circle may be constructed (see Fig. 13.10). The points D_1 and D_2 represent planes of maximum shear stress, and the diagram shows that the maximum shear stress is 15 MN/m² and that it is accompanied by a direct stress of 45 MN/m² (tensile).

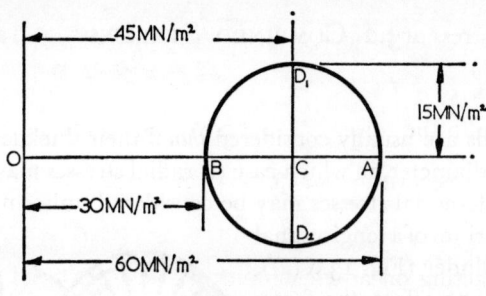

Fig. 13.10

Combined torsion and thrust

It may be shown that if a circular shaft is subjected to a torque T, shear stresses are produced, and

$$\tau/r = T/J = G\theta/l \qquad (13.14)$$

where τ = Shear stress at radius r,
   T = Torque transmitted,
   J = Polar second moment of area of the shaft cross-section ($=\pi d^4/32$ for a solid shaft of diameter d),
   G = Modulus of rigidity, and
   θ = Angle of twist over a length of shaft l.

Thus the shear stresses in a shaft vary from zero at the centre to a maximum at its surface. If there is also a longitudinal force on the shaft so that a uniform direct stress is combined with the torsional stresses, the combinations of stress will range from direct stress only at the centre to direct stress plus maximum shear stress at the surface. It is usually necessary to find maximum values of stress, so that only the stresses at the surface need be considered.

Example. *A ship's propeller shaft is* 400 mm *diameter and runs at* 120 rev/min. *The power transmitted is* 9 MW *and the thrust of the screw is* 850 kN. *Find the maximum compressive stress in the shaft.*

$$\text{Power transmitted by torque} = T\omega$$
$$\therefore 9 \times 10^6 = T \times (120 \times 2\pi)/60$$
$$\therefore T = 716 \times 10^3 \text{ Nm} = 0.716 \text{ MNm}$$

From Eq. (13.14),

$$\tau = Tr/J = 32\, Tr/\pi d^4$$

$$\therefore \text{ Shear stress at surface of shaft} = 32 \times 0.716 \times 0.2/\pi(0.4)^4$$
$$= 57.0 \text{ MN/m}^2$$

If the stress due to the thrust is evenly distributed over the cross-section,

$$\text{Compressive stress } \sigma = 850 \times 10^3/\pi(0.2)^2$$
$$= 6760 \times 10^3 \text{ N/m}^2 = 6.76 \text{ MN/m}^2$$

The stresses acting on an elementary block of material at the surface of the shaft are as shown in Fig. 13.11 (*a*) and from the Mohr stress circle, Fig. 13.11 (*b*), the maximum compressive stress (i.e., the principal stress of negative sign) is 60.5 MN/m².

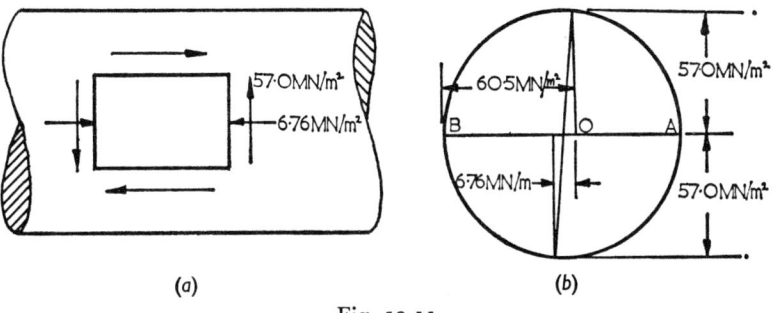

(*a*)   (*b*)

Fig. 13.11

Combined torsion and bending

The bending of a shaft involves (see Chapter 15) stresses ranging from maximum tensile (on the convex side) to maximum compressive (on the concave side) and if these are combined with the stresses due to torsion, various combinations will result. Maximum stresses will be found at the surface of the shaft and at the maximum distance from the neutral axis.

It is usually necessary to find the maximum tensile stress and maximum shear stress, which means an investigation along the line of intersection of the shaft surface with the plane of bending, on the convex side.

Example. *A shaft* 20 mm *diameter transmits a torque of* 100 Nm *and is acted on by transverse forces which give rise to a maximum bending moment of* 30 Nm. *Find (a) the maximum tensile stress, and (b) the maximum shear stress. Show by a diagram the positions of the principal planes at a point where these stresses occur.*

From Eq. (13.14),

$$\tau = Tr/J = 32\,Tr/\pi d^4$$

∴ at shaft surface,

$$\tau = 32 \times 100 \times 0 \cdot 01/\pi(0 \cdot 02)^4$$
$$= 6 \cdot 37 \times 10^7 \text{ N/m}^2 = 63 \cdot 7 \text{ MN/m}^2$$

From Eq. (15.1),

$$\sigma = My/I = 64My/\pi d^4$$

for a solid circular beam.

Therefore, at maximum distance from the neutral axis,

$$\sigma = 64 \times 30 \times 0 \cdot 01/\pi(0 \cdot 02)^4$$
$$= 3 \cdot 82 \times 10^7 \text{ N/m}^2 = 38 \cdot 2 \text{ MN/m}^2$$

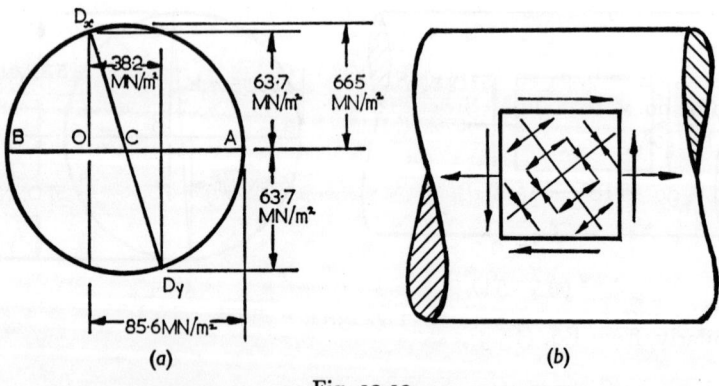

Fig. 13.12

From the Mohr stress circle (approximately to scale, Fig. 13.12 (*a*)), the maximum tensile stress is 85·6 MN/m² and the maximum shear stress 66·5 MN/m². The angle *DyCA* is 74°, so that the principal plane having tensile stress makes an angle of 37° with the plane on which the bending stress acts (Fig. 13.12 (*b*)).

Example. *A solid circular shaft is to transmit a torque of 1·6 kNm, and will be subjected to a maximum bending moment of 1 kNm. Calculate the required diameter (a) if the maximum tensile stress is not to exceed 90 MN/m², (b) if the maximum shear stress is not to exceed 45 MN/m².*

From Eq. (13.14), $\tau = Tr/J$. But $J = \pi d^4/32$, and at the shaft surface, $r = d/2$,

$$\therefore \tau = 16T/\pi d^3$$
$$= 16 \times 1\cdot6 \times 10^3/\pi d^3$$
$$= 8\cdot15 \times 10^3/d^3 \ \mathrm{N/m^2}$$

From Eq. (15.1),

$$\sigma = My/I$$

so that at maximum distance from the neutral axis ($y = d/2$),

$$\sigma = 32M/\pi d^3$$
$$= 32 \times 10^3/\pi d^3$$
$$= 10\cdot18 \times 10^3/d^3 \ \mathrm{N/m^2}$$

A Mohr stress circle could now be constructed to a scale in which a stress of $10^3/d^3 \ \mathrm{N/m^2}$ is represented by, say, 1 cm. Alternatively, from Eq. (13.6),

$$\sigma_{\theta \, (\mathrm{max})} = \tfrac{1}{2}(\sigma_x + \sigma_y) + \tfrac{1}{2}\sqrt{(\sigma_x - \sigma_y)^2 + 4\tau^2}$$

For (a), $\sigma_x = 0$, $\sigma_y = +10\cdot18 \times 10^3/d^3$, $\tau = 8\cdot15 \times 10^3/d^3$ and $\sigma_{\theta \, (\mathrm{max})}$ is to be $90 \times 10^6 \ \mathrm{N/m^2}$. Substituting these values and dividing throughout by 10^3,

$$90 \times 10^3 = \tfrac{1}{2}(10\cdot18/d^3) + \tfrac{1}{2}\sqrt{(-10\cdot18/d^3)^2 + 4(8\cdot15/d^3)^2}$$
$$= 14\cdot7/d^3$$
$$\therefore d = 0\cdot0547 \ \mathrm{m} = 54\cdot7 \ \mathrm{mm}$$

Similarly, from Eq. (13.9),

$$\tau_{\theta \, (\mathrm{max})} = \tfrac{1}{2}\sqrt{(\sigma_x - \sigma_y)^2 + 4\tau^2}$$

For case (b),

$$45 \times 10^3 = \tfrac{1}{2}\sqrt{(-10\cdot18/d^3)^2 + 4(8\cdot15/d^3)^2}$$
$$= 9\cdot61/d^3$$
$$\therefore d = 0\cdot0598 \ \mathrm{m} = 59\cdot8 \ \mathrm{mm}$$

13.3 Relationships between Stress and Strain

Longitudinal and lateral strains

If a bar of material is stressed in simple tension, there will be two kinds of strain (Fig. 13.13).

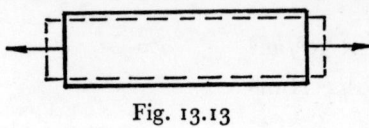

Fig. 13.13

1. The bar will extend in the direction of the applied stress σ so that there will be a *longitudinal strain* ε. Provided the stress is not greater than some limit, ε is proportional to σ and $\varepsilon = \sigma/E$.

2. The bar will contract in all directions at right angles to that of σ so that there will be a *lateral strain* which is a contraction when the longitudinal strain is an extension and vice versa.

Within the limit of proportionality, the ratio of lateral strain to longitudinal strain is a constant for each material and is known as *Poisson's ratio v*. The ratio is a positive number equal to the numerical ratio of the strains and ignoring the fact that they are of opposite signs. For most metals, σ lies between $\frac{1}{4}$ and $\frac{1}{3}$. In general, the lateral strain is given by $-v\sigma/E$, where the negative sign indicates that the lateral strain is in the opposite sense to that normally associated with the stress σ.

Principal stresses and principal strains

In any stress system, there are three mutually perpendicular planes on which only direct stresses act. If the three principal stresses are σ_1, σ_2 and v_3 and their effect on a cube of material with faces are parallel to the principal planes is considered (see Fig. 13.14), the strain in the direction of one principal stress will be the algebraic sum of the longitudinal strain due to that stress and the lateral strains due to the other two principal stresses. Hence the strain in the direction of σ_1,

$$\varepsilon_1 = \sigma_1/E - v\sigma_2/E - v\sigma_3/E$$
$$= [\sigma_1 - v(\sigma_2 + \sigma_3)]/E \qquad (13.15)$$

Similarly,
$$\varepsilon_2 = [\sigma_2 - v(\sigma_1 + \sigma_3)]/E \qquad (13.16)$$

and
$$\varepsilon_3 = [\sigma_3 - v(\sigma_1 + \sigma_2)]/E \qquad (13.17)$$

Although stresses in Fig. 13.14 are shown as tensile, Eqs. (13.15) to (13.17) apply to systems in which one (or more) of the principal stresses is compressive, so long as the convention is used that tensile stress and extension are positive, compressive stress and contraction negative.

The cube of material in Fig. 13.14 will be deformed but will remain rectangular, so that there is no shear strain in the directions of ε_1, ε_2 and ε_3 It will later be seen that, for a two-dimensional system, ε_1 and ε_2 represent maximum and minimum values of direct strain. There is an obvious analogy with the properties of the principal stresses, and the strains ε_1, ε_2 and ε_3 are termed *principal strains*.

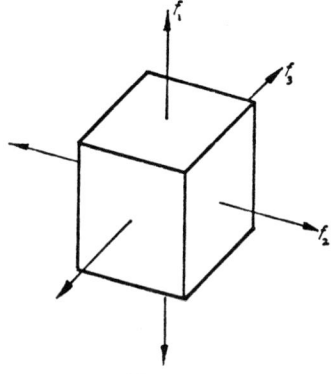

Fig. 13.14

Example. *In a previous example, a cylindrical shell was found to have a longitudinal stress* $\sigma_L = 30$ MN/m² *and a circumferential stress* $\sigma_C = 60$ MN/m². *If, for this material,* $E = 200$ GN/m² *and* $\nu = 0.3$, *find the proportional changes in length, diameter and thickness.*

The three principal stresses are $\sigma_1 = +30$ MN/m², $\sigma_2 = +60$ MN/m² and $\sigma_3 = 0$.

∴ Proportional change in length

$$= \varepsilon_1 = [\sigma_1 - \nu(\sigma_2 + \sigma_3)]/E$$
$$= [30 \times 10^6 - 0.3(60 \times 10^6 + 0)]/200 \times 10^9$$
$$= +6 \times 10^{-5} \text{ (i.e., an increase)}$$

Proportional change in diameter

$$= \text{Proportional change in circumference}$$
$$= \varepsilon_2 = [\sigma_2 - \nu(\sigma_1 + \sigma_3)]/E$$
$$= [60 \times 10^6 - 0.3(30 \times 10^6 + 0)]/200 \times 10^9$$
$$= +25.5 \times 10^{-5} \text{ (i.e., an increase)}$$

Proportional change in thickness

$$= \varepsilon_3 = [\sigma_3 - \nu(\sigma_1 + \sigma_2)]/E$$
$$= [0 - 0.3(30 \times 10^6 + 60 \times 10^6)]/200 \times 10^9$$
$$= -13.5 \times 10^{-5} \text{ (i.e., a decrease)}$$

13.4 COMPLEX STRAINS

Two-dimensional strain analysis

A small block of material ABCD forms part of a component subjected to two-dimensional stresses, the face ABCD being in the plane of the stress system and the faces AB and BC being parallel to the principal planes. Principal strains occur in the directions of the principal stresses or at right angles to the faces of the block. If these strains are ε_1 horizontally and ε_2 vertically (both positive), the block becomes $A_1B_1C_1D$ (Fig. 13.15 (a)) and the objective is to find the strain ε_θ at an angle θ with the horizontal, that is the proportional change in the diagonal DB. It is also important to find the rotation β of this diagonal relative to the principal planes, for this can be related to the shear strain γ_θ.

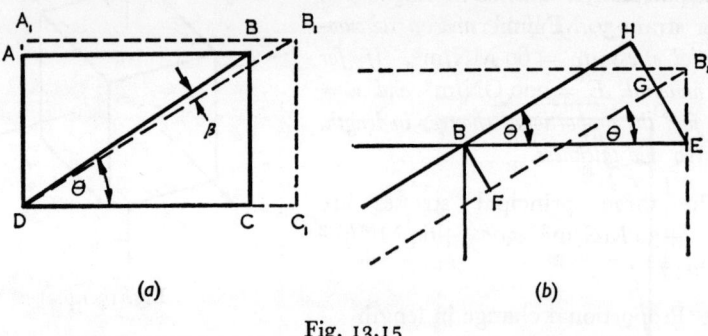

(a) (b)

Fig. 13.15

At the corner B (Fig. 13.15 (b)), for most materials, the strains are small so that B_1F may be considered parallel to BD. (This analysis does not apply to materials such as rubber.) Then,

$$BE = \varepsilon_1 \times AB \quad \text{and} \quad EB_1 = \varepsilon_2 \times CB$$

$$FB_1 = FG + GB_1 = BH + GB_1$$

$$= BE \cos \theta + EB_1 \sin \theta$$

$$= \varepsilon_1 . AB \cos \theta + \varepsilon_2 . CB \sin \theta$$

$$\therefore \varepsilon_\theta = FB_1/DB = \varepsilon_1 \cos^2 \theta + \varepsilon_2 \sin^2 \theta$$

$$= \tfrac{1}{2}(\varepsilon_1 + \varepsilon_2) + \tfrac{1}{2}(\varepsilon_1 - \varepsilon_2) \cos 2\theta \tag{13.18}$$

$$BF = HG = HE - GE$$

$$= BE \sin \theta - EB_1 \cos \theta$$

$$= \varepsilon_1 . AB \sin \theta - \varepsilon_2 . CB \cos \theta$$

$$\therefore \beta = BF/DB = \varepsilon_1 \cos \theta \sin \theta - \varepsilon_2 \sin \theta \cos \theta$$

$$= \tfrac{1}{2}(\varepsilon_1 - \varepsilon_2) \sin 2\theta \tag{13.19}$$

Relationship between β and γ. The shear strain in a material is usually regarded (Fig. 13.16 (*a*)) either as the ratio AA_1/BA or the angle γ. If, however, it is stipulated that the principal planes (i.e. the diagonals AC and DB) must not rotate, the situation is that shown in Fig. 13.16 (*b*). Both figures show the same deformation. The only difference is that in (*a*), the line BC is fixed while in (*b*) the point C is fixed and the diagonals do not rotate. The angle γ is now seen as the sum of the two angles β which correspond to the rotations of the sides of the cube relative to the principal planes. Hence in Eq. (13.19), the angle β represents half the shear strain γ_θ. Furthermore, the rotation of the sides BC and DA (on

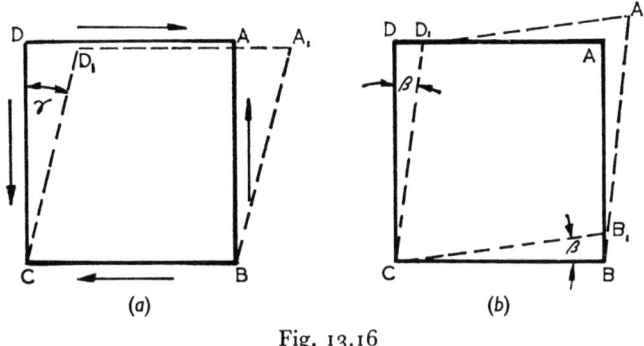

Fig. 13.16

which, by the usual convention, shear stress is positive) is seen to be anti-clockwise. In Fig. 13.15 (*a*), β represents a clockwise rotation and so must correspond to a negative shear strain. In other words,

$$\beta = -\tfrac{1}{2}\gamma_\theta = \tfrac{1}{2}(\varepsilon_1 - \varepsilon_2)\sin 2\theta \qquad (13.20)$$

Mohr's strain circle

For a principal plane where $\sigma_x = \sigma_1$, $\sigma_y = \sigma_2$ and $\tau = 0$, Eqs. (13.1) and (13.2), become

$$\sigma_\theta = \tfrac{1}{2}(\sigma_1 + \sigma_2) + \tfrac{1}{2}(\sigma_1 - \sigma_2)\cos 2\theta$$

and $$\tau_\theta = \tfrac{1}{2}(\sigma_1 - \sigma_2)\sin 2\theta$$

Eqs. (13.18) and (13.20) are seen to be similar in form, principal strains replacing principal stresses. Just as Eqs. (13.1) and (13.2) may be represented by a geometrical construction, so can be Eqs. (13.18) and (13.20). *Mohr's strain circle* is constructed in the same way as Mohr's stress circle, except that horizontal distances now represent direct strains and vertical distances represent *half* the corresponding shear strains. Furthermore, the negative sign in Eq. (13.20) means that the *lower* half of the strain circle represents positive shear strain.

Example. *In a material acted on by stresses only in one plane, the principal strains are $\varepsilon_1 = 5 \times 10^{-4}$, $\varepsilon_2 = 3 \times 10^{-4}$ (both being positive, i.e., extensions). Construct the Mohr strain circle and find the strains in a direction making an angle of 30° with the direction of ε_1. What is the maximum shear strain and where does it occur?*

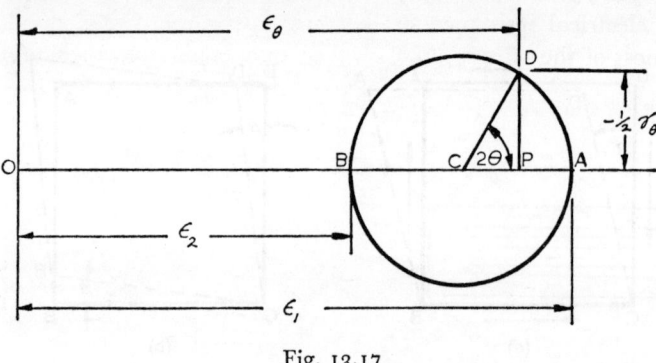

Fig. 13.17

Fig. 13.17 shows the strain circle, approximately to scale. If the 30° angle is measured anticlockwise from the direction of ε_1, the strains in the required direction are

$$\varepsilon_\theta = +4 \cdot 5 \times 10^{-4}$$

$$\gamma_\theta = -1 \cdot 73 \times 10^{-4} \text{ radians}$$

Thus, ε_θ represents an extension and γ_θ a clockwise rotation. Maximum shear strain corresponds to points on the circle vertically above and below its centre. From the diagram, the maximum shear strain is $\pm 2 \times 10^{-4}$ radians and occurs in directions making 45° with those of the principal strains. In other words, a cube of material with faces at 45° with the principal planes would have the maximum shear deformation.

Experimental strain analysis

Strains can be measured mechanically (for example, by an *extensometer* during a simple tensile test) or by attaching an *electrical resistance strain gauge* to the surface of the material.

Such a strain gauge (Fig. 13.18 (*a*)) consists of a length of thin wire bonded to an insulating backing which is firmly cemented in use to the surface of the material so that any strain is faithfully transmitted to the gauge. If the gauge is strained in the direction of its length, the wire is stretched and its electrical resistance is increased. Three factors contribute to this increase. The wire becomes longer, its cross-section is reduced and, because it is in a state of stress, its electrical resistivity increases. A strain of, say, 10^{-3} or 0·1 per cent thus causes a change in resistance somewhat in excess of 0·1 per cent. The ratio $(\delta R/R)/\varepsilon$ is called the *gauge factor* and is usually between 2 and 2·5.

If an electrical resistance strain gauge is strained transversely, or in shear, most of the wire is unaffected so that variations in resistance are

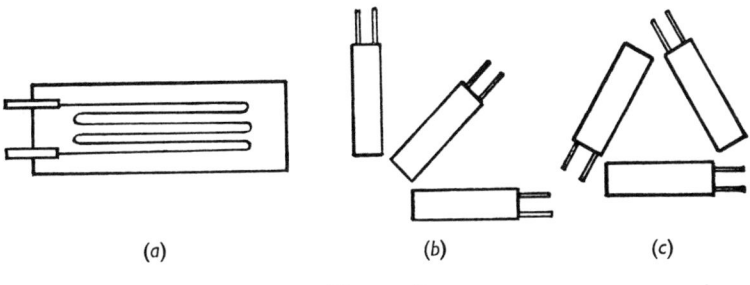

(a) (b) (c)

Fig. 13.18

negligible. Thus the gauge measures only the direct strain in the direction of its length. Changes in resistance are measured using a form of Wheatstone bridge circuit and the resistance of the "active" gauge is usually compared with that of a "dummy" gauge—an identical but unstrained gauge placed so that temperature variations will affect both equally.

Strain gauges can be used to find the principal strains in cases of complex stress. In two-dimensional systems, the magnitudes of the principal strains and the positions of the principal planes will both be unknown, but may be determined by measuring the direct strains in a number of different directions. The smallest such number is three and a *strain rosette* is an assembly of three strain gauges. Theoretically, measurements may be taken in any three directions but those directions used in practice are either the "45° strain rosette" (Fig. 13.18 (*b*)) or the "60° strain rosette" (Fig. 13.18 (*c*)). From the three strains measured by a rosette, the Mohr strain circle may be constructed and, from the principal strains, the principal stresses may be deduced.

Example. *A 45° strain rosette gave the following results: Gauge A, 7×10^{-4} (extension); Gauge B, at 45° (anticlockwise) to A, 3×10^{-4} (extension); Gauge C, at 90° (anticlockwise) to A, 2×10^{-4} (extension). Find the principal strains, the principal stresses and the positions of the principal planes relative to the strain rosette. Assume that $E = 200$ GN/m² and $v = 0.3$.*

Since the strains are measured in directions separated by 45°, they will be represented on the axis of Mohr's strain circle by the projections of three points on the circle separated by 90°. The construction of the circle is as follows:

(i) The axis of the diagram is drawn and, to a suitable scale, ε_A, ε_B and ε_C are represented by OP_a, OP_b and OP_c. Vertical lines are drawn through P_a, P_b and P_c.

(ii) The centre C of the circle is midway between points P_a and P_c. (Since planes A and C are 90° apart, D_aD_c will be a diameter of the circle.)

(iii) The point D_a is found by making P_aD_a equal to CP_b—triangles D_bCP_b and CD_aP_a are congruent.

The strain circle is now drawn (Fig. 13.19 (*a*)), whence the principal strains are $\varepsilon_1 = +742 \times 10^{-6}$ and $\varepsilon_2 = +158 \times 10^{-6}$. The principal stresses may be found by substituting these values in Eqs. (13.15) and (13.16) as follows:

$$742 \times 10^{-6} = (\sigma_1 - 0.3\sigma_2)/200 \times 10^9$$

$$158 \times 10^{-6} = (\sigma_2 - 0.3\sigma_1)/200 \times 10^9$$

Solving these simultaneous equations, $\sigma_1 = +174$ MN/m² and $\sigma_2 = +84$ MN/m² (i.e., both are tensile). In Fig. 13.19 (*a*), the angle ACD_a is 31° so that the angle between strains ε_1 and ε_A (measured anticlockwise) is 15·5°.

The principal stresses will be in the directions of ε_1 and ε_2, so that the principal planes make angles of 74·5° (anticlockwise) and 15·5° (clockwise) with the axis of strain gauge A. Fig. 13.19 (*b*) shows the relative positions of the strain rosette and a block of material having faces parallel to the principal planes.

Example. *A 60° strain rosette gave the following results: Gauge A: 7×10^{-4} (extension), Gauge B, at 60° (anticlockwise) to A: 2×10^{-4} (extension), Gauge C, at 120° (anticlockwise) to A: 10^{-4} (contraction). Find the principal strains and stresses, and the direction of the greater principal stress relative to gauge A if, for the material, $E = 70$ GN/m² and $v = 0.25$.*

The points P_a, P_b and P_c are constructed as in the previous example (P_c being to the left of O since ε_C is negative) and, as before, vertical lines are drawn. It is now necessary to construct a circle such that P_a, P_b and P_c are the projections of three points equally spaced around its circumference. One method (Fig. 13.20) is as follows.

(i) The centre C of the circle is located by

$$OC = \tfrac{1}{3}(OP_a + OP_b + OP_c)$$

(These quantities must be treated algebraically.)

(ii) With centre C, a circle is drawn tangential to one of the verticals, that through P_a in Fig. 13.20.

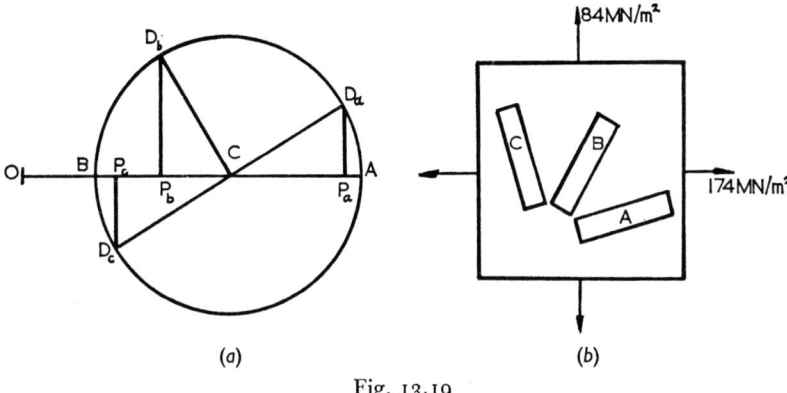

(a) (b)

Fig. 13.19

(iii) With this vertical as base, an equilateral triangle is constructed so that it encloses the circle. The points at which the sides of this triangle intersect the other verticals are points on the strain circle.

From Fig. 13.20, the principal strains are $\varepsilon_1 = +733 \times 10^{-6}$ and $\varepsilon_2 = -198 \times 10^{-6}$. Substituting in Eqs. (13.15) and (13.16),

$$733 \times 10^{-6} = (\sigma_1 - 0.25\sigma_2)/70 \times 10^9$$
$$-198 \times 10^{-6} = (\sigma_2 - 0.25\sigma_1)/70 \times 10^9$$

Solving these simultaneous equations,

$$\sigma_1 = +51.0 \text{ MN/m}^2 \quad \text{and} \quad \sigma_2 -1.1 \text{ MN/m}^2$$

This means that the principal stresses are 51·0 MN/m² (tensile) and 1·1 MN/m² (compressive). In Fig. 13.20, the angle D_aCA is 22°, so that the angle between strains ε_A and ε_1 (or between the axis of gauge A and the direction of the principal stress σ_1) is 11° (measured anticlockwise).

13.5 Relationships Between the Elastic Constants

Relationship between E, G and ν

In simple shear, the principal stresses are equal and opposite. From a consideration of the Mohr stress circle or from Eqs. (12.10) and (12.11),

$$\sigma_1 = \tau \quad \text{and} \quad \sigma_2 = -\tau$$

From Eq. (13.20),

$$-\tfrac{1}{2}\gamma_\theta = \tfrac{1}{2}(\varepsilon_1 - \varepsilon_2) \sin 2\theta$$

or

$$\gamma_\theta = -(\varepsilon_1 - \varepsilon_2) \sin 2\theta$$

On planes subjected to pure shear stress, however, $\gamma = \tau/G$ and in Eq. (13.20), the angle θ is measured (anticlockwise) from the direction of the

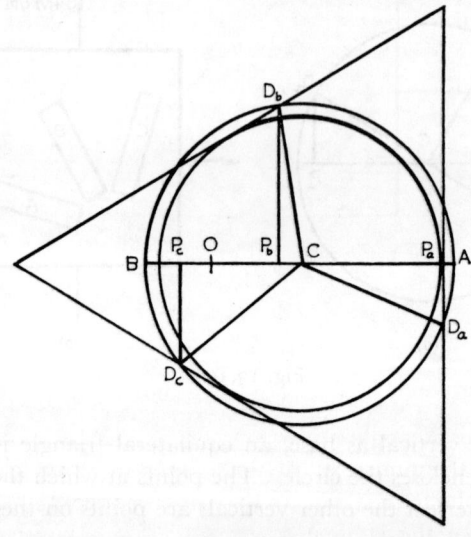

Fig. 13.20

strain ε_1. Fig. 13.21 shows that in a simple shear system, the angle between the positive principal strain and the plane on which the stress is pure shear (and, conventionally, positive) is 135°, so that

$$\gamma_{135°} = -(\varepsilon_1 - \varepsilon_2) \sin 270°$$
$$\therefore \ \tau/G = \varepsilon_1 - \varepsilon_2$$

From Eq. (13.15),

$$\varepsilon_1 = [\sigma_1 - \nu(\sigma_2 + \sigma_3)]/E$$

Here

$$\sigma_1 = \tau, \ \sigma_2 = -\tau \quad \text{and} \quad \sigma_3 = 0$$
$$\therefore \ \varepsilon_1 = (\tau + \nu\tau)/E$$
$$= \tau(1 + \nu)/E$$

Similarly, from Eq. (13.16),

$$\varepsilon_2 = (-\tau - \nu\tau)/E$$

$$= -\tau(1 + \nu)/E$$

Hence $\quad \tau/G = \tau(1 + \nu)/E - [-\tau(1 + \nu)/E]$

$$= 2\tau(1 + \nu)/E$$

$$\therefore \ E = 2G(1 + \nu) \tag{13.21}$$

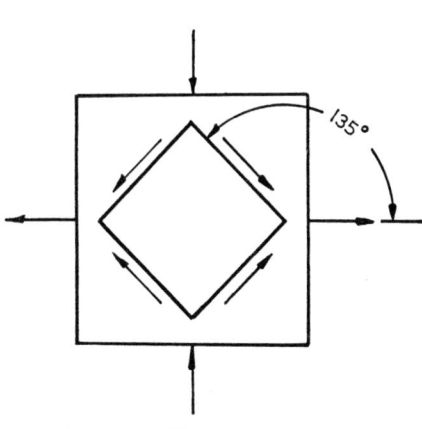

Fig. 13.21

Relationship between *E*, *K* and ν

The bulk modulus K applies only to bodies under hydrostatic pressure, and is defined as the numerical ratio of volumetric strain and hydrostatic pressure. The volumetric strain is the ratio change in volume/original volume and, since the volume V decreases for a positive pressure p, the strain is negative. Thus

$$K = \frac{-p}{(\delta V/V)}$$

Hydrostatic pressure brings about a stress system with all three principal stresses equal and of the same sign, so that the principal strains will be equal.

Let $\varepsilon = \varepsilon_1 = \varepsilon_2 = \varepsilon_3$ and consider an element of material consisting of a cube of side l. Under pressure, each side will become $l(1 + \varepsilon)$, where ε is negative for a body under pressure. Then

$$\delta V = [l(1 + \varepsilon)]^3 - l^3$$

Neglecting products of small quantities,

$$\delta V = 3\varepsilon l^3$$

and

$$\delta V/V = 3\varepsilon$$

But $\sigma_1 = \sigma_2 = \sigma_3 = -p$, so that Eqs. (13.15) to (13.17), give

$$\varepsilon = [-p - \nu(-p - p)]/E$$
$$= -p(1 - 2\nu)/E$$
$$\therefore \ \delta V/V = 3\varepsilon = -3p(1 - 2\nu)/E$$

$$\therefore \ K = \frac{-p}{(\delta V/V)} = E/3(1 - 2\nu)$$

or

$$E = 3K(1 - 2\nu) \tag{13.22}$$

Derived relationships

Transposition of Eqs. (13.21) and (13.22) gives the following relationships:

$$\nu = (E - 2G)/2G \tag{13.23}$$

$$\nu = (3K - E)/6K \tag{13.24}$$

Hence

$$(E - 2G)/2G = (3K - E)/6K$$

from which

$$E = 9KG/(G + 3K) \tag{13.25}$$

and

$$K = GE/(9G - 3E) \tag{13.26}$$

Example. *Typical values of the moduli of elasticity and rigidity for steel are* $E = 200 \ GN/m^2$ *and* $G = 80 \ GN/m^2$. *Calculate the values of the bulk modulus and Poisson's ratio.*

From Eq. (13.26),

$$K = GE/(9G - 3E)$$
$$= 200 \times 80/(9 \times 80 - 3 \times 200)$$
$$= 133.3 \ GN/m^2$$

From Eq. (13.23),

$$\nu = (E - 2G)/2G$$
$$= (200 - 2 \times 80)/2 \times 80$$
$$= 0.25$$

PROBLEMS

For tutorials

1. For a two-dimensional stress system, the stresses and strains in a material may be represented graphically by Mohr's stress circle and strain circle. The material concerned usually obeys Hooke's law and the strains are small. Are these conditions essential? State, giving reasons, whether each of the constructions would be valid for (a) a material in which strains were not proportional to stresses, (b) a material (such as rubber) in which the strains can exceed unity.

2. Materials of exceptional strength have been made by embedding "whiskers" (that is, long thin crystals of sapphire and other materials) in a metal matrix. A similar composite material is glass-fibre reinforced plastic, and in all such materials it may be assumed that the fibres are strong in tension while the matrix is strong in compression but weak in tension. A thin-walled tube is to be made of such a material. Show by diagrams the directions in which you think the fibres should be laid if the tube is to be subjected to (a) internal pressure, (b) torsion in one direction only, (c) torsion in either direction.

3. You have been given the task of determining the constants E, G, v and K for a new alloy. The only available specimen is a rod of circular section, 200 mm long and 4 mm diameter, and this must be returned intact. (In other words, you are not allowed to alter its shape or even to use an extensometer which indents its surface.) The elastic limits in tension and in shear are known to be, approximately, 300 MN/m² and 150 MN/m² respectively. Devise a procedure which will enable the elastic constants to be determined with reasonable accuracy and describe, in detail, the tests to be performed. Include the limits to be imposed so that stresses are kept well within the elastic range. (*Hint:* One procedure makes use of Eqs. (7.8), (16.14), (13.23) and (13.26).)

General

1. A small cube of material has no stress on two of its faces. On the remaining four, there are complementary shear stresses of 20 MN/m² and on the faces with conventionally positive shear stress, there is also a tensile stress of 36 MN/m². By consideration of the forces acting on a prism of material, find (a) the stresses on a plane making an angle of 30° (measured anticlockwise) from the plane of the tensile stress; (b) the positions of the principal planes; (c) the principal stresses.

Ans. (a) 9·7 MN/m² (tensile) and 25·6 MN/m² (shear); (b) principal planes make angles of 24° (clockwise) and 66° (anticlockwise) with that of the tensile stress; (c) 44·9 MN/m² (tensile) and 8·9 MN/m² (compressive).

2. At a point in a material, the principal stresses are 130 MN/m² (tensile) and 80 MN/m² (compressive). Construct the Mohr stress circle and find the stresses on planes making (a) 20° and (b) 60° with that of the tensile principal stress. Find also (c) the maximum shear stress, and (d) the positions of the planes on which the stress is pure shear.

Ans. (a) 105·4 MN/m² (tensile) and 67·5 MN/m² (shear); (b) 27·5 MN/m² (compressive) and 90·9 MN/m² (shear); (c) 105 MN/m²; (d) planes making angles of 52°, and 128° with that of the tensile principal stress.

3. The cylindrical shell of a steam boiler is made from 15 mm thick steel plates. Its diameter is 3 m and the working pressure is 850 kN/m^2 (8·5 bar). Construct a Mohr circle and find the maximum shear stress in the material (*a*) if its ends are supported only by the cylindrical shell, (*b*) if in addition they are connected by stays which exert a total inward force of 4 MN on each end.

Ans. (*a*) 21·25 MN/m^2; (*b*) 35·4 MN/m^2

4. During a test on a 45 mm twist drill, the following dynamometer readings were recorded: thrust, 4 kN; torque, 60 Nm. Find (*a*) the maximum compressive stress, and (*b*) the maximum shear stress in the plain cylindrical part of the drill.

Ans. (*a*) 4·84 MN/m^2; (*b*) 3·58 MN/m^2

5. A 100 mm diameter shaft transmits 500 kW at 500 rev/min. Transverse forces produce a maximum bending moment of 4 kNm. Find (*a*) the maximum shear stress, and (*b*) the maximum tensile stress in the shaft. Also, (*c*) show by a diagram the position of a plane on which maximum tensile stress acts.

Ans. (*a*) 52·9 MN/m^2; (*b*) 73·2 MN/m^2; (*c*) maximum tensile stress acts on planes making 56$\frac{1}{2}$° with the shaft axis.

6. A 45° strain rosette attached to a brass component ($E = 90$ GN/m^2 and $v = 0·3$) gave the following values of strain:

Gauge A: 432×10^{-6} (extension).
Gauge B: at 45° (anticlockwise) to A: 21×10^{-6} (extension).
Gauge C: at 90° (anticlockwise) to A: 138×10^{-6} (contraction).

Construct the Mohr strain circle and find (*a*) the principal strains, (*b*) the principal stresses, (*c*) the direction of the greater principal strain relative to the strain rosette.

Ans. (*a*) $\varepsilon_1 = 459 \times 10^{-6}$ (extension), $\varepsilon_2 = 165 \times 10^{-6}$ (contraction);

(*b*) 40·5 MN/m^2 (tensile) and 2·7 MN/m^2 (compressive); (*c*) the direction of strain ε_1 makes an angle of 12° (clockwise) with the axis of gauge A.

7. The following strains, all extensional, were measured by a 60° strain rosette:

Gauge A: 716×10^{-6},
Gauge B: at 60° (anticlockwise) to A: 539×10^{-6},
Gauge C: at 120° (anticlockwise) to A: 155×10^{-6}.

Construct the Mohr strain circle and find (*a*) the principal strains, (*b*) the principal stresses if $E = 200$ GN/m^2 and $v = 0·3$.

Ans. (*a*) 801×10^{-6} and 139×10^{-6} (both extensional); (*b*) 185·2 MN/m^2 and 83·4 MN/m^2 (both tensile).

8. Tests on a ceramic gave values of 284 GN/m^2 and 117 GN/m^2 for the moduli of elasticity and rigidity respectively. Calculate the values of (*a*) the bulk modulus, and (*b*) Poisson's ratio.

Ans. (*a*) 165 GN/m^2; (*b*) 0·214

CHAPTER 14

Shearing Force and Bending Moment

14.1 BASIC RELATIONSHIPS

A *beam* is a member acted on by transverse forces. It is usually horizontal and, when unloaded, straight, and the forces comprise downward *loads* and upward *reactions*. Loads may be *concentrated* or applied over such a small part of the beam that they may be considered to act at a point. Alternatively, they may be *distributed* over part or the whole of the beam, *uniformly* or otherwise. *Reactions* are the forces exerted on a beam by its supports, and are usually considered as concentrated. Supports on which the beam merely rests are termed *simple* and the phrase "a simply supported beam" also implies that there are only two supports. A beam with more than two supports is called a "continuous beam." If a beam is rigidly connected to its supports by building a short length into masonry, for example, or bolting it to a relatively rigid structure, the beam is said to be "built-in." As well as upward reactions, the supports will now apply couples called *fixing moments* to the beam which prevent (completely or partially) the departures from the horizontal which would otherwise occur under load. Built-in and continuous beams are both said to be *redundant* or *statically indeterminate* because the conditions of static equilibrium are not sufficient for their solution (see Chapter 16). A beam "built-in" only at one end is called a *cantilever* and has both a reaction and a fixing moment. It is statically determinate provided there are no other supports.

Equilibrium of a loaded beam

Fig. 14.1 (*a*) shows a loaded cantilever divided arbitrarily into two parts, A and B, by a line ZZ. When the equilibrium of each part is examined separately (Fig. 14.1 (*b*)), equilibrium for B requires an upward force F and an anticlockwise moment M. For part A, a downward force and a clockwise moment will be required and, since the cantilever as a whole is in equilibrium, these must be numerically equal to those acting on B. Thus at section ZZ, the parts A and B exert on each other equal and opposite forces F, and equal and opposite moments M, and the material of the cantilever is subjected to a combination of shearing and bending. Forces F are known as the shearing force and the moments M as the bending moment.

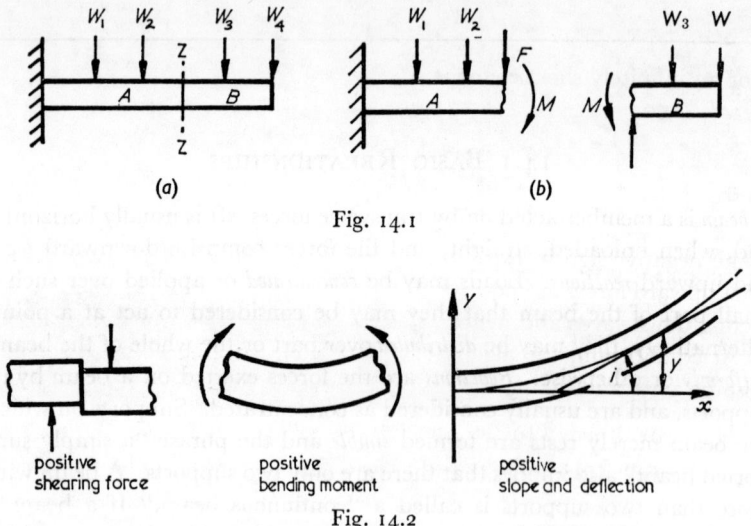

Fig. 14.1

Fig. 14.2

Sign conventions. There are several existing conventions for fixing the algebraic sign of forces, moment and deflections. All that matters is that conventions should be used consistently, and the set adopted here is shown in the table, which refers to a horizontal beam. See also Fig. 14.2.

Quantity	Positive direction
Loading (i.e. vertical force)	Upwards.
Shearing force	Upward force to the left of the section considered (or a tendency to rotate an element clockwise).
Bending moment	Clockwise moment to the left of the section considered (or, a tendency to bend the beam concave upwards or to cause "sagging").
Slope	Increasing deflection with increasing distance from the reference point (see Chapter 16).
Deflection	Upwards (see Chapter 16).

Loading, bending moment and shearing force

Fig. 14.3 shows a short length of a beam with a distributed load (not necessarily uniform) w per unit length. This element is distant x from some point of reference and has length δx. The shearing forces are F and $(F + \delta F)$ and the bending moments M and $(M + \delta M)$. In equilibrium, the algebraic sum of the vertical forces must be zero, so that

$$(F + \delta F) + w\delta x = F$$
$$\therefore \ \delta F/\delta x = -w$$

For an infinitely short element,

$$dF/dx = -w \tag{14.1}$$

and
$$(F_2 - F_1) = -\int_1^2 w \, dx \tag{14.2}$$

(The negative sign in Eqs. (14.1) and (14.2) comes from the convention that upward forces are positive. In Fig. 14.3 the distributed load is shown acting downwards and hence applies *negative loading* to the beam.)

At equilibrium, the algebraic sum of moments about any point must also be zero, and taking the right-hand end as reference point,

$$M + F\delta x + w\delta x \cdot \tfrac{1}{2}\delta x = M + \delta M$$
$$\therefore \ F\delta x + \tfrac{1}{2}w(\delta x)^2 = \delta M$$
$$\therefore \ F + \tfrac{1}{2}w\delta x = \delta M/\delta x$$

or, for an infinitely short element $(\delta x \to 0)$,

$$F = dM/dx \tag{14.3}$$

Fig. 14.3

and $(M_2 - M_1) = \int_1^2 F \, dx$ $\qquad$ (14.4)

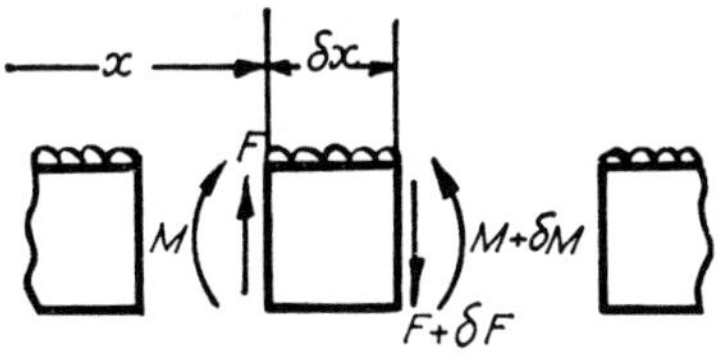

Thus if the loading (i.e., upward force per unit length) is integrated with respect to x, the result is the shearing force, and if the shearing force is integrated, the result is the bending moment. (In Chapter 16 it will be shown that two further integrations provide the slope and the deflection.)

14.2 FORCE AND MOMENT DIAGRAMS

Concentrated loading

Eq. (14.1) shows that $dF/dx = 0$ if $w = 0$, so that the shearing force is constant between loads and will be the algebraic sum of all forces to one side of any section. This fits in with the method usually adopted when drawing shearing force diagrams of proceeding from left to right, moving vertically up (or down) by the magnitude of each reaction (or load) as it is reached.

An attempt to apply Eq. (14.2) to a concentrated load results in an attempt to integrate an infinitely large load over an infinitely small distance. But concentrated loads exist only in theory, and their practical equivalent is a load applied over a short but measurable length, so that for a real beam, Eq. (14.2) applies and the line on the shearing force diagram is not vertical in the region where the load acts.

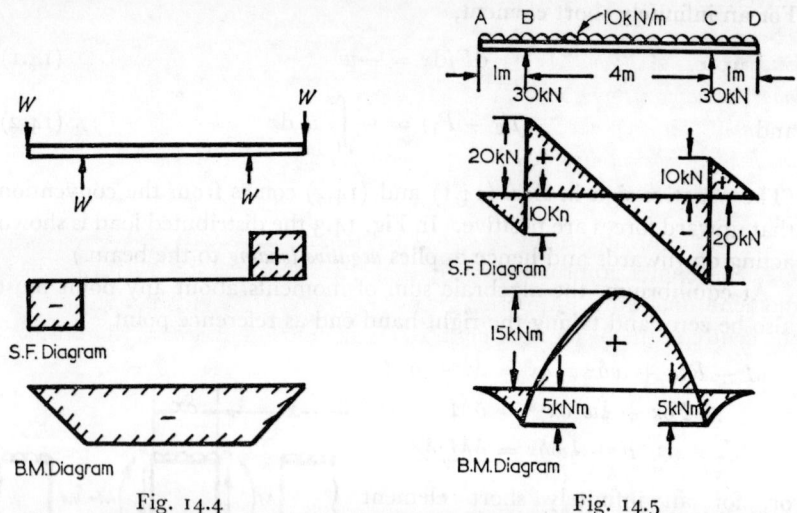

Fig. 14.4   Fig. 14.5

Eq. (14.3) shows $\mathrm{d}M/\mathrm{d}x$ is constant if F is constant, so that the bending moment diagram has constant slope between loads. The diagram is a series of straight lines only if the loads are concentrated and can be constructed once the bending moments at the loading points are calculated. A particularly interesting case is that shown by Fig. 14.4, where the shearing force is zero between the supports. From Eq. (14.3), if F is zero then $\mathrm{d}M/\mathrm{d}x$ is also zero, so that this arrangement gives a uniform bending moment over the central section of the beam.

Uniformly distributed loading

If the loading is uniformly distributed, Eq. (14.1) shows that $\mathrm{d}F/\mathrm{d}x$ is constant and the shearing force diagram is a straight line with a slope equal to the loading w. That is,

$$F \int w = \mathrm{d}x$$
$$= wx + A$$

From Eq. (14.4),

$$M = \int F \, dx$$
$$= \tfrac{1}{2}wx^2 + Ax + B$$

This equation represents a parabola so that for uniformly distributed loading, the bending moment diagram is parabolic. If there are concentrated loads as well, the shearing force diagram will consist of sloping lines between the concentrated loads and vertical lines at their points of application, and the bending moment diagram will be a series of parabolic curves.

Example. *A beam ABCD is 6 m long and carries a uniformly distributed load (including its own weight) of 10 kN per metre. It is supported at B and C so that it overhangs the supports by 1 m at each end. Sketch the shearing force and bending moment diagrams and state the position and magnitude of the maximum bending moment.*

Since the arrangement is symmetrical, each reaction will be 30 kN. Fig. 14.5 shows, approximately to scale, the shearing force and bending moment diagrams. Maximum bending moment evidently occurs at the centre of the beam and, considering forces to the right of this section,

$$M_{max} = +(30 \times 2) - (30 \times 1\cdot5)$$
$$= +15 \text{ kNm}$$

Non-uniformly distributed loading

If (as in the following example) it is possible to express the loading as a function of x, Eqs. (14.2) and (14.4) lead to expressions for shearing force and bending moment. If the loading varies irregularly, however, the problem may be solved graphically by stating Eqs. (14.2) and (14.4) in the form

$(F_2 - F_1) =$ Area of loading diagram between sections 1 and 2

and

$(M_2 - M_1) =$ Area of shearing force diagram between sections 1 and 2

Example. *A cantilever consists of a timber beam 20 cm wide and 5 m long. Its thickness varies uniformly from 10 cm at the free end to 40 cm at the fixed end. Taking its density as 700 kg/m³, construct the shearing force and bending moment diagrams if it is loaded only by its own weight. State the maximum bending moment.*

Consider a section distant x from the fixed end (and indicated by ZZ on Fig. 14.6).

The thickness of the beam at this section is

$$(40 - x \cdot 30/5) = (40 - 6x) \text{ cm}$$

Hence loading at this point (i.e., weight per unit length) is
$20 \times (40 - 6x) \cdot 10^{-4} \cdot 700 \cdot 9{\cdot}81 = 13{\cdot}73(40 - 6x) \text{ N/m}$ (downwards)
From Eq. (14.2),

$$F = \int w \, dx$$

$$= -13{\cdot}73\int(40 - 6x) \, dx$$

$$= -13{\cdot}73(40x - 3x^2 + A)$$

At the free end of the cantilever (when $x = 5$), $F = 0$,

$$\therefore \quad 0 = 13{\cdot}73(200 - 75 + A) \quad \text{or} \quad A = -125$$

Hence the expression for shearing force at any distance x from the fixed end is

$$F = -13{\cdot}73(40x - 3x^2 - 125)$$

From Eq. (14.4),

$$M = \int F \, dx$$

$$= -13{\cdot}73\int(40x - 3x^2 - 125) \, dx$$

$$= -13{\cdot}73(20x^2 - x^3 - 125x + B)$$

At the free end of the cantilever (when $x = 5$), $M = 0$,

$$\therefore \quad 0 = -13{\cdot}73(500 - 125 - 625 + B) \quad \text{or} \quad B = +250$$

Hence the expression for bending moment is

$$M = -13{\cdot}73(20x^2 - x^3 - 125x + 250)$$

The shearing force and bending moment diagrams are shown, approximately to scale, in Fig. 14.6 and it is evident that maximum bending moment occurs at the fixed end (i.e., $x = 0$).

$$\therefore \quad M_{\text{max}} = -13{\cdot}73(0 - 0 - 0 + 250)$$

$$= -3433 \text{ Nm} = -3{\cdot}433 \text{ kNm}$$

(The negative sign indicates "hogging," which means that at this point the cantilever will bend convex upwards.)

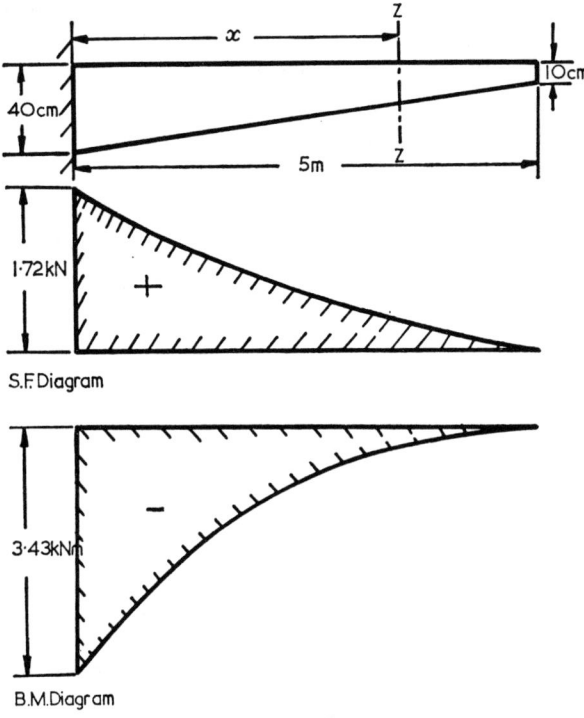

Fig. 14.6

Maximum bending moment

Eq. (14.3) shows that dM/dx is zero when F is zero, so that maximum bending moment must correspond with zero shearing force and points of maximum bending moment may be found from the shearing force diagram. In Fig. 14.5, for example, there are three points at which the shearing force diagram crosses the axis, and each corresponds to a possible point of maximum bending moment. In general, the bending moment diagram will indicate more than one such point and it will be necessary to calculate all these values of bending moment to find out which is the absolute maximum. The sign is usually unimportant and the "maximum bending moment" will be that having the greatest numerical value.

Points of contraflexure

With a beam loaded as in Fig. 14.5, two different kinds of bending will occur: positive bending or "sagging" at the centre and negative bending or "hogging" at the supports, and the points on the beam where the curvature changes are known as *points of contraflexure*. (See points CC in Fig. 14.7.) Since curvature is determined by bending moment, a point of contraflexure will occur wherever the bending moment changes sign, or wherever the bending moment diagram crosses its axis. For the example represented by the bending moment diagram in Fig. 14.5, they are situated 1·27 m from each end of the beam.

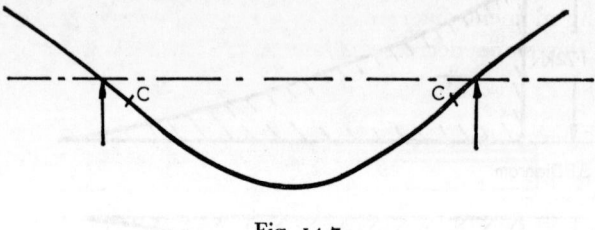

Fig. 14.7

Complex load systems

For any load system, however complex, Eqs. (14.1) to (14.4) apply and maximum bending moment will always correspond to zero shearing force (or, in the case of a concentrated load, to a point where the shearing force diagram crosses its axis).* Often it is unnecessary to construct accurate diagrams and, if only a sketch is required, the form of the bending moment diagram may be deduced from the relationship $dM/dx = F$, which means that the shearing force at any point gives the slope of the bending moment diagram.

Example. *A horizontal beam* 10 m *long is simply supported at its left-hand end* A *and at a point* B, 2 m *from its right-hand end* C. *It carries a uniformly distributed load (including its own weight) of* 16 kN/m *over its entire length and, in addition, a concentrated load of* 40 kN *at* C. *Sketch the shearing force and bending moment diagrams and find (a) the position and magnitude of the maximum bending moment, and (b) the position of the point of contraflexure.*

Taking moments about A in Fig. 14.8,

$$8R_B = (160 \times 5) + (40 \times 10)$$

$$\therefore R_B = 150 \text{ kN}$$

---

* For a cantilever and often for a built-in beam, maximum bending moment occurs at a fixed end, where the shearing force is not zero. In reality, however, this is not the end of the beam—there is a further section on which the fixing forces act and which is disregarded when drawing S.F. and B.M. diagrams. If this section is considered, the shearing force and the slope of the bending moment diagram change sign as it is entered.

For equilibrium of vertical forces,

$$R_A = 160 + 40 - R_B = 50 \text{ kN}$$

Fig. 14.8 shows the shearing force diagram (approximately to scale) on which the principal values have been inserted, and from this the form of the bending moment diagram may be deduced as follows.

The bending moment is zero at A, the shearing force is positive and so the bending moment diagram slopes upwards. From the shearing force diagram, this slope decreases to zero and then becomes increasingly

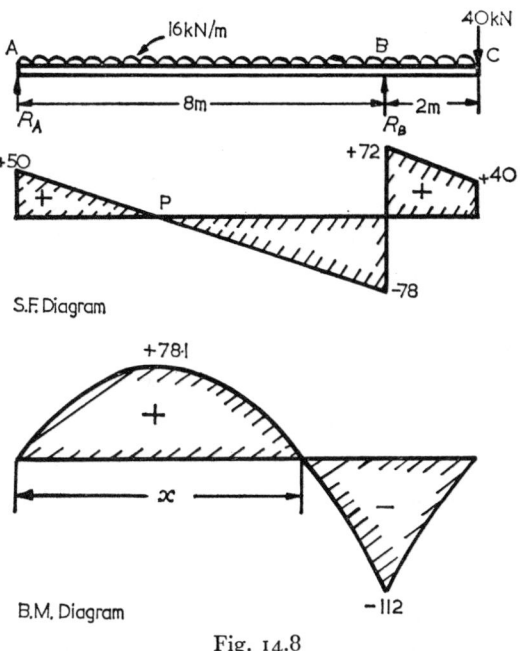

Fig. 14.8

negative until the point B is reached. (The bending moment diagram is a parabola and its apex is at the point where the shearing force is zero.) At B, the shearing force changes sign, and from B to C, the bending moment diagram slopes upwards with a slope which decreases from B to C but is not zero at C. (From B to C the bending moment diagram is part of a parabola with apex somewhere to the right of C.) Since C is at the end of the beam, the bending moment at this point must be zero. The general shape of the bending moment diagram is now known, and if the two maxima are calculated, a well-proportioned sketch may be made. Fig. 14.8 shows the bending moment diagram (approximately to scale) on which calculated values have been inserted.

Position of maximum bending moment. The shearing force diagram crosses the axis at two points, one at B and one at a point P between A and B. From the geometry of the diagram,

$$AP/AB = 50/128$$
$$\therefore AP = 8 \times 50/128 = 3{\cdot}125 \text{ m}$$

B.M. at P (considering moments to the left)

$$= (50 \times 3{\cdot}125) - (\tfrac{1}{2} \times 16 \times 3{\cdot}125^2) = +78{\cdot}1 \text{ kNm}$$

B.M. at B (considering moments to the right)

$$= -(40 \times 2) - (\tfrac{1}{2} \times 16 \times 2^2) = -112 \text{ kNm}$$

Hence the (numerically) maximum bending moment is at B.

Position of the point of contraflexure. From the bending moment diagram, the point of contraflexure is a little to the left of B, say x metres from A. Between A and B, the bending moment at any distance x from A is given by

$$M = 50x - \tfrac{1}{2}16x^2$$

At the point of contraflexure, $M = 0$, so that

$$\therefore \ 0 = 50x - 8x^2$$
$$\therefore \ x = 6{\cdot}25 \text{ m}$$

Example. *A beam AB is 6 m long and is built-in at both ends. Its mass is 510 kg and it supports a uniformly distributed load of 10 kN/m, together with a concentrated load of 36 kN at C, a point 4 m from A. The reactions and fixing moments are found to be:* $R_A = 42{\cdot}3$ kN, $M_A = 49$ kNm, $R_B = 59{\cdot}7$ kN, $M_B = 65$ kNm. *Construct, to scale, the shearing force and bending moment diagrams for the beam and find (a) the position and magnitude of the maximum bending moment and (b) the positions of the points of contraflexure.*

$$\text{Weight of beam} = 510 \times 9{\cdot}81 = 6000 \text{ N} \quad \text{or} \quad 6 \text{ kN}$$
$$\therefore \text{ Total distributed loading} = (10 + 6/6) = 11 \text{ kN/m}$$

In the shearing force diagram (Fig. 14.9) the principal values have been inserted. In order to construct the bending moment diagram, a series of values must be calculated at, say, intervals of 0·5 m.

For example, for a point 2 m from A, considering moments to the left,

$$M = -49 + (42{\cdot}3 \times 2) - (\tfrac{1}{2} \times 11 \times 2^2)$$
$$= +13{\cdot}6 \text{ kNm}$$

The shearing force diagram crosses its axis at a point 3·85 m from A, and at this point the bending moment reaches its maximum positive value of 32·3 kNm. This is, however, numerically smaller than the fixing moments, hence the maximum bending moment is 65 kNm at point B.

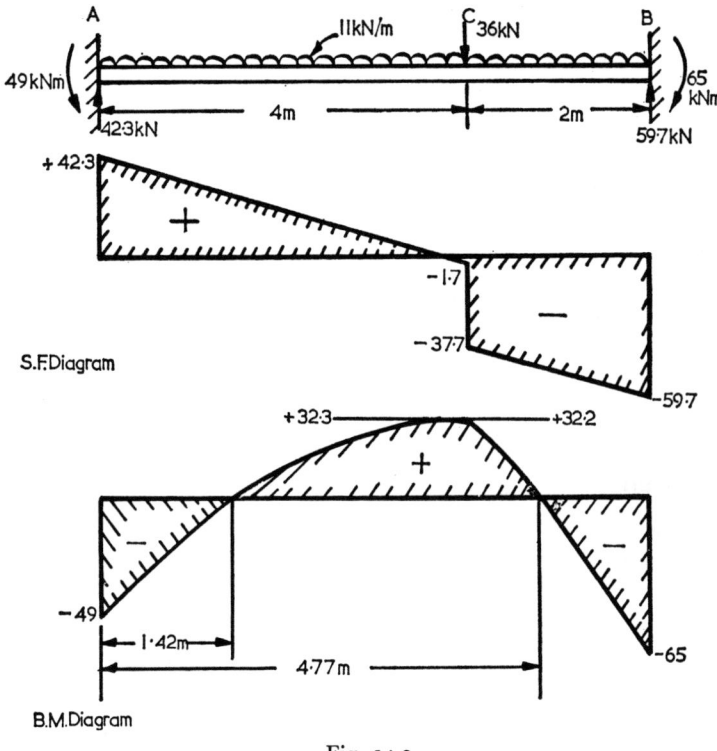

Fig. 14.9

Points of contraflexure. From the bending moment diagram, there are two points of contraflexure, one between A and C and one between C and B. For a point between A and C and distant *x* metres from A, considering moments to the left,

$$M = -49 + 42 \cdot 3x - 5 \cdot 5x^2$$

∴ For a point of contraflexure,

$$5 \cdot 5x^2 - 42 \cdot 3x + 49 = 0$$

Solving this quadratic equation,

$$x = 1 \cdot 42 \quad \text{or} \quad 6 \cdot 27$$

The value 6·27 is not within the specified limits (it represents the point where the parabola AC, if continued, would intersect the axis of the B.M. diagram) so that the point of contraflexure is situated 1·42 m from A.

For a point between C and B and distant x from A, again considering moments to the left,

$$M = -49 + 42 \cdot 3x - 5 \cdot 5x^2 - 36(x - 4)$$

$$= 6 \cdot 3x - 5 \cdot 5x^2 + 95$$

∴ For a point of contraflexure,

$$5 \cdot 5x^2 - 6 \cdot 3x - 95 = 0$$

$$∴ x = 4 \cdot 77 \quad \text{or} \quad -3 \cdot 62$$

Hence the second point of contraflexure is 4·77 m from A.

PROBLEMS

For tutorials

1. In Fig. 14.4, the central part of the beam is subjected to pure bending. To what is a beam subjected at a point of contraflexure? Prove that it is impossible to subject a finite length of a beam to pure shearing force, and consider the practical implications of this in the case of (i) bolts and rivets in shear, (ii) metal shearing machines.

2. End standards are often marked at the "Airy points" (named after Sir George Airy, who first calculated their position). If the length of the standard is L, their spacing is 0·577L and it may be shown that if a horizontal bar of uniform section is supported at these points its ends will remain horizontal although it sags at its centre. In Chapter 16 it will be shown that if two points on a uniform beam have the same slope, the total area of the bending moment diagram between them (that is, the algebraic sum of positive and negative areas) is zero. Construct, to a suitable scale, the bending moment diagram for a bar 1 m long and of mass 5·1 kg, supported (symmetrically) at points 577 mm apart and loaded only by its own weight. Measure the positive and negative areas and show that their algebraic sum is zero.

3. A horizontal beam 10 m long carries a uniformly distributed load of 20 kN/m and is simply supported at two points equidistant from its ends. What should be their spacing if (a) the maximum positive and maximum negative bending moments are to be numerically equal, (b) the bending moment at the centre of the beam is to be zero? In each case sketch the shearing force diagram, the bending moment diagram and (very approximately) the shape of the deflected beam. Find (c) the positions of the points of contraflexure in case (a).

Ans. (a) 5·86 m; (b) 5 m; (c) 2·93 m from each end

General

1. A beam 8 m long is simply supported at its ends. It carries a distributed load (including its own weight) which increases uniformly from 5 kN/m at the left-hand end to 29 kN/m at the right-hand end. Construct, to suitable scales, the shearing force and bending moment diagrams and find the magnitude and position of (a) the maximum shearing force, (b) the maximum bending moment.

Ans. (a) −84 kN, at the right-hand end; (b) +132 kN m, 4·45 m from the left-hand end.

2. A beam ABCD is simply supported at A and C. AB = 5 m, BC = 5 m and CD = 3 m. The beam supports a uniformly distributed load of 6 kN/m between B and D, together with concentrated loads of 25 kN at B and 10 kN at D. Neglecting the weight of the beam, sketch the shearing force and bending moment diagrams and calculate (a) the magnitude and position of the maximum bending moment, (b) the position of the point of contraflexure.

Ans. (a) +71·5 kN m at B; (b) 8·42 m from A

3. A cantilever of uniform section is 15 m long and of mass 1530 kg. In addition to its own weight, it supports a central load of 14 kN. It is propped at its free end, and a consideration of slope and deflection shows that this prop supplies a reaction of 10 kN. Sketch the shearing force and bending moment diagrams and find (a) the magnitude and position of the maximum bending moment, (b) the position of the point of contraflexure.

Ans. (a) −67·5 kN m at the fixed end; (b) 3·97 m from the fixed end.

4. A beam ABCD is built-in at A and D, and supports concentrated loads of 40 kN at B and 60 kN at C. AB = 5 m, BC = 12 m and CD = 3 m. At A there is a reaction of 37·3 kN and an anticlockwise fixing moment of 134·5 kN m, and at D there is a reaction of 62·7 kN and a clockwise fixing moment of 168·5 kN m. Calculate the bending moments at B and C, sketch the bending moment diagram and deduce from it the positions of the points of contraflexure.

Ans. M_B = +52 kN m, M_C = +19·6 kN m; 3·61 m from A and 2·29 m from D.

CHAPTER 15

Stresses in Beams

15.1 COMBINED BENDING AND DIRECT STRESS

Theory of simple bending

For a beam which is initially straight, it may be shown that

$$M/I = \sigma/y = E/R \qquad (15.1)$$

where M = Moment of resistance = Applied bending moment,

$\quad\quad I$ = Second moment of area of beam cross-section about the neutral axis,

$\quad\quad \sigma$ = Stress at distance y from the neutral axis,

$\quad\quad E$ = Modulus of elasticity, and

$\quad\quad R$ = Radius of curvature of neutral surface.

It may also be shown that the *neutral axis* (i.e., the intersection of the unstrained layer of material or *neutral surface* with a transverse section) is a straight line, perpendicular to the plane of bending and passing through the centroid of the beam cross-section.

The principle of superposition

In dealing with combinations of forces, the principle of superposition has a wide field of application. In general terms, it is as follows. *Any effect produced by two or more causes acting together is the algebraic sum of the effects which each would produce when acting alone.* In the present context, "causes" are the forces or moments applied to a beam and "effects" are the stresses (or, in Chapter 16, the deflections) produced.

The principle can be applied only if the relationship between "causes" and "effects" is always one of *direct proportionality*. For a beam, and for most problems involving stress and strain, this means that the limit of proportionality must not be exceeded. The principle cannot be applied to struts, for then neither stresses nor deflections are proportional to the applied loading.

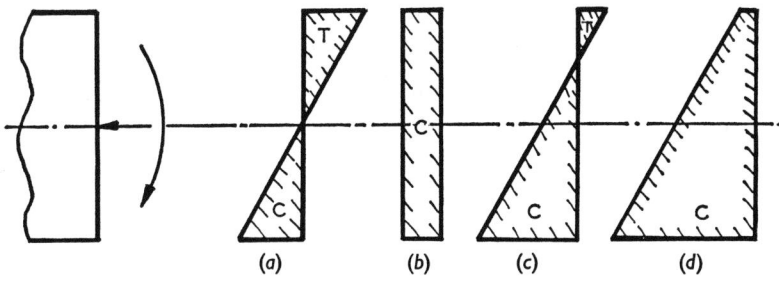

Fig. 15.1

Combined bending and direct stress

Eq. (15.1) for bending shows that $\sigma = y(M/I)$, so that bending produces stresses which are directly proportional to distances from the neutral axis and which vary (Fig. 15.1 (a)) from maximum tensile (on the convex side) to maximum compressive (on the concave side). If a longitudinal force acts on a beam, there will also be direct stresses and, if the force is applied at the centroid of the cross-section, these will be uniform and, for a compressive load, will be as in Fig. 15.1 (b). If a beam is subjected to both bending and axial loading, the principle of superposition may be applied, and the stress at any point will be the algebraic sum of direct and bending stresses (Fig. 15.1 (c)). A larger compressive load would produce the stress distribution shown in Fig. 15.1 (d).

Example. *A column is* 20 cm *high and of rectangular cross-section* 10 cm $\times$ 5 cm. *It is acted on by a load of* 5 kN *applied centrally but at an angle of* 20° *to the column axis, as shown by Fig. 15.2. Find the maximum tensile and compressive stresses at the base of the column.*

The load may be resolved into two components:

(i) (5 cos 20°) kN acting axially, producing direct compressive stress.

(ii) (5 sin 20°) kN acting transversely, producing bending.

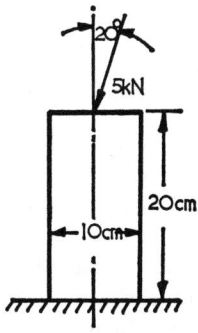

Fig. 15.2

$$\text{Direct stress} = F/A = (5 \times 10^3 \cos 20°)/(0.05 \times 0.1)$$

$$= 94 \times 10^4 \, \text{N/m}^2 = 0.94 \, \text{MN/m}^2$$

Bending moment due to (ii),

$$M = (5 \times 10^3 \sin 20°) \times 0.2$$

$$= 342 \, \text{Nm}$$

From Eq. (15.1),

$$\sigma = My/I$$

For a rectangular section,

$$I = bd^3/12 = (0.05 \times 0.1^3)/12 = 4.167 \times 10^{-6} \, \text{m}^4$$

On both sides of the neutral axis, y has a maximum value of 0.05 m, so that the maximum bending stress

$$\sigma = 342 \times 0.05/4.167 \times 10^{-6}$$

$$= 4.10 \times 10^6 \, \text{N/m}^2 = 4.10 \, \text{MN/m}^2$$

Thus at the right-hand edge of the base, the stress will be the algebraic sum of 0.94 MN/m² (compressive) and 4.10 MN/m² (tensile) or 3.16 MN/m² (tensile). Similarly, the maximum compressive stress will be at the left-hand edge of the base and is

$$(0.94 + 4.10) = 5.04 \, \text{MN/m}^2 \, \text{(compressive)}$$

The stress distribution would be similar to that in Fig. 15.1 (c).

15.2 ECCENTRIC LOADING

Fig. 15.3 (a) shows the end of a short column to which is applied a load F parallel to the column axis but displaced by x from its centroid. In Fig. 15.3 (b), two equal and opposite forces have been introduced (producing no effect on the equilibrium of the column), and in Figs. 15.3 (c) and (d), the forces have been re-grouped as an axial load at the centroid together with a couple. Thus the eccentric load is equivalent to a central load F plus a bending moment Fx, and will result in a combination of direct stress and bending.

Example. *A short concrete column of solid circular cross-section 40 cm diameter supports a mass of 15 tonne. The line of action of the load is displaced 4 cm from the column axis. Find the maximum compressive stress in the column and show by a diagram how the stress varies across the section. Concrete is weak in tension and, if cracking is to be avoided, no part of the column must be subjected to tensile stress. In view of this, what is the maximum eccentricity at which the load may be applied?*

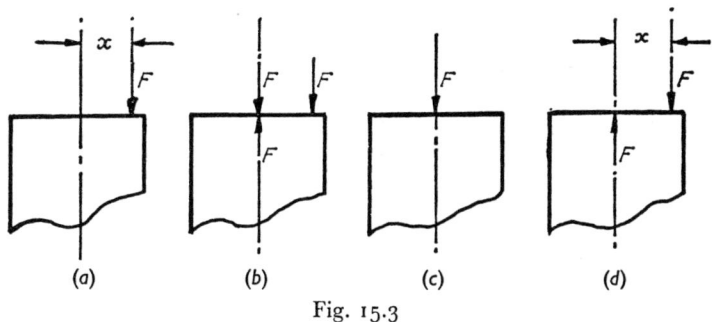

Fig. 15.3

$$\text{Load } F = mg = 15000 \times 9{\cdot}81$$
$$= 147{\cdot}15 \times 10^3 \text{ N}$$
$$\text{Bending moment } M = Fx = 147{\cdot}15 \times 10^3 \times 0{\cdot}04$$
$$= 5886 \text{ Nm}$$
$$\text{Direct compressive stress} = F/A = 147{\cdot}15 \times 10^3/\pi(0{\cdot}2)^2$$
$$= 1{\cdot}17 \times 10^6 \text{ N/m}^2 = 1{\cdot}17 \text{ MN/m}^2$$

From Eq. (15.1),

$$\sigma = My/I$$

For a solid circular section,

$$I = \pi d^4/64$$
$$= \pi(0{\cdot}4)^4/64 = 1{\cdot}257 \times 10^3 \text{ m}^4$$

and

$$\text{Maximum value of } y = \tfrac{1}{2}d = 0{\cdot}2 \text{ m}$$

∴ Maximum bending stress σ

$$= 5886 \times 0{\cdot}2/1{\cdot}257 \times 10^{-3}$$
$$= 936 \times 10^3 \text{ N/m}^2 \quad \text{or} \quad 0{\cdot}936 \text{ MN/m}^2$$

∴ Maximum compressive stress in the column

$$= (1{\cdot}17 + 0{\cdot}936) = 2{\cdot}106 \text{ MN/m}^2$$

The stress in the column will vary between this value and

$$(1 \cdot 17 - 0 \cdot 936) = 0 \cdot 234 \, \text{MN/m}^2 \text{ (compressive)}$$

The stress distribution is shown, approximately to scale in Fig. 15.4 (a).

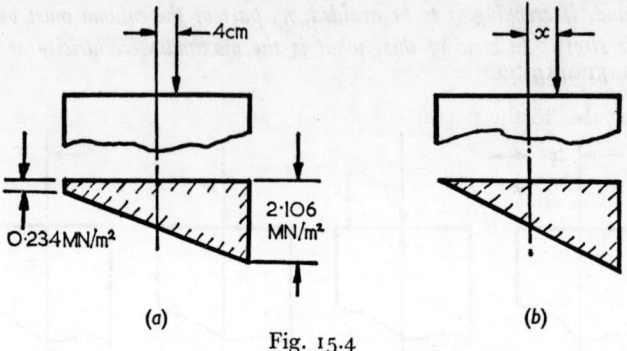

(a)   (b)

Fig. 15.4

The maximum bending moment which can be applied without giving rise to tensile stresses is that producing the stress distribution shown in Fig. 15.4 (b). This condition is that the maximum (tensile) bending stress shall not exceed the direct compressive stress. Hence

$$\text{Maximum bending stress } \sigma = 1 \cdot 17 \, \text{MN/m}^2$$

From Eq. (15.1),

$$M = \sigma I / y$$
$$= 1 \cdot 17 \times 1 \cdot 257 \times 10^3 / 0 \cdot 2 = 7357 \, \text{NM}$$
$$M = Fx$$
$$\therefore x = M/F$$
$$= 7357/147 \cdot 15 \times 10^3$$
$$= 0 \cdot 05 \, \text{m} = 5 \, \text{cm}$$

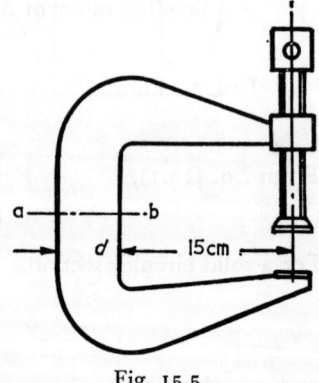

Fig. 15.5

Example. *The body of the simple clamp shown in* Fig. 15.5 *is to be cut from steel plate* 20 mm *thick. It is to exert a maximum clamping force of* 4 kN *and must be capable of applying this force* 15 cm *from the edge of the material being gripped. If, at the section* ab, *the tensile stress is not to exceed* 70 MN/m², *what is the minimum width d at this point?*

$$\text{Direct stress} = F/A = 4 \times 10^3 / 0 \cdot 02d$$
$$= 2 \times 10^5 / d \, \text{N/m}^2$$
$$= 0 \cdot 2/d \, \text{MN/m}^2 \text{ (tensile)}$$
$$\text{Bending moment } M = Fx = 4 \times 10^3 (\tfrac{1}{2}d + 0 \cdot 15) \, \text{Nm}$$

Maximum bending stress $\sigma = My/I$

and here

$$I = 0.02d^3/12 = d^3/600 \text{ m}^4$$

$$\therefore \sigma = 4 \times 10^3(\tfrac{1}{2}d + 0.15) \times \tfrac{1}{2}d \times 600/d^3 \text{ N/m}^2$$

$$= 1.2(\tfrac{1}{2}d + 0.15)/d^2 \text{ MN/m}^2$$

$\therefore$ Maximum tensile stress $= (0.2/d) + 1.2(\tfrac{1}{2}d + 0.15)/d^2 \text{ MN/m}^2$

Equating this to the permitted stress,

$$70 = (0.2/d) + 1.2(\tfrac{1}{2}d + 0.15)/d^2$$

$$\therefore 70d^2 = 0.2d + 1.2(\tfrac{1}{2}d + 0.15)$$

$$\therefore 70d^2 - 0.8d - 0.18 = 0$$

Solving this quadratic equation and disregarding the negative root,

$$d = 0.0567 \text{ m} = 56.7 \text{ mm}$$

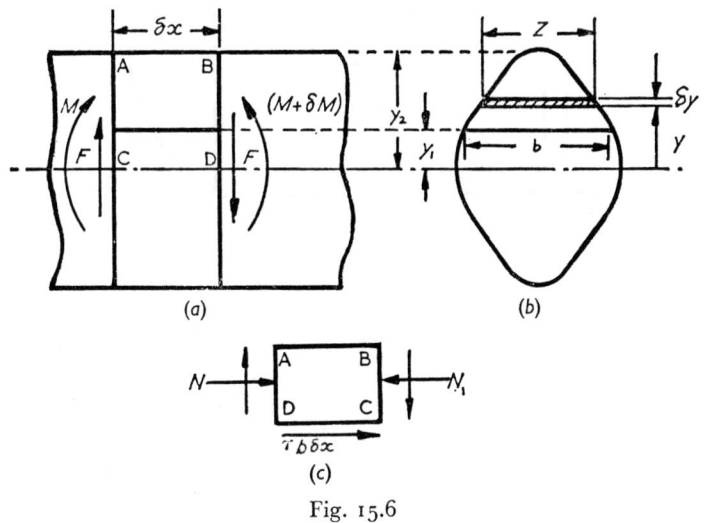

Fig. 15.6

15.3 SHEAR STRESS IN BEAMS

Consider a short element of a beam, of length δx, at a point where the shearing force is F. As shown in Chapter 14, if there is a shearing force, the bending moment must vary along the beam and, if F is positive, will increase from left to right. Let the bending moment be M at the left-hand side of the element and $(M + \delta M)$ at the right-hand side (Fig. 15.6 (a)). Fig. 15.6 (c) shows the forces acting on the block of material ABCD. Because of the difference in bending moments, the total compressive force N_1 acting on the face BC will be greater than the corresponding force N acting on AD, and the difference between these forces will be balanced by the only other horizontal force acting on the block—that due to the longitudinal shear stress on its base.

Although shear stress varies over a beam section, it is usually almost constant across its width, and will be assumed constant in this analysis. Thus if the shear stress on the face DC is τ and the width b,

$$\tau b \delta x = N_1 - N$$

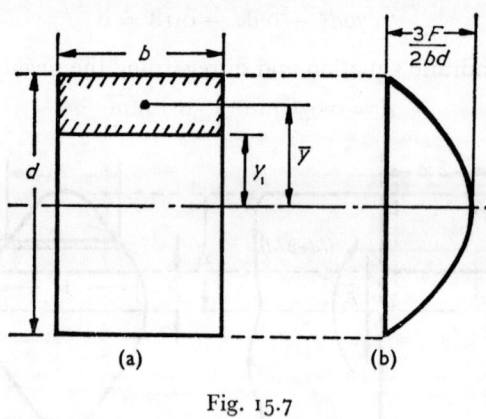

Fig. 15.7

Fig. 15.6 (*b*) shows the transverse section containing the face AD. From Eq. (15.1), the stress on a strip at a distance y from the neutral axis is My/I. Hence, if the width of the strip is z and its thickness δy, the force on the strip is $(My/I)z\delta y$ and, integrating over the face AD,

$$N = \int_{y_1}^{y_2} \frac{My}{I} z \, \mathrm{d}y$$

$$= \frac{M}{I} \int_{y_1}^{y_2} zy \, \mathrm{d}y$$

Similarly,

$$N_1 = \frac{M + \delta M}{I} \int_{y_1}^{y_2} zy \, \mathrm{d}y$$

Hence $$\tau b \delta x = N_1 - N = \frac{\delta M}{I} \int_{y_1}^{y_2} zy \, \mathrm{d}y$$

From Eq. (14.3), it follows that $\delta M/\delta x = F$, so that

$$\tau = \frac{F}{Ib} \int_{y_1}^{y_2} zy \, dy \qquad (15.2)$$

The expression $\int_{y_1}^{y_2} zy \, dy$ is the second moment of area of the face AD about the neutral axis, and if its area is A and its centroid is distant $\bar{y}$ from the neutral axis,

$$\tau = FA\bar{y}/Ib \qquad (15.3)$$

Although τ is the longitudinal shear stress at any point distant y_1 from the neutral axis, it is also the transverse shear stress at this point because

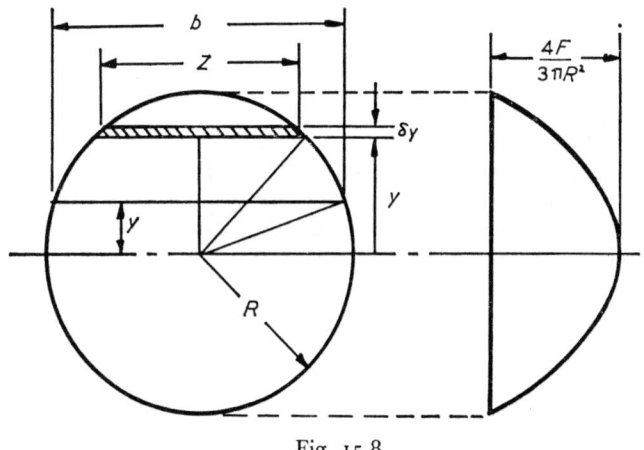

Fig. 15.8

complementary shear stresses are equal. This analysis may, for example, be used to calculate the shear stresses in beams of specified cross-section. For a rectangular beam, for example, with breadth b and depth d, the part of the section above a line drawn at a distance y_1 from the neutral axis is a rectangle (Fig. 15.7 (a)). Its height is $(\frac{1}{2}d - y_1)$ and the distance from its centroid to the neutral axis $y = \frac{1}{2}(\frac{1}{2}d + y_1)$.

Applying Eq. (15.3),

$$\tau = FAy/Ib$$
$$= 12Fb(\tfrac{1}{2}d - y_1)\tfrac{1}{2}(\tfrac{1}{2}d + y_1)/(bd^3)b$$
$$= \frac{6F}{bd^3} (\tfrac{1}{4}d^2 - y_1^2) \qquad (15.4)$$

The shear stress distribution is thus parabolic, as shown by Fig. 15.7 (b).

Maximum shear stress occurs at the neutral axis and, substituting $y_1 = 0$ in Eq. (15.4),

$$\tau_{(max)} = 3F/2bd$$

$$= 1 \cdot 5\tau_{(mean)}$$

For a beam of circular cross section (Fig. 15·8), the application of Eq. (15·2) gives

$$\tau = \frac{F}{Ib} \int_{y_1}^{y_2} zy \, dy = \frac{4F}{3\pi R^4} (R^2 - y_1^2) \qquad (15\cdot5)$$

The shear stress distribution is thus parabolic and

$$\tau_{(max)} = 4F/3\pi R^2$$

$$= \tfrac{4}{3} \times \tau_{(mean)}$$

PROBLEMS

For tutorials

1. A vertical brick wall of uniform thickness is to withstand a wind pressure of 800 N/m². If the density of the brickwork is 2100 kg/m³, what must be the relationship between its height h and its thickness d if tensile stresses at its base are to be avoided?

Ans. $h \not> 8 \cdot 58d^2$

2. A cantilever 2 m long and of square cross-section is made by bolting together two timber beams, the upper one having a cross-section 150 mm × 50 mm and the lower one a cross-section 150 mm × 100 mm. Working from first principles (i.e., considering the equilibrium of the upper beam), find the total shear force to be transmitted by the bolts when a downward load of 1·5 kN is applied to the free end of the cantilever and hence, if each bolt can safely transmit a shear force of 4 kN, the minimum number of bolts required. Show that the same result may be obtained using Eq. 15.4.

Ans. 26·67 kN; 7 bolts

General

1. A bar of circular section, 80 mm diameter, is subjected simultaneously to an axial compressive force of 20 kN and a bending moment of 500 Nm. Show by a diagram how the stress varies over a transverse section, and state the maximum tensile and compressive stresses.

Ans. 5·97 MN/m² tensile; 13·93 MN/m² compressive

2. In a link mechanism, a certain link transmits a maximum tensile force of 5 kN. In order to provide clearance for another part of the mechanism, a bent link is used, its central section having an axis parallel to the line of pull but displaced 60 mm from it. The link is of rectangular section, 20 mm × 50 mm, and the plane in which it is bent is parallel to a 50 mm side. Find the maximum tensile and compressive stresses at a central transverse section of the link.

Ans. 41 MN/m² tensile; 31 MN/m² compressive

3. A concrete dam 4 m high has vertical faces and is of uniform thickness 3 m. If the density of the concrete is 2400 kg/m³, find the maximum and minimum compressive stresses at its base when the water level is 0·4 m from the top of the dam. What is the minimum thickness of this dam if there are to be no tensile stresses at its base when the water level reaches the top of the dam?

(Shearing forces at the base of the dam may be ignored.)

Ans. 145·1 kN/m² and 43·3 kN/m²; 2·58 m

4. A beam is of solid circular section, 200 mm diameter. If the maximum transverse shear stress is to be 4 MN/m², find (*a*) the shearing force which may be applied, (*b*) the mean shear stress, (*c*) the shear stress at a point 40 mm from the neutral axis, assuming it to be constant across the width of the section.

Ans. (*a*) 94·25 kN; (*b*) 3 MN/m²; (*c*) 3·36 MN/m²

5. A horizontal cantilever of rectangular section, 120 mm deep and 80 mm wide, supports a downward load of 20 kN at its free end. Find the principal stresses at a point in the cantilever 150 mm from its free end and 40 mm below its upper surface.

Ans. 6·41 MN/m² (tensile), 1·20 MN/m² (compressive)

CHAPTER 16

Slope and Deflection of Beams

16.1 BENDING MOMENT, SLOPE AND DEFLECTION

When a beam is subjected to transverse loads, it bends and it is this bending action which accounts for nearly all the measurable vertical deflection along the length of the beam. There is also some deflection due to shear, but this is usually negligible.

For a beam which is initially straight and horizontal, it is convenient to specify the position of any point along the neutral axis of the beam by rectangular co-ordinates x and y, the origin being placed at the left-hand end of the beam (Fig. 16.1). The neutral axis may then be represented mathematically. Initially, this equation will be $y = 0$, since the vertical deflection y is zero for all values of x. When the beam is loaded, however, bending will occur, and the neutral axis will assume some other shape (Fig. 16.1). Clearly there must be a relationship between the equation of the deflected neutral axis and the equation for the bending moment M along the beam, and this may be found by considering the curvature of the neutral axis at some point along its length.

Curvature is defined as the reciprocal of radius of curvature. Thus, from the simple theory of bending,

$$\text{Curvature} = 1/R = M/EI \qquad (16.1)$$

where $R = $ Radius of curvature of the neutral axis.

It may be shown that the curvature at a point on a line is given by:

$$1/R = (\mathrm{d}^2y/\mathrm{d}x^2)/\{1 + (\mathrm{d}y/\mathrm{d}x)^2\}^{3/2} \qquad (16.2)$$

where $\mathrm{d}y/\mathrm{d}x = $ Slope of the line, and

$\mathrm{d}^2y/\mathrm{d}x^2 = $ Rate of change of the slope with x.

In most practical situations, the slope of a loaded beam at any point is small so that $(dy/dx)^2$ is negligible. Eq. (16.2) then reduces to

$$\text{Curvature} = 1/R = d^2y/dx^2 \tag{16.3}$$

It follows from Eqs. (16.1) and (16.3) that

$$\text{Curvature} = d^2y/dx^2 = M/EI$$

$$\therefore \ EI(d^2y/dx^2) = M \tag{16.4}$$

The bending moment M can in general be expressed in terms of x, and integration of Eq. (16.4) will therefore yield an expression for (dy/dx),

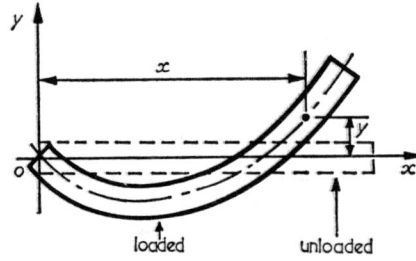

Fig. 16.1

the slope of the neutral axis, in terms of x. Integration of this equation will similarly yield an equation for the deflection y. Each integration will introduce a constant of integration, these can normally be obtained from known end conditions.

16.2 SIMPLE CASES

Cantilever beam with single concentrated load

At a distance x from the free end of the cantilever shown in Fig. 16.2

$$M = EI(d^2y/dx^2) = -Wx$$

$$\therefore \ EI(dy/dx) = -\tfrac{1}{2}Wx^2 + A$$

$$\therefore \ EIy = -\tfrac{1}{6}Wx^3 + Ax + B$$

where A and B are constants of integration. (It has, of course, been assumed that neither E nor I varies with x.)

When $x = L$, $dy/dx = 0$, so that

$$0 = -\tfrac{1}{2}WL^2 + A$$

i.e. $$A = \tfrac{1}{2}WL^2$$

When $x = L, y = 0,$

$$\therefore \; 0 = -\tfrac{1}{6}WL^3 + AL + B$$

i.e. $$B = \tfrac{1}{6}WL^3 - (\tfrac{1}{2}WL^2)L = -\tfrac{1}{3}WL^3$$

Thus: $$EI\,(dy/dx) = -\tfrac{1}{2}Wx^2 + \tfrac{1}{2}WL^2 \tag{16.5}$$

$$EIy = -\tfrac{1}{6}Wx^3 + \tfrac{1}{2}WL^2x - \tfrac{1}{3}WL^3 \tag{16.6}$$

Eqs. (16.5) and (16.6) enable the slope (dy/dx) and the deflection y to be calculated at any point along the length of the beam. In particular at the free end where $x = 0,$

$$\text{Slope} = +WL^2/2EI \tag{16.7}$$

$$\text{Deflection} = -WL^3/3EI \tag{16.8}$$

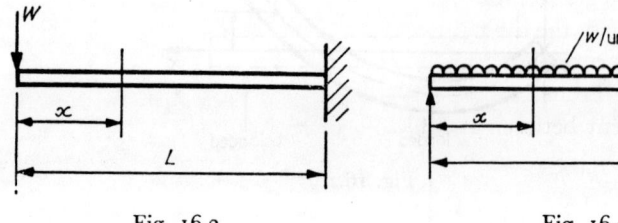

Fig. 16.2   Fig. 16.3

Simply supported beam with uniformly distributed load

Fig. 16.3 represents a simply supported beam with uniformly distributed load. By symmetry, the reaction at each end will be $\tfrac{1}{2}wL$. At some distance x from the left hand end,

$$M = EI\,(d^2y/dx^2) = -\tfrac{1}{2}wx^2 + \tfrac{1}{2}wLx$$

$$\therefore \; EI\,(dy/dx) = -\tfrac{1}{6}wx^3 + \tfrac{1}{4}wLx^2 + A$$

$$\therefore \; EIy = -\tfrac{1}{24}wx^4 + \tfrac{1}{12}wLx^3 + Ax + B$$

where A and B are constants of integration.

When $x = 0, y = 0$ and therefore $B = 0$. When $x = L, y = 0$, so that

$$0 = -\tfrac{1}{24}wL^4 + \tfrac{1}{12}wL^4 + AL$$

$$\therefore \; A = -\tfrac{1}{24}wL^3$$

Thus the equations for slope and deflection are:

$$EI\,(dy/dx) = -\tfrac{1}{6}wx^3 + \tfrac{1}{4}wLx^2 - \tfrac{1}{24}wL^3 \tag{16.9}$$

$$EIy = -\tfrac{1}{24}wx^4 + \tfrac{1}{12}wLx^3 - \tfrac{1}{24}wL^3x \tag{16.10}$$

The deflection y will be a maximum when (dy/dx) is zero. In general, the point of maximum deflection may be found by equating the expression for the slope to zero, but in this example, symmetry alone shows that (dy/dx) is zero at the midpoint of the beam. Thus, putting $x = L/2$ in Eq. (16.10),

$$EIy_{max} = -\tfrac{1}{24} w(L/2)^4 + \tfrac{1}{12} wL(L/2)^3 - \tfrac{1}{24} wL^3(L/2)$$

i.e. $$y_{max} = -5wL^4/384EI \qquad (16.11)$$

Simply supported beam with central concentrated load

Concentrated loads cause discontinuities in the bending moment diagram, so that the bending moment equation applicable to one section of a beam does not apply to others. In general, this hinders the determination of constants of integration, but for a simply supported beam and a central concentrated load, the difficulty may be avoided by considering only half of the beam. From Fig. 16.4, the bending moment between A and C is given by:

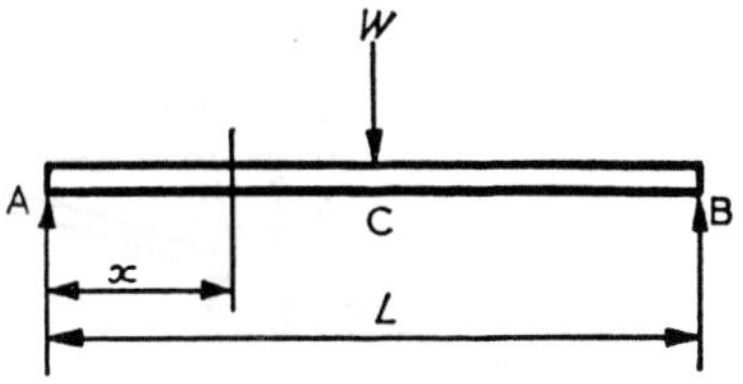

Fig. 16.4

$$M_{AC} = \tfrac{1}{2}Wx \quad \text{where} \quad 0 \leqslant x \leqslant L/2$$

$$\therefore \ EI(d^2y/dx^2) = \tfrac{1}{2}Wx$$

Thus:

$$EI(dy/dx) = \tfrac{1}{4}Wx^2 + A$$
$$EIy = \tfrac{1}{12} Wx^3 + Ax + B$$

where A and B are constants of integration.

By symmetry, $dy/dx = 0$ when $x = L/2$,

$$\therefore \ \tfrac{1}{4}W(L/2)^2 + A = 0$$

i.e. $$A = -WL^2/16$$

Also, when $x = 0$, $y = 0$, so that $B = 0$.

Thus the equations for slope and deflection between A and C are:

$$EI(dy/dx) = \tfrac{1}{4}Wx^2 - WL^2/16 \qquad (16.12)$$
$$EIy = \tfrac{1}{12} Wx^3 - \tfrac{1}{16} WL^2x \qquad (16.13)$$

The slope and deflection for AC will be those for BC at corresponding positions. The maximum deflection is obviously at the midpoint of the beam, where the slope is zero. Putting $x = L/2$ in Eq. (16.13),

$$EIy_{max} = \tfrac{1}{12}(WL/2)^3 - \tfrac{1}{16} WL^2(L/2)$$

i.e. $$y_{max} = -WL^3/48EI \qquad (16.14)$$

Principle of superposition

For linear systems in which the deflection is directly proportional to the applied load, the principle of superposition may be applied. This means that if more than one load is applied to a structure, the deflection at some point is equal to the sum of the deflections at that point due to each load acting alone.

This is illustrated by a simply supported beam carrying a uniformly distributed load w per unit length over its entire span, together with a central concentrated load W. The deflections at the mid-point due to w

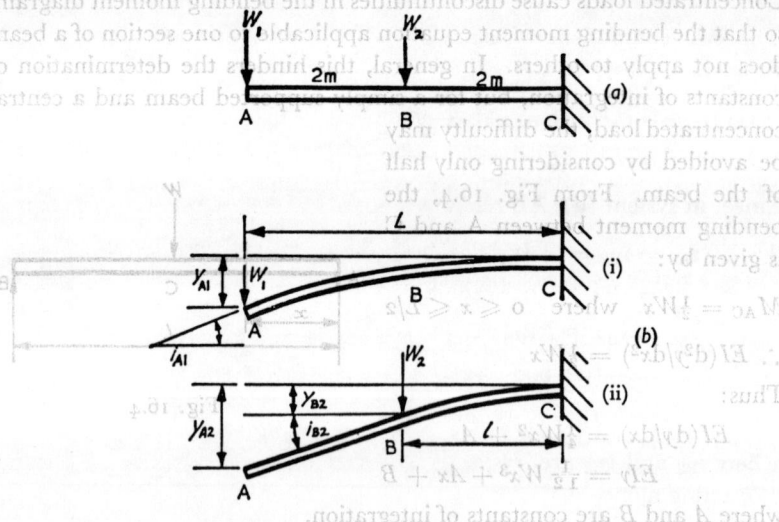

Fig. 16.5

and to W acting separately are given by Eqs. (16.11) and (16.14), so that, by the principle of superposition, the deflection at the mid-point due to the simultaneous application of w and W is

$$-\{(5wL^4/38EI) + (WL^3/48EI)\}$$

This result may be verified by the more conventional method used in the previous case.

Example. *A cantilever beam ABC is 4 m long and supports masses of 1 tonne at the free end A, and 2 tonne at its mid-point B (Fig. 16.5 (a)). The second moment of area of the section of the beam is 50×10^{-6} m^4 and $E = 200$ GN/m^2. Determine the slope and deflection of the beam at A.*

The calculation of W_1 and W_2 is simplified by assuming $g = 10$ m/s^2, so that:

$$W_1 = m_1 g \doteqdot 1000 \times 10 = 10 \text{ kN}$$
$$W_2 = m_2 g \doteqdot 2000 \times 10 = 20 \text{ kN}$$

Also, $EI = 200 \times 10^9 \times 50 \times 10^{-6} = 10^7 \text{ Nm}^2$

Let

y_{A1} = Deflection at A due to W_1 acting alone

y_{A2} = Deflection at A due to W_2 acting alone

i_{A1} = Slope at A due to W_1 acting alone

i_{A2} = Slope at A due to W_2 acting alone

The two component systems (i) and (ii) are shown in Fig. 16.5 (*b*). From Eqs. (16.7) and (16.8),

$$i_{A1} = +W_1 L^2/2EI = +10\,000 \times 4^2/(2 \times 10^7) = +8 \times 10^{-3} \text{ radian}$$
$$y_{A1} = -W_1 L^3/3EI = -10\,000 \times 4^3/(3 \times 10^7) = -21 \cdot 3 \times 10^{-3} \text{ m}$$

Since, in system (ii), AB remains straight,

$$i_{A2} = i_{B2} = +W_2 l^2/2EI = +20\,000 \times 2^2/(2 \times 10^7)$$
$$= +4 \times 10^{-3} \text{ radian}$$
$$y_{A2} = y_{B2} + AB i_{B2} = (-W_2 l^3/3EI) + (-2)(+4 \times 10^{-3})$$
$$= -\{(20\,000 \times 2^3/3 \times 10^7) + 8 \times 10^{-3}\}$$
$$= -13 \cdot 3 \times 10^{-3} \text{ m}$$

where i_{B2} and y_{B2} are, respectively, the slope and deflection at B due to W_2 acting alone.

By the principle of superposition,

Total slope at A $= i_A = i_{A1} + i_{A2} = 8 \times 10^{-3} + 4 \times 10^{-3}$
$$= 12 \times 10^{-3} \text{ radian}$$

Total deflection at A $= y_A = y_{A1} + y_{A2} = -21 \cdot 3 \times 10^{-3} - 13 \cdot 3 \times 10^{-3}$
$$= -34 \cdot 6 \times 10^{-3} \text{ m}$$
$$= -34 \cdot 6 \text{ mm}$$

16.3 MACAULAY'S METHOD

In general, concentrated loads or discontinuous distributed loads on a beam create difficulties in the derivation of slope and deflection equations, for a series of bending moment equations is required, each applicable only to one section of the beam between adjacent discontinuities. The substitution of known conditions of slope and deflection is thus impossible and the constants of integration cannot be found.

One way round this difficulty is to find an equation for bending moment which is applicable to the full length of the beam. This is not mathematically possible, but Macaulay has devised an artifice which provides this universal equation and gives correct results provided certain rules are obeyed:

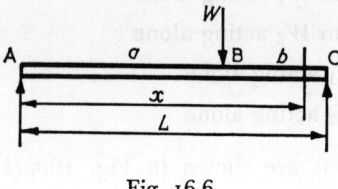

Fig. 16.6

1. The equation for the bending moment at a section just short of the right-hand end of the beam is written down in terms of x and the loads on the beam, x being measured from the extreme left-hand end of the beam. This equation will contain terms such as $W_1[x - a]$, $W_2[x - b]$, $\frac{1}{2}w[x - c]^2$, etc. (It is customary to use square brackets for these terms, as shown.)

2. The expressions in the square brackets are treated like x and are integrated intact. For example, $W_1[x - a]$ becomes $\frac{1}{2}W_1[x - a]^2$, which on integrating again becomes $\frac{1}{6}W_1[x - a]^3$. On no account must these binomial expressions be expanded. In this way equations for slope and deflection are derived. Any irregularity caused by this method of integration is automatically corrected by the constants of integration.

3. When particular values of x are substituted in the equations for bending moment, slope or deflection, any term inside the square brackets which becomes negative is ignored.

4. Distributed loads *must* continue to the extreme right-hand end of the beam. If this is not the case, then the load is extended to the right-hand end and the "non-existent" portion balanced out as shown in Fig. 16.7 (*b*).

5. When bending moment, slope or deflection equations are equated to zero with a view to solving for x, the appropriate binomial expressions —those which would be negative—must be ignored. It is therefore necessary to have an approximate idea of the roots of the equation. This seldom presents difficulties.

Example. *A beam of uniform section ABC is simply supported at A and C, and carries a single concentrated load W at B (Fig. 16.6). Determine the slope at A, the deflection at B and the position of the point of maximum deflection.*

By the principle of moments, $R_A = Wb/L$ and $R_C = Wa/L$. Consider a section just short of C and distance x from A.

$$M = EI(\mathrm{d}^2y/\mathrm{d}x^2) = R_A x - W[x - a]$$

$$\therefore\ EI(dy/dx) = \tfrac{1}{2}R_A x^2 - \tfrac{1}{2}W[x - a]^2 + A$$
$$\therefore\ EIy = \tfrac{1}{6}R_A x^3 - \tfrac{1}{6}W[x - a]^3 + Ax + B$$

where A and B are constants of integration.

When $x = 0, y = 0$, so that

$$0 = 0 - \tfrac{1}{6}W[-ve]^3 + 0 + B$$

Since terms having negative binomial expressions are ignored, $B = 0$.

When $x = L, y = 0$, and

$$0 = \tfrac{1}{6}R_A L^3 - \tfrac{1}{6}W[L - a]^3 + AL$$

i.e.
$$AL = \tfrac{1}{6}Wb^3 - \tfrac{1}{6}(Wb/L) \cdot L^3 = \tfrac{1}{6}Wb(b^2 - L^2)$$
$$= \tfrac{1}{6}Wb(b - L)(b + L) = -\tfrac{1}{6}Wab(L + b)$$
$$\therefore\ A = -Wab(L + b)/6L$$

Thus, the equations for slope and deflection are

$$EI(dy/dx) = (Wb/2L)x^2 - \tfrac{1}{2}W[x - a]^2 - \{Wab(L + b)/6L\} \quad (16.15)$$
$$EIy = (Wb/6L)x^3 - \tfrac{1}{6}W[x - a]^3 - \{Wab(L + b)/6L\} \cdot x \quad (16.16)$$

The slope at A is therefore given by:

$$EI(dy/dx)_A = 0 - \tfrac{1}{2}W[-ve]^2 - \{Wab(L + b)/6L\}$$

i.e.
$$(dy/dx)_A = -Wab(L + b)/6EIL$$

Deflection at B is given by:

$$EIy_B = (Wb/6L)a^3 - 0 - \{Wab(L + b)/6L\}a$$

i.e.
$$y_B = -Wa^2b^2/3EIL \quad (16.17)$$

This equation for the deflection at the load point is most useful, and falls into the same category as Eqs. (16.8), (16.11) and (16.14). All are standard cases with which engineers become familiar through constant use.

Maximum deflection occurs when dy/dx is zero but before Eq. (16.15) is equated to zero, it must be decided in which part of the beam the point of maximum deflection will occur. In this case the maximum deflection is obviously to be found in AB (if $a > b$) or, mathematically, $x < a$ Thus $[x - a]$ is negative, the second term of Eq. (16.15) is ignored and the point of maximum deflection is given by:

$$(Wb/2L)x^2 - \{Wab(L + b)/6L\} = 0$$

i.e.
$$x = \sqrt{\{a(L + b)/3\}} = \sqrt{\{(L^2 - b^2)/3\}} \quad (16.18)$$

The negative root has, of course, been neglected.

It is interesting to see how the position of the point of maximum deflection varies with b. From Eq. (16.18), as b varies from $L/2$ to zero, x varies from $L/2$ to $L/(\sqrt{3})$. In other words, as the load is moved from the mid-point towards one of the supports, the point of maximum deflection never moves more than $0.0773L$ from the mid-point.

Example. *A beam* ABCD *is supported at* A *and* C, *and carries a uniformly distributed load* w *per unit length between* B *and* C *(Fig. 16.7 (a)).* AB = BC = CD = l. *Calculate the deflection at the overhanging end* D.

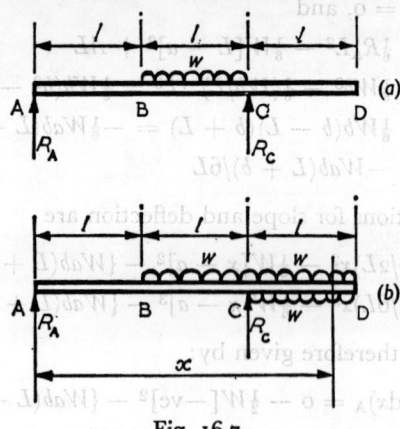

Fig. 16.7

First, the support reactions must be found. Take moments about A,

$$R_c \times 2l = wl \times 1.5l$$
$$\therefore R_c = \tfrac{3}{4}wl$$
$$\therefore R_A = \tfrac{1}{4}wl$$

In accordance with the rules for Macaulay's method, the distributed load between B and C must be continued to D and, to maintain the original condition, it is necessary to introduce an equal and opposite distributed load along CD (Fig. 16.7 (b)).

For a section of the beam distant x from A, just short of the right-hand end D,

$$M = EI(d^2y/dx^2) = R_Ax - \tfrac{1}{2}w[x - l]^2 + R_c[x - 2l] + \tfrac{1}{2}w[x - 2l]^2$$
$$\therefore EI(dy/dx) = \tfrac{1}{2}R_Ax^2 - \tfrac{1}{6}w[x - l]^3 + \tfrac{1}{2}R_c[x - 2l]^2$$
$$+ \tfrac{1}{6}w[x - 2l]^3 + A$$
$$\therefore EIy = \tfrac{1}{6}R_Ax^3 - \tfrac{1}{24}w[x - l]^4 + \tfrac{1}{6}R_c[x - 2l]^3$$
$$+ \tfrac{1}{24}w[x - 2l]^4 + Ax + B$$

where A and B are constants of integration.

When $x = 0, y = 0$, so that

$$0 = 0 - [-\text{ve}] + [-\text{ve}] + [-\text{ve}] + 0 + B$$

Since negative binomial terms in square brackets are ignored, $B = 0$.

When $x = 2l, y = 0$, and

$$0 = \tfrac{1}{6}R_A(2l)^3 - \tfrac{1}{24}w[2l - l]^4 + 0 + 0 + A(2l) + 0$$

$$\therefore A = \tfrac{1}{48}wl^3 - \tfrac{2}{3}R_Al^2 = \tfrac{1}{48}wl^3 - \tfrac{2}{3}(\tfrac{1}{4}wl)l^2$$

i.e. $\qquad A = -\tfrac{7}{48}wl^3$

At D, $x = 3l$, so that the deflection at D is given by:

$$EI y_D = \tfrac{1}{6}R_A(3l)^3 - \tfrac{1}{24}w[3l - l]^4 + \tfrac{1}{6}R_c[3l - 2l]^3$$

$$+ \tfrac{1}{24}w[3l - 2]l^4 + A(3l)$$

$$= \tfrac{1}{6}(\tfrac{1}{4}wl)27l^3 - \tfrac{1}{24}w \cdot 16l^4 + \tfrac{1}{6}(\tfrac{3}{4}wl)l^3 + \tfrac{1}{24}wl^4 - \tfrac{7}{48}wl^3 \cdot 3l$$

$$= +\tfrac{3}{16}wl^4$$

i.e. $y_D = +3wl^4/16EI$

16.4 MOHR'S THEOREMS

These mathematical arguments can be interpreted graphically and the results are two important theorems, due to Mohr, which are alternative ways of obtaining slope and deflection.

1. If A and B are two points on the neutral axis of a loaded beam, the change of slope of the neutral axis from A to B is equal to the area of M/EI diagram on AB. Thus in Fig. 16.8, the

Change of slope from A to B

$$= i_B - i_A = A_{AB}/EI \qquad (16.19)$$

where $A_{AB} = $ Area of B.M. diagram on AB, and EI is constant.

2. If A and B are two points on the neutral axis of a loaded beam, *the deflection of A from the tangent at B is equal to the* first moment of area *about A* of the M/EI diagram on AB. In Fig. 16.8,

Deflection of A from tangent B

$$= y_{AB} = A_{AB}\bar{x}/EI \qquad (16.20)$$

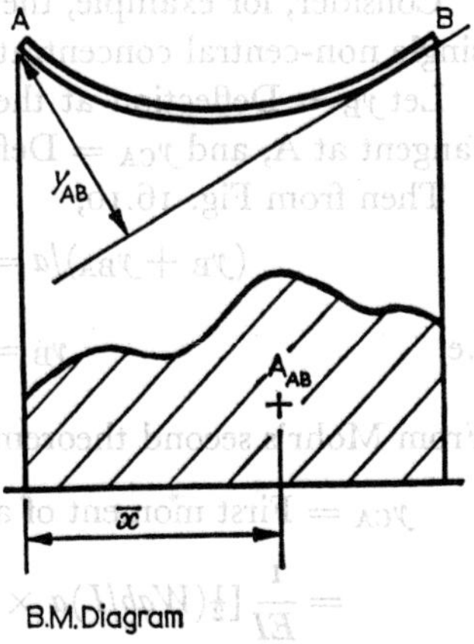

B.M. Diagram

Fig. 16.8

where $\bar{x} =$ Distance of centroid of A_{AB} from A, and again, EI is assumed
constant.

Mohr's theorem often referred to as the area-moment method, reduces
the calculation of slope and deflection to elementary geometry so long
as the bending moment diagram is relatively simple. In this connection,
the principle of superposition is often a great help. The bending moment
diagram can usually be reduced to a series of triangles and, since $\bar{x}$ is
measured horizontally along the beam, these triangles should be regarded
as lying on their sides with the "base" vertical and the "perpendicular

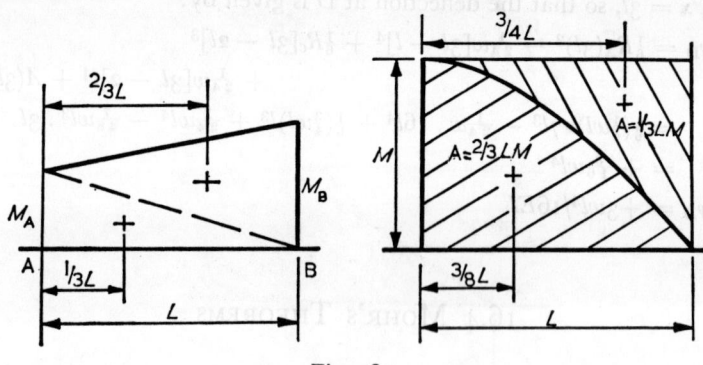

Fig. 16.9

height" horizontal. With uniformly distributed loads, the bending
moment diagram is parabolic and the information given in Fig. 16.9
will therefore prove useful.

Consider, for example, the case of the simply supported beam with the
single non-central concentrated load.

Let $y_B =$ Deflection at the load point B, $y_{BA} =$ Deflection of B from
tangent at A, and $y_{CA} =$ Deflection of C from tangent at A.

Then from Fig. 16.10,

$$(y_B + y_{BA})/a = y_{CA}/L \qquad \text{(similar triangles)}$$

i.e.
$$v_B = \left(\frac{a}{L}\right) y_{CA} - y_{BA} \qquad (16.21)$$

From Mohr's second theorem:

$y_{CA} =$ First moment of area of M/EI diagram on AC about C

$$= \frac{1}{EI} \left[\tfrac{1}{2}(Wab/L)a \times (\tfrac{1}{3}a + b) + \tfrac{1}{2}(Wab/L)b \times \tfrac{2}{3}b \right]$$

$$= \frac{Wab}{6EIL} [a^2 + 3ab + 2b^2]$$

y_{BA} = First moment of area of M/EI diagram on AB about B

$$= \frac{1}{EI} [\tfrac{1}{2}(Wab/L)a \times \tfrac{1}{3}a]$$

$$= \frac{Wab}{6EIL} a^2$$

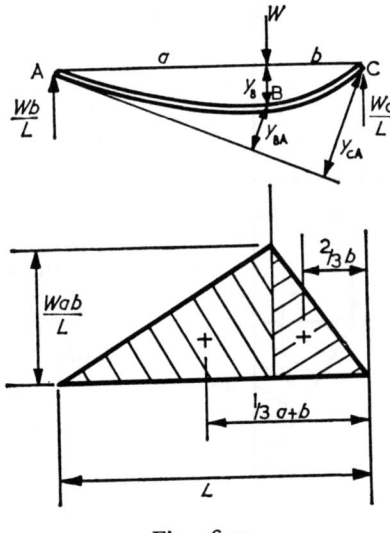

Fig. 16.10

Substitution in Eq. (16.21) gives:

$$y_B = \frac{a}{L} \frac{Wab}{6EIL} [a^2 + 3ab + 2b^2] - \frac{Wab}{6EIL} a^2$$

$$= \frac{Wa^2b}{6EIL^2} [a^2 + 3ab + 2b^2 - aL]$$

$$= \frac{Wa^2b}{6EIL^2} [a^2 + 3ab + 2b^2 - a(a + b)]$$

$$= \frac{Wa^2b}{6EIL^2} [2ab + 2b^2] = \frac{Wa^2b^2}{3EIL}$$

Alternatively, the bending moment may be split up using the principle of superposition (Fig. 16.11).

From Mohr's second theorem:

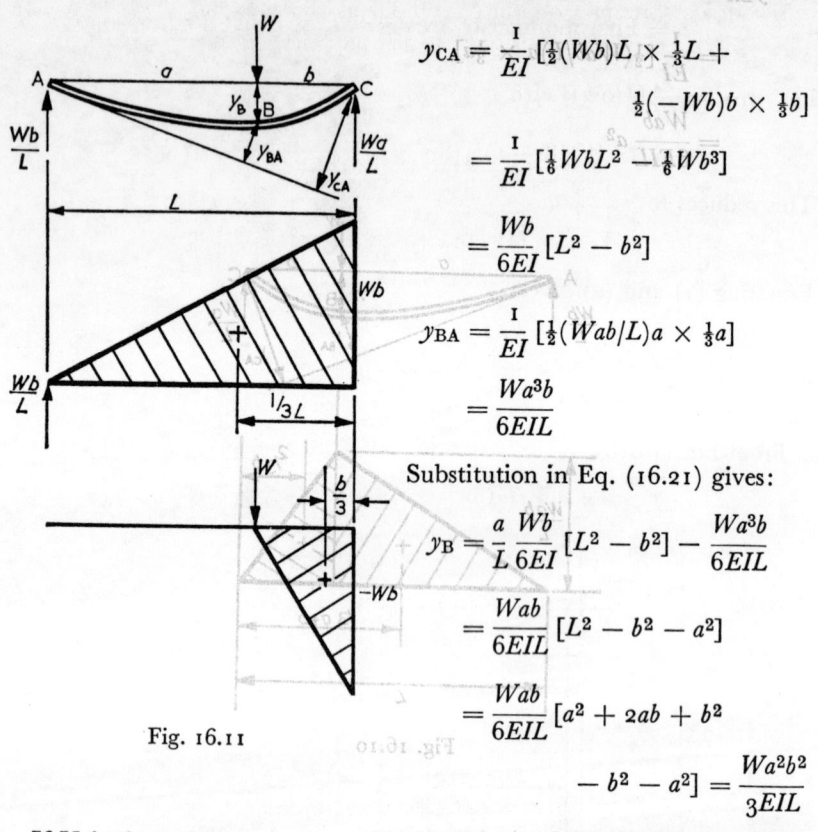

Fig. 16.11

$$y_{CA} = \frac{1}{EI}\left[\tfrac{1}{2}(Wb)L \times \tfrac{1}{3}L + \tfrac{1}{2}(-Wb)b \times \tfrac{1}{3}b\right]$$

$$= \frac{1}{EI}\left[\tfrac{1}{6}WbL^2 - \tfrac{1}{6}Wb^3\right]$$

$$= \frac{Wb}{6EI}[L^2 - b^2]$$

$$y_{BA} = \frac{1}{EI}\left[\tfrac{1}{2}(Wab/L)a \times \tfrac{1}{3}a\right]$$

$$= \frac{Wa^3b}{6EIL}$$

Substitution in Eq. (16.21) gives:

$$y_B = \frac{a}{L}\frac{Wb}{6EI}[L^2 - b^2] - \frac{Wa^3b}{6EIL}$$

$$= \frac{Wab}{6EIL}[L^2 - b^2 - a^2]$$

$$= \frac{Wab}{6EIL}[a^2 + 2ab + b^2 - b^2 - a^2] = \frac{Wa^2b^2}{3EIL}$$

If X is the point of maximum deflection, then the tangent at X will be horizontal. It follows that y_{AX} and y_{CX} are equal. By equating y_{AX} and y_{CX}, an equation in terms of $x(=AX)$ is produced which may then be solved to produce the same result as Eq. (16.18).

Example. *A uniform beam ABC is simply supported at A and C and carries a concentrated load at B. If* $AB = 8$ m *and* $BC = 2$ m, *find the position of the point of maximum deflection.*

The arrangement is shown in Fig. 16.12. X is the point of maximum deflection. Let $AX = x$.

Deflection of A from tangent at X

     = First moment of area *about A* of M/EI diagram on AX

i.e. $$y_{AX} = \frac{1}{EI}\left[\tfrac{1}{2}(0{\cdot}2Wx)x\tfrac{2}{3}x\right] = (W/3EI)0{\cdot}2x^3 \qquad (1)$$

Deflection of C from tangent at **X**

= First moment of area *about C* of M/EI diagram on **CX**

i.e. $\quad y_{CX} = \dfrac{1}{EI}\left[\tfrac{1}{2}(0{\cdot}2Wx)(8-x)\{\tfrac{2}{3}(8-x)+2\}\right.$

$\left. + \tfrac{1}{2}(1{\cdot}6W)(8-x)\{\tfrac{1}{3}(8-x)+2\} + \tfrac{1}{2}(1{\cdot}6W)\times 2\times \tfrac{2}{3}\times 2\right]$

This reduces to

$$Y_{CX} = (W/3EI)[0{\cdot}2x^3 - 3x^2 + 96] \qquad (2)$$

Equating (1) and (2) gives:

$$0{\cdot}2x^3 = 0{\cdot}2x^3 - 3x^2 + 96$$

i.e. $\qquad\qquad\qquad x^2 = 96/3$

i.e. $\qquad\qquad\qquad x = 5{\cdot}657 \text{ m}$

From Eq. (16.18),

$$x = \sqrt{\{(10^2 - 2^2)/3\}} = 5{\cdot}657 \text{ m}$$

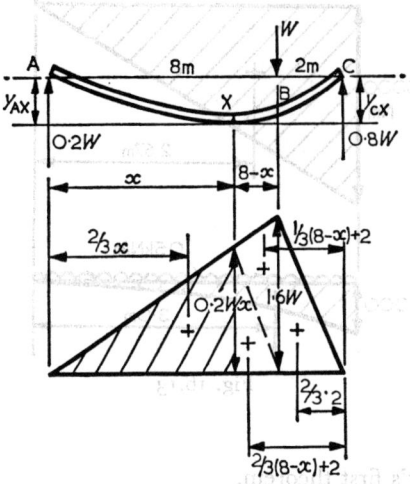

Fig. 16.12

Example. *A cantilever beam* ABC *is built-in horizontally at* A *and carries concentrated loads of* 10 kN *at* B *and* 5 kN *at* C. *In addition, the beam carries a uniformly distributed load of* 0·5 kN/m *along its entire length.* AB = 1·0 m; BC = 3 m; I = 50 × 10⁻⁶ m⁴ *and* E = 200 GN/m². *Determine the slope and deflection at the free end* C.

$$EI = 200 \times 10^9 \times 50 \times 10^{-6} = 10^7 \text{ Nm}^2$$

The system is shown in Fig. 16.13, together with its bending moment diagram, which, for convenience, has been split into 3 separate components in accordance with the principle of superposition.

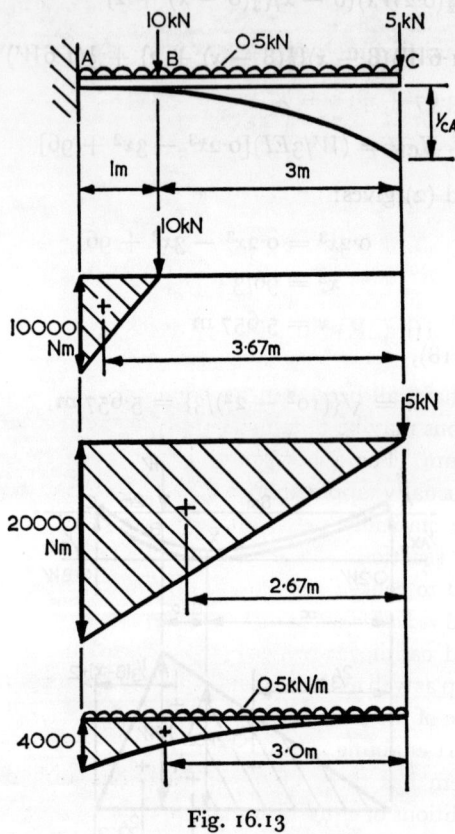

Fig. 16.13

Applying Mohr's first theorem,
   Change in slope from A to C = Area of M/EI diagram on AC

$$\therefore i_C - i_A = \frac{1}{EI} [\tfrac{1}{2} \times 10\,000 \times 1\cdot0 + \tfrac{1}{2} \times 20\,000 \times 4 + \tfrac{1}{3} \times 4000 \times 4]$$

But $i_A = 0$, so that

$$\text{Slope at C} = 10^{-7}[5000 + 40\,000 + 5333]$$

$$= 0\cdot0050 \text{ radian}$$

Since the beam is built-in horizontally at A, the tangent at A remains horizontal.

∴ Deflection at the free end C = Deflection of C from tangent at A

i.e.

$$y_C = y_{CA} = \text{Total first moment of area of } M/EI \text{ diagram about C}$$

$$= [5000 \times 3 \cdot 67 + 40\ 000 \times 2 \cdot 67 + 5333 \times 3]/EI$$

$$= (18\ 330 + 106\ 670 + 16\ 000)10^{-7}$$

$$= 0 \cdot 0141\ \text{m} = 14 \cdot 1\ \text{mm}$$

16.5 REDUNDANT SUPPORTS

So far the beams have all been statically determinate, which means that the support reactions may be obtained simply by applying the conditions of static equilibrium. For a horizontal beam, only two conditions o equilibrium are usually applicable—i.e. $\Sigma V = 0$ and $\Sigma M = 0$. The beam is therefore statically determinate provided there are only two unknown support reactions. In practice, however, additional supports are often provided to relieve stress or to give greater rigidity, and these supports are called *redundancies*.

Thus a propped cantilever has one redundant support for there is a reaction at the prop as well as the reaction and fixing moment at the built-in end. The presence of the prop considerably reduces the bending moment at the fixed support enabling a beam of lighter section to be used. However, such a system has three unknown reactions and, since there are still only two conditions of equilibrium applicable, none of these reactions can be obtained by normal methods. The structure is in fact statically indeterminate. Other examples of statically indeterminate structures are the continuous beam (a simply supported beam with more than two supports) and a beam which is built-in (or encastré) at both ends.

For design purposes it is clearly necessary to be able to calculate the reactions in redundant supports. The problem is that the conditions of equilibrium do not provide sufficient equations. But, although each redundant support introduces an additional unknown, it also introduces a known slope or deflection. Thus, the slope and deflection equations, as obtained for example, by Macaulay's method, can provide additional equations for the determination of redundant reactions. In general, let n = Number of supports.

$\therefore$ Unknowns $= n$ reactions $+$ 2 constants of integration $= n + 2$

Equations $= n$ conditions of slope or deflection

$+$ 2 conditions of equilibrium $= n + 2$

$\therefore$ Number of equations $=$ Number of unknowns.

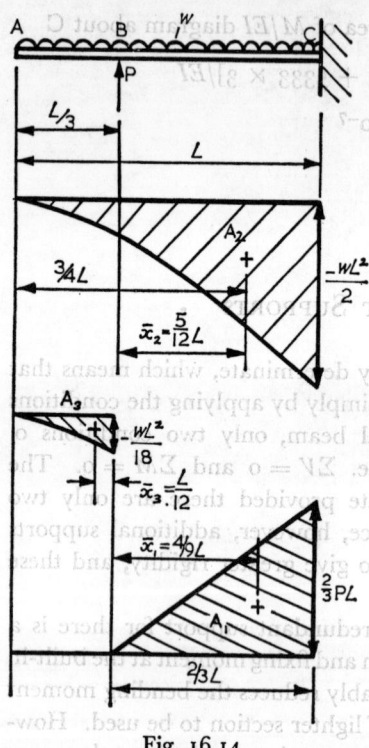

Any method of obtaining expressions for the known conditions of slope or deflection is, of course, acceptable, so that Mohr's theorems or strain energy methods may also be used (see Chapter 17).

Propped cantilever

For convenience, the bending moment diagram of a typical propped cantilever system (Fig. 16.14) has been split into two components in accordance with the principle of superposition. The system is redundant because there are three unknown reactions, P, R_C and M_C. It is advisable always to assume a positive sense for fixing moments in accordance with the bending moment sign convention.

Fig. 16.14

Using Macaulay's method

$$EI(\mathrm{d}^2y/\mathrm{d}x^2) = -\tfrac{1}{2}wx^2 + P[x - (L/3)]$$

$$EI(\mathrm{d}y/\mathrm{d}x) = -\tfrac{1}{6}wx^3 + \tfrac{1}{2}P[x - (L/3)]^2 + A$$

$$EIy = -\tfrac{1}{24}wx^4 + \tfrac{1}{6}P[x - (L/3)]^3 + Ax + B$$

There are five unknowns, P, R_C, M_C, A and B, but five equations may be obtained as follows:

1. Slope at C is zero.
2. Deflection at B is zero.
3. Deflection at C is zero.
4. Algebraic sum of vertical forces is zero.
5. Algebraic sum of the moments of the forces about any point is zero.

1. When $x = L$, $(dy/dx) = 0$,
$$\therefore \ 0 = -\tfrac{1}{6}wL^3 + \tfrac{1}{2}P[L - (L/3)]^2 + A$$
$$\therefore \ A = \tfrac{1}{6}wL^3 - \tfrac{2}{9}PL^2$$

2. When $x = L/3$, $y = 0$,
$$\therefore \ 0 = -\tfrac{1}{24}w(L/3)^4 + \tfrac{1}{6}P[0]^3 + A(L/3) + B$$
$$\therefore \ B = (wL^4/1944) - (L/3)\{\tfrac{1}{6}wL^3 - \tfrac{2}{9}PL^2\}$$
i.e. $$B = -(107wL^4/1944) + (2PL^3/27)$$

3. When $x = L$, $y = 0$,
$$\therefore \ 0 = -\tfrac{1}{24}wL^4 + \tfrac{1}{6}P[2L/3]^3 + AL + B$$
i.e. $$0 = -\tfrac{1}{24}wL^4 + \tfrac{4}{81}PL^3 + (\tfrac{1}{6}wL^3 - \tfrac{2}{9}PL^2)L - (107wL^4/1944)$$
$$+ (2PL^3/27)$$

From which:

$$P = \tfrac{17}{24}wL$$

4. Resolving vertically,
$$P + R_C = wL$$
$$\therefore \ R_C = wL - \tfrac{17}{24}wL = \tfrac{7}{24}wL$$

5. Taking moments about C,
$$M_C + \tfrac{1}{2}wL^2 = P \times \tfrac{2}{3}L$$
$$\therefore \ M_C = \tfrac{17}{24}wL \times \tfrac{2}{3}L - \tfrac{1}{2}wL^2 = -wL^2/36$$

All the reactions are known and it is a simple matter to draw the bending moment diagram for the beam.

Area-moment method. Since the beam is built-in at C, the tangent at C will remain horizontal. Therefore the deflection of B from the tangent at C = 0, and the first moment of area about B of the M/EI diagram on BC is also zero. From Fig. 16.14,

$$(A_1\bar{x}_1 + A_2\bar{x}_2 - A_3\bar{x}_3)/EI = 0$$

From the diagram:

$$A_1\bar{x}_1 = (\tfrac{1}{2} \times \tfrac{2}{3}PL \times \tfrac{2}{3}L)(\tfrac{4}{9}L) = 8PL^3/81$$
$$A_2\bar{x}_2 = \{\tfrac{1}{3} \times (-\tfrac{1}{2}wL^2) \times L\}(\tfrac{5}{12}L) = -5wL^4/72$$
$$A_3\bar{x}_3 = \{\tfrac{1}{3}(-wL^2/18)\tfrac{1}{3}L\}(-L/12) = +wL^4/1944$$
$$\therefore \ (8PL^3/81) + (-5wL^4/72) - (wL^4/1944) = 0$$

From which:

$$P = \tfrac{17}{24}wL$$

$$\therefore \ R_C = \tfrac{7}{24}wL \quad \text{and} \quad M_C = -wL^2/36, \text{ as before}$$

The area-moment method tends to be more direct, but care must be taken in the evaluation of areas, moment arms and their *signs*. The areas and moment arms for the parabolic figures in Fig. 16.14 have been calculated using the information given in Fig. 16.9.

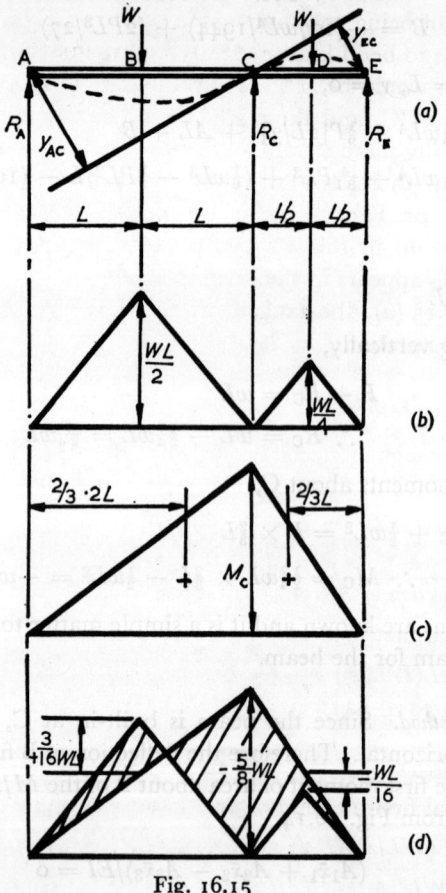

Fig. 16.15

Continuous beam

The simply supported beam shown in Fig. 16.15 (a) qualifies as a continuous beam since it has three supports and covers two spans. Again the redundant reaction can be found using Macaulay's method precisely as for the propped cantilever. There are three known positions of zero deflection and two available conditions of equilibrium giving five equations, from which the two constants of integration and the three unknown reactions may be found. Extension of the beam to cover more spans does not alter the problem, since each additional support introduces another unknown reaction but also another known point of zero deflection.

In what follows the problem of Fig. 16.15 (*a*) will be solved by the area-moment method, but the result should be checked using Macaulay's method.

If, instead of being continuous, the beam were *hinged* at C, each span would be able to bend independently and the bending moment diagram for the system would be as shown in Fig. 16.15 (*b*). However, the beam is continuous at C, and there will be interaction between the spans so that the bending moment at C is not zero but some unknown value M_C. It is necessary, therefore, to superimpose the bending moment diagram Fig. 16.15 (*c*) on Fig. 16.15 (*b*). It is convenient to split the bending moment diagram in this way, since, by doing so, only one unknown dimension M_C appears in the diagrams.

In Fig. 16.15 (*a*), the tangent at C is drawn and using the normal area-moment notation:

$$y_{AC}/2L = -y_{EC}/L \qquad \text{(similar triangles)}$$

(The negative sign is necessary because the deflections are in opposite directions relative to the tangent.)

$$\therefore \; y_{AC} + 2y_{EC} = 0$$

$y_{AC} =$ *Deflection of A from the tangent at C*

$\quad =$ Total first moment of area *about A* of the M/EI diagram on AC

$\quad = [\{\tfrac{1}{2}2L(WL/2)\}L + (\tfrac{1}{2}M_c \times 2L)\tfrac{4}{3}L]/EI$

$\quad = [\tfrac{1}{2}WL^3 + \tfrac{4}{3}M_cL^2]/EI$

$y_{EC} =$ *Deflection of E from the tangent at C*

$\quad =$ Total first moment of area *about E* of the M/EI diagram on EC

$\quad = [\{\tfrac{1}{2}L(WL/4)\}(L/2) + (\tfrac{1}{2}M_cL)\tfrac{2}{3}L]/EI$

$\quad = [\tfrac{1}{16}WL^3 + \tfrac{1}{3}M_cL^2]/EI$

Substituting,

$$(\tfrac{1}{2}WL^3 + \tfrac{4}{3}M_cL^2) + 2(\tfrac{1}{16}WL^3 + \tfrac{1}{3}M_cL^2) = 0$$

from which:

$$M_c = -\tfrac{5}{8}WL$$

This is as far as the calculation need be taken since the combined bending moment diagram may now be drawn (Fig. 16.15 (*d*)). The values obtained from this combined diagram are usually sufficient for design purposes, but if the support reactions are required, these may be easily calculated.

$$M_C = R_A \times 2L - WL = -\tfrac{5}{8}WL$$
$$\therefore \ R_A = \tfrac{3}{16}W$$

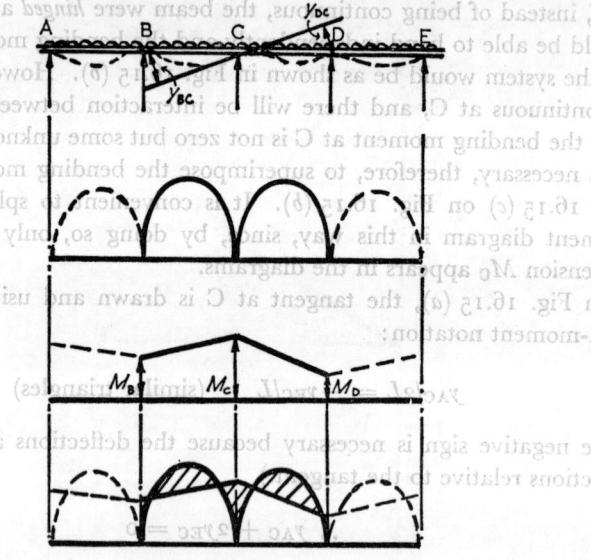

Fig. 16.16

Taking moments about E:

$$R_A 3L + R_C L = W2L + WL/2$$
$$\therefore \ R_C = \tfrac{31}{16}W$$

Resolving vertically:

$$R_A + R_C + R_E = 2W$$
$$\therefore \ R_E = -\tfrac{1}{8}W$$

The general case for a continuous beam is shown in Fig. 16.16 and may be dealt with as above. The spans are taken in pairs, as shown, and for each pair of spans an equation such as:

$$y_{BC}/L_{BC} = -y_{DC}/L_{CD} \qquad (16.22)$$

may be written. This equation will involve the bending moments M_B, M_C and M_D and is known as Clapeyron's equation of three moments.

With 2 spans, 1 equation of 3 moments may be written.

With 4 spans, 3 equations of 3 moments may be written.

With n spans $(n - 1)$ equations of 3 moments may be written to obtain fixing moments at $(n + 1)$ supports. The 2 additional equations are derived from known end conditions. In the above worked example the known end conditions were $M_A = 0$ and $M_E = 0$ (Fig. 16.15).

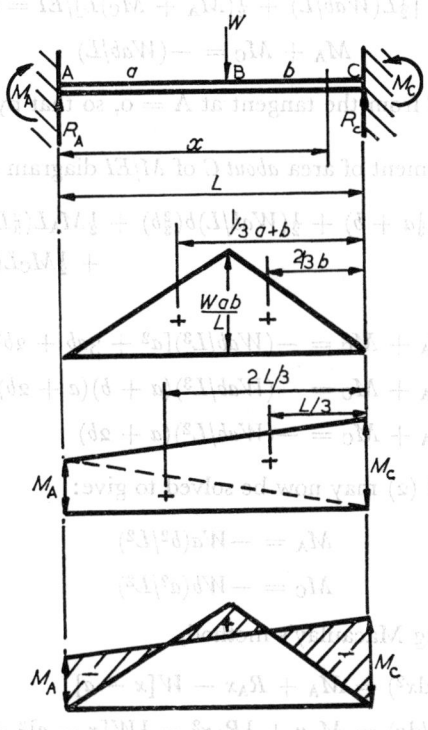

Fig. 16.17

Built-in beam

A built-in beam is usually taken as a beam which is built-in at *both* ends. The simplest case is shown in Fig. 16.17. This is just another example of a redundant structure and the methods already discussed are applicable.

If Mohr's theorems are used, the system should be split into two component systems, as shown. The component bending moment diagrams are shown in Fig. 16.17 together with the combined bending moment diagram.

Since the beam is built-in at both A and C,

$$\text{Slope at A} = \text{Slope at C} = 0$$

$$\therefore \text{ Change of slope from A to C} = 0$$

By Mohr's first theorem,

$$\text{Total area of } M/EI \text{ diagram on AC} = 0$$

$$\therefore [\tfrac{1}{2}L(Wab/L) + \tfrac{1}{2}(M_A + M_C)L]/EI = 0$$

i.e. $$M_A + M_C = -(Wab/L) \qquad (1)$$

Also, *deflection of C* from the tangent at A $= 0$, so that by Mohr's second theorem:

$$\text{Total first moment of area } about\ C \text{ of } M/EI \text{ diagram on AC} = 0$$

$$\therefore [\tfrac{1}{2}(Wab/L)a(\tfrac{1}{3}a + b) + \tfrac{1}{2}(Wab/L)b(\tfrac{2}{3}b) + \tfrac{1}{2}M_A L(\tfrac{2}{3}L)$$
$$+ \tfrac{1}{2}M_C L(\tfrac{1}{3}L)]/EI = 0$$

which reduces to:

$$2M_A + M_C = -(Wab/L^3)[a^2 + 3ab + 2b^2]$$

i.e. $$2M_A + M_C = -(Wab/L^3)(a + b)(a + 2b)$$

i.e. $$2M_A + M_C = -(Wab/L^2)(a + 2b) \qquad (2)$$

Equations (1) and (2) may now be solved to give:

$$M_A = -Wa(b^2/L^2)$$

$$M_C = -Wb(a^2/L^2)$$

Alternatively, using Macaulay's method,

$$M = EI(\mathrm{d}^2y/\mathrm{d}x^2) = M_A + R_A x - W[x - a]$$

$$\therefore EI(\mathrm{d}y/\mathrm{d}x) = M_A x + \tfrac{1}{2}R_A x^2 - \tfrac{1}{2}W[x - a]^2 + A$$

$$EIy = \tfrac{1}{2}M_A x^2 + \tfrac{1}{6}R_A x^3 - \tfrac{1}{6}W[x - a]^3 + Ax + B$$

When $x = 0$, $(\mathrm{d}y/\mathrm{d}x) = 0$,

$$\therefore A = 0$$

When $x = 0$, $y = 0$,

$$\therefore B = 0$$

When $x = L$, $(\mathrm{d}y/\mathrm{d}x) = 0$,

$$\therefore 0 = M_A L + \tfrac{1}{2}R_A L^2 - \tfrac{1}{2}W[L - a]^2$$

i.e. $$2M_A + R_A L = Wb^2/L \qquad (1)$$

When $x = L, y = 0$,

$$\therefore\; 0 = \tfrac{1}{2}M_A L^2 + \tfrac{1}{6}R_A L^3 - \tfrac{1}{6}W[L - a]^3$$

i.e.
$$3M_A + R_A L = Wb^3/L^2 \tag{2}$$

Subtracting (1) from (2) gives:

$$M_A = (Wb^3/L^2) - (Wb^2/L)$$
$$= (Wb^2/L^2)[b - L]$$

i.e.
$$M_A = -Wa(b^2/L^2) \tag{16.23}$$

It is convenient to memorise Eq. (16.23), since it provides a simple rule for the calculation of the fixing moments for a built-in beam carrying concentrated loads. For more than one load the principle of superposition may be used.

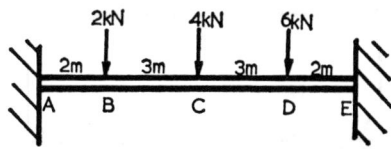

Fig. 16.18

Example. *A uniform beam is built-in horizontally at both ends (Fig. 16.18). The beam is* 10 m *long and carries loads of* 2, 4 *and* 6 kN *at distances of* 2 m, 5 m *and* 8 m, *respectively, from one end. Calculate the fixing moments at the supports.*

With reference to Fig. 16.18,

$$M_A = -2000 \times 2(8^2/10^2) - 4000 \times 5(5^2/10^2)$$
$$- 6000 \times 8(2^2/10^2)\ \text{N m}$$
$$= -(2560 + 5000 + 1920)\ \text{N m}$$

i.e.

$$M_A = -9480\ \text{N m} \qquad = -9{\cdot}48\ \text{kN m}$$

Similarly:

$$M_E = -6000 \times 2(8^2/10^2) - 4000 \times 5(5^2/10^2)$$
$$- 2000 \times 8(2^2/10^2)\ \text{N m}$$
$$= -(7680 + 5000 + 640)\ \text{N m}$$
$$= -13\,320\ \text{N m} \qquad = -13{\cdot}32\ \text{kN m}$$

Knowing M_A and M_E, the vertical reactions at A and E may be calculated and a complete bending moment diagram drawn.

Example. *A built-in beam ABC of length 2L is propped at its mid-point B (Fig. 16.19 (a)). A single concentrated load W is supported at D which is the mid-point of AB. Calculate the fixing moments at A and C and also the support reaction in the prop. Sketch the bending moment diagram for the beam.*

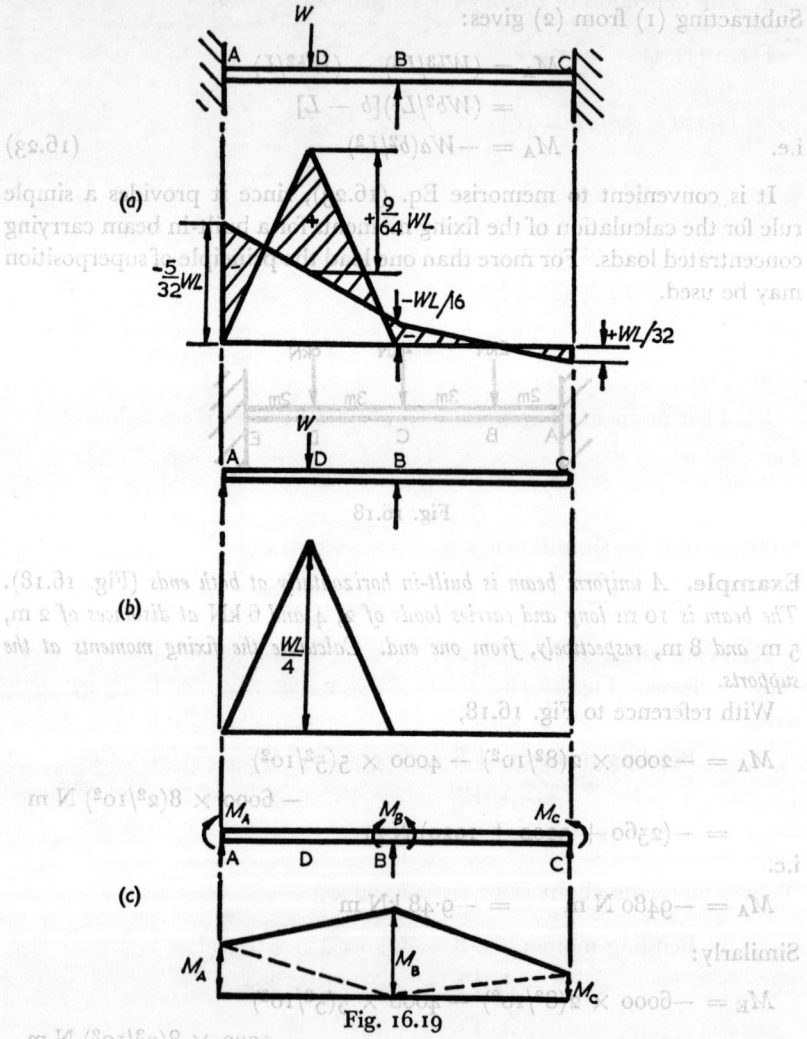

Fig. 16.19

The redundant system is split into two component systems. In the first (Fig. 16.19 (b)) all fixing moments are relaxed. The second component system is shown in Fig. 16.19 (c) and consists essentially of the fixing moments required to restore the actual conditions. Three equations are required to obtain the three unknown bending moments M_A, M_B

and M_C and these are provided by the following known conditions:

1. The change of slope from A to C is zero.
2. The deflection of B from the tangent at A is zero.
3. The deflection of B from the tangent at C is zero.

From (1),

Area of M/EI diagram on AC $= 0$

i.e. $\{(WL/4)(L/2) + \frac{1}{2}(M_A + M_B)L + \frac{1}{2}(M_B + M_C)L\}/EI = 0$

i.e. $WL/4 + M_A + 2M_B + M_C = 0$ (1)

From (2),

First moment of area about B of M/EI diagram on AB $= 0$

$\therefore \{(WL/4)(L/2)L/2 + (\frac{1}{2}M_AL)\frac{2}{3}L + (\frac{1}{2}M_BL)\frac{1}{3}L\}/EI = 0$

i.e. $\frac{3}{8}WL + 2M_A + M_B = 0$ (2)

From (3),

First moment of area about B of M/EI diagram on BC $= 0$

i.e. $\{(\frac{1}{2}M_BL)\frac{1}{3}L + (\frac{1}{2}M_CL)\frac{2}{3}L\}/EI = 0$

i.e. $M_B + 2M_C = 0$ (3)

Solution of these simultaneous equations gives:

$$M_A = -\tfrac{5}{32}WL; \quad M_B = -\tfrac{1}{16}WL; \quad M_C = +\tfrac{1}{32}WL$$

Knowing these moments, the combined bending moment diagram may be drawn (Fig. 16.19 (a)). Taking moments about B for left-hand side,

Bending moment at B $= M_B = M_A + R_AL - WL/2$

i.e. $-\tfrac{1}{16}WL = -\tfrac{5}{32}WL + R_AL - WL/2$

$$\therefore R_A = +\tfrac{19}{32}W$$

Taking moments about B for right-hand side,

Bending moment at B $= M_B = M_C + R_CL$

i.e. $-\tfrac{1}{16}WL = +\tfrac{1}{32}WL + R_CL$

$$\therefore R_C = -\tfrac{3}{32}W$$

Resolving vertically,

$$R_A + R_B + R_C = W$$

$\therefore$ Load in prop $= R_B = W - (19W/32) - (-3W/32)$

i.e. $R_B = W/2$

PROBLEMS

For tutorials

1. Discuss the relative merits of Macaulay's method, Mohr's method and the strain energy methods given in Chapter 17 for the determination of beam deflections and redundant reactions.

2. An accurate straight-edge, of length L, is simply supported at two points along its length, the supports being at the same level.

(a) Show that if the maximum deflection of the straight-edge due to its own weight is to be kept as small as possible, the supports should be $0.554L$ apart.

(b) Show that if the two ends of the straight-edge are to remain horizontal when it deflects under its own weight, the supports should be $0.577L$ apart.

3. A cantilever beam of uniform thickness t and length L has a rectangular section which varies in breadth linearly from zero at the free end to a maximum value of B at the fixed support. Show that when a concentrated load W is applied at the free end:

(a) the maximum bending stress is constant along the length of the beam,

(b) the beam bends into an arc of a circle,

(c) the deflection at the free end is $6WL^3/EBt^3$.

4. One of the disadvantages of a continuous beam is that slight subsidence of a support can bring about large changes in bending moment values and in the position of points of contraflexure. Comment on the advisability of introducing hinged joints at the points of contraflexure to combat this disadvantage, and quote an example of a structure of which this is a feature.

General

1. A uniform beam of length 4 m is placed symmetrically upon simple supports 2 m apart. A single concentrated load of 2 kN is applied at the mid-point of the beam. Neglecting the weight of the beam, calculate the slope and deflection of the ends of the beam if $EI = 50$ kNm².

Ans. 0.01 radian; 10 mm

2. A steel cantilever beam of uniform square cross-section throughout carries a uniformly distributed load along its entire length, which is 4 m. The second moment of area of the cross-section of the beam is 40×10^{-6} m⁴ and $E = 200$ GN/m². Calculate the maximum slope and deflection of the beam if the maximum bending stress induced by the loading is 180 MN/m².

Ans. 0.0162 radian; 48.65 mm

3. A uniform beam 10 m long, is simply supported at its ends, and carries a uniformly distributed load of 2 kN/m along its entire length. In addition, the beam carries a concentrated load of 30 kN at its mid-point. The second moment of area of the section of the beam is 150×10^{-6} m⁴ and $E = 200$ GN/m². Calculate the deflection at the mid-point of the beam and the slope, in degrees, at its ends.

Ans. 29.5 mm; 0.5172 degrees

4. A uniform beam ABCDE is simply supported at A and D and carries loads of 2 kN, 3 kN and 1 kN, respectively, at B, C and E:

$$AB = BC = CD = DE = 2 \text{ m}$$

Assuming that $EI = 400 \text{ kN m}^2$, calculate the slope of the beam at the supports, and the deflection at each load point.

Ans. $i_A = -0{\cdot}0194 \text{ rad}$; $i_D = +0{\cdot}0156 \text{ rad}$; $y_B = -32{\cdot}2 \text{ mm}$; $y_C = -31{\cdot}1 \text{ mm}$; $y_E = +24{\cdot}4 \text{ mm}$.

5. A beam, simply supported at its ends, is 10 m long and carries a central concentrated load of 2 kN and another concentrated load of 3 kN at a distance of 2 m from one end. The second moment of area of the section is $2 \times 10^{-6} \text{ m}^4$, and $E = 200 \text{ GN/m}^2$. Calculate the position and magnitude of the maximum deflection of the beam.

Ans. 4·735 m from one end; 194 mm

6. A uniform cantilever ABCD is built-in at D and carries a uniformly distributed load of 0·2 kN/m between B and C. Concentrated loads, each of 2 kN, are applied at A and B. AB = 2 m; BC = 6 m; CD = 2 m. The second moment of area of the section is $100 \times 10^{-6} \text{ m}^4$ and $E = 200 \text{ GN/m}^2$.

(*a*) Calculate the slope and deflection at A.

(*b*) What upward force is required at A to reduce the deflection to zero?

Ans. (*a*) 0·00904 radian, 63·5 mm, (*b*) 3·81 kN

7. A cantilever beam carries a uniformly distributed load w per unit length over its full length, together with a concentrated load W at the free end. The cantilever is propped at a point $L/4$ from the free end, where L is the length of the cantilever. Prove that the load in the prop is given by P, where

$$P = (19wL/32) + (3W/2)$$

8. A continuous beam of uniform section throughout has two spans, each of length 4 m, and concentrated loads of 12 kN and 6 kN, respectively, are carried at the mid-point of each span. Calculate the reaction in each of the three simple supports. Sketch the bending moment diagram and state the value of the maximum bending moment in the beam.

Ans. 4·31 kN; 12·38 kN; 1·31 kN; +8·63 kN m (at the 12 kN load)

9. (*a*) A uniform beam is built-in horizontally at both ends and carries a uniformly distributed load of 1 kN/m along its entire length. The beam is 12 m long. Calculate the fixing moment at the ends and the value of the bending moment at the mid-point.

(*b*) Concentrated loads, each of 20 kN, are now placed at 2 m and 4 m from one end. Calculate the fixing moments at each end.

Ans. (*a*) −12 kN m; +6 kN m; (*b*) −75·3 kN m; −35·3 kN m

10. A continuous beam ABCD is encastré at A and simply supported at B and D. A uniformly distributed load of w per unit length extends from B to C. The supports are all at the same level, and AB = BC = CD = L. The flexural rigidity is EI throughout. Determine the fixing moment at A and the vertical reactions at B and D.

Ans. $M_A = +9wL^2/88$; $R_B = +51wL/44$; $R_D = +13wL/88$

CHAPTER 17

Strain Energy

17.1 DEFINITIONS

Strain energy is the energy which a body possesses by virtue of its elastic deformation.

When forces or moments are applied to a piece of elastic material in equilibrium, the material deforms and, provided the elastic limit is not exceeded, the material will return to its original size and shape when applied forces or moments are removed. In the process the energy stored within the material due to elastic strain can be recovered. A familiar example of the use of strain energy as a source of power is the clockwork motor.

When a body is deformed, the forces responsible move their points of application and the work done by each force may be evaluated from the area under the force-displacement diagram. But from the principle of conservation of energy, the strain energy stored is equal to the total work done in deforming the body. This of course assumes that there is no loss of energy due to friction or hysteresis. Thus

Strain energy of a body = Total work done in deforming it   (17.1)

This equation may be used to obtain expressions for strain energy either in terms of the external loading on the body or in terms of the internal stresses in the body.

17.2 STRAIN ENERGY FOR VARIOUS TYPES OF LOADING

Direct loading

In a simple tension member (Fig. 17.1) the force F is applied gradually, so that the material is in equilibrium throughout the whole of the straining period (i.e. $F = \sigma A$). The load-extension diagram is then linear.

Work done in straining the member $= \tfrac{1}{2} F \delta L$

From Eq. (17.1),

$$\text{Strain energy stored} = \tfrac{1}{2}F\delta L$$

From Hooke's law,

$$\delta L = FL/EA \quad \text{where} \quad A = \text{Area of cross-section}$$
$$\text{Strain energy} = \tfrac{1}{2}F\delta L = \tfrac{1}{2}F(FL/EA)$$
$$= F^2L/2EA \qquad (17.2)$$

In general, allowing for a variation of F along the length x of a member,

$$\text{Strain energy in direct loading } U_D = \int_0^L \frac{F^2\,dx}{2EA} \qquad (17.3)$$

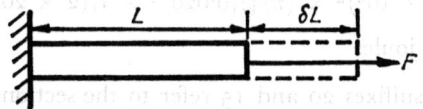

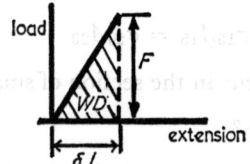

Fig. 17.1

The strain energy can be expressed in terms of stress by using Hooke's law in the form,

$$\delta L = \sigma L/E$$

if the material is in equilibrium,

$$F = \sigma A$$

so that

$$\text{Strain energy} = \tfrac{1}{2}F\delta L = \tfrac{1}{2}(\sigma A)(\sigma L/E)$$
$$= \frac{\sigma^2}{2E} \times AL$$
$$= \frac{\sigma^2}{2E} \text{ per unit volume} \qquad (17.4)$$

The capacity of a material to absorb energy by mechanical deformation is often referred to as resilience, and in this connection the terms "strain energy" and "resilience" are synonymous. Thus the term "proof resilience" refers to the energy absorbed per unit volume when a test piece is strained to a specified proof stress.

Example. *Two mild steel bars, A and B, are each 1 m long. A has a diameter of 20 mm over its entire length, while B has a diameter of 20 mm over 0·75 m and is turned down to 15 mm for the remainder of its length. If the stress in each bar is not to exceed 100 MN/m², calculate the strain energy capacity of each bar. Assume E = 200 GN/m².*

For bar A,

$$\text{Maximum strain energy} = (\sigma^2_{max}/2E) \times \text{Volume}$$

i.e. $U_{max} = (100 \times 10^6)^2 \times (\pi/4)(0\cdot020)^2 \times 1/(2 \times 200 \times 10^9)$ joules

$$= 7\cdot854 \text{ joules}$$

For bar B, let suffixes 20 and 15 refer to the sections of the bar with diameters 20 mm and 15 mm respectively. Then

$$\sigma_{15}A_{15} = \sigma_{20}A_{20}$$

The maximum stress will occur in the section of smaller diameter, so that

$$\sigma_{15} = 100 \text{ MN/m}^2$$

$$\therefore \sigma_{20} = \sigma_{15} \times (A_{15}/A_{20}) = 100(15/20)^2 \text{ MN/m}^2$$

$$= 56\cdot2 \text{ MN/m}^2$$

$$U_{15} = (100 \times 10^6)^2 \times (\pi/4)(0\cdot015)^2 \times 0\cdot25/(2 \times 200 \times 10^9) \text{ joules}$$

$$= 1\cdot105 \text{ joules}$$

$$U_{20} = (56\cdot2 \times 10^6)^2 \times (\pi/4)(0\cdot020)^2 \times 0\cdot75/(2 \times 200 \times 10^9) \text{ joules}$$

$$= 1\cdot861 \text{ joules}$$

$$\therefore U_{max} = U_{20} + U_{15}$$

$$= 1\cdot861 + 1\cdot105 = 2\cdot966 \text{ joules}$$

Example. *Obtain an expression for the strain energy per unit volume a rectangular block of material subjected to principal stresses σ_1 and σ_2.*

The block is shown in Fig. 17.2, and for convenience is assumed to be of unit thickness. The force corresponding to σ_1 is $\sigma_1 d$ and the

$$\text{Strain in direction } \sigma_1 = (\sigma_1 - \nu\sigma_2)/E = \delta b/b$$

$$\text{Extension in direction } \sigma_1 = b \times (\sigma_1 - \nu\sigma_2)/E = \delta b$$

$\therefore$ Work done in direction $\sigma_1 = \tfrac{1}{2}F_1\delta b$

$$= \tfrac{1}{2}(\sigma_1 d) \times \{b \times (\sigma_1 - \nu\sigma_2)/E\}$$

i.e. $\qquad\qquad\qquad U_1 = bd\sigma_1(\sigma_1 - \nu\sigma_2)/2E$

Similarly,

Work done in direction $\sigma_2 = U_2 = db\sigma_2(\sigma_2 - \nu\sigma_1)/2E$

Total strain energy $= U = U_1 + U_2$

i.e. $\qquad\qquad\qquad U = (bd/2E)\{\sigma_1(\sigma_1 - \nu\sigma_2) + \sigma_2(\sigma_2 - \nu\sigma_1)\}$

$$= (bd/2E)\{\sigma_1{}^2 + \sigma_2{}^2 - 2\nu\sigma_1\sigma_2\}$$

$\therefore$ Total strain energy/Unit volume

$$= (1/2E)\{\sigma_1{}^2 + \sigma_2{}^2 - 2\nu\sigma_1\sigma_2\}$$

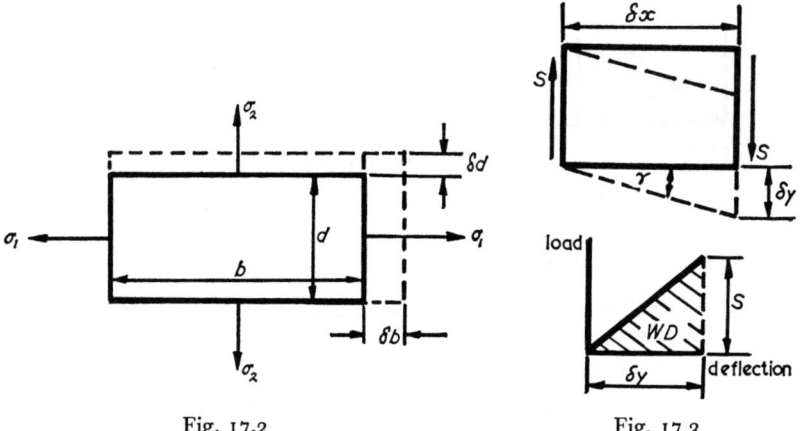

Fig. 17.2   Fig. 17.3

Shear loading

If a shear force S is applied gradually to an element of length δx of a member subjected to shear, so that the material is in equilibrium, the load/deflection diagram is linear as shown (Fig. 17.3). Then the work done in deforming the element is $\tfrac{1}{2}S\delta y$, and from Eq. (17.1),

Strain energy stored in element $= \tfrac{1}{2}S\delta y$

$\therefore$ Shear strain $\gamma = \delta y/\delta x$ or $\delta y = \gamma\delta x$

From Hooke's law,

Shear stress $= G \times \gamma = S/A$

$\therefore \gamma = S/GA$

where A = Area of cross-section.

∴ Strain energy stored in element $= \delta U_s = \tfrac{1}{2}S\gamma\delta x = \tfrac{1}{2}S(S/GA)\delta x$

i.e. $\delta U_s = S^2\delta x/2GA$

∴ Total strain energy due to shear in member $= U_s = \displaystyle\int_0^L \frac{S^2\,\mathrm{d}x}{2GA}$ (17.5)

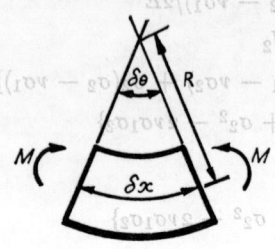

Note : Often the shear stress is not uniformly distributed across the section, in the distribution of shear due to bending in beams, for example, and Eq. (17.5) may then be multiplied by a distribution factor k_s.

Bending

Similar methods can be used to calculate the strain energy in bending. Consider an element of a beam, initially straight, of length δx. A bending moment M will cause this to bend as shown in Fig. 17.4 and if the bent element subtends an angle $\delta\theta$ at the centre of curvature, then

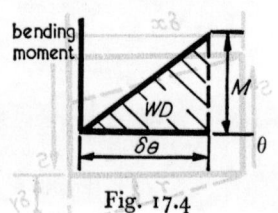

Fig. 17.4

Work done in bending element $= \tfrac{1}{2}M\delta\theta$

∴ Strain energy stored in element $= \tfrac{1}{2}M\delta\theta = \tfrac{1}{2}M(\delta x/R)$

But from the theory of bending,

$$1/R = M/EI$$

∴ Strain energy in element $\delta U_M = \tfrac{1}{2}M(M/EI)\delta x$

$$= M^2\delta x/2EI$$

∴ Total strain energy due to bending $= U_M = \displaystyle\int_0^L \frac{M^2\,\mathrm{d}x}{2EI}$ (17.6)

Torsion

To calculate the strain energy of torsion, consider an element of a shaft of length δx subjected to a gradually applied torque T (Fig. 17.5). Let the twist in the shaft over the length δx be $\delta\theta$. Then strain energy stored in element is

Work done in twisting the element $= \tfrac{1}{2}T\delta\theta$

From the theory of torsion,

$$T/J = G\delta\theta/\delta x$$

i.e.

$$\delta\theta = T\delta x/GJ$$

$$\therefore \; \delta U_T = \tfrac{1}{2}T(T\delta x/GJ) = T^2\delta x/2GJ$$

$$\therefore \; \text{Total strain energy due to torsion} = U_T = \int_0^L \frac{T^2 \, dx}{2GJ} \qquad (17.7)$$

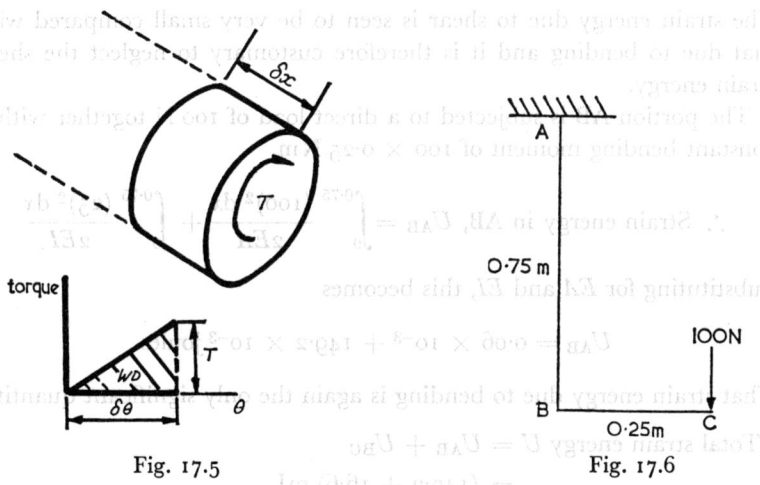

torque

Fig. 17.5

0·75 m

100N

B

0·25m C

Fig. 17.6

Example. *A mild steel bar, 20 mm diameter, is bent into an L-shape, ABC. AB = 0·75 m; BC = 0·25 m; and angle ABC = 90°. The bar is clamped at A and a vertical load of 100 N is applied at C. Calculate the total strain energy stored in the bar and hence the vertical deflection at C if (a) ABC lies in a vertical plane with AB vertical, (b) ABC lies in a horizontal plane. Assume E = 200 GN/m² and G = 80 GN/m².*

Cross-sectional area of bar $= \pi 0.01^2 = 3.142 \times 10^{-4}$ m²

Second moment of area in bending $= \pi 0.02^4/64 = 7.854 \times 10^{-9}$ m⁴

Polar second moment of area $= \pi 0.02^4/32 = 15.71 \times 10^{-9}$ m⁴

(a) The portion BC (Fig. 17.6) is subjected to shear loading and bending, and at a distance x from C,

$$S_{BC} = 100 \text{ N}$$

$$M_{BC} = 100x \text{ N m}$$

$\therefore$ Strain energy in BC,

$$U_{BC} = \int_0^{0\cdot25} \frac{100^2 \, dx}{2GA} + \int_0^{0\cdot25} \frac{(100x)^2 \, dx}{2EI}$$

Substituting for *GA* and *EI*, this becomes

$$U_{BC} = 0\cdot05 \times 10^{-3} + 16\cdot58 \times 10^{-3} \text{ joules}$$
$$= 16\cdot63 \text{ mJ}$$

The strain energy due to shear is seen to be very small compared with that due to bending and it is therefore customary to neglect the shear strain energy.

The portion AB is subjected to a direct load of 100 N together with a constant bending moment of 100 × 0·25 N m.

$$\therefore \text{ Strain energy in AB, } U_{AB} = \int_0^{0\cdot75} \frac{(100)^2 \, dx}{2EA} + \int_0^{0\cdot75} \frac{(25)^2 \, dx}{2EI}$$

Substituting for *EA* and *EI*, this becomes

$$U_{AB} = 0\cdot06 \times 10^{-3} + 149\cdot2 \times 10^{-3} \text{ joules}$$

That strain energy due to bending is again the only significant quantity.

Total strain energy $U = U_{AB} + U_{BC}$
$$= (149\cdot3 + 16\cdot6) \text{ mJ}$$
$$= 165\cdot9 \text{ mJ}$$

Work done by load = Area under the load-deflection diagram
$$= \tfrac{1}{2}100\delta \qquad \text{where } \delta = \text{Vertical deflection}$$

But

Work done by load = Total strain energy

$$\therefore \tfrac{1}{2}100\delta = 165\cdot9 \times 10^{-3} \text{ joules}$$
$$\therefore \delta = 3\cdot318 \times 10^{-3} \text{ m} = 3\cdot32 \text{ mm}$$

(*b*) When the plane of the bar is horizontal (Fig. 17.7), the portion BC is subjected to bending and shear, and the calculation of the strain energy in BC is identical with that carried out above.

$$\therefore U_{BC} = 16\cdot63 \text{ mJ}$$

The portion AB is subjected to shear, bending and torsion, for transference of the load to B introduces a twisting moment of 100 × 0·25 N m. As before, the shear strain energy may be neglected.

At a distance x from B,

$$M_{AB} = 100x \text{ N m}$$
$$T_{AB} = 25 \text{ N m (constant)}$$

$\therefore$ Strain energy in AB,

$$U_{AB} = \int_0^{0.75} \frac{(100x)^2 \, dx}{2EI} + \int_0^{0.75} \frac{(25)^2 \, dx}{2GJ}$$

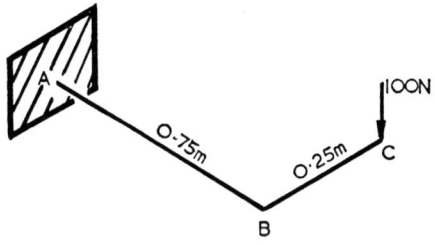

Fig. 17.7

Substituting for EI and GJ, this becomes

$$U_{AB} = 447 \cdot 6 \times 10^{-3} + 186 \cdot 5 \times 10^{-3} \text{ joules}$$
$$= 634 \cdot 1 \text{ mJ}$$

Total strain energy $U = U_{AB} + U_{BC}$
$$= (634 \cdot 1 + 16 \cdot 63) \text{ mJ}$$
$$= 650 \cdot 7 \text{ mJ}$$

Work done by load = Total strain energy

$$\therefore \tfrac{1}{2}100\delta = 650 \cdot 7 \times 10^{-3} \text{ joules}$$
$$\text{(where } \delta = \text{Vertical deflection)}$$

$$\therefore \delta = 13 \cdot 01 \times 10^{-3} \text{ m} = 13 \cdot 01 \text{ mm}$$

Example. *A beam 10 m long is simply supported at each end and supports a mass of 10 kg which may be assumed to be concentrated at a distance of 6 m from one end. The beam is of mild steel for which $E = 200 \text{ GN/m}^2$, and has a rectangular cross-section 30 mm wide by 40 mm deep. Calculate the vertical deflection of the load. Assume $g = 10 \text{ m/s}^2$ and neglect the deflection due to shear.*

$$\text{Weight of load} = 10g = 100 \text{ N}$$
$$\text{Reaction at A} = 40 \text{ N}$$
$$\text{Reaction at C} = 60 \text{ N}$$

The beam and load-deflection diagram are shown in Fig. 17.8.

$$I = bd^3/12 = 3 \times 4^3/12 = 16\,\text{cm}^4 = 16 \times 10^{-8}\,\text{m}^4$$

$$\therefore EI = 200 \times 10^9 \times 16 \times 10^{-8} = 32\,000\,\text{N m}^2$$

At a distance x from A,

$$M_{AB} = 40x\,\text{N m}$$

At a distance x from C,

$$M_{CB} = 60x\,\text{N m}$$

Neglecting the strain energy due to shear,

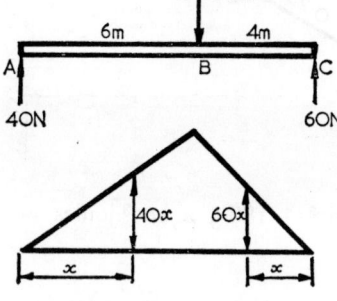

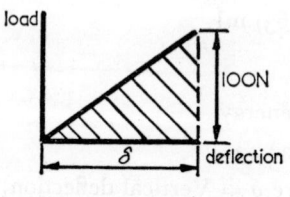

Fig. 17.8

Total strain energy

$$= U_{AB} + U_{CB}$$

$$= \int_0^6 \frac{(40x)^2\,\mathrm{d}x}{2EI} + \int_0^4 \frac{(60x)^2\,\mathrm{d}x}{2EI}$$

$$= 1{\cdot}8 + 1{\cdot}2\,\text{joules} = 3{\cdot}0\,\text{joules}$$

From the load-deflection diagram,

Work done by load

$$= \text{Area of load-deflection diagram}$$

$$= \tfrac{1}{2}\delta 100 = 3{\cdot}0\,\text{joules}$$

$$\delta = 0{\cdot}06\,\text{m} = 60\,\text{mm}$$

This result coincides with that obtained by more conventional methods (Chapter 16)

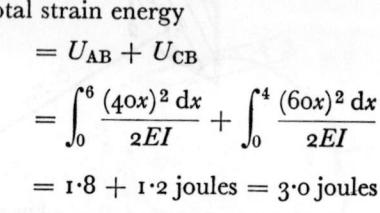

$$\delta = \frac{Wa^2b^2}{3EIL} = \frac{100 \times 6^2 \times 4^2}{3 \times 32\,000 \times 10}$$

$$= 0{\cdot}06\,\text{m} = 60\,\text{mm}$$

17.3 Castigliano's Theorem

If the total strain energy of a body or structure is known in terms of the various forces and torques acting on it, the partial differential coefficient of the strain energy with respect to a particular force or torque is equal to the displacement *corresponding* to that force or torque.

The term "corresponding displacement" is intended to mean the component of the displacement *in the direction of the force*. Thus, if a vertical force is applied to a structure and the structure deforms in such a way that the displacement of its point of application is other than vertical, then the application of this theorem gives only the vertical component of the resultant displacement.

$$\text{Total strain energy, } U = f(F_1, F_2, F_3, \text{ etc.})$$

$$\partial U/\partial F_1 = \delta_1, \quad \partial U/\partial F_2 = \delta_2, \quad \partial U/\partial F_3 = \delta_3, \text{ etc.} \qquad (17.8)$$

where δ_1, δ_2, δ_3, etc. are the displacements corresponding to F_1, F_2, etc.

Sign convention: δ is positive if it is in the same sense as F.

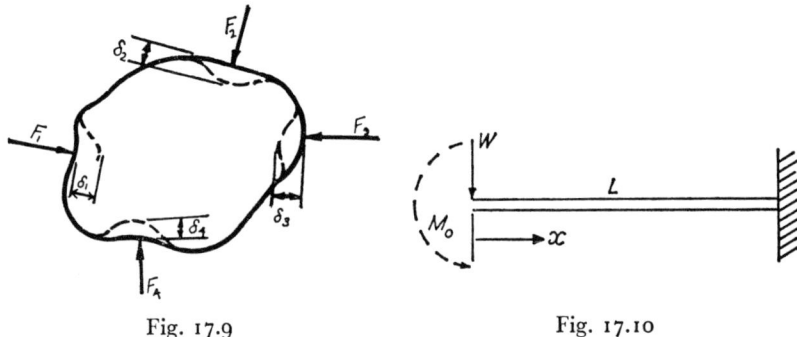

Fig. 17.9   Fig. 17.10

Applications

Castigliano's theorem is a powerful tool for dealing with problems relating to the analysis of forces and displacements in all types of structures, but especially those with redundant members. The theorem may also be applied to obtain temperature stress and stresses in members due to an initial lack of fit.

(*a*) *Simple tension member.* Consider a simple tension member of length L subjected to a tensile force F. Then

$$U = F^2L/2EA$$

and extension of member

$$\delta = \partial U/\partial F = FL/EA$$

(*b*) *Simple cantilever beam.* A cantilever of length L is subjected to a single concentrated load W at its free end (Fig. 17.10). Then at a distance x from the free end,

$$\text{Shearing force } S = -W$$
$$\text{Bending moment } M = -Wx$$

$$\text{Total strain energy} = \int_0^L S^2 \, dx/2GA + \int_0^L M^2 \, dx/2EI$$

$$= \int_0^L W^2 \, dx/2GA + \int_0^L W^2 x^2 \, dx/2EI$$

i.e. $U = (W^2 L/2GA) + (W^2 L^3/6EI)$

Deflection at the free end $\delta_W = \partial U/\partial_W = (WL/GA) + (WL^3/3EI)$

The first term of this expression is the deflection due to shear (usually negligible), and the second term is the deflection due to bending.

The slope of the beam at any point can also be obtained in this way. At the free end, for example, it is first necessary to calculate the strain energy with an imaginary moment M_0 at the free end, so that the deflection corresponding to M_0, is

$$i_0 = \partial U/\partial M_0$$

Next, M_0 is put equal to zero, when i_0 becomes the inclination of the beam at the free end.

At a distance x from the free end,

$$\text{Shearing force } S = -W$$

and   Bending moment $M = -M_0 - Wx$

Neglecting the strain energy due to shear

$$U = \int_0^L (-M_0 - Wx)^2 \, dx/2EI$$

$$= \int_0^L (M_0^2 + 2M_0 Wx + W^2 x^2) \, dx/2EI$$

$$= \{M_0^2 L + M_0 WL^2 + W^2 L^3/3\}/2EI$$

$$i_0 = \partial U/\partial M_0 = \{2M_0 L + WL^2 + 0\}/2EI$$

But $M_0 = 0$, so that

$$i_0 = WL^2/2EI$$

This expression is that obtained by conventional methods, but the sign is positive which indicates that i_0 is in the same sense as that assumed for M_0.

Example. *A cantilever mild steel beam* 10 m *long carries a single concentrated load of* 100 N *at its free end.* $E = 200$ GN/m² *and the beam has a rectangular section measuring* 30 mm *wide by* 40 mm *deep. Calculate the slope and deflection at the mid-point. Assume the shear strain energy negligible.*

Assume an imaginary load F_0 and moment M_0 to be applied at the mid-point (Fig. 17.11).

$$I = bd^3/12 = 3 \times 4^3/12 = 16 \text{ cm}^4$$

$$= 16 \times 10^{-8} \text{ m}^4$$

$$\therefore EI = 200 \times 10^9 \times 16 \times 10^{-8} = 32\,000 \text{ N m}^2$$

Total strain energy $U = \int_0^5 M_{AB}^2 \, dx/2EI + \int_5^{10} M_{BC}^2 \, dx/2EI$

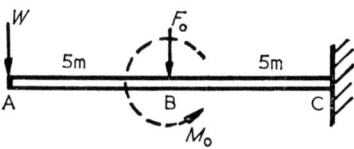

Fig. 17.11

In problems of this type, unnecessary calculations can frequently be avoided by differentiating *before* integrating. Thus

Deflection at mid-point $= \partial U/\partial F_0$

$$= \int_0^5 2M_{AB} \frac{\partial M_{AB}}{\partial F_0} \, dx/2EI + \int_5^{10} 2M_{BC} \frac{\partial M_{BC}}{\partial F_0} \, dx/2EI$$

$$= \int_0^5 M_{AB}(\partial M_{AB}/\partial F_0) \, dx/EI + \int_5^{10} M_{BC}(\partial M_{BC}/\partial F_0) \, dx/EI$$

$M_{AB} = Wx$

$$\therefore \partial M_{AB}/\partial F_0 = 0$$

$$M_{BC} = Wx + F_0(x - 5) + M_0$$

$$\therefore \partial M_{BC}/\partial F_0 = (x - 5)$$

Substituting,

Deflection at mid-point $= \partial U/\partial F_0$

$$= 0 + \int_5^{10} \{Wx + F_0(x - 5) + M_0\}$$
$$\times (x - 5) \, dx/EI$$

$$= \int_5^{10} \{W(x^2 - 5x) + F_0(x - 5)^2$$
$$+ M_0(x - 5)\} \, dx/EI$$

But M_0 and F_0 are both zero.

$$\therefore \text{ Deflection at mid-point} = \int_5^{10} W(x^2 - 5x) \, dx/EI$$

$$= W\{(1000 - 125)/3 - 5(100 - 25)/2\}/EI$$

$$= 100(292 - 187)/32\,000$$

$$= 0{\cdot}328 \text{ m} = 328 \text{ mm}$$

Since this is positive, the deflection is in the assumed direction of F_0, i.e. downwards.

$$\text{Slope at mid-point} = \partial U/\partial M_0$$

$$= \int_0^5 M_{AB}(\partial M_{AB}/\partial M_0) \, dx/EI$$

$$+ \int_5^{10} M_{BC}(\partial M_{BC}/\partial M_0) \, dx/EI$$

$$M_{AB} = Wx$$

$$\therefore \partial M_{AB}/\partial M_0 = 0$$

$$M_{BC} = Wx + F_0(x - 5) + M_0$$

$$\therefore \partial M_{BC}/\partial m_0 = 1{\cdot}0$$

$$\therefore \text{ Slope at mid-point} = \partial U/\partial M_0$$

$$= 0 + \int_5^{10} \{Wx + F_0(x - 5) + M_0\}1{\cdot}0 \, dx/EI$$

But M_0 and F_0 are both zero.

$$\therefore \text{ Slope at mid-point} = \int_5^{10} Wx \, . \, dx/EI = W(100 - 25)/2EI$$

$$= (100 \times 75)/(2 \times 32\,000) = 0{\cdot}1172 \text{ radian}$$

Since this is positive, the sense is again the same as that assumed for M_0.

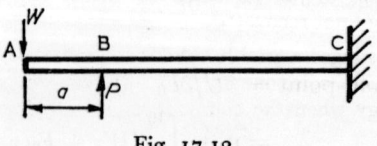

Fig. 17.12

(c) *Propped cantilever.* Castigliano's theorem may be used to obtain redundant support reactions such as occur in the propped cantilever. Consider the propped cantilever of length L (Fig. 17.12). If the prop is rigid, its deflection is zero, and hence if P is the force in the prop,

$\partial U/\partial P = 0$, and this condition yields the value of P. Thus

$$U = \int_0^a M_{AB}^2 \, dx/2EI + \int_a^L M_{BC}^2 \, dx/2EI \quad \text{(neglecting shear)}$$

$$\partial U/\partial P = \int_0^a M_{AB}(\partial M_{AB}/\partial P) \, dx/EI + \int_a^L M_{BC}(\partial M_{BC}/\partial P) \, dx/EI = 0$$

$$M_{AB} = -Wx$$

$$\therefore \; \partial M_{AB}/\partial P = 0$$

$$M_{BC} = -Wx + P(x - a)$$

$$\therefore \; \partial M_{BC}/\partial P = (x - a)$$

Substituting:

$$0 + \int_a^L \{-Wx + P(x - a)\}(x - a) \, dx/EI = 0$$

$$\therefore \; \int_a^L \{-W(x^2 - ax) + P(x^2 - 2ax + a^2)\} \, dx = 0$$

$$\therefore \; P[\tfrac{1}{3}(L^3 - a^3) - a(L^2 - a^2) + a^2(L - a)]$$
$$= W[\tfrac{1}{3}(L^3 - a^3) - \tfrac{1}{2}a(L^2 - a^2)]$$

Simplification then gives

$$P = W(2L + a)/2(L - a)$$

The same method may be applied to continuous beams and built-in beams.

PROBLEMS

For tutorials

1. A tensile test specimen loaded beyond its elastic limit is permanently deformed and will not return to its original length. In consequence, the energy released when the load is removed is less than the energy expended in deformation. Account for the energy not recovered.

2. Forgings and castings invariably contain internal stresses due to their method of manufacture. Such components therefore contain strain energy. What happens to this energy when the components are subjected to a stress relieving process?

3. It has been demonstrated that in the majority of cases the deflection due to shear in a beam is negligible compared with that due to bending. Show that if the deflection of a concentrated load at the free end of a mild steel cantilever beam is due to shear and bending in equal proportions, the slenderness ratio L/k of the beam must be as little as 2·74. Assume that for mild steel $E = 200$ GN/m² and $G = 80$ GN/m².

4. If energy in the form of heat is supplied to a metal bar which is not free to expand, the bar is subjected to compressive temperature stresses. Some of the energy supplied will therefore appear as strain energy. Will the final temperature of the bar therefore be reduced by the presence of the strain energy? Is the proportion of energy appearing as strain energy likely to reach a significant level?

5. One theory suggests that it is the total amount of strain energy absorbed by a piece of material that causes it to fail. According to this theory, unit volume of a material can absorb only so much energy before yielding occurs. Discuss the implications of this theory with regard to complex stress systems. In particular, show that for the system of Fig. 17.2, failure will occur when:

$$\sigma_{Y.P.}^2 = \sigma_1^2 + \sigma_2^2 - 2\nu\sigma_1\sigma_2$$

where $\sigma_{Y.P.}$ is the yield point stress for the material as determined by a simple tensile test.

6. Eq. 17.4 expresses the strain energy for direct loading in terms of the induced stress. Obtain a similar expression for the strain energy per unit volume for simple shear. Show also that for a solid round shaft subjected to torsion the strain energy stored is given by:

$$U = (\tau^2/4G) \times \text{Volume of the shaft}$$

where τ is the maximum shear stress at the outer surface of the shaft.

7. The strain energy of a body is equal to the total work done in deforming it. Explain how this statement may be used to obtain stresses induced by impact or shock loading.

General

1. The work done in inducing a direct tensile stress in a mild steel bar is 8 joules. If the bar is 500 mm long and has a diameter of 10 mm, calculate the stress induced.

Assuming the bar is in equilibrium throughout the whole of the straining period calculate the magnitude of the load on the bar. For mild steel $E = 200$ GN/m^2.

Ans. 285 MN/m^2; 22·4 kN

2. A piece of mild steel is subjected to principal stresses σ_1 and σ_2. Yielding is thought to occur when the total strain energy absorbed by the material reaches 264 kJ/m^3. If σ_1 is a tensile stress of 200 MN/m^2, calculate the value of σ_2 necessary to cause the material to yield (a) if σ_2 is tensile, (b) if σ_2 is compressive. Assume $E = 200$ GN/m^2 and $\nu = 0·25$.

Ans. (a) 311 MN/m^2; (b) 211 MN/m^2

3. A cantilever beam is 2 m long and has a rectangular cross-section 20 mm wide by 200 mm deep. The beam carries a single concentrated load 10 kN at its free end. Calculate (a) the strain energy due to shear; (b) the strain energy due to bending; (c) the deflection of the load due to shear; (d) the deflection of the load due to bending; and (e) the total deflection of the load. Take $E = 200$ GN/m^2 and $G = 80$ GN/m^2.

Ans. (a) 0·3125 J; (b) 50 J; (c) 0·0625 mm; (d) 10 mm; (e) 10·0625 mm

4. A solid shaft has a diameter of 120 mm and transmits 154 kW at 140 rev/min. Calculate the strain energy stored per unit length of the shaft. Hence calculate the angle of twist in the shaft over a length of 1 m. For the material of the shaft $G = 80$ GN/m^2.

Ans. 33·83 J/m; 0·00644 rad/m

5. A rectangular strip of mild steel is 200 mm long, 40 mm wide and 2 mm thick. A slot 30 mm wide is cut, symmetrically, from one end down the length of the strip to leave two parallel arms connected at one end by a piece of metal measuring 5 mm in the longitudinal direction of the arms. A bar 32 mm long, which may be assumed to be rigid, is inserted in the open end of the slot causing the free ends to move apart. Calculate the compressive force in this bar. Neglect the strain energy due to shear and assume $E = 200$ GN/m².

Ans. 1·276 N

6. ABCD is a piece of 5 mm diameter steel wire bent so that AB = 200 mm; BC = 100 mm; CD = 100 mm; $\angle$ABC = 90°; $\angle$BCD = 90°. ABCD is coplanar. The wire is clamped at A with ABCD lying in a horizontal plane, and a load of 10 N is applied vertically at D. Calculate the vertical deflection at D, if $E = 200$ GN/m² and $G = 80$ GN/m².

Ans. 8·3 mm

7. A uniform beam, 2 m long, is built-in horizontally at one end, and carries a concentrated transverse load of 1 kN at the other end. The section of the beam is such that $EI = 20000$ Nm².

(*a*) Using strain energy methods, calculate the slope and deflection at the mid-point of the cantilever.

(*b*) If the cantilever is propped at its mid-point, determine the load in the prop and the deflection at the free end.

Ans. (*a*) 41·67 mm; 0·075 radian; (*b*) 2·5 kN; 29·17 mm

CHAPTER 18

Theory of Struts

18.1 EULER'S THEORY FOR LONG SLENDER STRUTS

A strut is a member subjected to a direct compressive stress. The load carrying capacity of relatively short struts with large cross-section area is limited by the crushing strength of the material. Long and slender struts, however, can become unstable and tend to buckle.

A small transverse load applied to the mid-point of a slender strut will produce a lateral deflection which disappears when the transverse load is removed. As the compressive load is increased, however, a point is reached at which the lateral deflection does not disappear. At this point the strut is in a state of unstable equilibrium and the slightest lateral disturbance will cause it to buckle. Such a strut has clearly reached the limit of its load carrying capacity, and the load is said to have reached its critical value.

The critical load for a strut may be found using Euler's theory, which is based on a number of assumptions listed below.

Assumptions

1. The material is homogeneous.
2. The load is applied axially at the centroid of the section.
3. The cross-section is uniform.
4. The strut is initially straight.
5. The direct stresses due to the compressive load are negligible compared with the bending stresses induced by buckling.

Consider a strut of length L carrying its critical load F_E (Fig. 18.1). Let y be the lateral deflection at a distance x from one end. (If y is drawn positive (Fig. 18.1 (a)), then d^2y/dx^2 is negative and vice versa, Fig. 18.1 (b)).

Consider a section distance x from o. Then from Fig. 18.1 (a),

$$\text{Applied bending moment} = F_E \cdot y$$

$$\text{Resisting moment of section} = -EI(\mathrm{d}^2y/\mathrm{d}x^2)$$

(The negative sign is necessary since $(\mathrm{d}^2y/\mathrm{d}x^2)$ is negative.)

$$\therefore \text{ For equilibrium } -EI(\mathrm{d}^2y/\mathrm{d}x^2) = F_E y$$

i.e.
$$(\mathrm{d}^2y/\mathrm{d}x^2) + a^2y = \text{o} \qquad (18.1)$$

where $a^2 = F_E/EI = \text{a constant.}$

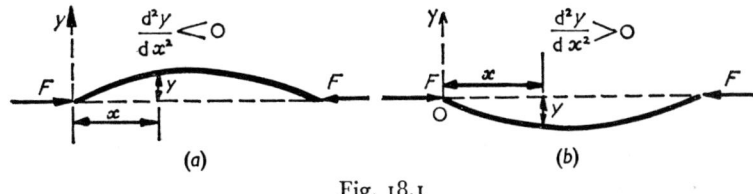

Fig. 18.1

Eq. (18.1) is a second order differential equation similar to that of Eq. (7.5). The solution may be expressed in the form

$$y = A \sin ax + B \cos ax$$

where A and B are constants.

When $x = \text{o}, y = \text{o}$,

$$\therefore B = \text{o}$$

When $x = L, y = \text{o}$,

$$\therefore A \sin aL = \text{o}$$

A cannot be zero, otherwise the strut would not be buckled.

$$\therefore \sin aL = \text{o}$$
$$\therefore aL = \text{o}, \pi, 2\pi, 3\pi, \text{ etc.} \qquad (18.2)$$

The first of these possible solutions, $aL = \text{o}$, is trivial since this would imply that there is no load on the strut. In general, therefore:

$$aL = n\pi, \qquad \text{where } n \neq \text{o}$$
$$\therefore a^2L^2 = n^2\pi^2$$

i.e.
$$(F_E/EI)L^2 = n^2\pi^2$$

i.e.
$$F_E = n^2\pi^2EI/L^2 \qquad (18.3)$$

The value of n will depend on the manner or mode of buckling. It is theoretically possible for a particular strut to buckle in any one of a number of ways, as shown in Fig. 18.2 (a), but clearly it is the *least* value of F_E (and therefore of n) which will cause the strut to fail.

The buckling mode of a strut will depend on the imposed constraints and, in particular, on the nature of the end-fixings. Fig. 18.2 (b) shows a strut with pin-jointed ends, while Fig. 18.2 (c) and (d) show, respectively, a strut built-in at one end and free at the other, and a strut built-in at both ends. It is clear from the theory that a strut deflects into the shape of a sine wave and evidently n is equal to the number of complete half-waves.

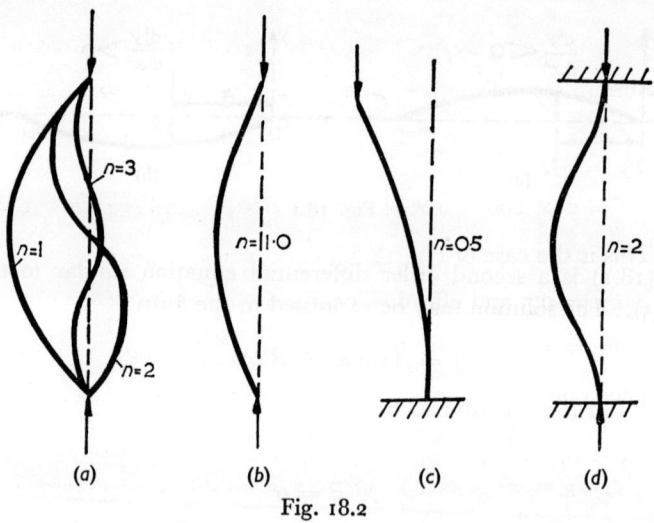

Fig. 18.2

Effective length

If a strut, by reason of its end-fixings or some lateral support, is compelled to deflect into 2 half-waves instead of one, then effectively it becomes two struts in series, each half as long as the original strut, so that the critical load is correspondingly greater. In general, if a strut is compelled to deflect into n half-waves, then the strut effectively becomes n struts in series each of length L/n, where L is the length of the original strut. Thus

$$\text{Effective length } l = L/n \qquad (18.4)$$

The Euler formula for the critical load of a strut Eq. (18.3) may now be written:

$$F_E = \pi^2 EI/l^2 \qquad (18.5)$$

where l = the *effective* length of the strut, and

I = the *minimum* second moment of area of the section in bending.

Example. *A strut is 2 m long and has a rectangular cross-section 30 mm ×
20 mm. Determine the Euler critical load for the strut if, (a) each end is pinned
and constrained to move axially in guides, (b) the strut is clamped at one end only,
and (c) each end is pinned and the strut is laterally supported at 0·5 m intervals
along its length.*

Assume E = 200 GN/m².

(*a*) For pin-jointed ends, the strut can deflect into a single half-wave,
so that

$$\text{Effective length} = \text{Actual length}$$

$$\text{Minimum second moment of area of section} = 0{\cdot}030 \times (0{\cdot}020)^3/12$$

$$= 2 \times 10^{-8}\ \text{m}^4$$

$$F_E = \pi^2 EI/l^2$$

$$= 2 \times 200 \times 10^9 \times 2 \times 10^{-8}/2^2 = 9868\ \text{N} = 9{\cdot}87\ \text{kN}$$

(*b*) This is the case of Fig. 18.2 (*c*),

$$n = 0{\cdot}5 \text{ and effective length} = 2L = 4\ \text{m}$$

$$\therefore F_E = \pi^2 \times 200 \times 10^9 \times 2 \times 10^{-8}/4^2 = 2467\ \text{N} = 2{\cdot}47\ \text{kN}$$

(*c*) Since points at 0·5 m intervals can have no lateral deflection,

$$\text{Effective length} = 0{\cdot}5\ \text{m} \qquad (\text{i.e. } n = 4)$$

$$\therefore F_E = \pi^2 \times 200 \times 10^9 \times 2 \times 10^{-8}/0{\cdot}5^2 = 157{\cdot}9\ \text{kN}$$

Slenderness ratio

The Euler equation for the critical load may be written

$$F_E = \pi^2 EAk^2/l^2$$

where A = Cross-sectional area of strut, and

k = Minimum radius of gyration of the section, so that

$Ak^2 = I$ = Minimum second moment of area of section.

$$\therefore F_E = \frac{\pi^2 EA}{(l/k)^2} \qquad\qquad (18.6)$$

This form of the Euler equation shows that for struts of a given cross-
sectional area, the critical load is inversely proportional to the square of

the ratio l/k. This quantity is clearly the criterion governing the onset of instability, and is called the *slenderness ratio*.

$$\text{Slenderness ratio} = l/k \qquad (18.7)$$

Euler's hyperbola

When the critical load is applied to a strut on the point of buckling, the nominal direct compressive stress in the strut is σ_E where $\sigma_E = F_E/A$. Thus from Eq. (18.6),

$$\sigma_E = \frac{\pi^2 E}{(l/k)^2} \qquad (18.8)$$

The graph of direct stress σ_E against slenderness ratio l/k (Fig. 18.3) is known as Euler's hyperbola. For small values of l/k, the stress corresponding to the critical load becomes very large, and obviously the relationship does not apply when the crushing strength of the material is exceeded. This happens for mild steel when the slenderness ratio is

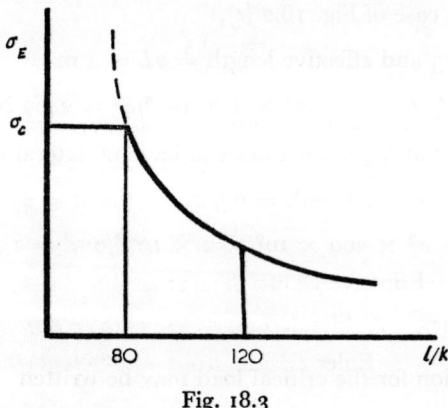

Fig. 18.3

about 80. Another limitation on the use of the Euler theory arises because of the assumption that the direct stress is small compared with the bending stresses induced by buckling, and in practice this formula is used only for struts having a slenderness ratio greater than about 120.

Example. *A vertical stanchion consists of an R.S.J. of length 16 m. The flanges of the R.S.J. each measure 25 cm × 1 cm and the web 30 cm × 1 cm. Assuming the stanchion to be built-in at each end calculate its slenderness ratio and its critical load. $E = 200 \text{ GN/m}^2$.*

The section is shown in Fig. 18.4,

$$I_{xx} = (25 \times 32^3/12) - (24 \times 30^3/12) = 14\,270 \text{ cm}^4$$
$$I_{yy} = (2 \times 25^3/12) + (30 \times 1^3/12) = 2607 \text{ cm}^4$$

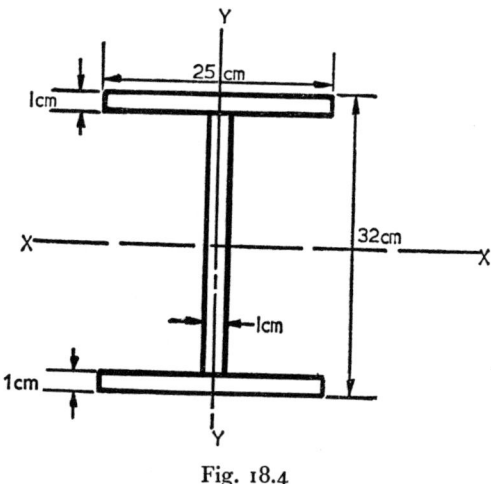

Fig. 18.4

$$\therefore \text{ Minimum second moment of area} = 2607 \text{ cm}^4$$
$$\text{Area of cross-section} = 80 \text{ cm}^2$$
$$\therefore k^2 = 2607/80 = 32 \cdot 59 \text{ cm}^2$$
$$\therefore k = 5 \cdot 71 \text{ cm} = 0 \cdot 0571 \text{ m}$$
$$\text{Effective length of strut} = L/2 = 8 \text{ m}$$
$$\therefore \text{ Slenderness ratio} = l/k = 8/0 \cdot 0571 = 140$$
$$\text{Euler critical stress} = \pi^2 E/(l/k)^2$$

i.e.

$$\sigma_E = \pi^2 \times 200 \times 10^9/140^2 = 100 \cdot 7 \times 10^6 \text{ N/m}^2 = 100 \cdot 7 \text{ MN/m}^2$$

$$\therefore \text{ Euler critical load} = \sigma_E \times A = 100 \cdot 7 \times 80 \times 10^{-4} = 0 \cdot 8056 \text{ MN}$$

i.e.
$$F_E = 805 \cdot 6 \text{ kN}$$

18.2 RANKINE-GORDON FORMULA

Many columns and stanchions have slenderness ratios between 80 and 120 and experimental evidence suggests that these members will fail at loads less than the Euler critical load. In this intermediate range, an empirical formula first suggested by Gordon and later modified by Rankine may be used.

Let F_R = Rankine critical load,

F_E = Euler critical load, and

F_C = Crushing load = $\sigma_c \cdot A$,

where σ_c = Yield point stress for the material. Then:

$$\frac{1}{F_R} = \frac{1}{F_E} + \frac{1}{F_C} \quad \text{or} \quad F_R = \frac{F_E \cdot F_C}{F_E + F_C} \qquad (18.9)$$

From this formula, F_R is smaller than either F_E or F_C. Furthermore, as the slenderness ratio is increased and F_C becomes large in comparison with F_E, F_R will tend towards F_E. Conversely, for small values of slenderness ratio, F_E will be very large so that F_R tends to F_C. Substituting for F_E and F_C Eq. (18.9) becomes

$$F_R = \frac{A\sigma_c}{1 + a(L/k)^2} \qquad (18.10)$$

where σ_C and a are constants.

Because the Rankine–Gordon formula is empirical, these constants must be obtained from test results. The slenderness ratio is based on the actual length of the strut, and the mode of buckling is accounted for by the constant a. The following are recommended values of σ_c and a for struts with pin-jointed ends ($n = 1$):

Material	σ_c MN/m²	a
Mild steel	325	1/7500
Wrought iron	247	1/9000
Cast iron	557	1/1600

For other modes of buckling, these values of a may be divided by n^2, where n is the number of half-waves into which the strut deflects. The stress corresponding to the Rankine critical load is plotted against slenderness ratio and compared with the Euler hyperbola in Fig. 18.5.

Example. *Tubular struts of various lengths are tested in order to find the constants in the Rankine–Gordon formula. All the struts have outside diameters of 20 mm and internal diameters of 15 mm. A strut having a length of 0·5 m* *was found to become unstable when the load reached 22·05 kN; and a strut of length 0·75 m was only able to support a load of 14 kN.*

Use this information to estimate the crippling load of a tubular strut 5 m long having an outside diameter of 150 mm and wall thickness 10 mm. Assume the material and end-fixings are the same as those under test.

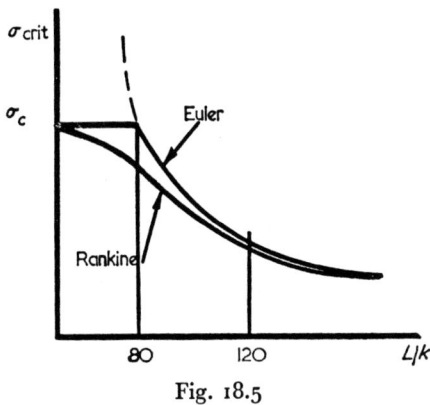

Fig. 18.5

Cross-sectional area of test struts

$$= (\pi/4)(20^2 - 15^2) = 137·4 \text{ mm}^2 = 137·4 \times 10^{-6} \text{ m}^2$$

Radius of gyration of section in bending

$$= \sqrt{\{(20^2 + 15^2)/16\}} = 6·25 \text{ mm}$$

For the 0·5 m strut,

$$L/k = 500/6·25 = 80$$

From Rankine–Gordon formula

$$22·05 \times 10^3 = \sigma_c \times 137·4 \times 10^{-6}/\{1 + a(80)^2\}$$

i.e. $$1 + 6400a = 6·232 \times 10^{-9}\sigma_c \qquad (1)$$

Consider the 0·75 m strut,

$$L/k = 750/6·25 = 120$$

From Rankine–Gordon formula

$$14 \times 10^3 = \sigma_c \times 137·4 \times 10^{-6}/\{1 + a(120)^2\}$$

i.e. $$1 + 14\,400a = 9·814 \times 10^{-9}\sigma_c \qquad (2)$$

Dividing (1) by (2),

$$(1 + 6400a)/(1 + 14\,400a) = 6·232/9·814$$

From which:

$$a = 1/7513$$

Substitution in (1) gives:

$$\sigma_c = 297 \cdot 3 \times 10^6 \text{ N/m}^2 = 297 \cdot 3 \text{ MN/m}^2$$

For the 5 m strut,

Cross-sectional area

$$= (\pi/4)(15^2 - 13^2) = 44 \text{ cm}^2 = 4 \cdot 4 \times 10^{-3} \text{ m}^2$$

Radius of gyration of section in bending

$$= \sqrt{\{(15^2 + 13^2)/16\}} = 4 \cdot 96 \text{ cm}$$

$\therefore$ Slenderness ratio $= L/k = 500/4 \cdot 96 = 100 \cdot 7$

$$F_R = \sigma_c A/\{1 + a(L/k)^2\}$$

$$= 297 \cdot 3 \times 10^6 \times 4 \cdot 4 \times 10^{-3}/\{1 + (100 \cdot 7)^2/7513\}$$

i.e. $$F_R = 556 \cdot 4 \times 10^3 \text{ N} = 556 \cdot 4 \text{ kN}$$

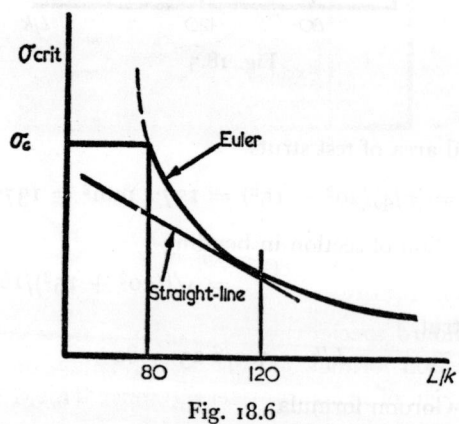

Fig. 18.6

18.3 OTHER EMPIRICAL FORMULAE

Straight-line formula

For struts of slenderness ratio less than 120, a linear relationship may be used:

$$\sigma_{cr} = \sigma_0 - C(L/k) \qquad (18.11)$$

where $\sigma_{cr} =$ Critical load per unit area of cross-section.

The constants σ_0 and C may be selected so that the line is tangential to the Euler hyperbola at the point where $L/k = 120$ (Fig. 18.6). In practice, these constants frequently include a load factor or factor of safety.

Johnson's parabolic formula

A formula due to Johnson is particularly suitable for struts of small slenderness ratio:

$$\sigma_{cr} = \sigma_0 - C(L/k)^2 \qquad (18.12)$$

where σ_{cr} = Critical load per unit area of cross-section.

Again the constants σ_0 and C may be chosen so that the parabola is tangential to the Euler hyperbola at the point where $L/k = 120$ (Fig. 18.7).

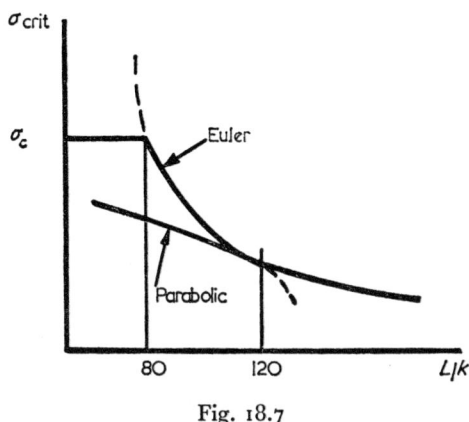

Fig. 18.7

Perry-Robertson formula

The British Standard specification BS 449:1959 recommends the use of the Perry–Robertson formula for the determination of allowable stress values for the design of structural steel columns. This is

$$\sigma_{cr} = K\sigma = \tfrac{1}{2}\{\sigma_y + (\eta + 1)\sigma_E\} - \sqrt{\left[\left\{\frac{\sigma_y + (\eta+1)\sigma_E}{2}\right\}^2 - \sigma_y\sigma_E\right]} \qquad (18.13)$$

where σ_{cr} = Critical load per unit area of section,
$\quad K$ = Load factor,
$\quad \sigma$ = Permissible average stress,
$\quad \sigma_y$ = Minimum yield stress of the material,
$\quad \sigma_E$ = Euler critical stress = $\pi^2 E/(l/k)^2$, and
$\quad \eta$ = $0.003(l/k)$.

In using this formula, the load factor is taken as 2 for values of slenderness ratio between 80 and 350. For slenderness ratios between 0 and 80, the permissible stress σ decreases linearly. Fig. 18·8 shows a comparison of this formula with the Euler curve.

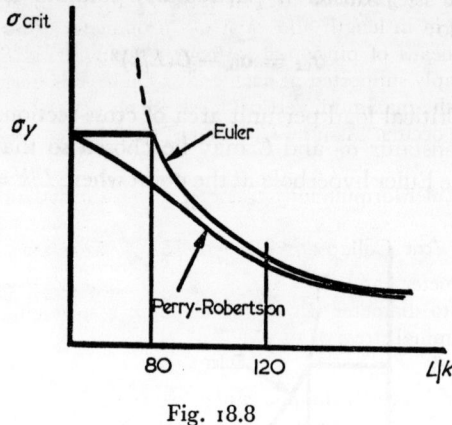

Fig. 18.8

PROBLEMS

For tutorials

1. A slender strut of circular section carries its Euler critical load and is deflected so that the slope at each end is θ radian. If the strut is of length L and diameter d, obtain a formula for the stress due to bending in terms of E, θ, d and L.

2. A long slender mild steel strut is pin-jointed at each end. If the critical load of the strut is inadequate, can this be rectified by specifying a steel of greater strength? If the section of the strut cannot be changed, what alternative methods exist for strengthening the strut?

3. A water tank is supported by steel tubular struts. The capacity of the tank is to be increased and it is proposed to strengthen the struts by pouring concrete into them via holes just below the top of each strut. Assuming no adhesion between steel and concrete, is this a reasonable proposition and, if so, what function does the concrete have?

4. A vertical strut is pinned at each end by parallel bolts passing through lugs so that, although the ends are free to rotate in one vertical plane, in the vertical plane at right angles the ends are fixed in direction. The strut is rectangular in section. Determine the conditions necessary for the crippling load to be the same for each of the principal axes of bending.

5. Investigate the possibility of strengthening a tubular strut by internal pressurisation.

A platform is supported by a hydraulically operated telescopic strut. The strut is in two sections and the platform is raised by pumping oil into the strut so that the upper section is forced upwards. Discuss the behaviour of this strut when fully extended. What type of stress is likely to occur in the lower section?

General

1. Three mild steel bars of circular section AB, BC and CD are, respectively, 5 m, 4 m and 3 m in length and each has a diameter of 40 mm. The bars are connected by means of pin-joints to form a triangular articulated framework, ABC. AB is simply supported at each end so that AB is horizontal with C above AB. Calculate the maximum vertical force which may be applied downwards at C before failure occurs. Assume $E = 200$ GN/m².

Ans. 25·83 kN

2. State the Euler formula for pin-ended struts, and show that for solid circular sections:

$$\text{Collapse stress} = (d^2/L^2) \times \text{Constant}$$

where $d =$ diameter, and $L =$ length. Given that $E = 200$ GN/m², find the ratio of length to diameter which would cause a solid circular section strut to collapse at a nominal stress of 140 MN/m².

Ans. 29·7

3. A mild steel strut with pin-jointed ends has a slenderness ratio of 120. If the strut is required to support a load of 200 kN, determine the minimum cross-sectional area of the strut based on (*a*) the Euler theory, (*b*) the Rankine–Gordon formula. Assume $E = 200$ GN/m² and that the constants for the Rankine formula are given in the table in Section 18.2.

Ans. (*a*) 1460 mm²; (*b*) 1797 mm²

4. A mild steel strut is 2 m long and has a cross-section of area 10 cm² and minimum radius of gyration of 20 mm. Using the constants for the Rankine formula as given in Section 18.2, estimate the load carrying capacity of this strut (*a*) if the ends are pin-jointed; (*b*) if the ends are built-in.

Ans. (*a*) 139·3 kN; (*b*) 243·8 kN

5. Compare the buckling strengths of two long slender struts, one having a circular section and the other a square section. Both struts have the same length, the same cross-sectional area, and are made of the same material.

Ans. Buckling strength of circular section $= 0·955 \times$ Buckling strength of square section.

CHAPTER 19

Stresses in Thick-Walled Cylinders

19.1 PRINCIPAL STRESSES AND STRAINS

Because a cylinder under pressure will deform symmetrically, there can be no shear on the radial face of an element in the wall (Fig. 19.1) and, as a result, there can be no complementary shear on the tangential face of the element. Evidently, therefore, the direct stresses which act on these faces are principal stresses, and these will produce principal strains. If the cylinder has closed ends and if the force on these ends is reacted by the cylinder walls, there will in general be three principal stresses acting on the element shown in Fig. 19.1.

  1. A longitudinal stress σ_L.
  2. A circumferential, tangential, or "hoop," stress σ_c.
  3. A radial stress σ_R.

For thin shells (Chapter 13), the radial stress is neglected, but when the wall thickness t becomes appreciable compared with the mean diameter d, the radial stress must be taken into account $\{(t/d) > (1/20)\}$. The principal strains corresponding to these principal stresses given in Chapter 13 are:

$$\varepsilon_L = \frac{1}{E}\{\sigma_L - \nu(\sigma_c + \sigma_R)\}$$

$$\varepsilon_c = \frac{1}{E}\{\sigma_c - \nu(\sigma_R + \sigma_L)\}$$

$$\varepsilon_R = \frac{1}{E}\{\sigma_R - \nu(\sigma_L + \sigma_c)\}$$

where E is Young's modulus and ν is Poisson's ratio for the material.

These expressions are, of course, applicable to any stress system. In accordance with usual convention, the stresses σ_L, σ_c and σ_R are all assumed positive when tensile. Other expressions for the principal strains in the cylinder wall can be obtained from the geometry of this particular

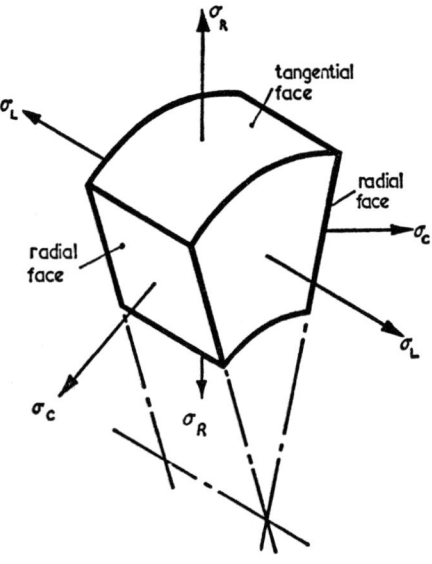

Fig. 19.1

system (Fig. 19.2). For an elemental annulus of material (radius r and thickness δr) the radial deformation at the inner radius is y and at the outer radius is $y + \delta y$. From Fig. 19.2,

Final thickness of element $= \delta r + (y + \delta y) - y = \delta r + \delta y$

Original thickness of element $= \delta r$

$\therefore$ Increase in wall thickness $= (\delta r + \delta y) - \delta r = \delta y$

$$\text{Radial strain of the element} = \frac{\text{Increase in wall thickness}}{\text{Original wall thickness}} = \frac{\delta y}{\delta r}$$

Thus, in the limit as δr tends to zero,

$$\text{Radial strain} = \varepsilon_R = dy/dr \qquad (19.1)$$

The circumferential strain (which is equal to diametral strain) may be found by considering the change in circumference at the radius r.

Change in circumference $= 2\pi(r + y) - 2\pi r = 2\pi y$

Original circumference $= 2\pi r$

$\therefore$ Circumferential strain $\varepsilon_c = 2\pi y / 2\pi r = y/r$ (19.2)

The expression for longitudinal strain is based on the assumption which is the basis of Lamé's theory for thick-walled cylinders, that plane sections of the cylinder remain plane after deformation. The longitudinal strain must then be the same at all radii so that

Longitudinal strain $\varepsilon_L = $ a constant (19.3)

This constant may sometimes be zero.

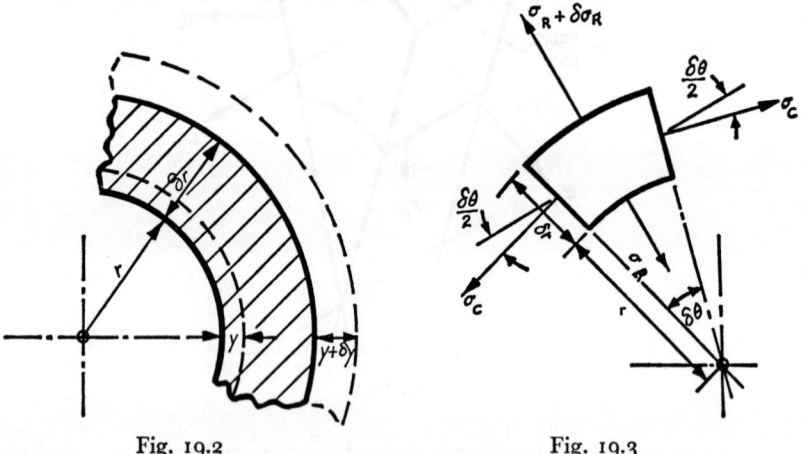

Fig. 19.2   Fig. 19.3

19.2 Lamé's Theory

A theory due to Lamé for the principal stresses in the cylinder wall, can be constructed from the following assumptions:

1. The material is perfectly homogeneous and isotropic.
2. The material obeys Hooke's law.
3. Young's modulus of elasticity is the same in tension and compression.
4. Plane sections of the cylinder perpendicular to its longitudinal axis remain plane after deformation.
5. The stresses are remote from the ends of the cylinder, where complicated stress patterns would be produced by the joints and change of geometry.

Assume for convenience that a small element taken from the cylinder wall (Fig. 19.3) is of unit thickness. Resolving forces radially,

$$(\sigma_R + \delta\sigma_R)(r + \delta r)\delta\theta - \sigma_R r\delta\theta - 2\sigma_c\delta r \sin(\delta\theta/2) = 0$$

Since $\delta\theta$ is small,

$$\sin(\delta\theta/2) \doteqdot \delta\theta/2$$

$$\therefore (\sigma_R r + r\delta\sigma_R + \sigma_R\delta r + \delta\sigma_R\delta r)\delta\theta - \sigma_R r\delta\theta - \sigma_c\delta r\delta\theta = 0$$

Dividing by $\delta\theta$ and neglecting products of small quantities, this becomes:

$$r\delta\sigma_R + \sigma_R\delta r - \sigma_c\delta r = 0$$

Dividing by δr and letting δr tend to zero,

$$r(d\sigma_R/dr) = \sigma_c - \sigma_R \qquad (19.4)$$

Since plane sections of the cylinder remain plane,

$$\varepsilon_L = \text{a constant for all values of } r$$

$$\therefore (d\varepsilon_L/dr) = 0$$

But

$$\varepsilon_L = \frac{1}{E}\{\sigma_L - \nu(\sigma_c + \sigma_R)\}$$

$$\frac{1}{E}\left\{\frac{d\sigma_L}{dr} - \nu\left(\frac{d\sigma_c}{dr} + \frac{d\sigma_R}{dr}\right)\right\} = 0$$

$$\therefore (d\sigma_c/dr) = \frac{1}{\nu}(d\sigma_L/dr) - (d\sigma_R/dr)$$

$$\therefore r(d\sigma_c/dr) = \frac{r}{\nu}(d\sigma_L/dr) - r(d\sigma_R/dr)$$

i.e.

$$r(d\sigma_c/dr) = \frac{r}{\nu}(d\sigma_L/dr) - (\sigma_c - \sigma_R) \qquad (19.5)$$

From Eq. (19.2),

$$\varepsilon_c = y/r = \frac{1}{E}\{\sigma_c - \nu(\sigma_R + \sigma_L)\}$$

$$\therefore y = \frac{1}{E}\{\sigma_c r - \nu(\sigma_R r + \sigma_L r)\}$$

Differentiating,

$$(dy/dr) = \frac{1}{E}[\sigma_c + r(d\sigma_c/dr) - \nu\{\sigma_R + r(d\sigma_R/dr) + \sigma_L + r(d\sigma_L/dr)\}]$$

Equating this with the value of (dy/dr) in Eq. (19.1) and simplifying,

$$\sigma_R - \nu\sigma_c = \sigma_c + r(d\sigma_c/dr) - \nu\sigma_R - \nu r(d\sigma_R/dr) - \nu r(d\sigma_L/dr)$$

Substitution for $r(d\sigma_R/dr)$ and $r(d\sigma_c/dr)$ from Eqs. (19.4) and (19.5) respectively, gives:

$$\sigma_R - \nu\sigma_c = \sigma_c + \frac{r}{\nu}(d\sigma_L/dr) - (\sigma_c - \sigma_R) - \nu\sigma_R - \nu(\sigma_c - \sigma_r)$$
$$- \nu r(d\sigma_L/dr)$$

This reduces to

$$r(d\sigma_L/dr)\left(\frac{1}{\nu} - \nu\right) = 0$$

$$\therefore (d\sigma_L/dr) = 0$$

$$\therefore \sigma_L = \text{a constant} \tag{19.6}$$

Longitudinal strain $\varepsilon_L = \{\sigma_L - \nu(\sigma_c + \sigma_R)\}/E$

But because both ε_L and σ_L are constant for all values of r, $\sigma_c + \sigma_R$ is a constant for all values or r, say $2a$. Then

$$\sigma_c = 2a - \sigma_R \tag{19.7}$$

Substituting for σ_c in Eq. (19.4),

$$r(d\sigma_R/dr) = (2a - \sigma_R) - \sigma_R$$

$$\therefore 2\sigma_R + r(d\sigma_R/dr) = 2a$$

Multiplying through by r, this becomes:

$$2r\,\sigma_R + r^2(d\sigma_R/dr) = 2ar$$

i.e.
$$\frac{d}{dr}(r^2\,\sigma_R) = 2ar$$

$$\therefore r^2\sigma_R = \int 2ar\,dr = ar^2 + (-b)$$

where $(-b)$ is a constant of integration.

$$\therefore \sigma_R = a - \frac{b}{r^2} \tag{19.8}$$

Substituting for σ_R in Eq. (19.7) gives

$$\sigma_c = a + \frac{b}{r^2} \tag{19.9}$$

Eqs. (19.8) and (19.9) provide a means of computation for the radial and hoop stresses in the cylinder wall at any radius r. The constants a and b are usually obtainable from known boundary conditions.

The other principal stress σ_L has been shown to be uniform across the section (Eq. (19.6)), and may therefore be calculated by dividing the total force due to the internal pressure on one end plate by the cross-sectional area of the cylinder. For a cylinder subjected only to internal pressure, it can be shown that the longitudinal stress σ_L and the constant a are the same.

Example. *A hydraulic cylinder has an internal diameter of* 100 mm *and a wall thickness of* 20 mm. *Determine the radial and hoop stresses at the inside and outside surfaces of the cylinder when subjected to an internal gauge pressure of* 100 MN/m². *Find, also, the change in internal diameter of the cylinder. Assume E =* 200 GN/m² *and* $v = 0.25$.

Because of the hydraulic pressure, the inside surface must be subjected to a direct compressive stress of 100 MN/m².

Therefore, when $r = 0.050$ m,

$$\sigma_R = -100 \times 10^6 \text{ N/m}^2$$

$$\therefore \ -a \, (b/0.050^2) = -10^8 \tag{1}$$

At the outside surface there is zero hydraulic pressure.

Therefore, when $r = 0.070$ m,

$$\sigma_R = 0$$

$$\therefore \ a - (b/0.070^2) = 0 \tag{2}$$

From (1) and (2) the constants a and b can be found.

Substituting in (1) from (2):

$$b[(1/0.05^2) - (1/0.07^2)] = 10^8$$

$$\therefore \ b = 510\ 000 \text{ N}$$

From (2):

$$a = b/0.0049 = 104 \times 10^6 \text{ N/m}^2$$

(Note the units of the constants a and b.) Thus

$$\sigma_R = 104 \times 10^6 - (510\ 000/r^2) \text{ N/m}^2 \tag{19.10}$$

$$\sigma_c = 104 \times 10^6 + (510\ 000/r^2) \text{ N/m}^2 \tag{19.11}$$

where r is the radius in metres.

At the inside surface,

$$r = 0.050 \text{ m}$$

$$\therefore \ \sigma_R = -10^8 \text{ N/m}^2 = -100 \text{ MN/m}^2$$

$$\sigma_c = +308 \times 10^6 \text{ N/m}^2 = +308 \text{ MN/m}^2$$

At the outside surface,

$$r = 0.070 \text{ m}$$

$$\therefore \sigma_R = 0$$

$$\therefore \sigma_c = +208 \times 10^6 \text{ N/m}^2 = +208 \text{ MN/m}^2$$

The positive sign indicates a tensile stress and the negative sign a compressive stress.

Eqs. (19.10) and (19.11) show that both the compressive radial stress and the tensile hoop stress are greatest when r is least, so that the greatest stresses occur at the inside surface of the cylinder.

Since

$$\text{Circumference} = \pi \times \text{Diameter}$$

$$\text{Circumferential strain} = \text{Diametral strain}$$

$$\therefore \varepsilon_D = \varepsilon_c = \frac{1}{E}\{\sigma_c - \nu(\sigma_R + \sigma_L)\}$$

In a hydraulic cylinder, the thrust on the ends would not normally be reacted by the cylinder walls, so that in this case $\sigma_L = 0$.

Therefore, at the inside surface:

$$\varepsilon_D = \varepsilon_c = \{(308 \times 10^6) - 0.25(-100 \times 10^6)\}/(200 \times 10^9)$$

$$= +1.667 \times 10^{-3}$$

$$\therefore \text{ Increase in internal diameter} = 1.667 \times 10^{-3} \times \text{Original diameter}$$

$$= 1.667 \times 10^{-3} \times 100 \text{ mm} = 0.1667 \text{ mm}$$

19.3 Stress Distributions Across the Cylinder Wall

It has been shown that the longitudinal stress, when it exists, is uniform across the section of the cylinder. The radial and hoop stresses, however, vary across the section according to Eqs. (19.8) and (19.9). Because many ductile materials are liable to fail in shear rather than in direct tension or compression, the maximum shear stress is also important. The maximum shear stress is given (Chapter 13) by

$$\tau_{max} = \pm\tfrac{1}{2}(\sigma_c - \sigma_R)$$

σ_c and σ_R being principal stresses. Thus:

$$\tau_{max} = \pm\tfrac{1}{2}[\{a + (b/r^2)\} - \{a - (b/r^2)\}]$$

$$= \pm b/r^2 \tag{19.12}$$

Eqs. (19.10) and (19.11) show how the radial and hoop stresses may vary across the section of a stressed cylinder, and given b, Eq. (19.12) produces a similar variation in maximum shear stress. The following

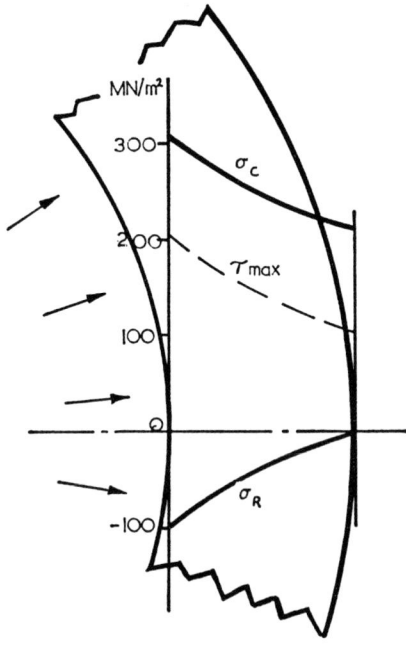

Fig. 19.4

table gives values of σ_R, σ_c and τ_{max} for various values of r (see also Fig. 19.4) where $a = 104 \times 10^6$ N/m², and $b = 0.51 \times 10^6$ N.

Radius r mm	$\tau_{max} = \pm b/r^2$ MN/m²	$\sigma_R = a - b/r^2$ MN/m²	$\sigma_c = a + b/r^2$ MN/m²
50	± 204	-100	$+308$
55	± 169	-65	$+273$
60	± 142	-38	$+246$
65	± 121	-17	$+225$
70	± 104	0	$+208$

Consider now a cylinder of similar dimensions to the previous one, but which is subjected to an *external* pressure of 100 MN/m² when the internal pressure is zero.

When $r = 0.070$ m,

$$\sigma_R = -100 \times 10^6 \text{ N/m}^2$$

$$\therefore \quad a - (b/0.07^2) = -10^8 \tag{1}$$

When $r = 0.050$ m,

$$\sigma_R = 0$$

$$\therefore \quad a - (b/0.05^2) = 0 \tag{2}$$

Solving for a and b gives

$$a = -204 \times 10^6 \text{ N/m}^2$$

and

$$b = -0.51 \times 10^6 \text{ N}$$

Thus

$$\sigma_R = -204 + (0.51/r^2) \text{ MN/m}^2 \tag{19.13}$$

$$\sigma_c = -204 - (0.51/r^2) \text{ MN/m}^2 \tag{19.14}$$

$$\tau_{max} = \pm 0.51/r^2 \text{ MN/m}^2 \tag{19.15}$$

Values of σ_R, σ_c and τ_{max} for various values of r are shown in the table, and the variation of stresses across the cylinder wall is shown in Fig. 19.5.

Radius r mm	τ_{max} MN/m^2	σ_R MN/m^2	σ_c MN/m^2
50	± 204	0	-408
55	± 169	-35	-373
60	± 142	-62	-346
65	± 121	-83	-325
70	± 104	-100	-308

The distribution of radial stress is reversed but the hoop stress, although compressive instead of tensile, has much the same distribution as before. The maximum direct stress is still at the inside surface, but is compressive. The distribution of shear stress is unchanged.

The case of a cylinder subjected to external pressure may be extended to include a solid cylindrical shaft or plug of circular section. For a solid cylinder, the internal radius is zero. Substitution of $r = 0$ into the equation for σ_R would make the radial stress infinite. This is obviously not true and the only conclusion to be drawn is that the constant b must also be zero. Thus for a solid shaft subjected to an external pressure p,

$$\sigma_R = a; \quad \sigma_c = a$$

Also, since at the outer skin $\sigma_R = -p$, it follows that $a = -p$.

$$\therefore \quad \sigma_R = \sigma_c = -p \tag{19.16}$$

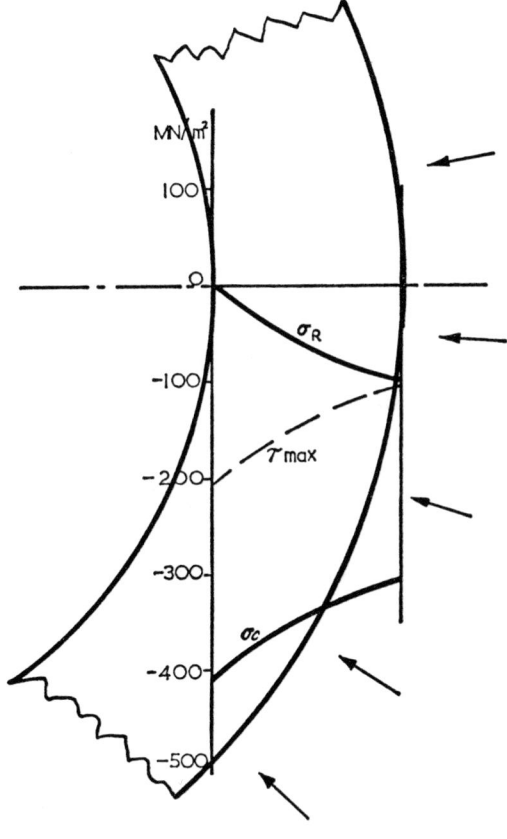

Fig. 19.5

Example. *A cylindrical pressure vessel has an internal diameter of 250 mm and is required to withstand an internal gauge pressure of 100 MN/m². The yield stress of the material is 824 MN/m². Using a factor of safety of 2 on the yield stress, calculate the minimum wall thickness required. (a) if the maximum principal stress is the criterion of failure, (b) if the maximum shear stress is the criterion of failure.*

(*a*) If the maximum principal stress is the criterion of failure, the maximum principal stress in the cylinder must be limited to $824 \div 2$ MN/m² $= 412$ MN/m². For a cylinder subjected to internal pressure, the greatest value of the maximum principal stress is the hoop stress σ_c at the inside surface.

Thus, when $r = 0 \cdot 125$ m,

$$\sigma_c = 412 \text{ MN/m}^2$$

$$\therefore 412 \times 10^6 = a + (b/0 \cdot 125^2)$$

$$= a + 64b \qquad (1)$$

Also, when $r = 0 \cdot 125$ m,

$$\sigma_R = -100 \text{ MN/m}^2$$

$$\therefore a - (b/0 \cdot 125^2) = a - 64b = -100 \times 10^6 \qquad (2)$$

Adding (1) and (2),

$$2a = 312 \times 10^6$$

i.e.

$$a = 156 \times 10^6 \text{ N/m}^2$$

Substitution in (2) gives:

$$b = 4 \times 10^6 \text{ N}$$

$$\therefore \sigma_R = 156 - (4/r^2) \text{ MN/m}^2$$

$$\sigma_c = 156 + (4/r^2) \text{ MN/m}^2$$

When $r = $ Radius at outer surface $= R$ (say),

$$\sigma_R = 0$$

$$\therefore 156 - (4/R^2) = 0$$

$$\therefore R^2 = 4/156 = 0 \cdot 02564 \text{ m}^2 = 25\,640 \text{ mm}^2$$

$$\therefore R = 160 \cdot 1 \text{ mm}$$

$$\therefore \text{ Wall thickness} = 160 \cdot 1 - 125 = 35 \cdot 1 \text{ mm}$$

Since the pressure vessel presumably has closed ends, there will be a longitudinal stress σ_L. Does this lie within the safe limit?

$$\sigma_L = \text{Force on one end/C.S.A.}$$

$$= 100 \times \pi \times 0 \cdot 125^2/\pi (0 \cdot 1601^2 - 0 \cdot 125^2) \text{ MN/m}^2$$

$$= 156 \text{ MN/m}^2$$

This is considerably less than 412 MN/m^2, so that the wall thickness is satisfactory. It should be noticed that σ_L has the same value as the constant a. This is no accident, for it can be proved that $\sigma_L = a$.

(b) In a simple tensile test, the maximum shear stress in the specimen is $\sigma/2$ on a plane at $45°$ to the direct stress σ. At yield $\sigma = 824 \text{ MN/m}^2$, so that

Maximum shear stress at yield $= \pm 824/2 = \pm 412 \text{ MN/m}^2$

Using the factor of safety, the maximum permissible shear stress in the cylinder is $\pm 412/2$ MN/m$^2 = \pm 206$ MN/m^2. The maximum shear stress in the cylinder occurs at the inside surface, so that

$$\pm b/0.125^2 = \pm 206 \times 10^6$$

or
$$b = 3.22 \times 10^6 \text{ N}$$

When $r = 0.125$ m,

$$\sigma_R = -100 \text{ MN/m}^2$$

$$\therefore \ a - (b/0.125^2) = -100 \times 10^6$$

$$\therefore \ a = -(100 \times 10^6) + 64b = (-100 + 206) \times 10^6 \text{ N/m}^2$$

i.e.
$$a = 106 \times 10^6 \text{ N/m}^2$$

$$\therefore \ \sigma_R = 106 - (3.22/r^2) \text{ MN/m}^2$$

$$\sigma_c = 106 + (3.22/r^2) \text{ MN/m}^2$$

When $r = R$ (the outer radius),

$$\sigma_R = 0$$

$$\therefore \ 106 - (3.22/R^2) = 0$$

$$\therefore \ R^2 = 3.22/106 = 0.03037 \text{ m}^2 = 30\ 370 \text{ mm}^2$$

$$\therefore \ R = 174.3 \text{ mm}$$

$$\therefore \ \text{Wall thickness} = 174.3 - 125 = 49.3 \text{ mm}$$

PROBLEMS

For tutorials

1. A cylindrical pressure vessel having closed ends is subjected to an internal pressure p. Prove that:

(a) The longitudinal stress σ_L is equal to the constant a in Lamé's equation;

(b) $\sigma_{c_{\max}} = p \left(\dfrac{n^2 + 1}{n^2 - 1} \right)$

where $n =$ External radius/Internal radius.

2. A thick-walled cylinder is subjected to an external pressure p while the internal pressure is zero. Prove that the maximum hoop stress is given by:

$$\sigma_{c_{\max}} = -2pn^2/(n^2 - 1)$$

where $n =$ External radius/Internal radius.

3. The distribution of hoop stress in the wall of a cylinder subjected to an internal pressure only is shown in Fig. 19.4. If n is the ratio of the external to the internal radius, show that

$$\sigma_{c_{max}}/\sigma_{c_{min}} = \tfrac{1}{2}(1 + n^2)$$

Plot a graph of the ratio $\sigma_{c_{max}}/\sigma_{c_{min}}$ against n given that the smallest possible value of n is unity. What significance can be attached to the value of $\sigma_{c_{max}}/\sigma_{c_{min}}$ when (a) $n = 1$, (b) when n is large?

4. A thick-walled cylinder with closed ends is subjected to an internal pressure p. Given that $1 - 2\nu = E/3K$ (Chapter 13), show that the longitudinal strain in the cylinder is given by:

$$\varepsilon_L = p/\{3K(n^2 - 1)\}$$

where n is the ratio of external to internal diameter.

Use this equation to devise an experiment for the measurement of the bulk modulus K of the material.

General

1. A closed cylindrical pressure vessel has an internal diameter of 1·5 m, a wall thickness of 100 mm and a length of 4 m. The vessel is subjected to an internal pressure of 16 MN/m², while the external pressure is zero. Determine:

(a) The maximum direct tensile stress and the maximum shear stress in the cylinder wall.

(b) The increase in internal diameter and the increase in length of the cylinder.

Assume $E = 200$ GN/m² and $\nu = 0.28$.

Ans. (a) 128·5 MN/m²; 72·25 MN/m²; (b) 0·879 mm; 0·495 mm

2. A hydraulic cylinder subjected to an internal pressure of 40 MN/m² has an external diameter of 200 mm. This pressure produces an increase in the external diameter of 0·05 mm. Assuming that the force on the ends of the cylinder is not reacted by the cylinder wall, estimate the internal diameter of the cylinder. Take $E = 200$ GN/m² and $\nu = 0.28$.

Ans. 124 mm

3. A thick-walled cylinder has to withstand an internal pressure of 60 MN/m². The internal diameter is 100 mm and the external diameter is 200 mm. There is no axial load reacted by the cylinder wall. Make the necessary calculations and sketch graphs to show the variations in hoop and radial stress, and maximum shear stress through the cylinder wall.

Ans. σ_c varies from 100 MN/m² at inner radius to 40 MN/m² at outer radius; τ_{max} varies from 80 MN/m² at inner radius to 20 MN/m² at outer radius.

4. An 80 mm internal diameter copper tube is 1·0 m long and the metal is 8 mm thick. The ends of the cylinder are closed and may be assumed not to distort. The tube is initially full of water. Determine the change of pressure when an additional 8 ml of water is pumped in. For copper, $E = 100$ GN/m² and $\nu = 0.3$. The bulk modulus of water is 2 GN/m².

Ans. 2·6 MN/m².

CHAPTER 20

Dimensional Analysis

20.1 NATURAL LAWS

The laws describing a physical process may often be derived analytically from first principles and expressed in the form of simple algebraic equations—in elementary mechanics, for example. In some instances, however, particularly in fluid mechanics and thermodynamics, processes are too complex or too little understood for laws to be derived analytically, and *determination* of the law by experiment—may then be necessary. A reasoned estimate must be made of the variables likely to be involved, and suitable experiments devised to determine their interdependence.

In problems relating to fluid flow or to heat transfer, the laws under investigation often relate not individual variables but non-dimensional *groups* of variables. Before, embarking on experiments, it is therefore essential to determine the nature of these groups. which may be done by dimensional analysis. Dimensional analysis is a process by which the physical dimensions of both sides of an equation are evaluated, and then equated. The method was first used by Osborne Reynolds in his investigations of the flow of fluids, but was later more fully developed by Lord Rayleigh and by Buckingham.

Principle of dimensional homogeneity

Every term in an equation describing a real physical process must be dimensionally the same when expressed in terms of fundamental dimensions. This principle is self-evident, since only physical quantities of the same dimensional form could ever be termed "equal." Similarly, no two physical quantities may be added together unless they have the same dimensional form.

For example, a force F applied to a body of mass m may result in the body acquiring kinetic and potential energy. Energy is acquired because the force does work, and

Work done by F = Kinetic energy gained + Potential energy gained

i.e. $\qquad\qquad Fs = \frac{1}{2}mv^2 + mgh$

where s = Displacement of the point of application of the force,
$\qquad h$ = Vertical displacement of the mass, and
$\qquad v$ = Velocity acquired by the mass.

Substitution of units reveals that:

$\quad Fs$ has the units: $\text{N} \times \text{m} = (\text{kg m/s}^2)\,\text{m} = \text{kg m}^2/\text{s}^2$
$\quad \frac{1}{2}mv^2$ has the units: $\text{kg} \times (\text{m/s})^2 = \text{kg m}^2/\text{s}^2$
$\quad mgh$ has the units: $\text{kg} \times (\text{m/s}^2) \times \text{m} = \text{kg m}^2/\text{s}^2$

As anticipated, each term in the equation has precisely the same fundamental units.

In dimensional analysis, the actual unit of measurement of each quantity is unimportant. Only the *dimension* need be considered so that instead of substituting the units m, kg and s, the dimensions length (L), mass (M), and time (T) are substituted.

20.2 FUNDAMENTAL DIMENSIONS

The fundamental dimensions of mechanics are length, mass and time. Every other physical quantity may be expressed in terms of these dimensions in accordance with some physical law or definition. For example,

$$\text{Velocity} = \text{Displacement/Unit time}$$
$$\text{Length/Time} = LT^{-1}$$
$$\text{Acceleration} = \text{Change in velocity/unit time}$$
$$= \text{Length/(time)}^2 = LT^{-2}$$
$$\text{Force} = \text{Mass} \times \text{Acceleration}$$
$$= \text{Mass . Length/(time)}^2 = MLT^{-2}$$

Temperature is defined in Chapter 25 as "the mean translational kinetic energy" of the molecules of a substance, and is theoretically expressable in terms of L, M and T. This, however, is not always convenient and similar devices are not practical for electricity and light. Consequently, there are three auxiliary fundamental dimensions—thermodynamic temperature, electric current and luminous intensity.

Fundamental and derived units

Units of measurement must be defined for each of the six fundamental dimensions. These are called *fundamental units*, and are listed in the table.

Dimension	Symbol	Unit	Unit abbreviation
Length	L	metre	m
Mass	M	kilogramme	kg
Time	T	second	s
Temperature	Θ	degree Kelvin	°K
Electric current	I	ampere	A
Luminous intensity	J	candela	cd

All other quantities may be expressed as combinations of these, and the compound units of measurement formed are called *derived units*. Some derived units occur so frequently that they are given special names, e.g. the derived unit of force is the kg . m/s² which is called the newton (N).

Dimensions of some physical quantities

Following is a selection of physical quantities and their corresponding dimensions.

Displacement $s = L$; Area $A = L^2$; Volume V $\qquad = L^3$

Mass density $\rho = $ Mass/Unit volume $= M/L^3$ $\qquad = ML^{-3}$

Velocity $v = $ Displacement/Unit time $= L/T$ $\qquad = LT^{-1}$

Momentum $= $ Mass $\times$ Velocity $= M(L/T)$ $\qquad = MLT^{-1}$

Acceleration $a = $ Velocity/Unit time $= (L/T)/T$ $\qquad = LT^{-2}$

Force $F = m \cdot a$ $\qquad = MLT^{-2}$

Angular displacement $= s/r = L/L$ $\qquad = [L]^\circ$

Angular velocity $\omega = v/r = LT^{-1}/L$ $\qquad = [L]^\circ T^{-1}$

Frequency $f = 1/T$ $\qquad = T^{-1}$

Moment of inertia $= mk^2$ $\qquad = ML^2$

Volume flow rate $\dot{V} = $ Volume/Unit time $= L^3/T$ $\qquad = L^3 T^{-1}$

Mass flow rate $\dot{m} = $ Mass/Unit time $= M/T$ $\qquad = MT^{-1}$

Angular acceleration $\alpha = a/r = LT^{-2}/L$ $\qquad = [L]^\circ T^{-2}$

Torque $= Fr = MLT^{-2}L$ $\qquad = ML^2 T^{-2}$

Linear strain $= \delta x/x = L/L$ $\hspace{3cm} = [L]^0$

Pressure, stress, modulus $=$ Force/Unit area

$$= MLT^{-2}/L^2 = ML^{-1}T^{-2}$$

Energy (U), Work $W = Fs = MLT^{-2} \cdot L$ $\hspace{2cm} = ML^2T^{-2}$

Power $=$ Work/Unit time $= ML^2T^{-2}/T$ $\hspace{2cm} = ML^2T^{-3}$

Dynamic viscosity $\eta =$ Shear stress/Velocity gradient

$$= ML^{-1}T^{-2} \div (LT^{-1}/L) = ML^{-1}T^{-1}$$

Kinetic viscosity $v = \eta/\rho = ML^{-1}T^{-1}/ML^{-3}$ $\hspace{1.5cm} = L^2T^{-1}$

Heat $Q = U + W =$ Energy $\hspace{3cm} = ML^2T^{-2}$

Heat flow rate $\mathring{Q} = dQ/dt = ML^2T^{-2}/T$ $\hspace{1.5cm} = ML^2T^{-3}$

Specific heat $c = Q/m\theta = ML^2T^{-2}/M\Theta$ $\hspace{1.5cm} = L^2T^{-2}\Theta^{-1}$

Enthalpy $=$ Energy (dimensionally) $\hspace{2.5cm} = ML^2T^{-2}$

Specific enthalpy $=$ Enthalpy/Unit mass

$$= ML^2T^{-2}/M = L^2T^{-2}$$

Entropy $= \delta Q/\theta = ML^2T^{-2}/\Theta$ $\hspace{2cm} = ML^2T^{-2}\Theta^{-1}$

Thermal conductivity $k = \mathring{Q}/\{A \cdot (d\theta/dx)\}$

$$= ML^2T^{-3}/\{L^2(\Theta/L)\} = MLT^{-3}\Theta^{-1}$$

20.3 DIMENSIONAL ANALYSIS AND FLUID FLOW

In problems relating to the flow of a fluid, the main consideration is the force of resistance offered by a body *around* which the fluid is flowing, or by the pipeline, orifice or channel *through* which the fluid is flowing. In general, the problem will be determined in terms of a number of variables which will fall into three categories:

1. A sufficient number of lengths to define the system geometrically, such as l, b, d.

2. Dynamic and kinematic variables such as (a) a velocity v or a quantity flow $\mathring{V}$; (b) a force of resistance F or a pressure p; (c) the acceleration due to gravity g.

3. Fluid properties, such as (a) mass density ρ; (b) dynamic viscosity η; (c) modulus of elasticity K, and (d) surface tension σ.

Thus, it would be reasonable to suppose that:

$$F = \phi(l, b, d, v, g, \rho, \eta, K, \sigma) \tag{20.1}$$

where the symbol ϕ indicates some mathematical function the form of which is not determined. The large number of variables on which F

might depend, suggests that any dimensional analysis of the problem would be a formidable task. However, Eq. (20.1) may be simplified by considering more specific cases in which it is obvious that some variables will have little or no effect.

The Reynolds number

If, for example, the problem relates specifically to a sphere deeply submerged in an incompressible fluid, the following conclusions may be drawn:

1. Only one dimension, the diameter of the sphere d, is required to define the system geometrically.

2. Since there is no free surface, gravitational effects on the fluid will be balanced by static buoyancy, so that g will have no effect on F.

3. Since the fluid is incompressible, there will be no elasticity effects and hence K may be ignored.

4. Since there is no free surface, there can be no surface tension effects. Eq. (20.1) thus reduces to:

$$F = \phi(d, v, \rho, \eta) \qquad (20.2)$$

F is now seen to depend on only four variables and it may be assumed tentatively that:

$$F = C d^r v^s \rho^t \eta^u \qquad (20.3)$$

where C is a dimensionless constant. Substitution of the fundamental dimensions of each variable gives:

$$MLT^{-2} = (L)^r (LT^{-1})^s (ML^{-3})^t (ML^{-1}T^{-1})^u$$

By the principle of dimensional homogeneity, the respective exponents of M, L and T on each side of the equation must be the same.

$$\therefore \text{ For } M: \quad 1 = t + u$$
$$\text{For } L: \quad 1 = r + s - 3t - u$$
$$\text{For } T: \quad -2 = -s - u$$

The four exponents cannot be determined from only three equations, but three may be expressed in terms of the fourth.

In general the exponent retained will be that be that of the variable whose effects are the main subject under investigation. In this case the viscous effects are likely to be of most interest, and thus:

$$t = 1 - u$$
$$s = 2 - u$$
$$r = 1 - (2 - u) + 3(1 - u) + u = 2 - u$$

Eq. (20.3) may be re-written:

$$F = Cd^{(2-u)}v^{(2-u)}\rho^{(1-u)}\eta^u$$

i.e.

$$F = C\rho v^2 d^2 (\rho v d/\eta)^{-u}$$

i.e.

$$\frac{F}{\rho v^2 d^2} = C\left(\frac{\rho v d}{\eta}\right)^{-u} \tag{20.4}$$

Eq. (20.4) relates two non-dimensional groups of variables by means of an unknown exponent $(-u)$. The non-dimensional group $(\rho v d/\eta)$ is called Reynolds number and is denoted by Re. This was first discovered by Osborne Reynolds, who found that its value determined the type of flow through round pipes. Rayleigh later showed that the number has a much wider significance and that it is the governing factor in problems relating to frictional resistance in fluid flow. Thus:

$$\text{Reynolds number } \text{Re} = (\rho v d/\eta) = (v d/\nu) \tag{20.5}$$

The other non-dimensional group $(F/\rho v^2 d^2)$ is known as the Newton number.

Eq. (20.4) may be written:

$$F = \rho v^2 d^2 \phi(\text{Re}) \tag{20.6}$$

The cross-sectional area of a sphere $A = \pi d^2/4$.

$$\therefore F = \rho v^2 . (4A/\pi) . \phi(\text{Re})$$
$$= \tfrac{1}{2}\rho A v^2 (8/\pi) . \phi(\text{Re})$$

i.e.

$$F = \tfrac{1}{2}C_D \rho A v^2 \tag{20.7}$$

where $C_D = (8/\pi)\phi(\text{Re}) = $ Drag coefficient and must be obtained by experiment.

The problem of the flow of a fluid through a pipeline is much the same. The major differences lie in the fact that a pressure difference p is required to overcome the viscous resistance, and an additional dimension is required to specify the geometry. Not only must the diameter d of the pipe be specified but also the length l. Thus:

$$p = \phi(l, d, v, \rho, \eta) \tag{20.8}$$

There are now five variables to be considered, and if

$$p = Cl^q d^r v^s \rho^t \eta^u \tag{20.9}$$

where C is a dimensionless constant.

Substituting dimensions, as before:

$$ML^{-1}T^{-2} = (L)^q(L)^r(LT^{-1})^s(ML^{-3})^t(ML^{-1}T^{-1})^u$$

Applying the principle of dimensional homogeneity:

For M: $1 = t + u$

For L: $-1 = q + r + s - 3t - u$

For T: $-2 = -s - u$.

Since there are now five unknowns and still only three equations, *two* exponents must be retained. For the same reason as before, the exponent of viscosity will be retained, and the logical choice for the other is the additional geometrical exponent q. Solving for, r, s and t:

$$t = 1 - u$$

$$s = 2 - u$$

$$r = -1 - q - (2 - u) + 3(1 - u) + u$$

$$= -q - u$$

Eq. (20.9) thus becomes:

$$p = Cl^q d^{(-q-u)} v^{(2-u)} (\rho^{(1-u)} \eta^u$$

$$= C\rho v^2 (l/d)^q (\rho v d/\eta)^{-u}$$

i.e. $$(p/\rho v^2) = C(l/d)^q (\rho v d/\eta)^{-u} \qquad (20.10)$$

Eq. (20.10) relates three non-dimensional groups by means of the two unknown exponents q and $-u$. The factor $(l/d)^q$ indicates the geometric proportions of the pipe. This can be removed from Eq. (20.10) by employing *geometrically similar* pipes for which (l/d) is a constant. However, it is useful to have some indication of the geometrical proportion factor in the equation, so that Eq. (20.10) is written:

$$(p/\rho v^2) = C(l/d) \cdot (l/d)^{q-1} (\rho v d/\eta)^{-u}$$

Thus, for geometrically similar pipes,

$$p = (\rho v^2 l/d) \cdot \phi(\text{Re}) \qquad (20.11)$$

The Froude number

For a sphere which is only partly submerged in an incompressible fluid, elasticity may be ignored since the fluid is incompressible, but there is now a free surface and the sphere will displace liquid upwards against the force of gravity on its upstream side. Thus, surface wave action will effect the force of resistance F and g must be retained in the equation. Although surface tension is unlikely to have a great effect, it should be included and so Eq. (20.1) reduces to:

$$F = \phi(d, v, g, \rho, \eta, \sigma) \qquad (20.12)$$

Thus: $$F = Cd^p v^q g^r \rho^s \eta^t \sigma^u \qquad (20.13)$$

The surface tension coefficient σ has dimensions of force/unit length, or $MLT^{-2}/L = MT^{-2}$.

Therefore, substituting dimensions:

$$MLT^{-2} = (L)^p(LT^{-1})^q(LT^{-2})^r(ML^{-3})^s(ML^{-1}T^{-1})^t(MT^{-2})^u$$

Applying the principle of dimensional homogeneity,

For M: $1 = s + t + u$

For L: $1 = p + q + r - 3s - t$

For T: $-2 = -q - 2r - t - 2u$

There are six unknowns and only three equations, and therefore three unknown exponents must be retained. As well as viscosity, the effects of gravity and surface tension are under investigation, so that the exponents of η, g and σ must be those retained. Solving for p, q and s:

$$s = 1 - t - u$$
$$q = 2 - 2r - t - 2u$$
$$p = 1 - (2 - 2r - t - 2u) - r + 3(1 - t - u) + t$$
$$= 2 + r - t - u$$

Eq. (20.13), thus becomes:

$$F = Cd^{(2+r-t-u)}v^{(2-2r-t-2u)}g^r\rho^{(1-t-u)}\eta^t\sigma^u$$

i.e. $F = C\rho v^2 d^2(v^2/gd)^{-r}(\rho vd/\eta)^{-t}(\rho v^2 d/\sigma)^{-u}$

This may be written:

$$\frac{F}{\rho v^2 d^2} = C\left(\frac{v}{\surd(gd)}\right)^{-2r}(\text{Re})^{-t}\left(\frac{\rho v^2 d}{\sigma}\right)^{-u} \qquad (20.14)$$

Eq. (20.14) relates four non-dimensional groups, two of which are recognised as Reynolds number and the Newton number. The other other two, $v/\surd(gd)$ and $\rho v^2 d/\sigma$, are called respectively the Froude number and the Weber number. Of these, the more important is the Froude number Fr, since in the majority of cases surface tension effects are negligible. Thus:

$$\text{Froude number} = \text{Fr} = v/\surd(gd) \qquad (20.15)$$

For a partly submerged sphere, if surface tension effects are negligible:

$$F = \rho v^2 d^2 \phi(\text{Fr, Re}) \qquad (20.16)$$

The Mach number

If the sphere is once more deeply submerged but in a fluid which is not incompressible and in which the effects of elasticity will be apparent. Eq. (20.1) will reduce to:

$$F = \phi(d, v, \rho, \eta, K) \qquad (20.17)$$

This may be written:

$$F = Cd^q v^r \rho^s \eta^t K^u \qquad (20.18)$$

where C is a dimensionless constant.

Substituting dimensions

$$MLT^{-2} = (L)^q (LT^{-1})^r (ML^{-3})^s (ML^{-1}T^{-1})^t (ML^{-1}T^{-2})^u$$

Applying the principle of dimensional homogeneity,

For M: $1 = s + t + u$

For L: $1 = q + r - 3s - t - u$

For T: $-2 = -r - t - 2u$

There are five unknowns and only three equations, and therefore the exponents of η and K must be retained.

Solving for q, r and s,

$$s = 1 - t - u$$

$$r = 2 - t - 2u$$

$$q = 1 - (2 - t - 2u) + 3(1 - t - u) + t + u$$

$$= 2 - t$$

Eq. (20.18) thus becomes:

$$F = Cd^{(2-t)} v^{(2-t-2u)} \rho^{(1-t-u)} \eta^t K^u$$

$$= C\rho v^2 d^2 (\rho vd/\eta)^{-t} (\rho v^2/K)^{-u}$$

But $\sqrt{(K/\rho)} = $ Velocity of sound in the fluid $= a$

$\therefore \rho v^2/K = v^2/a^2 = (v/a)^2$

Thus the above equation becomes

$$F/\rho v^2 d^2 = C(\mathrm{Re})^{-t}(v/a)^{-2u} \qquad (20.19)$$

Eq. (20.19) relates three non-dimensional groups, namely Reynolds number, the Newton number and the ratio of the fluid velocity to the local velocity of sound. The latter is called the Mach number, Ma.

$$\mathrm{Ma} = v/a \qquad (20.20)$$

Thus if elasticity effects are appreciable:

$$F = \rho v^2 d^2 \phi(\text{Re, Ma}) \qquad (20.21)$$

This equation and the Mach number are of considerable importance for bodies moving through air with sonic velocity.

20.4 Model Testing

Equations obtained by dimensional analysis can only be partial solutions, since there are always insufficient equations relating the exponents. The unknown exponents and the values of dimensionless constants such as C must be obtained by suitable experiments and it is often convenient and sometimes essential to rely on small scale model experiments. With ships, aircraft and other large structures, for example, information required in the design stage can only be obtained by model tests, the results of which will be valid only if the experiments are conducted under conditions of both geometrical similarity and dynamical similarity.

Principle of geometrical similarity

Two bodies with the same shape are geometrically similar if the ratio of corresponding dimensions is the same.

With the exception of Eq. (20.10), all the equations obtained so far by dimensional analysis have related to a sphere where only one dimension is required to define the geometry. The same is true of *geometrically similar* bodies. For example, a 1/10 scale model of an aircraft may be tested in a wind tunnel to obtain the exponents of an equation obtained by dimensional analysis and involving only one typical dimension, and the results will be valid for geometrically similar aircraft of any size.

The principle of geometrical similarity is illustrated by the reduction of Eq. (20.10) to that of Eq. (20.11).

Principle of dynamical similarity

Two systems with the same types of forces are dynamically similar if the ratio of corresponding forces is the same. In fluid flow, this means that the paths taken by corresponding fluid elements will be geometrically similar. In, for example, a system in which the flow of a fluid is governed by its inertia force and viscous resistance, there will be dynamical similarity between systems of this type only if the ratio of inertia force to viscous resistance is the same.

If l is a typical dimension, the

$$\text{Inertia force} \propto \text{Mass} \times \text{Acceleration}$$

$$\propto \rho l^3 \times l/t^2$$

$$\propto \rho l^2 \times v^2$$

since

$$v \propto l/t$$

$$\text{Viscous resistance} \propto \eta \times \text{Velocity gradient} \times \text{Area}$$

$$\propto \eta \times (v/l) \times l^2$$

$$\propto \eta v l$$

∴ The ratio inertia force/viscous resistance $\propto \rho l^2 v^2 / \eta v l$

$$\propto \rho v l / \eta$$

This, of course, is Reynolds number, Re.

It may similarly be shown that for systems in which surface wave resistance is involved

$$\text{Inertia force/Gravity force} \propto \text{Fr}$$

and, for compressibility effects,

$$\text{Inertia force/Elastic force} \propto \text{Ma}$$

Dynamical similarity is thus achieved in a model test by ensuring that the appropriate non-dimensional group of variables is the same in the model as in the full scale counterpart. If, for example, a model is to be tested to obtain the drag force on a deeply submerged body (Eq. (20.4)) then the Reynolds number for the model must be equal to that for the actual body. Thus using suffix "m" for the model,

$$\rho_m v_m l_m / \eta_m = \rho v l / \eta$$

$$\therefore \ v_m = (\rho/\rho_m)(l/l_m)(\eta_m/\eta)v \qquad (20.22)$$

If the model is tested in a different fluid, then (ρ/ρ_m) and (η_m/η) will not be unity. Eq. (20.22) gives the *corresponding speed* of the model.

Example. *A $1/10$ scale model is to be tested in a wind tunnel to determine the drag force on a diving bell when submerged in sea water running at 4 m/s. Determine the corresponding speed of the air in the wind tunnel. If the drag force on the model is found to be 2 N, determine the drag force on the actual diving bell. Assume that the density of sea water is 1030 kg/m³ and that its dynamic viscosity is 1·5 centipoise, while the density of air is 25·75 kg/m³ and has a dynamic viscosity of 0·0182 centipoise.*

Since the diving bell is submerged, only inertia and viscous forces will be involved and the drag force will be given by an equation of the form:

$$F = \rho l^2 v^2 \phi(\mathrm{Re})$$

For dynamical similarity:

$$\mathrm{Re_m} = \mathrm{Re}$$

i.e. $\qquad \rho_m v_m l_m / \eta_m = \rho v l / \eta$

i.e. $\qquad v_m = (\rho/\rho_m)(l/lm)(\eta_m/\eta)v$

$$= (1030/25\cdot75)(10)(0\cdot0182/1\cdot5)4 \text{ m/s}$$

$$= 19\cdot41 \text{ m/s}$$

Since $F/\rho l^2 v^2 = \phi(\mathrm{Re})$ and Re is the same for both model and diving bell, it follows that:

$$F/\rho l^2 v^2 = F_m/\rho_m l_m^2 v_m^2$$

i.e. $\qquad F/F_m = (\rho l^2 v^2)/(\rho_m l_m^2 v_m^2)$

$$= (1030/25\cdot75)(10)^2(4/19.41)^2 = 170$$

$$\therefore F = 170 \times 2 \text{ N} = 340 \text{ N}$$

Example. *A ship which is to have a speed of* 15 m/s *has a hull* 200 m *long. A scale model* 5 m *long is to be towed in a fresh water tank to determine the probable resistance to motion of the ship due to surface wave action. Determine the speed at which the model should be towed, and calculate the wave resistance of the ship if at this speed the wave resistance of the model is found to be* 16 N. *The density of sea water and fresh water are, respectively,* 1030 kg/m³ *and* 1000 kg/m³.

If viscous resistance is neglected, Eq. (20.16) reduces to

$$F = \rho v^2 l^2 \phi(\mathrm{Fr})$$

But for dynamical similarity,

$$\mathrm{Fr_m} = \mathrm{Fr}$$

i.e. $\qquad v_m/\sqrt{(gl_m)} = v/\sqrt{(gl)}$

$$\therefore v_m = v\sqrt{(l_m/l)}$$

$$= 15\sqrt{(5/200)} = 2\cdot37 \text{ m/s}$$

If the Froude numbers are equal,

$$F_m/\rho_m v_m^2 l_m^2 = F/\rho v^2 l^2$$

$$\therefore F = F_m (\rho v^2 l^2/\rho_m v_m^2 l_m^2)$$

$$= 16 \times 1\cdot03 \times 40 \times 40^2 = 1\cdot055 \text{ MN}$$

Problems

For tutorial

1. Distinguish between (a) dimensional homogeneity, (b) geometrical similarity, and (c) dynamical similarity.

2. Evaluate the dimensions of the following quantities in terms of M, L, T and Θ:

(a) The power transmitted by a shaft $T\omega$, where T is the torque applied to the shaft and ω is its angular velocity,

(b) the characteristic constant of a gas pv/T, where p is the pressure of the gas, v is its specific volume, and T is its absolute temperature.

3. Check the dimensional homogeneity of the equation

$$m(\mathrm{d}^2x/\mathrm{d}t^2) + F_v(\mathrm{d}x/\mathrm{d}t) + Sx = 0$$

where m is a mass, F_v a viscous damping force, S an elastic stiffness and x a linear displacement.

4. It may be shown that the resistance to motion of a partly submerged body is dependent on both the Reynolds number and the Froude number. To achieve dynamical similarity in a model test, both Re and Fr must therefore be the same for the model as for the full size body. Is it possible to satisfy both these conditions simultaneously?

Discuss the implications of this for a model test on the hull of a ship.

5. (a) It may be thought that the resistance to flow in a pipeline is dependent on the degree of roughness of the inside surface of the pipe, as well as on the length of pipe l, its diameter d, the density ρ and viscosity η of the fluid and the velocity v of the fluid. If ε is the average height of roughness projections, show by dimensional analysis that:

$$F \propto \rho v^2 d^2 \phi\{(l/d),\ (\rho v d/\eta),\ (\varepsilon/d)\}$$

(b) If it is shown by experiment that F is directly proportional to l, use the above expression to show that the frictional coefficient f in the Darcy formula is a function of both Reynolds number and the roughness factor (ε/d).

General

1. Assuming that the resistance of a hydroplane is due entirely to wave resistance, calculate its resistance at a speed of 30 m/s if that of a scale model whose linear dimensions are 1/20 of those of the hydroplane is 2 N at the corresponding speed. What is the speed of the model? Establish any formula used.

Ans. 16 kN; 6·71 m/s

2. A disc of diameter d is rotated at a speed n rev/min in a fluid of dynamic viscosity η and mass density ρ. Show that the torque required to overcome the frictional resistance is given by

$$T = \rho n^2 d^5 \phi(\rho n d^2/\eta)$$

3. Show by dimensional analysis that the thrust of a propeller is given by:

$$F = \rho v^2 d^2 \phi\{(nd/v), (\rho v d/\eta)\}$$

where d is the diameter of the propeller, n its speed of revolution, v its speed of advance, ρ the mass density of the fluid, and η the dynamic viscosity of the fluid. A propeller has a diameter of 3 m and rotates at a speed of 100 rev/min. A scale model of the propeller 300 mm in diameter is tested in sea water at a speed of 400 rev/min, the thrust developed by the model being 200 N and the speed of advance 4 m/s. Neglecting viscous resistance, determine the thrust and speed of advance of the full-size propeller.

Ans. 125 kN; 10 m/s

4. If the torque due to viscous resistance T_R of a lubricated bearing depends on the diameter d of the bearing, the dynamic viscosity η of the oil, the speed of rotation n of the shaft and the bearing pressure p, show by dimensional analysis that:

$$T_R = C\eta n d^3$$

where C is a function of the non-dimensional factor $(\eta n/p)$.

CHAPTER 21

Hydrostatics

21.1 FLUIDS

Fluids are substances which will flow under gravity. All gases and liquids are fluids, and certain solids such as silicone putty, lead and pitch will at normal temperatures flow under gravity but much more slowly. Flow takes place because the substances cannot resist shear indefinitely and completely. For a fluid in motion, there are viscous tangential (shear) forces which in fact tend to resist motion (see Chapter 22) but these are proportional to the velocity of flow and become zero when the fluid is at rest.

Liquids differ from gases in their molecular arrangement. In liquids the arrangement is such that a given mass will retain virtually the same volume under considerable change in pressure, so that a liquid poured into a vessel will have a well defined surface in contact with the atmosphere above which is known as a free surface. Gases, because of their freer molecular arrangement, tend to fill any vessel into which they are put and so do not possess a free surface. Moreover, they change considerably in volume when subjected to changes in pressure.

Laws of hydrostatics

Hydrostatics is concerned with the equilibrium of and the pressure exerted by fluids at rest. The whole or any part of a fluid at rest can be assumed to be in static equilibrium under the action of external forces. This assumption together with the knowledge that fluids at rest cannot resist shear have led to the formulation of certain laws relating to hydrostatics.

Fluid pressure

Fluid pressure acts normal to the surfaces in contact with the fluid.

If a fluid (Fig. 21.1) exerts a force p on unit area of surface SS at some angle θ to the surface then,

Component of force parallel to the surface SS

$$= \text{Tangential force (shear)} = p \cos \theta$$

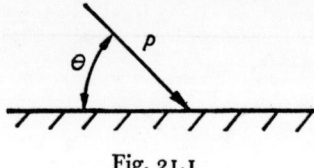

Fig. 21.1

But this component must be zero, since fluids at rest cannot exert shear forces. Therefore θ must equal 90° and hence the normal component, which is $p \sin \theta$, must equal p. That is, the fluid pressure p must always act normal to the surface.

Static head

Pressure is proportional to head of fluid.

Consider a vertical column of liquid (Fig. 21.2) of uniform cross section forming part of the fluid content of a tank. If the upper end of the column lies in the free surface of the liquid, the column will be in equilibrium under its own weight and the upward thrust of the surrounding liquid on the lower end. The fluid thrust on the vertical surface of the column will be horizontal, since it acts normal to the surface, and therefore need not be considered further.

Let h = Height of the column of fluid (m)

ρ = Density of the fluid (kg/m³)

ρg = Specific weight of fluid (N/m³)

p = Pressure at the lower end of the column due to head h (N/m²), and

a = Cross-sectional area of the column (m²). Then

$$\text{Weight of column} = \text{Volume} \times \text{Specific weight}$$
$$= ah\rho g$$

Upward thrust of surrounding fluid $= pa$

For equilibrium

$$pa = ah\rho g$$

or $$p = \rho g h \qquad (21.1)$$

Since ρ and g are constants,

$$p \propto h$$

The height h is known as the static head of fluid.

This relationship does not include the effect of atmospheric pressure p_a, which acts uniformly on the free surface and is transmitted uniformly throughout the liquid. When this pressure is taken into account, then

$$p = \rho gh + p_a \qquad (21.2)$$

Pressure at points of equal depth

Pressure is the same at all points at the same depth.
This follows directly from Eq. (21.1), since if h is constant then p must also be constant.

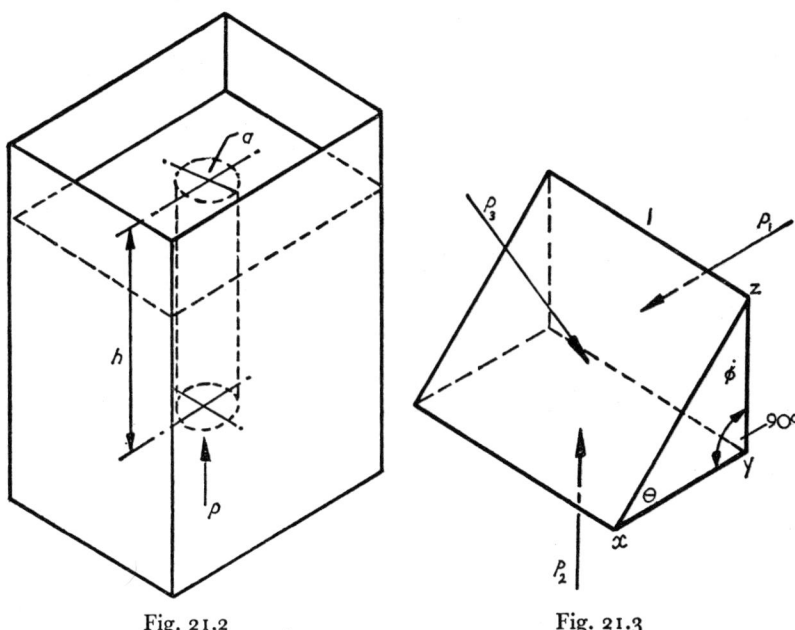

Fig. 21.2   Fig. 21.3

Pressure at a point

Pressure at any point in a fluid at rest is equal in all directions.
Consider an element of the fluid in the form of a right angled triangular prism of unit length (Fig. 21.3). Let the intensities of pressure on the rectangular faces be p_1, p_2 and p_3 as shown.

$$\text{Weight of the prism} = [(xy)(yz)/2]\rho g$$

Resolving forces horizontally and equating to zero for equilibrium,

$$p_1(yz) - p_3(xz)\cos\phi = p_1(yz) - p_3(yz) = 0$$

so that
$$p_1 = p_3 \qquad (a)$$

Resolving forces vertically and equating to zero.

$$p_2(xy) - p_3(xz) \cos \theta - [(xy)(yz)/2]\rho g = 0$$

$$p_2(xy) - p_3(xy) - [(xy)(yz)/2]\rho g = 0$$

$$p_2 - p_3 - \tfrac{1}{2}(yz)\rho g = 0$$

In the limit as $(yz) \rightarrow 0$, these pressures will act at a point, and

$$p_2 = p_3 \tag{b}$$

Hence, from Eqs. (a) and (b)

$$p_1 = p_2 = p_3$$

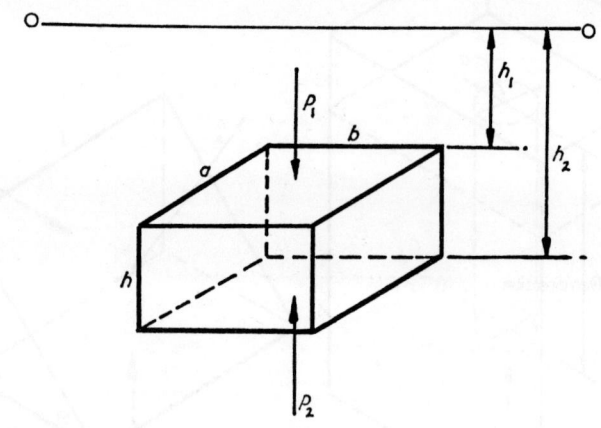

Fig. 21.4

Archimedes principle

If a body is immersed or partly immersed in a fluid at rest, the upthrust on the body due to fluid pressure is equal to the weight of the fluid displaced.

Suppose the body is removed from the fluid and that the "cavity" left behind is filled with additional fluid. Equilibrium is maintained, and since the forces due to the fluid surrounding the body now act unchanged on the replacement fluid, their resultant must act upwards to balance the weight of the additional fluid. But this resultant must be the upthrust on the body and must therefore equal the weight of the fluid displaced by the body.

The principle can be illustrated for a simple shape in the following manner. A solid rectangular prism (Fig. 21.4) length a, breadth b and thickness h is immersed in a fluid of density ρ. The upper horizontal surface is at depth h_1 and is subjected to an intensity of pressure p_1, while the lower horizontal surface is at depth h_2 and subjected to pressure p_2.

$$\text{Total force on upper surface} = p_1ab$$
$$= h_1\rho gab$$
$$\text{Total force on lower surface} = p_2ab$$
$$= h_2\rho gab$$

Therefore,

$$\text{Upthrust} = (h_2 - h_1)\rho gab$$
$$= (\text{Volume of displaced fluid})\,\rho g$$
$$= \text{Weight of displaced fluid}$$

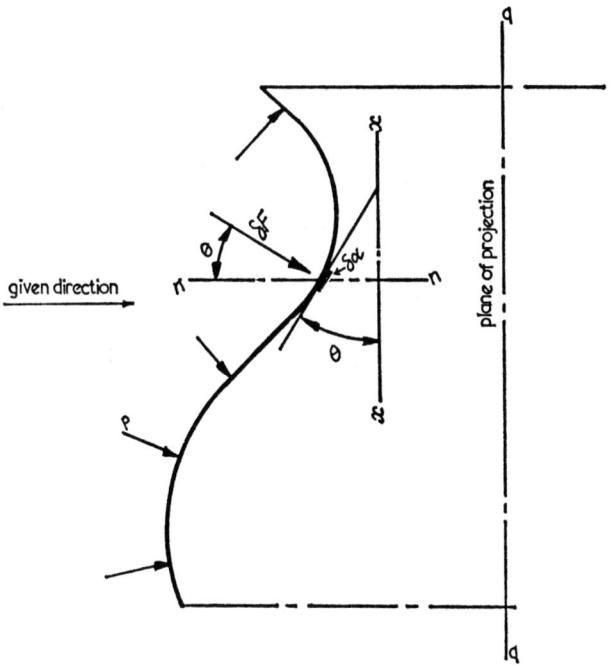

Fig. 21.5

Uniform pressure on curved surface

It can also be shown that the total force on a curved surface in a given direction equals the intensity of pressure multiplied by the area of the surface as projected onto a plane at right angles to the given direction.

Fig. 21.5 shows a curved surface subjected to a uniform pressure p. Consider an elemental area δa inclined at angle θ to the plane of projection.

Let x–x and n–n be parallel and normal to the plane of projection q–q respectively. Let δF be the force acting on area δa. Then δF will act at angle θ to n–n. Since

$$p = \delta F/\delta a, \quad \delta F = p\delta a$$

Component of force normal to plane of projection

$$= p\delta a \cos \theta$$
$$= p \times \text{Projection of } \delta a \text{ on q–q}$$

Total force in the given direction

$$= p\Sigma\delta a \cos \theta$$
$$= p \times \text{Projected area of the total surface}$$

Atmospheric pressure. The atmosphere is itself a fluid subjected to a gravitational pull which, because of its weight, exerts a pressure at the surface of the earth which must be taken into account when determining the absolute pressure acting upon a particular surface. Atmospheric pressure must be measured by some form of barometer and varies from day to day with changing meteorological conditions. The average value for atmospheric pressure is about $1 \cdot 013 \times 10^5 \text{ N/m}^2$ but it is convenient to assume a value of 10^5 N/m^2 or 1 bar for general purposes. Since the density of water is 1000 kg/m^3, the head of water equivalent to an atmospheric pressure of 1 bar is given by Eq. (21.1) as

$$h = 10^5/1000 \times 9 \cdot 81 = 10 \cdot 2 \text{ m}$$

This means that air at pressure of 1 bar will support a column of water approximately $10 \cdot 2$ metres in height. If mercury is substituted for water, the height of the column will be $10 \cdot 2/13 \cdot 6 = 0 \cdot 75$ metres. For the average value of atmospheric pressure of $1 \cdot 013 \times 10^5 \text{ N/m}^2$, the equivalent static heads are approximately $10 \cdot 36$ metres of water and $0 \cdot 76$ metres of mercury.

21.2 MEASUREMENT OF PRESSURE

Piezometer

If an open-ended vertical tube (Fig. 21.6) is connected to a vessel containing liquid under pressure, liquid will rise up the tube to form a column whose height equals the static head equivalent to the pressure of the liquid. That is,

$$p = h\rho g$$

where ρ = Density of liquid in the vessel (kg/m³),

    h = Height of the liquid column (m), and

    p = Gauge pressure of the liquid within the vessel (N/m²).

A tube used in this way to measure fluid pressure is known as a piezometer or pressure tube.

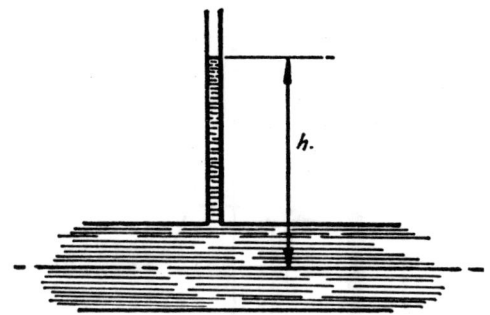

Fig. 21.6

The U-tube

U-tubes may be used for measuring the pressure at one point or the difference in pressure between two points in a fluid. The U-tube in Fig. 21.7 is connected at one end to the vessel containing the fluid F_1 under pressure p and is open at its other end. A second fluid F_2 (usually mercury) which is denser than that in the vessel is contained in the bend of the U-tube. The two fluids will have a common surface at a–a in the right-hand leg of the tube. Due to the pressure p, the fluid F_2 will be pushed down in the right hand leg and up in the left hand leg, producing a difference in levels. If ρ and ρ' are the densities of the fluids F_1 and F_2, the

Ratio of specific weights of the two fluids $= r = \rho'g/\rho g = \rho'/\rho$

Taking a–a as datum, since the pressures in both legs of the U-tube are equal at this level, then,

Pressure corresponding to difference in U-tube levels

$$= \text{(Pressure in Vessel)}$$

$$+ \text{ Pressure corresponding to head } h_2 \text{ of fluid } F_1$$

$$h\rho'g = p + h_2\rho g$$
$$p = h\rho'g - h_2\rho g$$
$$= (h_1 + h_2)\rho'g - h_2\rho g$$

because $h = h_1 + h_2$

Substituting r for ρ'/ρ,

$$p = [(h_1 + h_2)rg - h_2g]\rho$$
$$= [h_1rg + h_2rg - h_2g]\rho$$
$$= [(r - 1)h_2g + rh_1g]\rho \qquad (21.3)$$

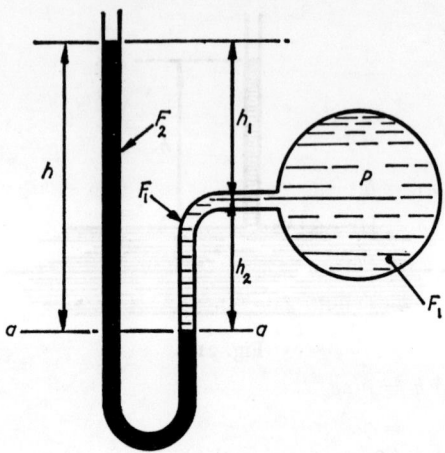

Fig. 21.7

In Fig. 21.8 the U-tube is shown connected to the entrance and throat of a Venturi-meter (see Chapter 22) in order to measure the difference in pressures at these two points. Taking a–a as datum,

Pressure corresponding to difference in U-tube levels

= Pressure difference of fluid F_1 at entrance and throat
+ Pressure corresponding to head h of fluid F_1

$$h\rho'g = (p_1 - p_2) + h\rho g$$
$$p_1 - p_2 = h\rho'g - h\rho g$$
$$= h(\rho'g - \rho g) \qquad (21.4)$$

The difference in pressure can be expressed as a head by dividing through by ρg, so that,

$$(p_1 - p_2)/\rho g = h(\rho'g/\rho g - 1)$$
$$= h(r - 1) \qquad (21.5)$$

Example. *What will be the gauge pressure in the pipe line shown in* Fig. 21.7 *if h = 600 mm and h_2 = 400 mm. The pipe is carrying oil having a specific* *gravity of 0·85 and the U-tube contains mercury (specific gravity 13·6). Also express the pressure as a head of oil.*

$$\text{Density of water} = 1000 \text{ kg/m}^3$$

Ratio r of specific weights of mercury and oil

$$= \rho'g/\rho g$$
$$= \rho'/\rho$$
$$= 13\cdot6 \times 1000/(0\cdot85 \times 1{,}000)$$
$$= 16$$

Also

$$h_1 = h - h_2 = 600 - 400 = 200 \text{ mm} = 0\cdot20 \text{ m}$$
$$p = [(r - 1)h_2g + rh_1g]\rho$$
$$= [(16 - 1)0\cdot40 \times 9\cdot81 + 16 \times 0\cdot2 \times 9\cdot81]850$$
$$= [58\cdot86 + 31\cdot40]850$$
$$= 76\,700 \text{ N/m}^2$$

Static head $h = p/\rho g$

$$= 76\,700/(850 \times 9\cdot81)$$
$$= 9\cdot2 \text{ metres of oil}$$

21.3 PRESSURE ON IMMERSED SURFACES

Intensity of pressure and total pressure on a surface

The pressure exerted by a fluid at rest varies with depth, but at a point or over an infinitely small area da of an immersed surface, the force dF may be considered to be constant so that

$$\text{Intensity of pressure at a point} = p = dF/da$$

The total thrust on the whole surface will equal the sum of the individual thrusts at the various points so that,

$$\text{Total pressure } F = \int p \, da$$

Fig. 21.9 shows a surface immersed in a liquid, and the force on an element of area da of the surface at depth h below o–o is given by

$$dF = p \, da$$
$$= h\rho g \, da$$

Therefore, the total pressure on the whole surface,

$$F = \rho g \Sigma h \, da$$
$$= \rho g \text{ (First moment of area of the surface about o–o)}$$
$$= \rho g A \bar{h} \qquad (21.6)$$

where A is the total wetted surface area and $\bar{h}$ is the vertical distance from the free surface o–o to the centre of area of the wetted surface. The right-hand side of Eq. (21.6) can be written as $A(\rho g \bar{h})$, which shows

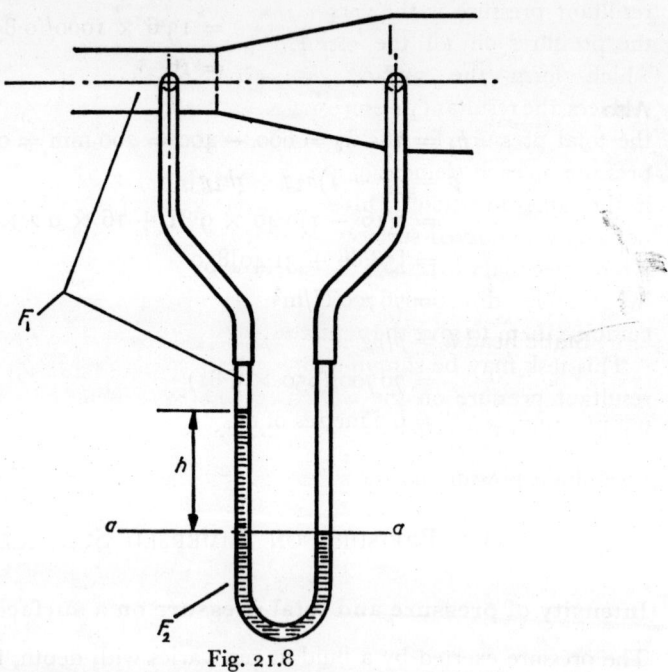

Fig. 21.8

that the total pressure F on an immersed surface is the product of the surface area and the intensity of pressure at the centre of area. Eq. (21.6) applies to both plane and curved surfaces.

Example. *A hollow metal sphere of 2 metres internal diameter is filled with water having a density of* 1000 kg/m³. *Determine* (a) *the maximum intensity of pressure, and* (b) *the total fluid pressure exerted on the interior surface of the sphere.*

$$\text{Maximum intensity of pressure} = \rho g h_{max}$$
$$= 1000 \times 9 \cdot 81 \times 2$$
$$= 19\,620 \text{ N/m}^2$$
$$\text{Total pressure } F = A \rho g \bar{h}$$

For a sphere $\quad A = 4\pi r^2 = 4\pi \times 1^2 = 12\cdot56 \text{ m}^2$

and $\qquad\qquad\quad h = d/2 = 1\cdot0 \text{ m}$

therefore $\qquad\quad F = 12\cdot56 \times 1000 \times 9\cdot81 \times 1\cdot0$

$$= 123\ 214 \text{ N}$$

Resultant pressure on a surface. The resultant pressure is the vector sum of the pressures on all the elements da which form the surface. For flat surfaces, the resultant pressure will equal the total pressure, for the intensities of pressure on each element of area all act in the same direction. This result does not apply to curved surfaces, so that it is often necessary to resolve the pressures in suitable directions and then to combine them to give the resultant.

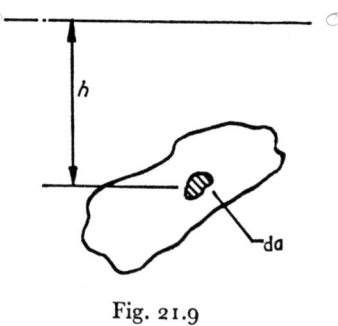

Fig. 21.9

This task may be simplified for vessels containing liquids because the resultant pressure on the wetted surface must equal the weight of the liquid within the vessel. Thus in the last worked example,

Resultant pressure on the spherical surface = Weight of fluid

$$= \text{Specific weight} \times \text{volume}$$

$$= \rho g(4\pi r^3/3)$$

$$= 1000 \times 9\cdot81(4\pi \times 1^3/3)$$

$$= 41\ 000 \text{ N}$$

For floating bodies, the resultant pressure on the wetted surface must be equal and opposite to the weight of the body.

Centre of pressure on a plane surface

When a thin plate is immersed in a liquid the intensity of pressure acting on each element of surface depends upon the static head of fluid above it. For a plate completely surrounded by liquid, the variation in pressure will be the same on both sides and the *resultant* pressure on the plate will be zero. If, however, the plate is in contact with the liquid only on one side the resultant pressure will equal the sum of the pressures on each element of the wetted surface.

Fig. 21.10 shows a lamina immersed in a liquid, inclined at angle θ to the free surface o–o and intersecting the free surface at q. To calculate the pressure on one side only of the lamina, divide the surface of the lamina into horizontal elemental strips each of area da and at vertical depth y below the free surface. Let l be the distance of each strip from q, measured in the plane of the lamina. Then $y = l \sin \theta$. The pressure p on each strip will act normal to the surface and will have an intensity equal to $\rho g y$.

$$\text{Thrust on each element} = \rho g y \,.\, \mathrm{d}a$$

$$= \rho g l \sin \theta \,.\, \mathrm{d}a$$

$$\text{Moment of this thrust about q} = \rho g l^2 \sin \theta \,.\, \mathrm{d}a$$

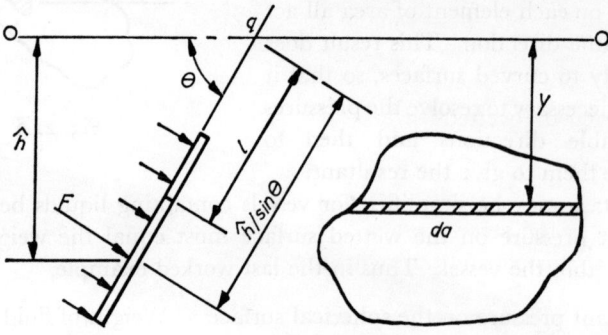

Fig. 21.10

Therefore,

$$\text{Sum of the moments of all thrusts} = \rho g \sin \theta \int l^2 \, \mathrm{d}a$$

But $\int l^2 \, \mathrm{d}a = I_q$, the second moment of area of the wetted surface about q. Therefore,

$$\text{Total moment of force on the whole surface} = \rho g I_q / \sin \theta$$

It has already been shown that the total pressure F on a surface is equal to $\rho g A \hat{h}$ (Eq. (21.6)), and for a plane surface, F will also be the resultant pressure. If F acts at a vertical depth of $\hat{h}$, then the total moment of pressure on the surface about q is given by $F\hat{h}/\sin \theta$ so that,

$$F\hat{h}/\sin \theta = \rho g I_q \sin \theta$$

or $$\rho g A \hat{h} \hat{h}/\sin \theta = \rho g I_q \sin \theta$$

from which $$\hat{h} = I_q \sin^2 \theta / A \hat{h} \qquad (21.7)$$

where $\hat{h}$ is the depth of the centre of pressure below o–o.

Example. *A plate covers a rectangular hole 300 mm wide by 600 mm high in a vertical side of a tank. The tank contains water which is level with the top edge of the hole. Determine the resultant pressure on the plate and the position of the centre of pressure relative to the top edge of the hole. If the water level is raised by one metre, what will be the new position of the centre of pressure?*

$$\text{Total pressure } F = \rho g A \bar{h}$$

(a) When the water is at the lower level

and
so that

$$\rho = 1000 \text{ kg/m}^3$$
$$A = 0.6 \times 0.3 = 0.18 \text{ m}^2$$
$$\bar{h} = 0.3 \text{ m}$$
$$F = 1000 \times 9.81 \times 0.18 \times 0.3$$
$$= 529.74 \text{ N}$$

Considering the wetted surface,

$$I_q = I_G + A\bar{h}^2/\sin^2 \theta$$

where I_G is the second moment of area of the wetted surface about a horizontal axis passing through the centre of area. Then,

$$I_q = (0.3 \times 0.6^3/12) + (0.18 \times 0.3^2/1)$$
$$= 0.0054 + 0.162 = 0.0216 \text{ m}^4$$
$$\hat{h} = I_q \sin^2 \theta / A\bar{h}$$
$$= 0.0216 \times 1/0.18 \times 0.3 = 0.4 \text{ m}$$

This means that the centre of pressure lies 0·4 m below the top edge of the hole.

(b) When the water is at the higher level,

Then

$$\bar{h} = 1.3$$
$$A = 0.18 \text{ m}^2 \text{ as before, and}$$
$$I_q = I_G + A\bar{h}^2/\sin^2 \theta$$
$$= 0.0054 + (0.18 \times 1.3^2)$$
$$= 0.0054 + 0.3042 = 0.3096 \text{ m}^4$$
$$\hat{h} = 0.3096 \times 1/0.18 \times 1.3$$
$$= 1.323 \text{ m}$$

The centre of pressure now lies 0·323 m below the top edge of the hole.

Centre of pressure on a curved surface

The centre of pressure is determined once the line of action of the resultant pressure is known and this can be done by determining the total pressures in suitable directions and combining these components to obtain the resultant pressure.

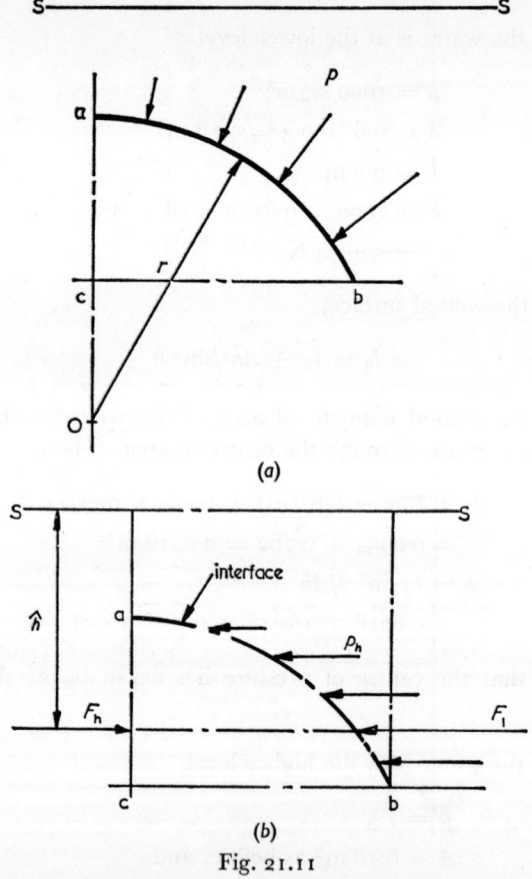

(a)

(b)

Fig. 21.11

Fig. 21.11 (a) shows a thin curved surface ab forming part of a cylinder (radius r) totally immersed in a liquid having a free surface s–s. The pressure p of the liquid on the upper side of the surface, will everywhere act normal to the surface. The surface ab is replaced by an interface ab and the pressure p is resolved into horizontal (Fig. 21.11 (b)) and vertical (Fig. 21.11 (c)) components p_h and p_v. For the equilibrium of a mass of liquid abc of unit thickness at right angles to the paper (Fig. 21.11 (b)), the total horizontal pressure F_h on the rectangular vertical interface ac can be found by using Eq. (21.6) and will act at a depth $\hat{h}_1$ given by Eq. (21.7). The resultant force F_h must balance F_1, the resultant of the intensities of pressure p_h, which must, of course, be equal and opposite to F_h and will therefore also act at depth $\hat{h}_1$ below the surface s–s.

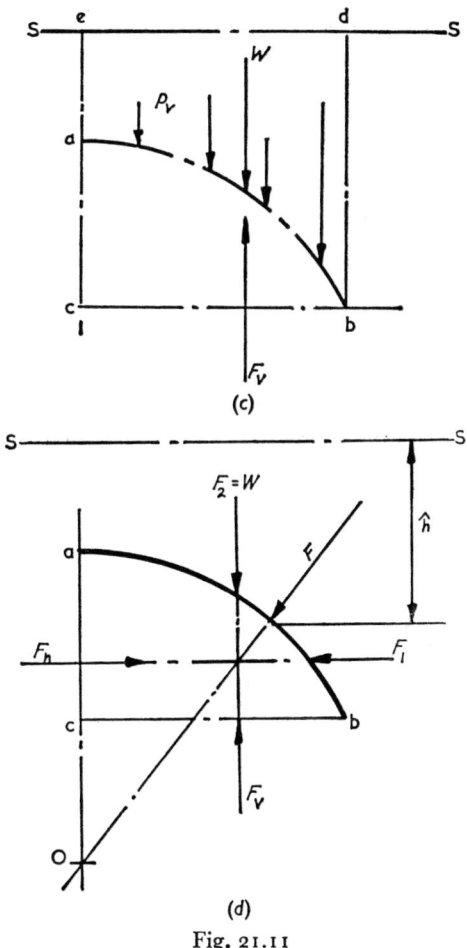

(c)

(d)

Fig. 21.11

The total vertical downward pressure must be the weight W of the mass of fluid abde of unit thickness lying above the interface ab, and will correspond to a resultant force $F_2 = W$ acting vertically downwards through the centre of gravity of the mass abde on the upper side of the interface. The force F_2 will be balanced by an upward pressure with an equal and opposite resultant F_v. The horizontal and vertical components

F_1 and F_2 can therefore be determined, and the resultant pressure F on the surface must pass through their point of intersection (Fig. 21.11 (d)). Since the intensities of pressure p act normal to the surface ab and pass through the centre of curvature O, the resultant F must also pass through O. The line of action of F is therefore determined and its point of intersection with ab forms the location of the centre of pressure, the depth of which below the free surface is $\hat{h}$.

Example. *The ends of a cylinder 2 metres internal diameter are in the form of hemispheres. The cylinder is placed with its longitudinal axis horizontal and filled with water. Determine the resultant pressure and the position of the centre of pressure for each cylinder end.*

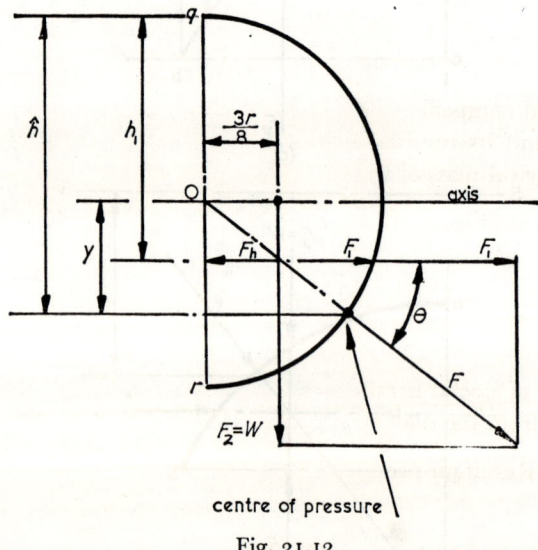

Fig. 21.12

The horizontal component of pressure F_1 equals the total force F_h on the projected surface of the hemisphere referred to the plane of projection qr (Fig. 21.12). This projected surface is a circle of 2 metres diameter. Then,

$$A = \pi r^2 = \pi \times 1^2 = 3 \cdot 14 \text{ m}^2$$

$$\bar{h} = 1 \text{ m}$$

$$\theta = 90°$$

$$\therefore \sin^2 \theta = 1$$

It follows that,

$$F_1 = F_h = \rho g A \bar{h}$$
$$= 1000 \times 9 \cdot 81 \times 3 \cdot 14 \times 1$$
$$= 30\,803 \text{ N}$$

To find the depth $\hat{h}_1$ of F_1,

$$I_q = I_G + A\bar{h}^2/\sin^2 \theta$$

where $\qquad I_G = \pi d^4/64 = \pi \times 2^4/64 = 0 \cdot 7854 \text{ m}^4$

Therefore, $\qquad I_q = 0 \cdot 7854 + 3 \cdot 14 \times 1^2/1$
$$= 3 \cdot 9254 \text{ m}^4$$

and $\qquad \hat{h}_1 = I_q \sin^2 \theta / A\bar{h}$
$$= 3 \cdot 9254 \times 1/(3 \cdot 14) \times 1$$
$$= 1 \cdot 25 \text{ m}$$

The vertical component F_2 is equal to the weight of water within the hemisphere and its resultant will act through the centre of gravity of the hemispherical mass of water.

Weight of water $W = F_2 =$ Specific weight $\times$ Volume
$$= \rho g \times 2\pi r^3/3$$
$$= 1000 \times 9 \cdot 81 \times 2\pi \times 1^3/3$$
$$= 20\,532 \text{ N}$$

The centroid of a solid hemisphere is at a distance of $3r/8$ along the axis of symmetry from the diametral plane. Hence,

Resultant pressure $F = \sqrt{(F_1{}^2 + F_2{}^2)}$
$$= \sqrt{(30\,803^2 + 20\,532^2)}$$
$$= 37\,010 \text{ N}$$

$$\text{Tan } \theta = 20\,532/30\,803 = 0 \cdot 6665$$

so that $\qquad \theta = 33° 41'$

Alternatively,

$$\text{Tan } \theta = (\hat{h}_1 - r)/(3r/8) = 0 \cdot 25/0 \cdot 375$$

which gives the same result.

The resultant F must pass through O, the centre of curvature of the hemispherical surface, so that

$$\hat{h} = r + y = r + r \sin \theta$$
$$= 1 + 1 \times \sin 33° 41' = 1 \cdot 5546 \text{ m}$$

Problems

For tutorials

1. A "dock" for a small model ship has been designed so that when the ship is floating within the dock, the space between the ship and the dock is almost negligible. The dock is closed at both ends and just sufficient water poured in to cause the level to rise to the plimsol line when the ship is lowered carefully into position. The ship is thus able to float but does so without having displaced a weight of water equal to its own weight. Explain this apparent violation of the principle of Archimedes'.

2. If a solid object made of, say, steel, is thrown overboard in deep water, will it sink to the bottom of the ocean or will it reach a depth at which upthrust equals the weight of the object, so preventing further fall.

3. A man floats in a rowing boat within a small open tank of water. He lifts a large weight from the floor of the boat carefully lowers it over the side and, releasing it, allows it to sink to the floor of the tank. (*a*) Does the boat rise a little or fall a little relative to the water line? (*b*) Does the water level rise, or fall, relative to the sides of the tank?

5. Suppose a piezometer is being installed in a pipe through which water is flowing. Describe what precautions must be taken in the installation if correct readings are to be obtained.

General

1. In a hydraulically powered machine tool, the cutting stroke is produced by means of a system of levers connected to a ram which slides within a cylinder 10 centimetres in diameter. The cylinder is connected to a pump by a pipeline, the working fluid being oil. If the desired maximum cutting force is 4 kilonewtons, the velocity ratio, through the system of levers, of the ram to the cutting tool 0·5, the mechanical efficiency 96 per cent and the loss of head in the pipeline is 2 metres of oil determine the pressure of the oil supplied by the pump. The density of the oil is 900 kilogrammes per cubic metre.

Ans. 1,076·4 kN/m²

2. A thin metal cylinder 0·5 metres in diameter, 1·0 metre in length and weighing 80 newtons when empty is half filled with oil and allowed to float in water with its longitudinal axis horizontal. Determine the density of the oil if the level of the oil within the drum is 0·04 metres higher than the level of water surrounding the drum. Assume that the density of water is 1000 kg/m³.

Ans. 714·7 kg/m³

3. A rectangular plate of a width *b* and a length *d* is immersed in a liquid with one of its narrower sides lying in the free surface. Show, working from first principles, that the centre of pressure lies at a point 2*d*/3 along the plate from the upper edge.

4. Fig. 21.13 shows a circular sluice gate, one metre in diameter, hinged at its uppermost point. The gate is of uniform thickness, weighs one kilonewton and when closed is inclined at 30° to the vertical. The gate is opened by exerting a pull on a chain attached to its lowest point and running at right angles to the plane of the opening. Determine the pull in the chain required to open the gate when the water level is 0·5 metres above the hinge.

Ans. 4268 N

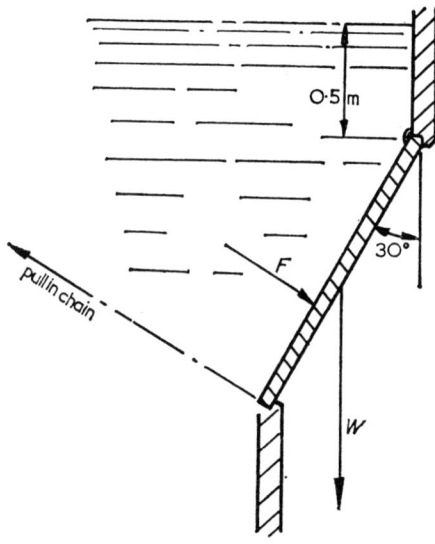

Fig. 21.13

5. A cylindrical tank with flat ends has an internal diameter of 0·8 metres and a length of 2 metres. It is filled with oil having a density of 850 kg/m³. Determine the total pressure on the curved surface, the resultant pressure and the centre of pressure when the cylinder is lying on its side and the total pressure on the curved surface when the cylinder is standing on one end.

Ans. 16 780 N, 8390 N, $\hat{h}$ = 0·8 m, 42 000 N

CHAPTER 22

Fluid Flow

22.1 VISCOSITY

When a layer of fluid is moved laterally relative to an adjacent layer, a force is set up within the fluid in opposition to the shearing action. This internal resistance, known as the viscosity of the fluid, is caused by molecular adhesion and acts along the common boundary of the fluid layers. Viscosity involves a relationship between the shear stress and the rate of shear.

A flat plate of area A is separated from a stationary surface o–o by a fluid film of thickness h (Fig. 22.1). If the plate has a uniform velocity v

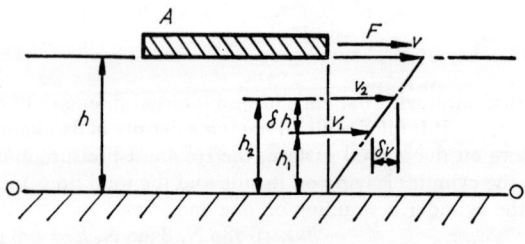

Fig. 22.1

relative and parallel to o–o, the layer of fluid in contact will move with the plate, being held to it by molecular forces, while the layer in contact with o–o will remain at rest for similar reasons. Intermediate layers will have velocities which are directly proportional to their distances from surface o–o, so that

$$v_1/h_1 = v_2/h_2 = v/h = \delta v/\delta h = \text{a constant}$$

The ratio $\delta v/\delta h$ is referred to as the velocity gradient or the rate of shear R.

The force per unit area causing motion is given by F/A and is known as the shear stress τ. For laminar (i.e. streamline or viscous) flow, the force F is proportional to the area A and the rate of shear, so that

$$F \propto A\delta v/\delta h = \eta A\delta v/\delta h$$

where the constant η is the absolute, or dynamic, viscosity of the fluid and is a measure of that property by virtue of which a fluid resists shear.

$$\therefore \; \eta = \frac{F/A}{\delta v/\delta h} \quad \text{or} \quad \eta = \tau/R \tag{22.1}$$

Units of dynamic viscosity

The units of η can be obtained by assigning appropriate units to the symbols in Eq. (22.1). Thus, the units of η are:

$$\frac{N}{m^2} \times \frac{ms}{m} = Ns/m^2$$

If A, δh and δv are equal to unity, the viscosity may thus be defined as that force in newtons which would produce unit speed (1 m/s) in a plate of unit area (1 m^2) at unit distance (1 m) from a parallel stationary plate.

S.I. and metric units of viscosity. In the metric (centimetre, gram, second) system, the unit of viscosity is the *poise* and is given by Eq. (22.1) as follows:

$$\eta \; (\text{poise}) = \frac{F \; (\text{dynes})/A \; (\text{cm}^2)}{\delta v \; (\text{cm/s})/\delta h \; (\text{cm})}$$

Now, 1 dyne $= 10^{-5}$ N, and 1 cm $= 10^{-2}$ m, so that substituting in Eq. (22.1),

$$1 \; \text{poise} = \frac{10^{-5} N}{10^{-4} \, m^2} \times \frac{10^{-2} \, ms}{10^{-2} \, m} = 0 \cdot 1 \; Ns/m^2$$

or
$$1 \; \text{centipoise (cP)} = 0 \cdot 001 \; Ns/m^2$$

Newtonian fluids

Newton, in his work on this topic, concluded that the viscosity of a fluid is independent of the rate of shear at any particular pressure and temperature. Another way of expressing this is that shear stress τ is directly proportional to shear rate R, and fluids which behave in this way are known as Newtonian fluids. They include water and many of the mineral oils.

Kinematic viscosity

Kinematic viscosity is the ratio of dynamic viscosity to mass density and is a measure of the fluidity of a substance. Thus

Kinematic viscosity = (Dynamic viscosity)/(Mass density)

or $$v = \eta/\rho \qquad\qquad (22.2)$$

The units of kinematic viscosity are given by

$$\frac{\text{Ns}}{\text{m}^2} \times \frac{\text{m}^4}{\text{Ns}^2} = \text{m}^2/\text{s}$$

Example. *A shaft (diameter 0·1 m) of a length 2 m slides vertically in a sleeve 0·1001 m in diameter and 0·25 m long. If the annular space between the sliding pair is filled with a lubricant having a viscosity of 1·5 poise, determine the force necessary to overcome the viscous resistance when the shaft is sliding with a velocity of 0·5 m/s.*

$$\eta = 1\cdot5 \text{ poise} = 0\cdot15 \text{ Ns/m}^2$$
$$A = \pi \times 0\cdot1 \times 0\cdot25 = 0\cdot025\pi \text{ m}^2$$
$$h = (0\cdot1001 - 0\cdot1000)/2 = 0\cdot00005 \text{ m}$$
$$v = 0\cdot5 \text{ m/s}$$

From Eq. (22.1),
$$\eta = (F/A)/(h/v)$$

$$\therefore F = \frac{0\cdot15 \times 0\cdot025\pi \times 0\cdot5}{0\cdot00005} = 117\cdot3 \text{ N}$$

Example. *A simple thrust bearing for a shaft is in the form of a flat collar integral with the shaft. When the latter is at its normal operating speed of 1,800 rev/min, the load is transmitted to an annular surface in the bearing housing through a film of oil 0·3 mm thick. If the oil has a viscosity of 0·10 Ns/m² and the inner and outer diameters of contact of the bearing surfaces are 80 mm and 140 mm respectively, determine the power required to overcome the viscous force of the lubricant.*

Consider an elemental area of the thrust face concentric with the bearing of width δx, radius x and area δA (Fig. 22.2).

Let   Bearing velocity at radius $x = v = \omega x$

Velocity gradient at radius $x = v/h = \omega x/h$

Viscous force on the element $= \eta \delta A v/h$

$$= \eta 2\pi x \delta x \omega x/h$$

∴ Torque required to overcome viscous resistances

$$= \eta 2\pi x^2 \delta x \omega x / h$$

∴ Total torque required at bearing $= \dfrac{2\pi\omega\eta}{h} \displaystyle\int_{x=r_2}^{x=r_1} x^3 \, dx$

$$= \frac{\pi\omega\eta}{2h} [r_1^4 - r_2^4]$$

∴ Power required $=$ Torque

$\times$ Angle turned/second

$$= \frac{\pi\omega^2\eta}{2h} [r_1^4 - r_2^4]$$

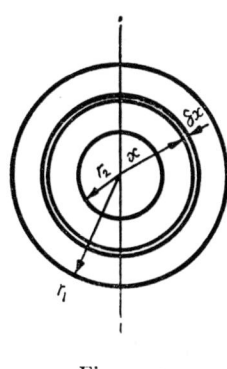

Fig. 22.2

In the example

$$\omega = 2\pi n/60 = 2\pi \cdot 1{,}800/60 = 60\pi \text{ rad/s}$$
$$\eta = 0\cdot 10 \text{ Ns/m}^2$$
$$h = 0\cdot 3 \text{ mm} = 0\cdot 0003 \text{ m}$$
$$r_1 = 70 \text{ mm} = 0\cdot 07 \text{ m}$$
$$r_2 = 40 \text{ mm} = 0\cdot 04 \text{ m}$$

Substituting,

$$\text{Power required} = \frac{\pi \times 3600\pi^2 \times 0\cdot 10}{2 \times 0\cdot 0003} [(0\cdot 07)^4 - (0\cdot 04)^4]$$

$$= 398\cdot 3 \text{ W}$$

22.2 Laminar and Turbulent Flow

The flow of a fluid in motion relative to surfaces with which it is in contact may be either laminar or turbulent. In laminar (streamline, viscous) flow, molecules of the fluid follow well defined paths, referred to as streamlines, which may converge or diverge. There is no interchange of molecules between adjacent paths, and, the relationships of Eq. (22.1) apply. The viscous forces are dependent upon temperature but independent of pressure and surface roughness.

Critical velocity

The flow of fluids around objects and within pipes was investigated by Osborne Reynolds, who discovered that if the velocity of streamlined flow is steadily increased, the flow becomes turbulent when the velocity reaches a particular value and remains turbulent at velocities beyond this point. Furthermore, if the velocity is now steadily decreased, there is a return to laminar flow but at a lower velocity than that at which turbulence sets in. There is a difference between these two critical velocities because the inertia of the fluid tends to preserve whichever happens to be the existing flow form.

Turbulence is characterised by the formation of vortices or eddies which interrupt and destroy orderly streamline motion. It is usually visible only when a free surface is distorted by the vortices, but turbulence can be observed by the introduction of thin streams of coloured dye which follow the erratic convolutions of the current. With laminar flow, the dye preserves its smooth thread-like form along the pipe or around the object in the path of the fluid.

The internal resistances in turbulent flow are proportional to the density of the fluid and to the square of the velocity. They are also proportional to the contact area and are affected by surface roughness, but are relatively independent of temperature.

Reynolds number

Reynolds found that the form of the flow through pipes could be determined from the number Re, called Reynolds number

$$\text{Re} = \rho v d / \eta = v d / \nu \qquad (22.3)$$

where $v =$ Velocity of flow (m/s),

$d =$ Bore of pipe (m),

$\nu =$ Kinematic viscosity (m²/s),

$\rho =$ Density of fluid (kg/m³ or Ns²/m⁴), and

$\eta =$ Absolute viscosity (Ns/m²).

If units are assigned to either form of the expression (see 21.2) these are seen to cancel, indicating that Reynolds number is non-dimensional.

Reynolds's chief conclusion was that the lowest critical velocity for any fluid flowing through a pipe is reached when the value of Re is between 2,000 and 2,500. Below this range, the flow will be viscous; above it, the flow will be turbulent.

More generally, Reynolds number can be expressed as $\rho v l / \eta$ or $v l / \nu$ where l represents one linear dimension of the body around or through which fluid is flowing. This is why it is important that comparisons in the flow of fluids around objects or through pipes be made only between objects or pipes which are geometrically similar. Reynolds number for the critical velocity of flow will therefore depend upon the particular linear dimension chosen to represent l.

Poiseuille's law

Laminar flow in pipes which are long in relation to their diameter is determined by Poiseuille's law, which can be expressed as:

$$V = (h_f \rho g \pi . r^4 t)/8 \eta l \qquad (22.4)$$

where V = Volume of liquid flowing in time t seconds (m³),
h_f = Head lost in overcoming friction (m),
ρ = Density of fluid (kg/m³),
g = Gravitational constant (m/s²),
r = Radius of pipe bore (m), and
l = Length of pipe (m).

The velocity of flow is related to the volume flowing per second by $v = V/\pi r^2 t$, so that Eq. (22.4) becomes

$$v = h_f \rho g r^2/8 \eta l \qquad (22.5)$$

But pressure = head × specific weight, so that the pressure difference p causing flow = $h_f \rho g$. Therefore, $h_f = p/\rho g$, and substituting for h_f in Eq. (22.4),

$$V = p \pi r^4 t/8 \eta l \qquad (22.6)$$

Eqs. (22.5) and (22.6) are alternative forms of Poiseuille's equation.

Example. *An oil which has a viscosity of 1·5 cP and a density of 750 kg/m³ is found to have a critical velocity of 0·04 m/s when flowing through a pipe of 0·1 m diameter. What will be the critical velocity of water in the pipe if the viscosity and density of water are 1·0 cP and 1,000 kg/m³ respectively?*

$$\text{Reynolds number Re} = (\rho v d)/\eta$$

Therefore, for the oil

$$\text{Re} = (750 \times 0·04 \times 0·1)/0·0015$$

$$= 2000$$

Reynolds number for water must also be 2,000, so that

$$2000 = (1,000 \times v \times 0·1)/0·001$$

and

$$\text{Critical velocity for water} = v = (2000 \times 0·001)/(1000 \times 0·1)$$

$$= 0·02 \text{ m/s}$$

22.3 CONTINUITY OF FLOW

Flow through a system is continuous when the masses of fluid entering and leaving in unit time are the same.

Liquids are virtually incompressible, so that the density of any liquid may be taken as constant. Since the volume per second passing any one point in a system equals the velocity multiplied by the cross-sectional area of flow, there is an equation relating flow at any two points and known as the continuity equation for liquids:

$$a_1 v_1 = a_2 v_2 \qquad (22.7)$$

where a_1, a_2 and v_1, v_2 refer to cross-sectional areas and velocities respectively at any two points in the system.

Gases are compressible fluids in which densities vary considerably with changes in temperature and pressure, so that Eq. (22.7) does not apply (see Chapter 26).

Example. *Three pipes feed into a single tapering pipe as shown in* Fig. 22.3. *If the system runs full of water determine* (a) *the area of cross-section a_4 if the velocity v_4 is to be* 0·5 m/s, *and* (b) *the velocity v_5 if the diameter d_5 is* 2·5 cm.

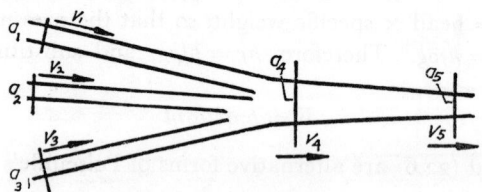

Fig. 22.3

The branch pipes have cross-sectional areas of $a_1 = 6$ cm², $a_2 = 12$ cm² and $a_3 = 20$ cm² respectively, the velocities of flow being $v_1 = 0·75$ m/s, $v_2 = 1$ m/s and $v_3 = 0·25$ m/s respectively.

For continuity of flow,

$$a_1 v_1 + a_2 v_2 + a_3 v_3 = a_4 v_4$$
$$(6 \times 0·75) + (12 \times 1) + (20 \times 0·25) = a_4 \times 0·5$$
$$\therefore a_4 = 21·5/0·5 = 43 \text{ cm}^2$$

Further,

$$a_4 v_4 = a_5 v_5$$

so that

$$43 \times 0·5 = (\pi \times 2·5^2/4) v_5$$
$$\therefore v_5 = 43 \times 0·5 \times 4/\pi \times 2·5^2$$
$$= 4·38 \text{ m/s}$$

22.4 ENERGY OF A LIQUID

The total energy of a liquid in motion is given by the sum of the potential energy and of the contributions due to pressure and velocity.

Potential energy

The potential energy (Fig. 22.4) is obtained by letting a mass m be lifted through a vertical distance Z.

Work done (energy supplied) in lifting $= mgZ$

$$\therefore \text{ Potential energy per unit mass of fluid} = \frac{\text{Total energy supplied}}{\text{Mass of fluid}}$$

$$= mgZ/m = Zg \qquad (22.8)$$

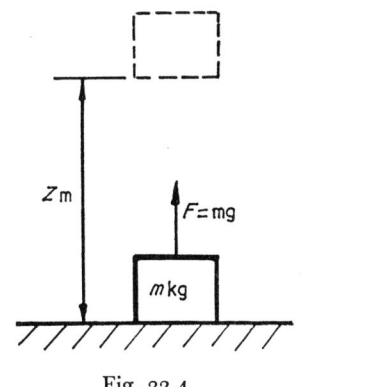

Fig. 22.4

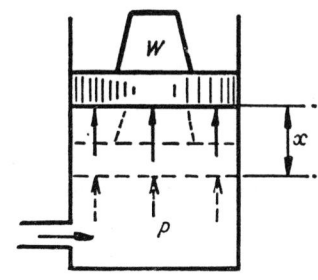

Fig. 22.5

Pressure energy

The energy due to pressure is obtained as follows (Fig. 22.5). Let a mass with a weight W be supported on a piston within a cylinder by fluid under a pressure p. Let the piston have an area of a. If additional fluid at pressure p is now supplied to the cylinder so that the piston is displaced a distance x,

$$\text{Energy supplied by fluid} = \text{Work done on the mass}$$
$$= Wx = pax$$

The volume of additional fluid supplied is ax, so that the mass of this fluid is $ax\rho$, where ρ is the density of the fluid. It follows that,

$$\text{Energy contained in unit mass of the fluid} = pax/ax\rho = p/\rho \quad (22.9)$$

Velocity energy

The energy per unit of mass of fluid due to its velocity is calculated by considering (Fig. 22.6) a mass of fluid of m initially at rest and acted upon by a resultant force of F through a distance of S. The mass will be accelerated at a metres/second^2 in accordance with Newton's second law.

Work done in moving the mass through $S = F \times S$

but $\qquad\qquad\qquad\qquad\qquad\qquad\qquad F = ma$

and $\qquad\qquad\qquad\qquad\qquad\qquad S = v^2/2a \quad \text{(since } u = 0\text{)}$

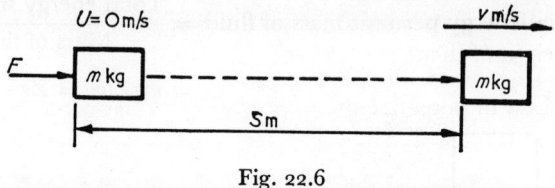

Fig. 22.6

From this, it follows that

$$\text{Work done on the fluid} = m \cdot av^2/2a$$

$$= mv^2/2$$

$\therefore$ Kinetic energy of unit mass of the fluid $= mv^2/2m$

$$= v^2/2 \qquad (22.10)$$

Adding these three separate contributions together,

$$\text{Total energy per unit mass of fluid} = Zg + p/\rho + v^2/2 \quad (22.11)$$

Using S.I. units, the energy is, of course, expressed in joules.

22.5 HEAD OF A LIQUID

It is sometimes convenient to describe the position, pressure and velocity of a liquid in terms of what are the equivalent static heads.

Potential head. If a quantity of fluid is situated at an instant at a height Z vertically above some arbitrarily chosen datum, then

$$\text{Potential head} = Z \text{ (metres)} \qquad (22.12)$$

Pressure head. Consider a column of liquid of height h, cross sectional area a and density ρ.

$$\text{Volume of the column} = ah$$

$$\text{Mass of the column} = ah\rho$$

$$\text{Weight of the column} = ah\rho g$$

Then

$$\text{Pressure on the base of the column due to its weight} = p = ah\rho g/a$$

$$= h\rho g$$

or
$$\text{Pressure head } h = p/\rho g \quad (22.13)$$

Velocity head. If a mass m of fluid with zero initial velocity falls freely through a height h, then

$$\text{Loss in potential energy} = \text{Gain in kinetic energy}$$

or
$$mgh = mv^2/2$$

Thus
$$\text{Velocity head } h = v^2/2g \quad (22.14)$$

Putting these results together, the

$$\text{Total head of a liquid} = Z + p/\rho g + v^2/2g \quad (22.15)$$

using S.I. units, the result is in metres.

Work done and power required in overcoming pressure. If in Fig. 22.5, the piston rises through x metres per second then

$$\text{Work done per second by the liquid} = p \cdot ax$$

$$= p \text{ (Volume of flow/second)}$$

But from Eq. (22.13),

$$p = \rho g h$$

$$\therefore \text{ Work done/second} = \rho g h \text{ (Volume of flow/second)}$$

$$= Wh$$

where $W =$ Weight of fluid flowing per second, and

$\quad h =$ Static head equivalent to pressure p.

$$\therefore \text{ Power required} = Wh = Wh \text{ watts} \quad (22.16)$$

Example. *Determine the power required to pump 360 cubic metres of oil per hour along a pipe line 400 metres long and 0·2 metres diameter. The oil has a dynamic viscosity of 400 cP and a density of 910 kg/m³.*

The first step is to determine whether the flow is laminar or turbulent.

Volume of oil flowing = Area of pipe × Velocity of flow

i.e. $360/3600 = 0.7854 \times (0.2)^2 \times v$

$$v = 0.1/(0.7854 \times (0.2)^2) = 3.19 \text{ m/s}$$

$$\text{Re} = \rho v d/\eta$$

$$= 910 \times 3.19 \times 0.2/0.4 = 1450$$

Reynolds number is below the range 2,000 to 2,500, so that the flow is laminar and Poiseuille's equation can be applied, so that V, the volume flowing per second, is given by

$$V = 0.1 \text{ m}^3/\text{s} = (h_f \rho g \pi r^4 t)/8\eta l$$

$$\therefore 0.1 = \frac{h_f \times 910 \times 9.81 \times \pi \times 0.1^4 \times 1}{8 \times 0.4 \times 400}$$

$$\therefore h_f = \frac{0.1 \times 8 \times 0.4 \times 400}{910 \times 9.81 \times \pi \times 0.1^4 \times 1}$$

$$= 45.6 \text{ m}$$

Power required to overcome friction = Wh_f

where W = Weight of oil flowing per second

$$= V\rho g$$

$$= \frac{360 \times 910 \times 9.81}{3,600} = 890 \text{ N/s}$$

$$\therefore \text{ Power required} = 890 \times 45.6 = 40\,800 \text{ W} = 40.8 \text{ kW}$$

Bernoulli's theorem (see Chapter 26)

In any system of fluid flow, the total energy in the system at any two points is the same provided that energy is neither given to nor extracted from the system. For liquids, this means that

$$Z_1 g + p_1/\rho + v_1^2/2 = Z_2 g + p_2/\rho + v_2^2/2 \qquad (22.17)$$

where the suffixes 1 and 2 refer to any two points in the system.

Eq. (22.17) is only true if there is no friction. When energy losses due to friction are included the equation must be modified to

$$Z_1 g + p_1/\rho + v_1^2/2 = Z_2 g + p_2/\rho + v_2^2/2 + h_f g \qquad (22.18)$$

where h_f = Head lost (in metres) due to friction.

Bernoulli's equation may also be expressed in terms of equivalent static heads, so that

$$Z_1 + p_1/\rho g + v_1^2/2g = Z_2 + p_2/\rho g + v_2^2/2g + h_f \qquad (22.19)$$

Example. *Water having a density of* 1000 kg/m³ *is under a pressure of* 8 kN/m² *and has a velocity of* 10 m/s *when passing a certain point in a tapering pipeline where the pipe diameter is* 0·1 m. *Determine the pressure at a second point where the pipe diameter is* 0·2 m *if this point is* 5 m *higher than the first point. Assume frictionless flow.*

For continuity of flow,

$$a_1v_1 = a_2v_2$$

so that
$$v_2 = (d_1^2/d_2^2)v$$
$$= (0\cdot1^2/0\cdot2^2)10 = 2\cdot5 \text{ m/s}$$

Taking the lower point as datum and applying Bernoulli's equation,

$$Z_1g + p_1/\rho + v_1^2/2 = Z_2g + p_2/\rho + v_2^2/2$$
$$0 + 8000/1000 + 10^2/2 = 5 \times 9\cdot81 + p_2/1000 + 2\cdot5^2/2$$
$$8 + 50 = 49\cdot05 + p_2/1{,}000 + 3\cdot125$$
$$p_2 = 5\cdot825 \times 1000$$
$$= 5\cdot825 \text{ kN/m}^2$$

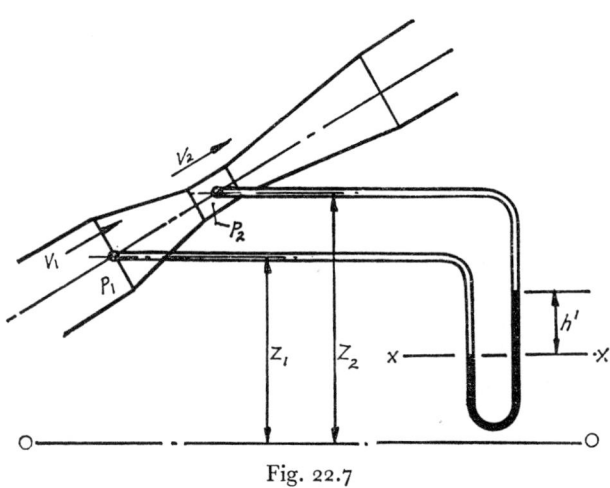

Fig. 22.7

The venturi meter

The flow of a fluid through a pipeline can be measured by a device which consists of a pipe which converges to a throat from the main pipeline diameter and then diverges to the full diameter again. Fig. 22.7 shows a

venturi tube through which water is assumed to be flowing. The convergence causes velocity and pressure changes in the water between inlet
and throat. The difference in pressure is measured by means of a U-tube
containing a fluid which is denser than, and will not mix with, the water.

Let p_1, v_1 and Z_1 be the pressure, velocity and potential head at the
inlet, and p_2, v_2 and Z_2 be the same quantities at the throat. For continuity
of flow, $a_1 v_1 = a_2 v_2$, so that

$$v_1 = (a_2/a_1)v_2 \qquad\qquad (22.20)$$

Applying Bernoulli's equation to inlet and throat,

$$Z_1 + p_1/\rho g + v_1{}^2/2g = Z_2 + p_2/\rho g + v_2{}^2/2g$$

or $$(v_2{}^2 - v_1{}^2)/2g = (p_1 - p_2)/\rho g + Z_1 - Z_2$$

Substituting for v_1 from Eq. (22.20),

$$(v_2{}^2/2g)(1 - a_2{}^2/a_1{}^2) = (p_1 - p_2)/\rho g + Z_1 - Z_2$$

$$\therefore\ v_2{}^2 = 2g[(p_1 - p_2)/\rho g + Z_1 - Z_2]/(1 - a_2{}^2/a_1{}^2)$$

But $$1 - a_2{}^2/a_1{}^2 = (a_1{}^2 - a_2{}^2)/a_1{}^2$$

$$\therefore\ v_2 = a_1\sqrt{\{2g[(p_1 - p_2)/\rho g + Z_1 - Z_2]\}}/\sqrt{(a_1{}^2 - a_2{}^2)}$$

But $(p_1 - p_2)/\rho g = h$, the pressure difference in head of *water* between
inlet and throat, and the
Theoretical discharge $= a_2 v_2$

$$= a_1 a_2 \sqrt{\{2gh + Z_1 - Z_2\}}/\sqrt{(a_1{}^2 - a_2{}^2)}$$

and

Actual discharge $= C_d a_1 a_2 \sqrt{\{2gh + Z_1 - Z_2\}}/\sqrt{(a_1{}^2 - a_2{}^2)}$

$$(22.21)$$

When the venturi meter is placed horizontally, $Z_1 = Z_2$, and

$$\text{Actual discharge} = C_d a_1 a_2 \sqrt{2gh}/\sqrt{(a_1{}^2 - a_2{}^2)} \qquad (22.22)$$

An average value for the meter is 0·97.

The difference of pressure head between inlet and throat is measured
at the U-tube. The pressure in the two arms will be equal at X–X, since
the liquid in that portion of the tube below X–X has the same density
throughout. It follows that the difference of head in metres of water is
(head of water equivalent to head h^1 of liquid in right hand limb) — (head
h^1 of water in the left hand limb).

If d = Relative density of the heavier liquid in the U-tube,

Difference of head = $(dh^1 - h^1)$ metres of water

If the heavy liquid is mercury, then $d = 13 \cdot 6$, and

Difference of head = $13 \cdot 6h^1 - h^1$

$= 12 \cdot 6h^1$ metres of water

22.6 ORIFICES AND MOUTHPIECES

Small orifices

A small orifice in a vertical surface is one which is small enough for its whole area to be considered at the same depth in determining the rate of discharge.

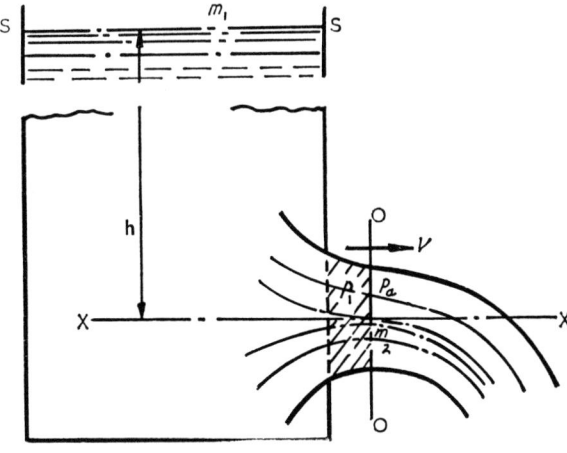

Fig. 22.8

In a tank (Fig. 22.8) of fluid with a small orifice in one of its vertical sides, assume that the head of water above the centre line XX is kept constant at h metres. Movement of fluid towards the orifice takes place because of the difference in pressure between the inside and outside of the tank on the datum line XX. Away from the orifice, there is a slow drift towards the opening, but as the particles of fluid approach the orifice, their potential and pressure energies are converted into velocity energy and they increase in speed until the fluid pressure drops to that of the surrounding atmosphere—that is, to p_a.

Within the shaded part of the jet, the average pressure across the jet, denoted by p_1, is still a little above p_a even though the fluid is now clear of the tank and surrounded by the atmosphere. This happens because of the centrifugal forces directed towards the jet axis and caused by the inertia of fluid particles as they follow the curved streamlines. At O–O, the streamlines have become parallel, the fluid pressure is at its lowest and the jet has attained its maximum velocity v m/s. The section of the jet at O–O is known as the *vena contracta* and its area will be less than that of the orifice itself.

The velocity of the jet can be obtained by comparing the energy in a particle of fluid of mass m first when the particle is at m_1, on the free surface S–S and later at m_2 on the *vena contracta* O–O. If there are no energy losses,

Gain in kinetic energy = Loss in potential energy

or
$$\tfrac{1}{2}mv^2 = m \cdot g \cdot h$$
so that
$$\text{Theoretical velocity } v = \sqrt{2gh}$$

Discharge from a sharp-edged orifice

Before the actual discharge from an orifice can be calculated, it is necessary to introduce coefficients which will allow for (1) the reduction in the velocity due to fluid friction up to the *vena contracta*, and (2) the contraction of the jet from the area of the orifice to that which it has at the *vena contracta*.

Coefficient of velocity (C_v). The ratio C_v of the actual velocity at *vena contracta* to the theoretical velocity there is called the coefficient of velocity.

$$\therefore C_v = \frac{V}{\sqrt{2gh}}$$

$$\therefore \text{Actual velocity } V = C_v\sqrt{2gh}$$

An average value for C_v is approximately 0·97.

Coefficient of contraction (C_c). The ratio C_c of the cross-sectional area of jet at *vena contracta* to the area of the orifice is the coefficient of contraction. If

a_0 = Area of the orifice, and

a_v = Area of the jet at the *vena contracta*

then
$$C_c = \frac{a_v}{a_0}$$

$$\therefore a_v = C_c a_0$$

An average value for C_c is 0·64.

Coefficient of discharge (C_d). The ratio C_d of the actual discharge to the theoretical discharge is called the coefficient of discharge, and can be related to the coefficients of velocity and contraction.

The theoretical discharge $= a_0\sqrt{2gh}$. The actual discharge is the product of the velocity and area at the *vena contracta*, and is given by

$$C_v\sqrt{2gh} \times C_c a_0 = C_c C_v a_0\sqrt{2gh}$$

$$\therefore C_d = \frac{C_c C_v a_0\sqrt{2gh}}{a_0\sqrt{2gh}}$$

$$= C_c C_v \qquad (22.23)$$

An average value for C_d is about 0·62.
Thus

Actual rate of discharge from a sharp edged orifice $= C_d a_0\sqrt{2gh}$ (22.24)

Example. *Water flows from a tank of uniform rectangular cross section 0·75 m wide by 0·5 m deep through a small orifice 40 mm diameter set in a vertical side. Determine the time taken for the level of water in the tank to fall from a height of 9 m to a height of 4 m above the centre line of the orifice. C_d for the orifice is 0·62.*

Let h = Height of the water level at time t,

dh = Fall in the water level in time dt, and

A = Area of cross-section of tank.

From Eq. (22.24),

$$\text{Rate of discharge} = C_d \cdot a_0\sqrt{2gh}$$

But

Loss in the volume of water in the tank in time dt

$$= \text{Quantity of water discharged from the orifice in time } dt$$

$$\therefore -A\,dh = C_d a_0\sqrt{2g} \cdot h^{\frac{1}{2}}\,dt$$

(As h is measured upwards, dh will be negative since it represents a downwards movement of the water level.)

$$\therefore dt = -\frac{A}{C_d a_0\sqrt{2g}}h^{-\frac{1}{2}}\,dh$$

Integrating

$$\int_{t_1}^{t_2} dt = -\frac{A}{C_d a_0\sqrt{2g}}\int_{h_1}^{h_2} h^{-\frac{1}{2}}\,dh$$

The time required to lower the level from h_1 to h_2, is therefore

$$T = -\frac{2A}{C_d a_0 \sqrt{2g}} [h^{\frac{1}{2}}]_{h_1}^{h_2}$$

$$= \frac{2A}{C_d a_0 \sqrt{2g}} [h_1^{\frac{1}{2}} - h_2^{\frac{1}{2}}] \qquad (22.25)$$

$$= \frac{2 \times 0{\cdot}75 \times 0{\cdot}5}{0{\cdot}62 \times 0{\cdot}7854 \times 0{\cdot}04^2 \times \sqrt{2 \times 9{\cdot}81}} [9^{\frac{1}{2}} - 4^{\frac{1}{2}}]$$

$$= \frac{2 \times 0{\cdot}75 \times 0{\cdot}5 \times 1}{0{\cdot}62 \times 0{\cdot}7854 \times 0{\cdot}04^2 \times 4{\cdot}43}$$

$$= 217{\cdot}3 \text{ seconds} \quad = 3 \text{ min } 37{\cdot}3 \text{ s}$$

Large rectangular orifices

With large orifices set in vertical surfaces, the effect on the velocity of flow of the variation in depth from top to bottom of the aperture may be too great to be ignored (Fig. 22.9).

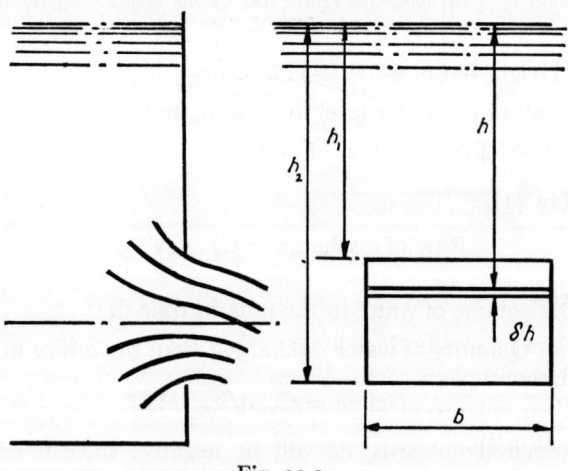

Fig. 22.9

Let h_1 and h_2 be the heads of fluid above the top and bottom edges of the orifice and b the breadth of the orifice. Consider an elemental area of the orifice having width δh and breadth b and lying at a uniform depth h below the free surface.

Theoretical velocity of flow through the strip $= \sqrt{2gh}$

Theoretical discharge through the strip

$$= \text{Area of strip} \times \text{Velocity}$$

$$= B \cdot \delta h \sqrt{2gh}$$

Actual discharge through the strip $= C_d \cdot B\sqrt{2gh} \cdot \delta h$

$\therefore$ Actual total discharge from orifice $= C_d B \sqrt{2g} \displaystyle\int_{h_1}^{h_2} h^{\frac{1}{2}}\, dh$

$$= \tfrac{2}{3} C_d B \sqrt{2g}(h_2^{3/2} - h_1^{3/2})$$

$$(22.26)$$

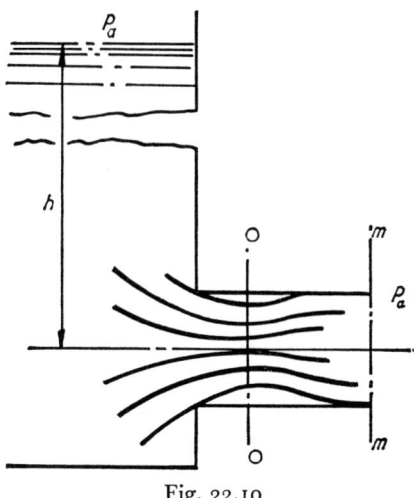

Fig. 22.10

Mouthpieces

Fig. 22.10 shows a small orifice fitted with a cylindrical extension to form an external mouthpiece. The jet will narrow to the *vena contracta* at O–O and then enlarge to fill the mouthpiece before emerging into the atmosphere at m–m.

If v_0 = Velocity of fluid at O–O,
$\quad v_m$ = Velocity of fluid at m–m,
$\quad a_0$ = Area of jet at O–O, and
$\quad a_m$ = Area of jet at m–m.

From the continuity equation

$$v_0 = a_m v_m / a_0$$

Assuming that $C_c = 0.64$, then

$$a_0 = 0.64 a_m \quad \text{and} \quad v_0 = v_m/0.64$$

It will be shown in Chapter 23, that

Loss of head due to enlargement of the jet between O–O and m–m

$$= (v_0 - v_m)^2/2g$$
$$= (v_m/0.64 - v_m)^2/2g$$
$$= 0.317 v_m^2/2g$$

Applying Bernoulli's equation to the free surface S–S and the mouthpiece at m–m

$$p_a/\rho g + h = p_a/\rho g + v_m^2/2g + 0.317 v_m^2/2g$$
$$h = 1.317 v_m^2/2g$$

$$C_v = \frac{\text{Actual velocity at exit}}{\text{Theoretical velocity at exit}}$$

$$= v_m/\sqrt{2gh}$$
$$= v_m/\sqrt{2g} \times 1.317 v_m^2/2g$$
$$= 0.874$$

The coefficient of contraction is unity.

$$\therefore \; C_d = C_c \cdot C_v = 0.874$$

Thus an external mouthpiece will cause an increase in flow over a plain orifice of

$$[(0.874 - 0.62)/0.2] = 41 \text{ per cent approximately}$$

Example. *Find the pressure at the vena contracta and discharge from an orifice fitted with an external mouthpiece if the head of water is 1 metre and the diameter of the mouthpiece is 30 mm. Assume that C_c is 0.64 and that the atmospheric pressure is 1 bar (10^5 N/m^2).*

Applying Bernoulli's equation to the free surface and to the *vena contracta* and taking the axis of the mouthpiece as the datum,

$$p_a/\rho + hg = p_0/\rho = + v_0^2/2$$

where p_a = Atmospheric pressure. But

$$v_0 = v_m/0.64$$
$$\therefore \; v_0^2 = 2.44 v_m^2$$
$$\therefore \; p_a/\rho + 1 \cdot 9.81 = p_0/\rho + 1.22 v_m^2$$

But $\qquad h = 1\cdot317 v_m{}^2/2g$

so that $\qquad v_m = \sqrt{2gh/1\cdot317}$

$$= \sqrt{(2 \times 9\cdot81 \times 1)/1\cdot317} = 3\cdot85 \text{ m/s}$$

$$\therefore (p_a - p_0)/\rho = 1\cdot22 \times 3\cdot85^2 - 9\cdot81$$

$$= 8\cdot29$$

$$\therefore p_0 = p_a - 8\cdot29\rho$$

$$= 100,000 - 8\cdot29 \times 1,000$$

$$= 91\,710 \text{ N/m}^2$$

$$= 0\cdot9171 \text{ bar}$$

Quantity flowing $= a_m v_m$

$$= 0\cdot7854 \times 30^2/1000^2 \times 3\cdot85$$

$$= 0\cdot00272 \text{ m}^3/\text{s}$$

Re-entrant mouthpiece. When a plain orifice is fitted internally with a cylindrical tube, a re-entrant mouthpiece is formed. The mouthpiece may run free as shown in Fig. 22.11 or full as in Fig. 22.12.

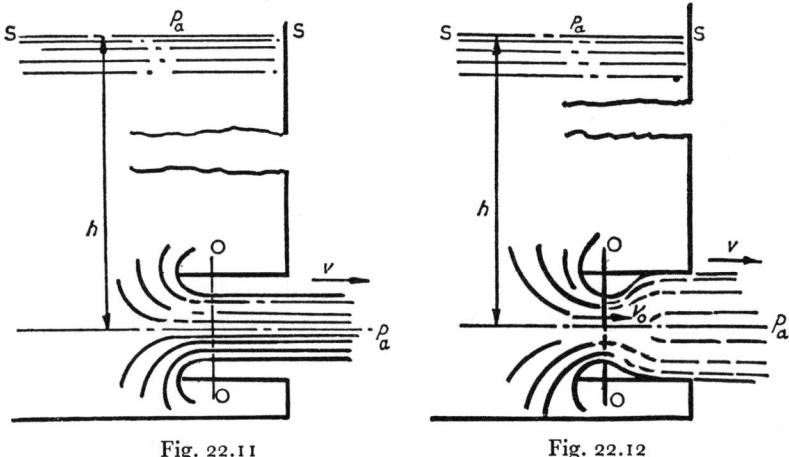

Fig. 22.11   Fig. 22.12

Let $a_0 = $ Area of jet at *vena contracta* O–O,

$\quad a = $ Area of mouthpiece,

$\quad v_0 = $ Velocity of flow at *vena contracta* (also the velocity at exit in Fig. 22.11),

$\quad p = $ Fluid pressure under head h.

When the mouthpiece is running free,

$$\text{Force} = \text{Rate of change of momentum}$$

so that
$$pa = mv_0$$

But
$$p = h\rho g$$

and
$$m = \text{Mass flowing/second} = a_0 v_0 \rho$$

Substituting these values for p and m in the force equation,

$$h\rho g a = a_0 v_0{}^2 \rho$$

But
$$h = v_0{}^2/2g$$

$$\therefore (v_0{}^2/2g)\rho g a = a_0 v_0{}^2 \rho$$

$$a_0 = 0.5a$$

which means that

$$C_c = 0.5$$

$$\therefore \text{Discharge/second} = a_0 v_0$$

$$= a_0 \sqrt{2gh}$$

$$= 0.5a\sqrt{2gh} \qquad (22.27)$$

When the mouthpiece is running full (Fig. 22.12), there will be losses due to enlargement of the jet beyond the *vena contracta*. If

$$v = \text{Velocity at exit}$$

For continuity of flow

$$va = v_0 a_0$$

But
$$a_0/a = 0.5$$

so that
$$v_0 = v/0.5$$

It will be shown in Chapter 23, that

$$\text{Loss of energy due to enlargement} = (v_0 - v)^2/2$$
$$= [(v/0.5) - v]^2/2$$
$$= v^2/2$$

Applying Bernoulli's equation to the free surface S–S and to the mouthpiece,

$$\text{Energy at the free surface} = \text{Energy at outlet} + \text{Losses}$$

$$\therefore hg = v^2/2 + v^2/2$$

$$h = v^2/g$$

or
$$v = \sqrt{gh}$$

Therefore

$$\text{Discharge/second} = av$$

$$= a\sqrt{gh} \qquad (22.28)$$

and is therefore greater than when the mouthpiece is running free.

Problems

For tutorials

1. Why does a jet of water issuing from a small rectangular or square orifice appear twisted? Is the jet really twisted? If not, what is happening along its length?

2. A tank with a small sharp-edged circular orifice in one of its vertical sides contains water maintained at a constant head above the orifice. The necessary equipment is available for measuring the quantity of water discharged from the orifice in a given time and for determining the heights fallen through by the jet at specified horizontal distances from the plane of the orifice.

Explain how it would be possible, using these measurements, to determine the coefficients of velocity and contraction for the orifice.

3. Why is the discharge from a re-entrant mouthpiece greater when running full than when running free? Is the pressure at the vena contracta different in the two cases? If so, how does this affect the flow?

4. The frictional resistances to which a shaft is subjected are often calculated by using the relationship:

$$\text{Friction torque} = \mu Wr$$

where μ is the coefficient of friction for the shaft and bearing metals, W is the load on the bearing and r is the radius of the shaft.

In which kinds of machinery and under what conditions should the friction torque be determined in this manner and when might it be advisable to assess the bearing resistance in terms of viscous drag?

General

1. A shaft having a nominal diameter of 50 mm and a length of 75 mm rotates in a bearing with a clearance of 0·16 mm. Lubrication of the shaft and bearing is by an oil having an absolute viscosity of 0·20 Ns/m². For a speed of 2000 rev/min determine (a) the viscous resistance of the oil; (b) the power required to overcome this viscous resistance.

Assume that the shaft and bearing are concentric and that the velocity gradient across the oil film between the shaft and the bearing is constant.

Ans. (a) 154 N; (b) 806 W

2. A multi-collar thrust bearing has six pairs of annular bearing surfaces each 100 mm outside diameter and 60 mm inside diameter. It is found that 60 W is required to overcome the viscous resistance of the lubricant when the shaft on which the bearing is mounted is rotating at 600 rev/min and the film of lubricant between each pair of bearing surfaces is 0·3 mm thick. Determine the viscosity of the lubricant in Ns/m². Work from first principles.

Ans. $\eta = 0.089$ Ns/m²

3. A machine component of 100 kg mass rests on a flat plate. The surface area of the component in contact with the plate is 0·5 m². If the plate is tilted until it makes an angle of 10° with the horizontal, what will be the maximum speed of descent of the component relative to the plate assuming that the two surfaces are separated by a film of oil 1 mm thick of viscosity 75 cP?

Ans. 4·55 m/s

4. A tank is of circular cross section 1 metre in diameter. It contains water to a height of 2 m above a small orifice set in the base of the tank. If the orifice has a diameter of 30 mm and a coefficient of discharge of 0·62 determine the time taken to empty the tank.

Ans. 19 min 4 s

5. A horizontal pipeline contains a venturi meter, the inlet and throat diameters of which are 250 mm and 150 mm respectively. The pipeline conveys water to a tank which has a "large" rectangular orifice in one of its vertical sides. If the head of water above the top edge of the orifice is 0·20 m when the flow into the tank equals the flow from the tank, determine the head of mercury, in millimetres, in the U-tube connected to the venturi meter. The orifice is 300 mm wide and 150 mm deep. C_d for the orifice is 0·62, C_d for the venturi meter 0·97.

Ans. 49·87 mm

CHAPTER 23

Turbulent Flow Through Pipelines

23.1 ENERGY LOSSES IN PIPES

Turbulence in pipelines

The factors which determine whether the flow through a pipeline shall be laminar or turbulent are embodied in Reynolds's equation $\mathrm{Re} = (\rho v d)/\eta$ (Eq. (22.3)). If the Reynolds number (Re) is less than 2,000 to 2,500, flow will be laminar. Above this range, the flow will be turbulent. For a relatively viscous fluid such as an SAE 30 oil, which has a dynamic (absolute) viscosity η of 300 cP (0·3 Ns/m²) at 20°C and a density ρ of 880 kg/m³, flowing along a pipe of diameter $d = 50$ mm, the critical velocity will be found by taking Reynolds number as 2,000

$$v = \mathrm{Re}\,\eta/\rho d$$

$$= 2000 \times 0{\cdot}3/880 \times 0{\cdot}05$$

$$= 15{\cdot}6 \text{ m/s approximately}$$

Most hydraulic systems using oil are designed so that the maximum velocity of flow lies between 3 and 5 metres per second, rising perhaps to 7 metres per second for short spells. Oils such as SAE 30 will therefore usually have laminar flow and be governed by Poiseuille's equation (see Section 22.2).

If, however, Reynolds's equation is applied to water flowing through pipe of the same diameter, the critical velocity is much lower. The dynamic viscosity of water at 20°C is only 1 cP (0·001 Ns/m²) and the density may be taken as 1,000 kg/m³. The critical velocity is now given by

$$v = (2000 \times 0{\cdot}001)/1000 \times 0{\cdot}05$$

$$= 0{\cdot}04 \text{ m/s}$$

This velocity is very low and would be lower still for larger diameter pipes. It follows that in pipelines carrying water and other low viscosity fluids such as petrol, the velocity of flow will be well above critical, so that the flow will be turbulent.

Pipeline problems involve the assessment of energy losses caused by resistance to the flow of the fluid. These include losses due to sudden enlargements and contractions, bends, entrances and exits to the pipe as well as those caused by the pipe itself. With turbulent flow, these losses occur partly because of the viscous forces exerted at the surfaces in contact with the fluid but mainly because of the eddies set up in the fluid as an indirect result of the viscous forces.

Reynolds showed that the fluid resistances are proportional to v^n where v is the velocity of flow and n is a factor depending upon the degree of roughness of the pipe surface. With turbulent flow, n varies from about 1·8 to 2·0, but for practical purposes is usually assumed to be equal to 2·0. This assumption is supported by investigations carried out by Froude on the resistance to motion of flat surfaces in water and which also showed that the resistances are directly proportional to the area of the surface in contact with the fluid. Thus,

$$R \propto A'v^n$$

or, for most purposes,

$$R = f'A'v^2 \tag{23.1}$$

where R = Frictional resistance,
A' = Area of surface in contact with the fluid,
v = Velocity of flow, and
f' = Frictional coefficient.

If R is measured in newtons, a in square metres and v in metres per second, the frictional coefficient $f' = R/A'v^2$ and represents the frictional resistance in newtons per square metre of wetted surface per unit velocity. The units of f' are found by substitution to be Ns^2/m^4.

Hydraulic gradient

Fig. 23.1 shows two water reservoirs connected by a horizontal pipe of uniform cross-sectional area A. The two free surfaces are at the same level and hence the pressure p is equal at each end of the pipe. The water within the pipe is therefore in static equilibrium under equal and opposite forces pA applied at each end of the pipe. The pressure p will be transmitted uniformly along the pipe in accordance with Pascal's law (see Chapter 21). This could be demonstrated by erecting piezometer tubes

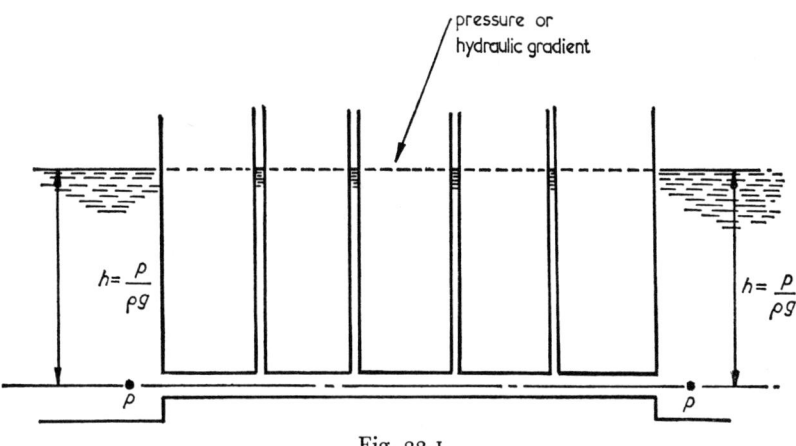

Fig. 23.1

along the pipe; the water would rise by equal amounts in each tube to reach the level of the water in the two reservoirs. The line joining the levels in the reservoirs (and the tubes) is of course, horizontal, and its vertical height above any point in the pipeline represents the pressure

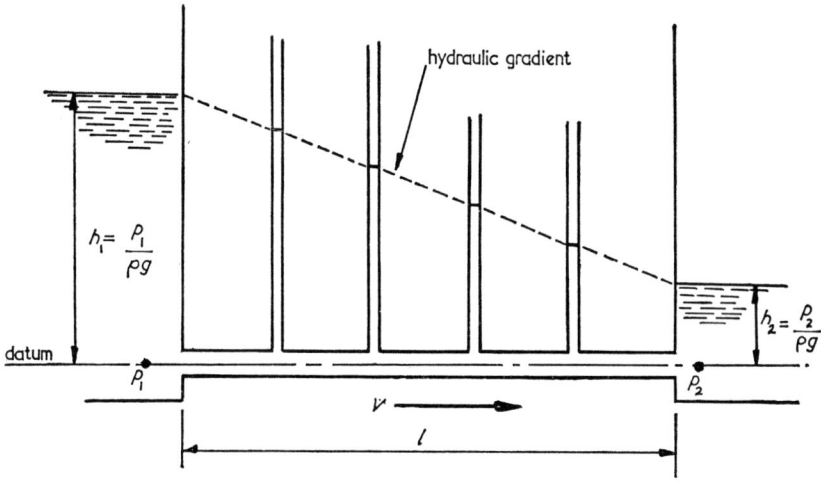

Fig. 23.2

head of fluid in the pipe at that point. The line itself indicates the height at which the fluid pressure is atmospheric and is called the pressure gradient line or hydraulic gradient.

If there is a constant difference of level between the two reservoirs (Fig. 23.2), the pressures p_1 and p_2 at the ends of the pipe will be different and liquid will flow along the pipe, creating frictional resistances which oppose motion and which increase approximately in proportion to v^2 (Eq. (23.1)). The flow becomes uniform when dynamic equilibrium is established between the forces caused by the different pressures and the frictional resistances, so that

$$p_1 A = p_2 A + R \tag{23.2}$$

The resistance R acts uniformly along the pipe, causing a uniform drop in pressure so that, as before, the pressure head at any point in the pipeline is given by the height above the pipe of the straight line joining the levels in the two reservoirs.

The difference of levels, above some horizontal datum, of any two points on the hydraulic gradient is therefore a measure of the pressure difference between two corresponding points in the pipeline causing fluid to flow between those points.

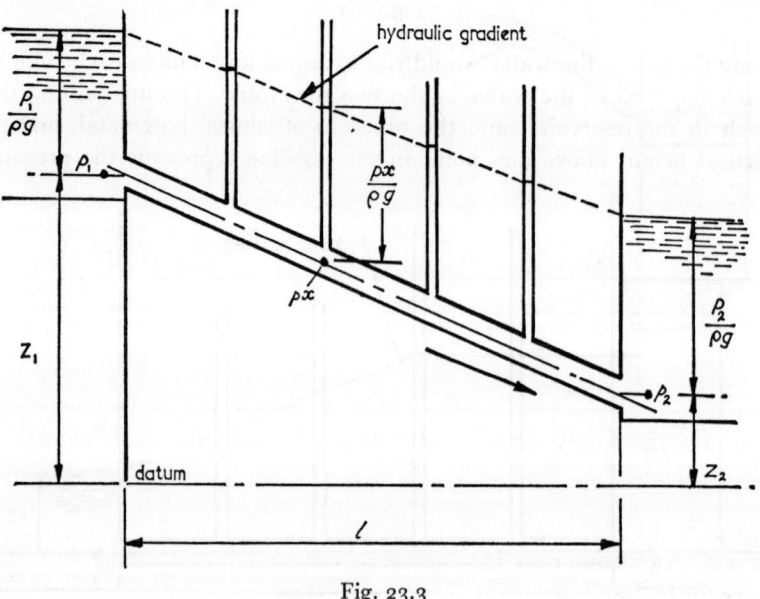

Fig. 23.3

Slope of the hydraulic gradient. If

h_f = Head lost due to friction between two points in the pipeline,
l = Distance between the points, and
i = Slope of the hydraulic gradient between the point, then

$$i = h_f/l \tag{23.3}$$

In Fig. 23.2, the slope is given by

$$i = (h_1 - h_2)/l$$
$$= [(p_1 - p_2)/\rho g]/l$$

where l is the length of the pipe. Provided the pipes are very long in relation to the difference in levels between inlet and outlet, which is usually the case in civil engineering, the horizontal projection of the pipe length may be taken as the actual length l (Fig. 23.3). If Bernoulli's equation is applied to the pipe inlet and exit (Fig. 23.3)

$$Z_1 + p_1/\rho g = Z_2 + p_2/\rho g + h_f$$

or
$$h_f = (Z_1 + p_1/\rho g) - (Z_2 + p_2/\rho g)$$

As before, this is the difference in the water levels.

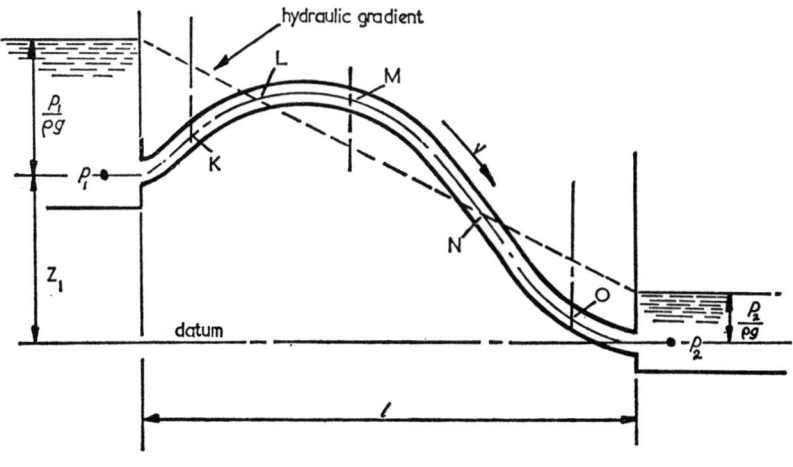

Fig. 23.4

Syphons. A pipeline may have to be laid over high ground, so raising it above the hydraulic gradient (Fig. 23.4). The piezometer tubes have been replaced by ordinates to indicate that fluid at points such as K and O below the hydraulic gradient will be under pressures greater than that of the atmosphere. Fluid at L and N will be at atmospheric pressure, while the fluid at those points in the pipe above the gradient such as M will be below atmospheric pressure. A pipeline laid in this manner forms a syphon through which flow will continue provided that the highest point does not lie more than about 7·9 metres above the hydraulic gradient. At greater heights, separation will take place, the water vapourising under the reduced pressure.

Pipeline losses

It has so far been assumed that the only energy losses are those due to friction between the fluid and the pipe itself, but sudden changes in the pipe diameter, bends, elbows, etc., also cause losses. These are fortunately small, and can be ignored when the pipe is long, so that the hydraulic gradients in Figs. 23.2, 23.3 and 23.4 approximately represent the variation in pressure in a long pipe length. For a short pipe, however, the other energy losses may be comparatively large and can no longer be ignored. The simplified versions in Figs. 23.2 to 23.4 are no longer appropriate and it is necessary to derive expressions for the various losses.

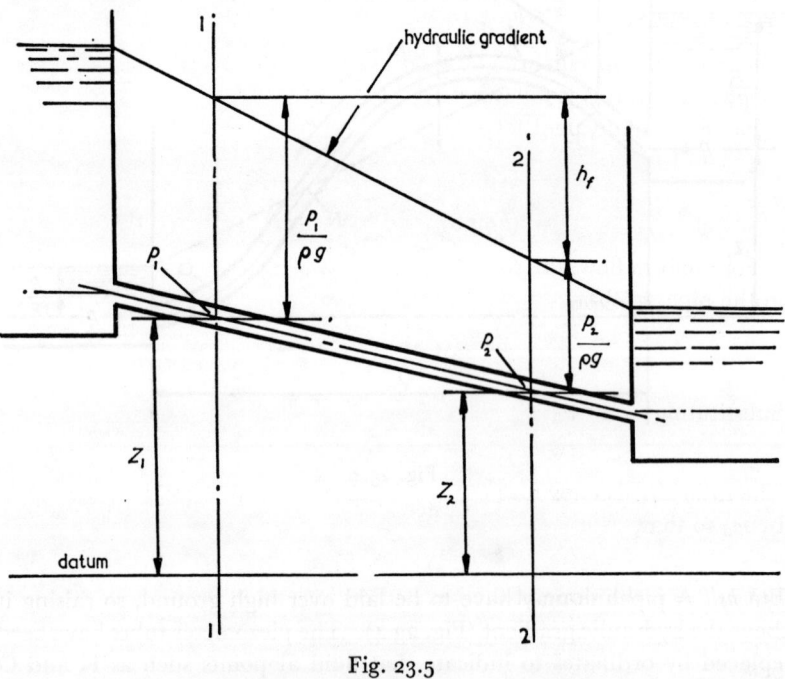

Fig. 23.5

The Chezy formula. For a liquid flowing with constant velocity along a pipe of uniform cross-section (Fig. 23.5), the flow will be due to the difference h_f in the heights of the hydraulic gradient at the Sections 1–1 and 2–2.

Let $l =$ Length of the pipeline between Sections 1–1 and 2–2 (m),
$\quad A =$ Area of cross-section of the pipe (m²),
$\quad v =$ Velocity of flow (m/s),

$p_1 =$ Pressure in the pipe at Section 1–1 (N/m^2),

$p_2 =$ Pressure in the pipe at Section 2–2 (N/m^2),

$f' =$ Frictional resistance per unit area at unit velocity (Ns/m^4),

$A' =$ Wetted area of the pipe (m^2), and

$w =$ Wetted perimeter of the pipe (m).

Applying Bernoulli's equation to Sections 1–1 and 2–2 and recalling that the velocity of flow is uniform, so that $v_1 = v_2$,

$$Z_1 + p_1/\rho g = Z_2 + p_2/\rho g + h_f$$

or, in terms of the equivalent static pressure with respect to the datum,

$$Z_1 \rho g + p_1 = Z_2 \rho g + p_2 + h_f \rho g$$

$$\therefore\ h_f \rho g = Z_1 \rho g + p_1 - Z_2 \rho g - p_2$$

where ρg is the specific weight of the fluid and pressure = head $\times$ specific weight. From this it follows that

Resultant equivalent force on the liquid between the sections

$$= (Z_1 \rho g + p_1 - Z_2 \rho g - p_2)A$$

$$= (h_f \rho g)A$$

For uniform flow, this force must be balanced by the frictional resistance in the pipe, so that

$$h_f \rho g A = f' A' v^2$$

or

$$h_f = f' A' v^2 / (A \rho g)$$

Substituting

$$A' = wl$$

$$h_f = f' w l v^2 / (A \rho g)$$

The ratio A/w is known as the hydraulic mean depth and is denoted here by m_d so that,

$$h_f = f' l v^2 / (m_d \rho g) \tag{23.4}$$

But $h_f/l = i$, the slope of the hydraulic gradient,

$$\therefore\ i = f' v^2 / (m_d \rho g)$$

or

$$v = c \sqrt{m_d i} \tag{23.5}$$

where

$$c = \sqrt{(\rho g / f')}$$

The units of the coefficient c can be determined by substitution as $s^{-1}\,m^{\frac{1}{2}}$. The value of c varies with the temperature and viscosity of the fluid.

Eq. (23.5) is known as the Chezy formula and can be applied to pipes running full or partly full. A pipe running partly full is equivalent to flow along an open channel, and the appropriate Chezy formula is derived in Section 23.3.

The Darcy formula. Pipes usually run completely full, in which case

$$m_d = A/w = (\pi d^2/4)/(\pi d) = d/4$$

Substituting for m_d in Eq. (23.4),

$$h_f = 4f'lv^2/\rho \, dg$$

If f' is replaced by $f\rho/2$, where f is as experimental coefficient, then

$$h_f = 4flv^2/2gd \qquad (23.6)$$

This equation is known as the Darcy formula.

Since the units of ρ are kg/m³ (or Ns²/m⁴ in derived units) and the units of f' are also Ns²/m⁴, substitution in the equation $f = 2f'/\rho$ shows that f is non-dimensional. The value of f is found experimentally and depends on the degree of roughness of the pipe surface and the viscosity of the fluid. Values for f vary from around 0·004 to 0·010. Since $Q = Av$, then $v^2 = Q^2/A^2 = 16Q^2/\pi^2 d^4$. Substituting for v^2 in Eq. (23.6) and evaluating the numerical constants,

$$h_f = flQ^2/3d^5 \text{ approximately} \qquad (23.7)$$

where Q is the discharge in metres³/second.

This rearrangement of the Darcy formula brings out more clearly than Eq. (23.6) the fact that the head lost due to friction varies directly as the square of the discharge and inversely as the fifth power of the pipe diameter.

Example. *A pipe* 200 mm *diameter and* 100 *metres long supplies* 0·1 *cubic metres of water per second. Determine the head lost in friction over the pipe length and the slope of the hydraulic gradient. Assume* $f = 0·005$.

$$\text{Velocity of flow } v = Q/A$$

$$= 0·1/(0·7854 \times 0·2)^2$$

$$= 3·18 \text{ m/s}$$

$$h_f = 4flv^2/2gd$$

$$= \frac{4 \times 0·005 \times 100 \times 3·18^2}{2 \times 9·81 \times 0·20}$$

$$= 5·14 \text{ m}$$

and

$$\text{Slope } i = h_f/l = 5·14/100 = 0·0514$$

Example. *Two reservoirs* (Fig. 23.6) *3 kilometres distant from each other have a difference in surface levels of* 40 *metres. They are connected by a pipe* 600 *mm diameter. Intervening ground necessitates laying the pipe above the hydraulic gradient. If the highest point in the pipe line is* 500 *metres from the inlet and the advisable minimum absolute pressure within the pipe is* 3 *metres of water, determine the rate of flow through the pipe and the maximum height above the water level in the upper reservoir to which the pipe may be taken in order to minimise excavation. Assume the atmospheric pressure to equal* 1 bar *and the coefficient for the pipe to be* 0·005.

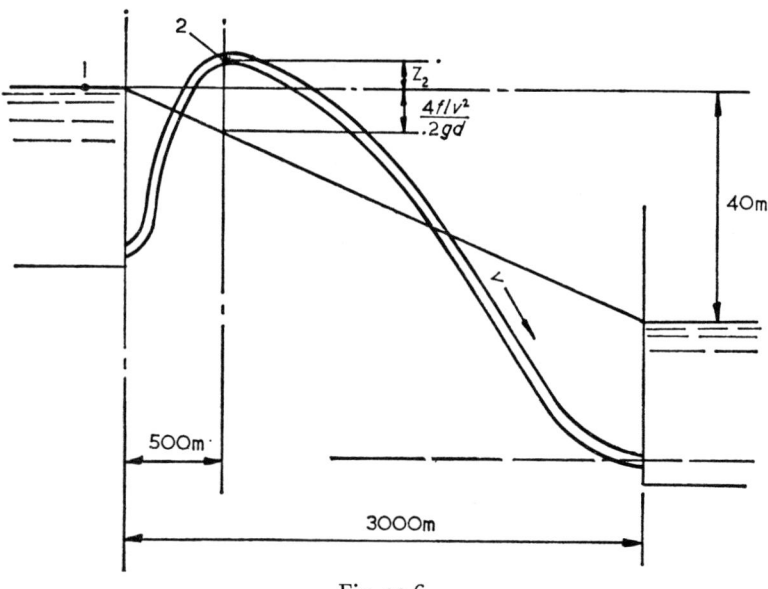

Fig. 23.6

The rate of flow is found by applying Bernoulli's equation to the two ends of the pipe, taking the datum through the lower end of the pipe.

Difference in levels = Velocity head + Head lost due to friction

$$\therefore \ 40 = v^2/2g + 4flv^2/2gd$$

$$= v^2/2g + \frac{4 \times 0 \cdot 005 \times 3000 \times v^2}{2g \times 0 \cdot 6}$$

$$= v^2/2g + 100v^2/2g$$

$$\therefore \ v^2 = (40 \times 2 \times 9 \cdot 81)/101$$

$$= 7 \cdot 78$$

$$\therefore \ v = \sqrt{7 \cdot 78} = 2 \cdot 79 \ \text{m/s}$$

Rate of flow $Q = Av$

$$= 0.7854 \times 0.6^2 \times 2.79$$

$$= 0.79 \text{ m}^3/\text{s}$$

To find the height of the pipeline,

$$\text{Atmospheric pressure } p_1 = 1 \text{ bar}$$

$$= 10^5 \text{ N/m}^2$$

$$\text{Density of water } \rho = 1000 \text{ kg/m}^3$$

$$\text{Head of water equivalent to atmospheric pressure} = p_1/\rho g$$

$$= 10^5/1000 \times 9.81$$

$$= 10.2 \text{ m}$$

Apply Bernoulli's equation to point 1, the free surface of the water in the upper reservoir, and to point 2, the highest point in the pipe line (Fig. 23.6) and take the datum through the free surface. Ignoring all losses other than pipe friction,

$$Z_1 + p_1/\rho g + v_1{}^2/2g = Z_2 + p_2/\rho g + v_2{}^2/2g + h_f$$

Z_1 and v_1 are both zero, so that

$$p_1/\rho g = Z_2 + p_2/\rho g + v_2{}^2/2g + h_f$$

or $\qquad 10.2 = Z_2 + 3 + \dfrac{2.79^2}{2g} + \dfrac{4 \times 0.005 \times 500 \times 2.79^2}{2 \times g \times 0.60}$

$$10.2 = Z_2 + 3 + 0.396 + 6.62$$

$$Z_2 = 10.2 - 10.016 = 0.184 \text{ m}$$

That means that the pipeline must not, at its highest point, be more than 0.184 m above the surface level of the upper reservoir.

Loss of head at enlargement. At a sudden enlargement of the cross-sectional area, the flowing liquid will experience a reduction in velocity and an increase in pressure as it diverges to fill the larger pipe. These changes result in the creation of eddies which form a region of increased turbulence distinct from the general turbulence of the liquid and existing only in the vicinity of the enlargement. The eddies transfer kinetic energy from the liquid through the turbulent region to the pipe walls, where it is ultimately dissipated in the form of heat.

This loss of energy can be determined by considering the mass of fluid between Sections 1–1 and 2–2 (Fig. 23.7) and equating the resultant force on this mass to the rate of change of the momentum. Let p_1 and v_1 be the pressure and velocity of the liquid at Section 1–1 before the enlargement, and p_2 and v_2 be the pressure and velocity respectively at Section 2–2 after the enlargement. A_1 and A_2 are the areas of cross-section of the pipe at Sections 1–1 and 2–2 respectively, and p the pressure within the turbulent region on the annular area $(A_1 - A_2)$.

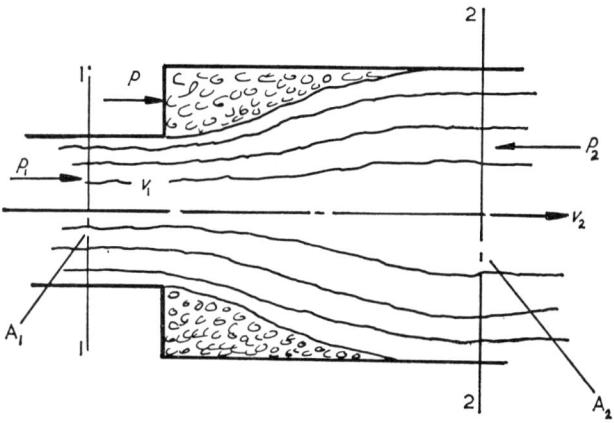

Fig. 23.7

Experiment suggests that $p = p_1$, which implies that the annular area will exert an equal but opposite pressure $p = p_1$ on the liquid in the direction of motion. The

Resultant force on the liquid between Sections 1 and 2

$$= p_2 A_2 - p_1 A_1 - p(A_2 - A_1)$$
$$= p_2 A_2 - p_1 A_2$$
$$= A_2(p_2 - p_1)$$

If m is the mass of fluid passing between Sections 1–1 and 2–2 each second, then

$$m = \rho A_1 v_1 = \rho A_2 v_2$$

and

Rate of change of momentum $= mv_1 - mv_2$

$$= \rho A_1 v_1 (v_1 - v_2)$$

Equating force to rate of change of momentum,

$$A_2(p_2 - p_1) = \rho A_1 v_1 (v_1 - v_2)$$

Dividing through by $A_2 \rho g$,

$$(p_2 - p_1)/\rho g = (A_1 v_1/A_2)(v_1 - v_2)/g$$
$$= v_2(v_1 - v_2)/g \qquad (23.8)$$

If h_f is the head lost by the sudden enlargement, Bernoulli's equation applied to Sections 1–1 and 2–2 gives

$$p_1/\rho g + v_1{}^2/2g = p_2/\rho g + v_2{}^2/2g + h_f$$

so that

$$h_f = (v_1{}^2 - v_2{}^2)/2g - (p_2 - p_1)/\rho g$$

Substituting from Eq. (23.8),

$$h_f = (v_1{}^2 - v_2{}^2)/2g - v_2(v_1 - v_2)/g$$
$$= (v_1{}^2 - v_2{}^2 - 2v_2v_1 + 2v_2{}^2)/2g$$
$$= (v_1{}^2 - 2v_2v_1 + v_2{}^2)/2g$$
$$= (v_1 - v_2)^2/2g \qquad (23.9)$$

Loss of head at contraction. Fig. 23.8 represents a sudden contraction in a pipeline. Because of its inertia, the fluid will continue to contract up to Section 1–1, where its cross-sectional area will have become less than that

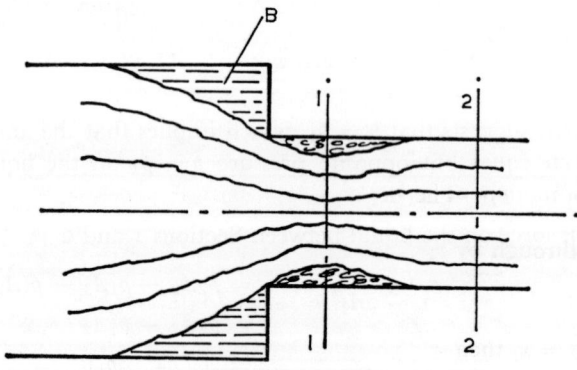

Fig. 23.8

of the smaller pipe. Expansion now takes place until the flow fills the pipe at Section 2–2. During convergence, the stream lines are stable, being contained by relatively quiet regions of fluid at B, but the subsequent divergence creates eddies as in sudden enlargement.

Let p_1 and v_1, p_2 and v_2, A_1 and A_2 represent pressure, velocity and cross-sectional areas at Sections 1–1 and 2–2 respectively, and let C_c be the coefficient of contraction A_1/A_2.

From Eq. (23.9),

$$\text{Loss of head due to expansion} = (v_1 - v_2)^2/2g$$

Continuity of flow gives

$$v_1 = (A_2/A_1)v_2 = v_2/C_c$$
$$\therefore \text{Loss of head} = (v_2/C_c - v_2)^2/2g$$
$$= (1/C_c - 1)^2 v_2^2/2g$$
$$= kv_2^2/2g \qquad (23.10)$$

The constant k is found by experiment to approximate to 0·5.

Example. *A reversible flow of water takes place along a pipeline in which there is a sudden change of diameter. Assuming that the quantity flowing remains constant irrespective of direction, what must be the ratios of the diameters in order that the loss at the change of section is the same for both directions of flow? Assume* $C_c = 0.61$.

Let v_1 and A_1 refer to the smaller diameter and v_2 and A_2 to the larger diameter of pipe. Then
Head lost due to sudden expansion

$$= \text{Head lost due to sudden contraction}$$

or $$(v_1 - v_2)^2/2g = (1/C_c - 1)^2 v_1^2/2g$$

Substituting $C_c = 0.61$ and then expanding,

$$v_1^2 - 2v_1v_2 + v_2^2 = 0.64^2 v_1^2$$
$$v_1^2 - 2v_1^2 A_1/A_2 + v_1^2(A_1/A_2)^2 = 0.64^2 v_1^2$$

Dividing through by v_1^2,

$$1 - 2A_1/A_2 + (A_1/A_2)^2 = 0.64^2$$

If $(A_1/A_2) = x$, then

$$x^2 - 2x + 0.591 = 0$$

from which $x = 1.64$ or 0.36,

$$\therefore A_1/A_2 = 0.36 = d_1^2/d_2^2$$
$$\therefore d_1/d_2 = \sqrt{0.36} = 0.60$$

or $$d_2/d_1 = 1/0.6 = 1.667$$

Loss of head at pipe entrance. When water enters a pipe from a tank or reservoir, there will first be a sudden contraction of the stream followed by expansion to fill the pipe. If v is the velocity of the liquid in the pipe after expansion, then from Eq. (23.10),

$$\text{Loss of head at entrance to pipe} = 0.5v^2/2g \text{ approximately} \quad (23.11)$$

Head loss on entering reservoir. If v is the velocity of the water in the pipe at exit and if the pipe remains cylindrical in form right to the end, the whole of this velocity will be dissipated in the form of eddies.

$$\text{Head lost on entering reservoir} = v^2/2g$$

If the pipe exit is expanded so that the flow becomes divergent most of the velocity energy will be converted to pressure energy, making the loss at exit almost negligible.

Loss in pipe bends and elbows. The head lost at pipe bends depends upon the radii of the bends and the pipe diameter. In general, the head loss is equal to $kv^2/2g$ where k varies from 0.3 to 0.4 for bends and 1.0 to 1.2 for elbows.

The loss of head in various circumstances are set out in the table.

Cause	Loss
Friction in pipeline =	$4flv^2/2gd$
Sudden enlargement =	$(v_1 - v_2)^2/2g$
Sudden contraction =	$0.5v^2/2g$
Entrance =	$0.5v^2/2g$
Pipe exit to reservoir =	$v^2/2g$
Bends and elbows =	$kv^2/2g$

Representation of pipeline losses

With short pipes, the energy losses at enlargements, bends, etc., become significant in relation to the losses in the pipe itself and must be included when drawing the hydraulic gradient. Fig. 23.9 represents two tanks or reservoirs joined by a short pipe of varying diameter and shows diagrammatically, in terms of head, how the total energy of the fluid varies with changes in potential, pressure and velocity energy between point L, just within the upper reservoir, and point T, just within the lower reservoir.

(a) At L, the fluid may be considered at rest so that the pressure head is given by $p_1/\rho g$ and the total head by $Z_1 + p_1/\rho g$. Hence the hydraulic gradient and the total energy line coincide with each other and lie within the free surface S_1–S_1 of the water in the upper reservoir.

(*b*) At M, the fluid has entered the pipe. There is an immediate loss of head at entrance of $0 \cdot 5 v_1^2/2g$ shown by a fall in the total energy line and a drop in pressure of $v_1^2/2g$.

(*c*) Between M and N, there is a uniform fall in the pressure head because of pipe friction of $4fl_1v_1^2/2gd_1$.

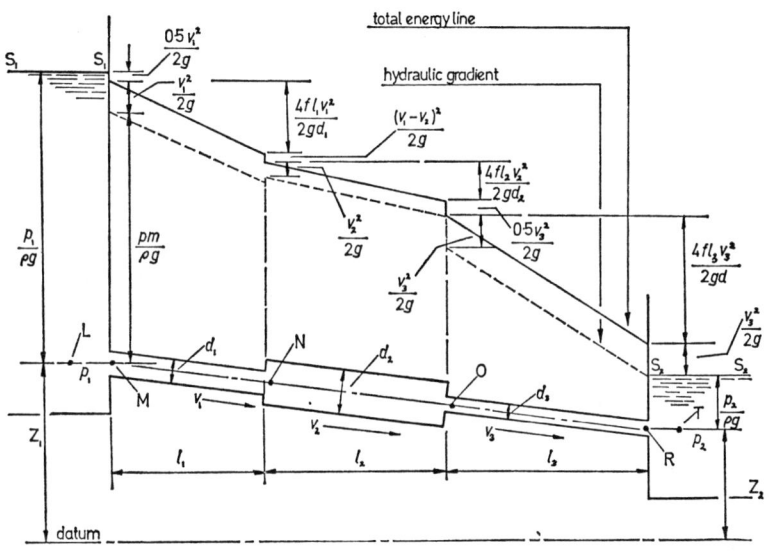

Fig. 23.9

(*d*) At N there is a further loss of head due to sudden expansion equal to $(v_1 - v_2)^2/2g$ and, as the velocity is reduced to v_2 in the larger section, some reconversion of velocity head to pressure head shown by a vertical rise in the hydraulic gradient.

(*e*) Between N and O, a second uniform fall of pressure and hence of total energy equal to $4fl_2v_2^2/2gd_2$ takes place.

(*f*) At O the sudden contraction causes a loss of $0 \cdot 5v^2/2g$ and a drop in the hydraulic gradient as additional pressure energy is converted to velocity energy, raising the latter to $v_3^2/2g$ and providing the higher velocity of flow v_3 required in the narrower pipe of diameter d_3 to preserve the continuity of flow.

(*g*) Between O and R, a further uniform loss of head equal to $4fl_3v_3^2/2gd_3$ takes place.

(*h*) At R, the total energy line drops suddenly to the free surface level S_2–S_2 as the velocity energy $v_3^2/2g$ is dissipated within the reservoir or reconverted to pressure energy. With reconversion, the hydraulic gradient will rise sharply by a small amount to S_2-S_2.

(*i*) At T, the fluid may once again be considered at rest, the total head being given by $Z_2 + p_2/\rho g$ so that the total energy line and the hydraulic gradient coincide with each other and with the surface S_2-S_2.

At any point in the pipeline, the total head is given, in general terms, by $Z + p/\rho g + v^2/2g$ where Z (the potential head) is the height of the pipe centre line above the datum, $p/\rho g$ (the pressure head) is the height of the hydraulic gradient above the pipe centre line and $v^2/2g$ (the velocity head) is the height of the total energy line above the hydraulic gradient.

23.2 Power Transmission Through Pipelines

If W is the weight of liquid flowing through a pipeline in unit time (N/s), h the total head of the liquid on entering the pipeline (m), and h_f the head lost due to friction in the pipeline (*m*), for a long pipe, the

$$\text{Power available at the pipe outlet} = W(h - h_f)$$

$$= W(h - 4flv^2/2gd)$$

but $$W = \rho gAv$$

$$\therefore \text{ Power available at outlet} = P = \rho gA(hv - 4flv^3/2gd) \quad (23.12)$$

where $A =$ Cross-sectional area (m^2),

$v =$ Velocity of flow (m/s),

$l =$ Length of the pipeline (m),

$d =$ Diameter (m), and

$f =$ Frictional coefficient.

Differentiating Eq. (23.12) with respect to v,

$$dP/dv = \rho gA[h - 3(4flv^2/2gd)]$$

Equating to zero for maximum conditions,

$$h_f = h/3 \quad (23.13)$$

The power transmitted therefore becomes a maximum when the flow is such that one third of the total head is lost in overcoming friction.

Example. *A pump supplies water at a pressure of* 5 MN/m² *to one end of a pipeline. If the pipeline is* 4 *kilometres long and has a diameter of* 200 mm, *what is the maximum available power at the other end of the pipeline? Assume that* $f = 0.005$.

$$\text{Pressure head } h = p/\rho g$$
$$= 5 \times 10^6/(1000 \times 9.81)$$
$$= 509 \text{ m}$$

When transmitting maximum power,

$$h_f = h/3 = 509/3 = 169.7 \text{ m}$$

and
$$h_f = 4flv^2/2gd$$

$$\therefore v^2 = \frac{169.7 \times 2 \times 9.81 \times 0.20}{4 \times 0.005 \times 4000} = 8.34$$

$$\therefore v = \sqrt{8.34} = 2.89 \text{ m/s}$$

$$\text{Power available} = W(h - h_f) = W(h - h/3)$$
$$= 2Wh/3 = 2\rho g Avh/3$$

$$= \frac{2 \times 1,000 \times 9.81 \times \pi \times 0.20^2 \times 2.89 \times 509}{3 \times 4}$$

$$= 302\,500 \text{ watts} = 302.5 \text{ kW}$$

PROBLEMS

General

1. Explain what is meant by the critical velocity of flow in a pipe. Determine whether the flow will be laminar or turbulent in each of the following cases. (*a*) A pipe of 60 mm diameter delivering olive oil at a rate of 0·1 m³/s. The oil has an absolute viscosity of 100 cP and a density of 920 kg/m³. (*b*) A pipe of 300 mm diameter delivering 0·05 m³/s of petrol having an absolute viscosity of 0·6 cP and a density of 880 kg/m³. (*c*) A pipe of 50 mm diameter supplying 0·001 m³/s of mercury having an absolute viscosity of 1·6 cP and a density of 13 600 kg/m³. *Ans.* (*a*) laminar, (*b*) turbulent, (*c*) turbulent.

2. A pipe supplying at 0·01 m/s has a sudden contraction which reduces the diameter from 160 mm to 100 mm. What will be the loss of head due to this contraction? Taking only this loss into account and applying Bernoulli's equation, find the change in pressure head caused by the contraction. State this pressure difference also in newtons per square metre. Assume that the coefficient of contraction $C_c = 0.62$.

Ans. 0·0308 m; 0·1008 m; 987 N/m²

3. Determine the head lost due to friction in a pipe 350 metres long and 200 mm in diameter when supplying water at 10 cubic metres/minute. Assume that the friction coefficient $f = 0.008$, that the pipe is flowing full and that the frictional resistance per unit area varies as the square of the velocity of flow.

Ans. 80·4 m

4. Two reservoirs 5 kilometres apart are joined by a pipe which is 300 mm diameter for the first 2 kilometres, when it enlarges suddenly to 500 mm diameter for the next kilometre and then reduces suddenly to 300 mm for the remaining 2 kilometres. The difference in head between the reservoirs is 100 metres. Determine the quantity flowing in cubic metres per second and the various losses in the pipe. Assume the coefficient of friction f to be 0·005. Draw the hydraulic gradient.

Ans. 0·19 m³/s

5. A pipe 5 kilometres long and 300 mm in diameter supplies water for driving equipment requiring 400 kW of power. If the velocity of flow through the pipe is 2·5 m/s, determine the necessary head at the entrance to the pipe. Assume $f = 0.005$.

Ans. 337 m

CHAPTER 24

Friction and Lubrication

24.1 FRICTION

Laws of friction

When two surfaces are in contact and slide relative to each other, a tangential force acting so as to resist the motion is set up along the plane of contact. This force, referred to as force of friction, is generally assumed to obey the laws of friction formulated by Coulomb (1736–1806). Coulomb's laws are approximately true for smooth surfaces subjected to what is known as boundary lubrication. Coulomb's laws for "dry" friction give the following properties of the friction force:

1. Independent of the area of contact.
2. Directly proportional to the normal force holding the surfaces together.
3. Independent of the velocity of sliding.

Coulomb also drew attention to the fact that the frictional resistance is greater at the moment when the surfaces are about to commence sliding than during the process of sliding. He distinguished between these two forms of resistance by referring to them as *static* and *kinetic* friction respectively. His second law embodies the well-known relationship

$$F = \mu R$$

where F = Tangential friction force resisting sliding,
R = Normal force holding the surfaces together, and
μ = Constant, referred to as the coefficient of friction,

which represents the static or kinetic coefficient of friction depending upon whether F is equal to the force required to cause or to maintain sliding.

Cause of friction

In Coulomb's first law, as given above, the area of contact means the nominal area of the surfaces. More recent research carried out at Cambridge by Dr. F. P. Bowden and others has shown that when two metal surfaces are brought together, the actual area of contact is much smaller than the nominal. Accurately ground steel surfaces having an area of 20 cm^2 were shown to be in actual contact over only 1/10,000 of the nominal area when a normal force of 200 N was applied to hold the plate, together. The explanation of this is that even when the metal surfaces are finely ground, there remain irregularities which cause variations in the surface level of as much as 1·5 microns. With lapping and polishing, this figure can be reduced to below 0·15 microns.

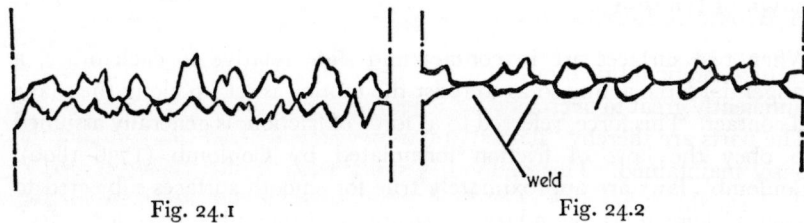

Fig. 24.1   Fig. 24.2

When the surfaces are brought together, the irregularities are the parts which come into actual contact (Fig. 24.1). The pressure at the points of contact will be very high (in the case cited above $200/(0·002/10,000) = 1$ GN/m^2), which causes the metal to yield so that energy is released in the form of heat. This energy, though not great in amount, is concentrated at the points of contact producing temperatures high enough to cause the metal to soften and melt. Plastic flow continues until the cross-sectional areas of the irregularities have enlarged sufficiently to withstand the load, when cooling and solidification take place (Fig. 24.2).

Under higher loads, this process is intensified. Asperities are further reduced in height, the area of weld is enlarged and so the area of contact is increased. If the surfaces are made to slide relative to each other, the welded junctions must shear. The highly localised ruptures cause fresh irregularities and further welds to occur, so that there is a continuous resistance to movement. This resistance is the force of friction.

Bearing seizure. Lubrication is a means of separating bearing surfaces so as to reduce or eliminate contact of surface asperities. When lubrication is insufficient and the bearing pressure is great enough, the welded areas may spread over the whole bearing resulting in complete seizure.

24.2 LUBRICATION

Lubrication implies the introduction of grease or oil between the bearing surfaces of machine components with the objectives of reducing friction (and hence the power required to overcome frictional resistances) and of minimising surface wear. The lubricant forms a thin film between the moving parts, separating or tending to separate them so as to prevent the asperities from coming into contact with each other. The way in which it does this depends very much upon the manner in which the film is introduced and maintained, some types of lubrication are listed below.

Hydrostatic lubrication

In hydrostatic lubrication, oil is pumped to the bearings under a pressure sufficiently great to overcome the average pressure at the bearing surfaces. The parts are thereby "floated" on a layer of the fluid which is continuously maintained. This method of lubrication is often incorporated in the large thrust bearings of hydroelectric generators for use when starting. (At low speeds, the hydrodynamic form of lubrication on which these bearings rely is not effective.) Hydrostatic lubrication is also used in machine tools which have a reciprocating motion such as shapers, planers and slotting machines, the oil being pumped to the slides carrying the reciprocating parts. This is necessary because the bearing surfaces are parallel to each other and the velocity of sliding is low towards the ends of the stroke of a reciprocating component.

Hydrodynamic lubrication

The bearing surfaces involved in the transmission of motion and power are usually so shaped that they do not lie parallel but are inclined or curved relative to each other. Examples are the anti-friction bearings in which ball or roller surfaces engage with races in a rolling motion associated with a small amount of slip, gear teeth which combine rolling and sliding motions, thrust bearings of the tilting pad kind involving sliding motion and journal bearings in which the surfaces slide over each other but which, because of the eccentricity of the journal and its bearing, are not parallel (see Figs. 24.3 (*a*), (*b*), (*c*) and (*d*)). In these and similar cases, the requisite thickness of oil film is maintained by what is known as hydrodynamic lubrication.

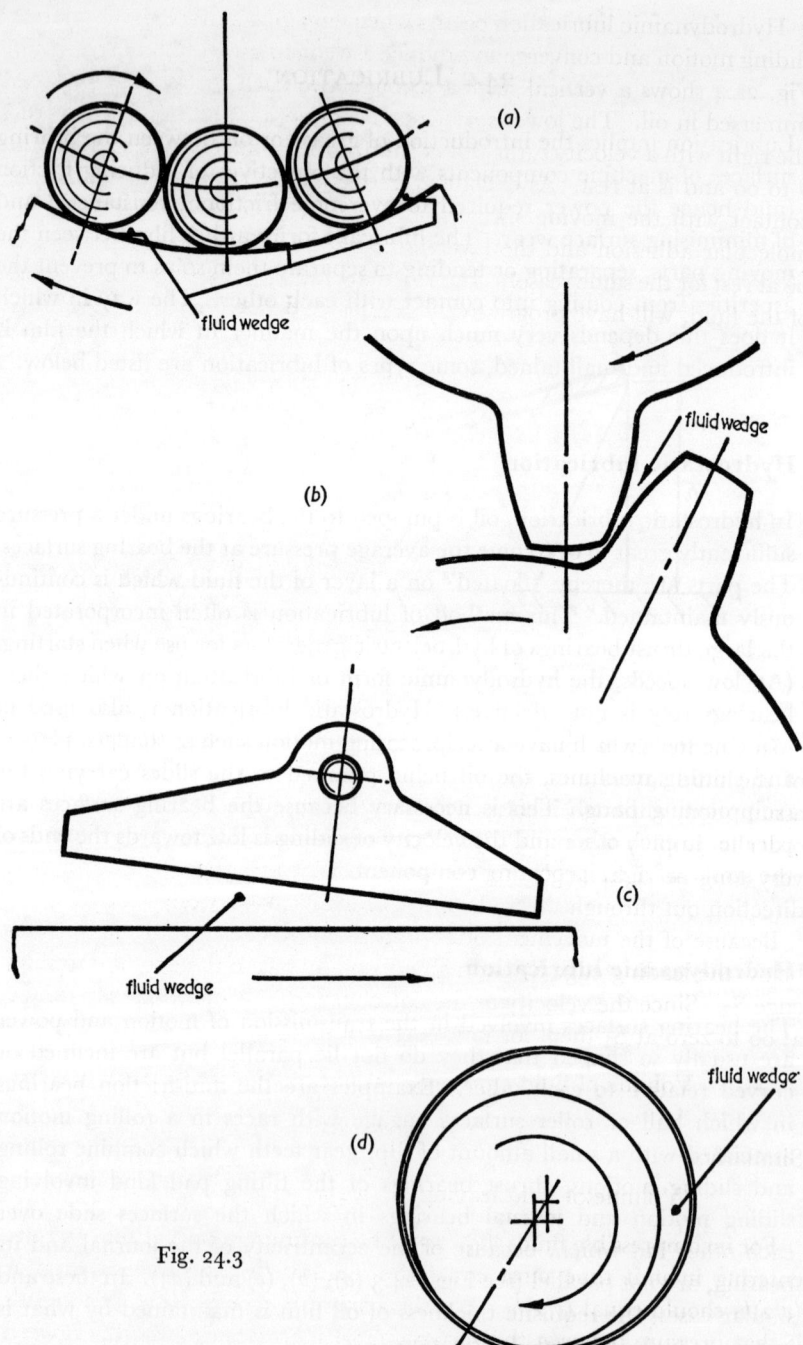

fluid wedge

fluid wedge

fluid wedge

fluid wedge

Fig. 24.3

Hydrodynamic lubrication occurs when engaging surfaces have relative sliding motion and converge to produce a wedge-shaped film of lubricant. Fig. 24.4 shows a vertical section through two such rectangular surfaces immersed in oil. The lower surface oo is horizontal and moving towards the right with a velocity v, the upper surface s_1s_2 is inclined at a small angle θ to oo and is at rest. As explained in Section 22.1, the layer of fluid in contact with the moving surface will move with it at velocity v due to molecular adhesion and the layer in contact with the tilted surface will be at rest for the same reason. Intermediate layers, because of the viscosity of the fluid, will have progressively smaller velocities towards the right.

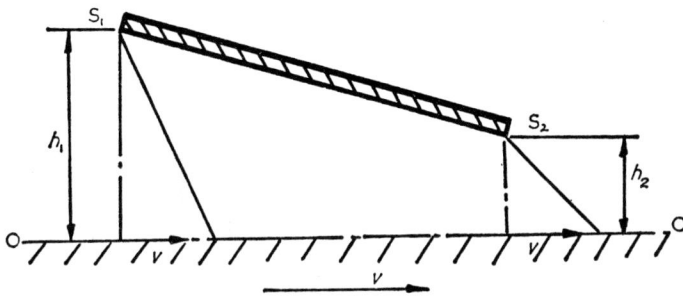

Fig. 24.4

In Chapter 22, a linear relationship was assumed between the velocity of the fluid layers and their distances from the fixed surface and this assumption, although incorrect for converging surfaces, is made here in order to simplify the explanation. It is also assumed that the surfaces are very long at right angles to the page so that any flow of fluid in that direction out through the ends of the bearing can be neglected.

Because of the movement of oo, fluid will be drawn into the opening below the leading edge S_1 and will eventually pass out below the trailing edge S_2. Since the velocity at the entrance varies proportionately from v at oo to zero at S_1 then, for unit length of surface,

$$\text{Volume of fluid entering} = \text{Average velocity} \times \text{Area}$$
$$= (v/2)h_1$$

Similarly,

$$\text{Volume of fluid leaving} = (v/2)h_2$$

For incompressible fluids, flow must be continuous, so that the quantity entering in unit time must equal the quantity leaving in unit time or $(v/2)h_1$ should equal $(v/2)h_2$. But this cannot be, for $h_1 \neq h_2$. The result is that pressure between the surfaces must increase, resisting the inward flow of fluid and reducing the velocity of the intermediate layers, and the velocity gradient is no longer constant across the entrance but curved (Fig. 24.5). The average velocity is less than $v/2$.

The heightened pressure between the surfaces also acts rearwards to assist the outgoing fluid, the effect here being to increase the velocity of the intermediate layers and curve the velocity gradient so that the average velocity is now greater than $v/2$. In this way, the continuity of flow is

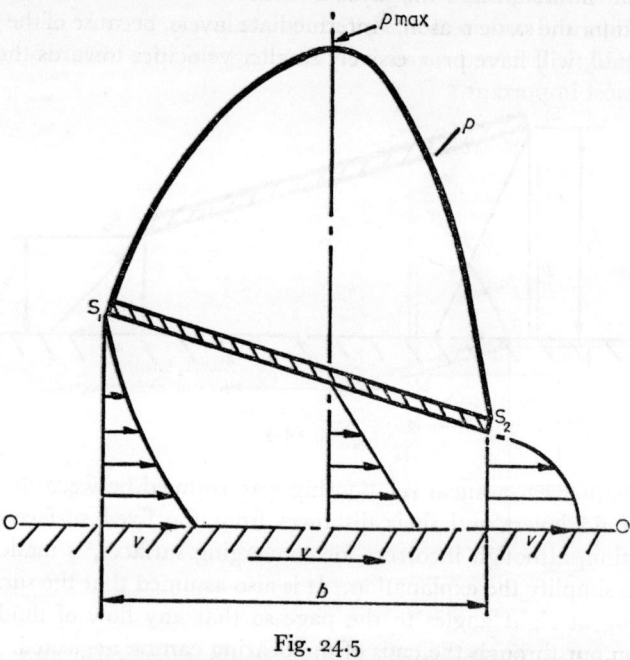

Fig. 24.5

preserved. The increase of pressure between the surfaces provides the necessary upward force to resist the load acting on the bearing and so preserves the requisite thickness of oil film.

Pressure distribution. The hydrodynamic pressure p (Fig. 24.5) rises from zero at the leading edge S_1 to a maximum and then falls away to zero at the trailing edge. The maximum value of p occurs at some point just to the rear of the geometrical centre of S_1S_2, and at this point the gradient of the pressure curve is zero and the velocity gradient is a straight line. The pressure curve shows the distribution of pressure along the breadth b of the surface for one vertical section only. For bearings of great length in which end flow of fluid is negligible, the same curve would represent the pressure distribution for all parallel sections.

Bearings are, however, usually quite short *l* being equal to or not much larger than *b*, so that the proximity of the edges limiting the length of the bearing causes the pressure to fall quickly towards these edges.

Rectangular slider bearing

The following analysis is a brief introduction to the hydrodynamic lubrication of plane rectangular slider bearings. Rigorous mathematical sections have been avoided, but the equations obtained will give some idea of the relationships and orders of magnitude of the parameters involved.

The analysis is based on several assumptions, of which the following are the most important.

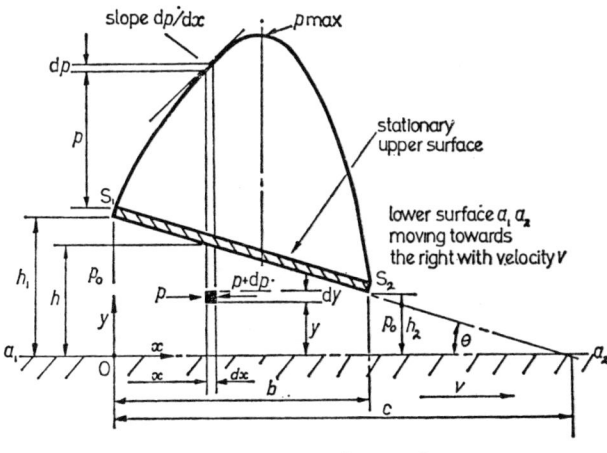

Fig. 24.6

1. The flow of lubricant is entirely in the direction of the width of the bearings; there is no end flow.
2. The flow is laminar between the sliding surfaces, or the lubricants are Newtonian fluids.
3. The viscosity of the fluid remains constant.
4. Fluid inertia is neglected.
5. Fluid pressure is constant through the film thickness.
6. External effects such as gravitational forces are neglected.

Fig. 24.6 shows a vertical section through the pad and runner.

Let θ = Angle of inclination of the pad $s_1 s_2$ to the runner aa,

b = Breadth or width of the pad,

h_1 = Height of the gap at the leading edge s_1,

h_2 = Height of the gap at the trailing edge s_2,

h = Height or thickness of the oil film at distance x from the origin O,

c = Distance from the leading edge to the point at which the plane of the pad intersects the plane of the runner,

V = Velocity of the runner,

v = Velocity of an element of the fluid at distance y above aa,

p_0 = Pressure of the oil before entering or after leaving the bearing,

p = Pressure of the oil within the bearing at distance x from O,

η = Dynamic viscosity of the oil,

τ = Shear stress within the oil, and

R = Rate of shear.

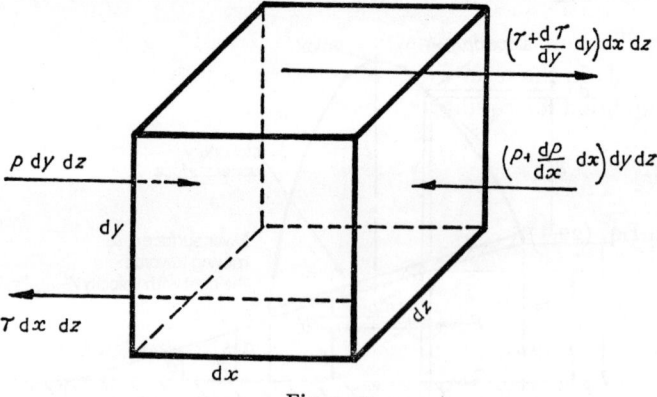

Fig. 24.7

The width of the bearing lies in the direction of the x axis, the thickness of the fluid film in the direction of the y axis and the length of the bearing along the z axis at right angles to the page. The origin O is considered to be situated in aa, below the leading edge s_1. The runner is assumed to be moving with velocity V towards the right and the pad to be stationary. Fluid is therefore being drawn into the bearing through the gap below the leading edge s_1 and is leaving the bearing through the gap below the trailing edge s_2. The upper part of the diagram shows the approximate pressure distribution across the width of the pad for the section under consideration. For pads of this kind, there would usually be end flow and the fluid pressure would vary with z as well as with x. In this case, end flow does not take place (assumption 1), so that the variation in pressure with respect to x can be written as dp/dx.

For an element of the fluid dx, dy and dz which is situated within the bearing at a point whose co-ordinates are x and y, at equilibrium the forces towards the right must equal the forces towards the left. These (Fig. 24.7) are as follows:

The total force to the right is the sum of the force due to pressure on the left-hand face and the shear force due to viscosity on the upper face, and is thus given by

$$p \, dy \, dz + \left(\tau + \frac{\partial \tau}{\partial y} \, dy \right) dx \, dz$$

Similarly, the total force towards the left is the sum of the force due to pressure on the right-hand face and the shear force due to viscosity on the lower face, and this works out as

$$\left(p + \frac{dp}{dx} \, dx \right) dy \, dz + \tau \, dx \, dy$$

At equilibrium, these quantities must be equal to one another. But they differ only in the term which involves the differential coefficients, so that the condition for equilibrium becomes

$$\frac{dp}{dx} = \frac{\partial \tau}{\partial y} \tag{24.1}$$

From Eq. (22.1),

$$\tau = \eta R = \eta \frac{\partial v}{\partial y} \tag{24.2}$$

Both τ and R must be written as partial derivatives since they vary with respect to x as well as with respect to y. Substituting the expression for τ given by Eq. (24.2) in Eq. (24.1),

$$dp/dx = \eta \partial^2 v / \partial^2 y$$

Integrating

$$(\partial v / \partial y) = (dp/dx)y + A$$

Integrating again

$$v = (dp/dx)y^2/2\eta + Ay/\eta + B/\eta \tag{a}$$

The constants of integration A and B can be found as follows. Here v represents the velocity of the element at distance y from aa. The total thickness of the oil film is equal to h, so that y can vary from o to h. When $y = 0$, $v = V$ so that, upon substituting in Eq. (a)

$$V = B/\eta \quad \text{or} \quad B = \eta V$$

Thus

$$v = (dp/dx)y^2/2\eta + Ay/\eta + V \tag{b}$$

When $y = h$, $v = 0$ so that upon substituting in Eq. (b)

$$o = (dp/dx)h^2/2\eta + Ah/\eta + V$$

or

$$A = -(dp/dx)h/2 - V\eta/h$$

Substituting for A in Eq. (b),

$$v = (dp/dx)y^2/2\eta + \{[-(dp/dx)h/2 - V\eta/h]y/\eta\} + V$$
$$= [(dp/dx)y/2\eta](y - h) - V/h(y - h)$$
$$= V/h(h - y) - [(dp/dx)y/2\eta](h - y) \qquad (24.3)$$

The quantity of oil passing through the cross-section of the element in the $y2$ plane is given by $v \, dy \, dz$.

If unit length of bearing is considered (i.e. $dz = 1$), the total quantity of oil passing through unit length of bearing in unit time

$$Q = \int_0^h v \cdot dy$$
$$= \int_0^h [V/h(h - y) - [(dp/dx)y/2\eta](h - y)] \, dy$$
$$= \int_0^h [V - Vy/h - (dp/dx)yh/2\eta + (dp/dx)y^2/2\eta] \, dy$$
$$= \left[Vy - Vy^2/2h - (dp/dx)y^2h/4\eta + (dp/dx)y^3/6\eta \right]_0^h$$
$$= Vh/2 - h^3/12\eta(dp/dx) \qquad (24.4)$$

Transposing, this equation can be written as

$$dp/dx = 12\eta(V/2h^2 - Q/h^3) \qquad (24.5)$$

Here Q is constant, since there is continuity of flow. V, the velocity of the pad, and η, the viscosity of the oil, are constant.

Fig. 24.6 shows that when θ is small,

$$h = (c - x)\theta \text{ approximately} \qquad (24.6)$$

An expression giving the pressure p at any distance x from the leading edge can now be obtained by integrating Eq. (24.5) as follows:

$$dp = \{12\eta V/2h^2 - 12\eta Q/h^3\} \, dx$$

so that

$$p = p_0 + 12\eta V \int_0^x dx/2h^2 - 12\eta Q \int_0^x dx/h^3$$

Substituting $h = (c - x)\theta$,

$$p = p_0 + \frac{12\eta V}{2} \int_0^x \frac{dx}{(c - x)^2\theta^2} - 12\eta Q \int_0^x \frac{dx}{(c - x)^3\theta^3}$$
$$= p_0 + \frac{12\eta V}{2} \left[\frac{1}{\theta^2}\left(\frac{1}{c - x} - \frac{1}{c} \right) \right] - 12\eta Q \left[\frac{1}{2\theta^3} \left(\frac{1}{(c - x)^2} - \frac{1}{c^2} \right) \right]$$
$$= p_0 + \frac{6\eta x}{c\theta^2(c - x)} \left[V - \frac{Q(2c - x)}{c\theta(c - x)} \right] \qquad (24.7)$$

Thus $p = p_0$ when $x = 0$. But p must also equal p_0 when $x = b$, so that it is necessary that

$$\left[V - \frac{Q(2c - b)}{c\theta(c - b)} \right] = 0$$

$$\therefore Q = \frac{Vc\theta(c - b)}{(2c - b)} \tag{24.8}$$

Substituting for Q, putting $h = (c - x)\theta$ in Eq. (24.7) and simplifying gives:

$$p = p_0 + \frac{6\eta Vx(b - x)}{h^2(2c - b)} \tag{24.9}$$

An approximate expression for the average pressure on the pad can be obtained by assuming that the maximum pressure occurs at the mid-point of the pad. This assumption is reasonable, for actual pressure plots indicate that the point of maximum pressure lies behind but near to the centre of the pad.

At the mid-point of the width of the pad $x = b/2$ and the fluid is of mean thickness h_m. Therefore, by Eq. (24.6),

$$h_m = \theta(c - b/2) \tag{24.10}$$

Substituting $b/2$ for x and h_m for h in Eq. (24.9), the

Pressure at the pad centre $= p_c = p_0 + \dfrac{(6\eta Vb/2)(b - b/2)}{h_m^2(2c - b)}$

or $\qquad p_c - p_0 = \dfrac{3\eta Vb^2}{2h_m^2(2c - b)}$

For a parabolic distribution of pressure (approximately true), the average pressure p_a above the inlet and outlet pressure p_0 is equal to $\frac{2}{3}(p_{max} - p_0)$ which is approximately equal to $\frac{2}{3}(p_c - p_0)$

$$\therefore p_a \simeq \frac{\eta Vb^2}{h_m^2(2c - b)} \tag{24.11}$$

Load carried by the bearing. These calculations can also provide the load carried by the bearing, which is equal to the average pressure multiplied by the bearing area, or

$$p_a lb \simeq \frac{\eta Vb^3 l}{h_m^2(2c - b)} \tag{24.12}$$

where l is the length of the bearing.

Shear stress at the runner. The frictional resistance at the runner is caused by the shear stress τ of the lubricant where, by Eq. (24.2),

$$\tau = \eta \,\partial v/\partial y$$

at $y = 0$ (the surface of the runner). But, by Eq. (24.3),

$$v = V/h(h - y) - [(\mathrm{d}p/\mathrm{d}x)y/2\eta](h - y)$$

$$= V - Vy/h - (\mathrm{d}p/\mathrm{d}x)yh/2\eta + (\mathrm{d}p/\mathrm{d}x)y^2/2\eta$$

so that, differentiating,

$$\partial v/\partial y = -V/h - (\mathrm{d}p/\mathrm{d}x)h/2\eta + (\mathrm{d}p/\mathrm{d}x)y/\eta$$

or, when $y = 0$ (for runner surface),

$$\partial v/\partial y = -(\mathrm{d}p/\mathrm{d}x)h/2\eta - V/h$$

If the viscous shear stress τ is taken as positive on the runner surface,

$$\tau = -\eta \,\partial v/\partial y$$

$$= (\mathrm{d}p/\mathrm{d}x)h/2 + V\eta/h \qquad\qquad (24.13)$$

In using Eq. (24.13), $\mathrm{d}p/\mathrm{d}x$ can be obtained from Eqs. (24.5) and (24.8).

Example. *A rectangular plane bearing is 100 mm wide and 200 mm long. The pad is inclined to the runner so that the distance c is twice the width b of the pad. If the load on the bearing is 30 kN, the viscosity of the oil 0·1 Ns/m² and the speed of the runner 5 m/s find the mean thickness of the oil film. The oil when clear of the bearing is at amospheric pressure.*

$$\text{Average pressure on the bearing } p_a = \text{Load/Bearing area}$$

$$= 30\,000/0\!\cdot\!1 \times 0\!\cdot\!2$$

$$= 1\,500\,000 \text{ N/m}^2$$

Also,

$$(2c - b) = (40\mathrm{a} - 100) = 300 \text{ mm} = 0\!\cdot\!3 \text{ m}$$

By Eq. (24.11),

$$p_a \simeq \eta V b^2/h_m{}^2(2c - b)$$

$$\therefore \ h_m{}^2 \simeq 0\!\cdot\!1 \times 5 \times 0\!\cdot\!1^2/(0\!\cdot\!3 \times 1\,500\,000)$$

$$\simeq 0\!\cdot\!000\,000\,011\,1$$

$$\therefore \ h_m \simeq 0\!\cdot\!000\,105\,3 \text{ m}$$

Example. *In the previous example, what will be the quantity of oil flowing through the bearing per unit length of bearing and the viscous shear stress at the centre of the bearing. Also estimate the friction force acting on the runner.*

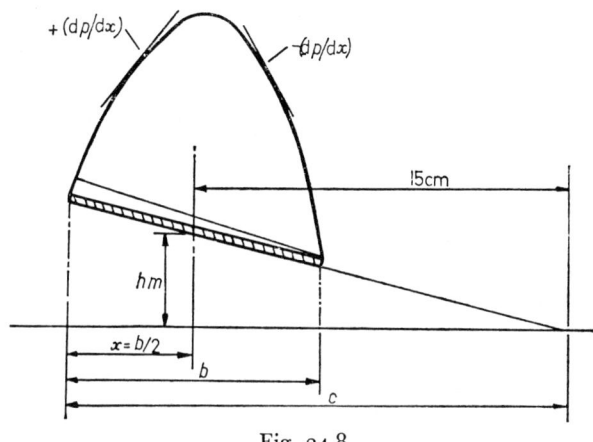

Fig. 24.8

By Eq. (24.8) (see Fig. 24.8),

$$Q = \frac{Vc\theta(c - b)}{(2c - b)}$$

where

$$\theta = h_m/(c - x)$$
$$= 0.0001053/0.15$$
$$= 0.000774 \text{ radian}$$

Therefore,

$$Q = (5 \times 0.2 \times 0.000774 \times 0.1)/0.3$$
$$= 0.000258 \text{ m}^3/\text{s per unit length of bearing}$$

Putting $h = h_m$ in Eq. (24.5)

$$(dp/dx) = 12\eta(V/2h_m^2 - Q/h_m^3)$$

or,

$$(dp/dx) = 12 \times 0.1[5/(2 \times 0.0001053^2) - (0.000258/0.0001053^3)]$$
$$= 1.2(225\,000\,000 - 220\,500\,000)$$
$$= 1.2 \times 4\,500\,000$$
$$= 5\,400\,000 \text{ N/m}^2/\text{m}$$

Then, by Eq. (24.13),

$$\tau = (\mathrm{d}p/\mathrm{d}x)h_m/2 + V\eta/h_m$$
$$= [(5{,}400{,}000 \times 0 \cdot 0001053)/2] + [5 \times 0 \cdot 1/0 \cdot 0001053]$$
$$= 284 + 4740$$
$$= 5024 \ \mathrm{N/m^2}$$

Since the pressure rises from p_0 at the leading edge to a maximum at a point just behind the centre of the bearing and then falls again to p_0 at the trailing edge, the slope $(\mathrm{d}p/\mathrm{d}x)$ of the pressure curve will be positive up to a maximum value of p when it becomes zero, and thereafter will have negative values as far as the trailing edge. The shear stress on the runner as given by Eq. (24.13) therefore varies from a maximum at the leading edge to a minimum at the trailing edge as the term $(\mathrm{d}p/\mathrm{d}x)h_m/2$ in the equation changes from positive to negative.

For design purposes it is usually sufficiently accurate to take the average shear stress as being that mid-way across the bearing at $h = h_m$. Therefore, in the above example, the average shear stress equals 5,024 $\mathrm{N/m^2}$ approximately as calculated for τ. Therefore,

$$\text{Friction force} \simeq \tau bl$$
$$\simeq 5024 \times 0 \cdot 1 \times 0 \cdot 2$$
$$\simeq 100 \cdot 48 \ \mathrm{N}$$

A further simplification is sometimes made by assuming that $(\mathrm{d}p/\mathrm{d}x)$ equals zero mid-way across the bearing (approximately true) when τ becomes equal to $V\eta/h_m$. That is,

$$\tau = 4740 \ \mathrm{N/m^2}$$
and
$$\text{Friction force} = 94 \cdot 8 \ \mathrm{N}$$

Journal bearings

The most common forms of bearings are probably journal bearings. Lubrication is hydrodynamic but the analysis is difficult, so that only a brief qualitative account is given here.

If a shaft is to run freely in its bearing, there must be clearance between the two parts and inevitably there must be shaft eccentricity and wedge-shaped clearance spaces when the shaft is under load. These are shown, very much exaggerated, in Fig. 24.9. Hydrodynamic pressure arises because of the rotation of the shaft within the bearing in the presence of the wedge shaped film of oil formed in the clearance space. When the shaft is at rest (Fig. 24.9 (a)) it presses down on to the bearing, forcing out most of the oil and the coefficient of friction between the two components is then relatively high. As the shaft begins to rotate, because of friction it first rolls along and up the bearing wall to reach the position shown in Fig. 24.9 (*b*). The shaft rotation now causes oil to be drawn

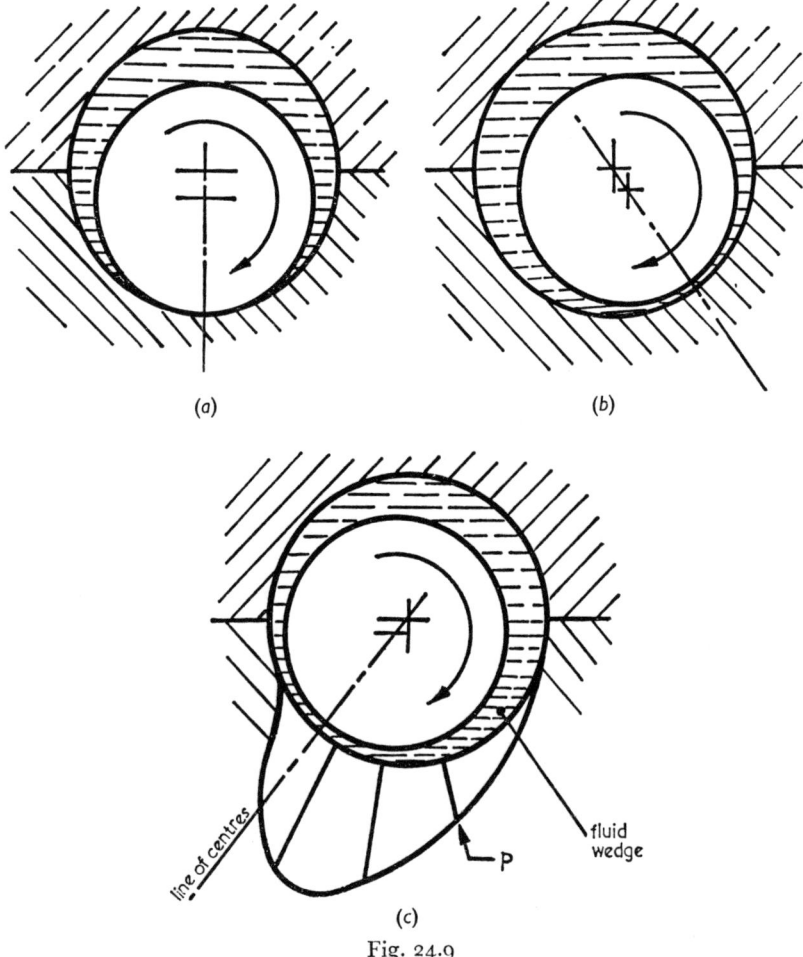

(*a*) (*b*)

(*c*)

Fig. 24.9

into the wedge shaped clearance space, hydrodynamic pressure builds up and the shaft is "floated" back through the central position and over to the opposite side of the bearing (Fig. 24.9 (*c*)) where equilibrium is established between the load on the shaft and the oil pressure. The distribution of oil pressure is shown in Fig. 24.9 (*c*) and is similar to that for the plane rectangular bearing.

It will be clear from the diagram in Fig. 24.9 that the least thickness of the oil film, located on the line of centres, is the critical thickness. This critical thickness is influenced by shaft load and speed and the viscosity of the oil. An increase in speed will increase the amount of oil drawn into the clearance space and enlarge the critical thickness. An increase in load will have the opposite effect. The viscosity of an oil is altered by temperature—the greater the temperature the lower the viscosity—and also by pressure—the greater the pressure, the greater the viscosity. The oil temperature which affects viscosity is now known to be influenced very much by the rate of oil flow to and from the surfaces so that adequate attention must be given to holes and grooves through which oil is admitted to the bearing.

Boundary lubrication

In the absence of hydrostatic lubrication and at low speeds or when the sliding surfaces are parallel, thus preventing the build up of hydrodynamic pressure, the film of lubricant may be squeezed out by the bearing pressure until it is only of molecular thickness. This film continues to adhere to the bearing surfaces because of the strong molecular attraction between the oil and the metal, but is so thin that there are insufficient intermediate layers to permit free fluid motion. Lubrication is now the result of the free sliding action of the few remaining layers of molecules. Because the film is so thin, it is possible for some surface irregularities to come into contact with each other so that the rate of wear and the frictional resistances will be increased, the coefficient of friction rising to about 0·1.

Boundary lubrication exists in bearings when machinery is being started, in oscillating members such as levers or, with reciprocating parts, towards the ends of the strokes as the velocity approaches zero. Many workshop processes involving metal drawing take place under boundary lubrication conditions due to the high pressures involved.

Mixed lubrication

In between hydrodynamic and boundary is the condition known as mixed lubrication. As the film of lubricant is made thinner by increase in bearing pressure or reduction in speed, the fluid conditions associated with hydrodynamic lubrication are lessened and the relationships derived earlier no longer fully apply. With mixed lubrication, the bearing runs partly under hydrodynamic and partly under boundary conditions, depending upon the extent to which the film of lubricant is interfered with by the surface irregularities.

PROBLEMS

For tutorials

1. Describe the manner in which fluid lubricants is dependent upon the property known as viscosity.

2. Assuming that a lubricant has the necessary viscous properties, what are the remaining essential conditions if a bearing is to function hydrodynamically?

3. Why are force-feed pumps often necessary with journal bearings even though the latter operate under hydrodynamic pressure?

4. What is the main function of a lubricant: to prevent wear of the moving parts, or to reduce friction? Define carefully the following terms: wear, pitting, scoring, scuffing, seizure, spalling, embedding, abrasion.

5. What is meant by the term "running in" as applied to a bearing? Describe what happens when bearing surfaces are "run in" and why this process is beneficial.

6. Draw diagrammatic arrangements of the lubrication systems for the following machine tools: lathe, shaping machine and planing machine.

7. At one time it was thought that the braking of a locomotive was more effective if the brakes were applied to lock the wheels, so causing them to slide along the lines. Comment on this.

General

1. A rectangular plane bearing is 100 mm wide and 200 mm long. The pad is inclined to the runner so that the film thickness at the leading edge is twice that at the trailing edge.
If the viscosity of the oil is 0.08 Ns/m², determine the speed of the runner in order that the bearing will support a load of 20 kN when the mean thickness of the oil film is 0.01 cm.
Ans. 3.75 m/s

2. A hydrodynamic bearing is comprised of five rectangular pads, each 100 mm wide by 100 mm long, and a runner. If the total load on the bearing is 100 kN, each pad has an equal share of the load, the kinematic viscosity of the lubricant 0.1 Ns/m² and the speed of the runner 5 m/s, what must be the angle of tilt of the pads if the mean thickness of the film of lubricant is not to exceed 0.0001 m? Also determine the oil thickness at the leading and trailing edges.
Ans. $\theta = 0.000893$ rad; $h_1 = 0.0001497$ m; $h_2 = 0.0000503$ m

3. Determine the total quantity of oil flowing through a plane rectangular bearing 50 mm wide and 100 mm long which is supporting a load of 7.5 kN. The oil has a viscosity of 50 centipoise, the runner speed is 10 m/s and the pad is tilted so that the thickness of the film at the leading edge is twice that at the trailing edge. Neglect end flow.
Ans. 0.0000332 m³/s

4. Explain the advantage of having bearing pads which are free to tilt while in operation.

5. In a rectangular plane bearing, the pad is tilted so that the oil film thickness at inlet is twice that at outlet. The mean thickness of the film is 0·00001 m. The dimensions of the pads are, width 100 mm, length 150 mm. Determine the angle at which the pads are tilted and the total intensity of pressure on the bearing at the leading and trailing edges and at points distant one quarter, one half and three quarters along the width of the bearing.

The oil has a viscosity of 1 poise and the runner speed is 5 m/s. The pressure p_0 at entry and upon leaving the bearing is approximately atmospheric, that is 1 bar or 10^5 N/m^2. Use the values obtained to sketch the pressure curve.

Ans. $\theta = 0·0000667$ radians

$p = 100; \ 137\,800; \ 250\,100; \ 270\,300; \ 100$ kN/m^2 respectively

CHAPTER 25

Thermodynamic Properties of Fluids

25.1 PROPERTIES OF FLUIDS

Thermodynamics is the science which deals with energy and its trans-
formations, so that the *thermodynamic properties* of a fluid are those concerned
with interchanges of energy. A fluid may have other measurable characteristics, such as colour, but these play no part in energy calculations and
are thus not involved in thermodynamics.

Intensive and extensive properties

Properties may be divided into two kinds: those which depend on the
mass of fluid considered (for example, volume and total internal energy)
which are known as *extensive properties,* and those which are independent
of mass and known as *intensive properties*. Intensive properties include
pressure and temperature and also those properties based on unit mass,
such as specific volume or specific internal energy.

In order to define the state of a fluid, it is unnecessary to specify all
of its properties. Indeed, *any two* properties are sufficient so long as the
following conditions are satisfied.

1. They must be intensive properties.

2. They must be independent, which means that it must be possible
to vary one while keeping the other constant. It is, for example, obvious
that it is not possible to define the state of a fluid by specifying density
and specific volume, for one property is simply the reciprocal of the other.
It is not quite so obvious that it is not always possible to define the state
of a fluid by specifying the temperature and pressure. If the fluid is
water, for example, the statement "pressure 1·01325 bar, temperature
100°C" could refer to liquid water, to dry saturated water vapour or to
any mixture of these. Further consideration shows that in this particular

case, pressure and temperature are not independent: if liquid water and water vapour are in contact and the pressure is 1·01325 bar, the temperature must be 100°C.

3. The fluid concerned must be a "pure substance," which may be defined as "a single substance of unvarying chemical composition or a solution, in fixed proportions, of such substances." If, for example, a chemical change takes place in a fluid it becomes, in effect, a different fluid and its properties no longer have the same relationships. Most of the fluids met with in practice are pure substances. Dry air, for example, is a pure substance except at very high temperatures (in which case molecular dissociation alters the chemical composition) and at very low temperatures (in which case the oxygen and nitrogen liquify at different rates so that the liquid and vapour will have different proportional compositions).

Thus the *two-property rule* may be stated as follows: *The state of a pure substance may be defined by any two independent intensive properties.*

From this it follows that any two independent intensive properties may form the axes of a plane diagram on which the state of a fluid will be represented by a point. If the fluid undergoes a *process* (that is, if some or all of its properties change), the corresponding change of state will be represented by the movement of this point. The line thus traced on the diagram is known as a *path*.

Thermodynamic properties

No further discussion is needed of the properties, volume (V), specific volume (v) and density (ρ). The first is an extensive property and the other two are intensive properties.

Pressure. Pressure (p) is defined as force per unit area, so that the unit of pressure is the N/m^2. This is rather a small unit and an alternative unit, the bar is often more convenient; 1 bar $= 10^5 \, N/m^2$.

In what follows, pressures are "true" or "absolute" unless otherwise stated. Pressure gauges are usually calibrated so that "zero" means atmospheric pressure, which means that absolute pressure = gauge pressure + atmospheric pressure. (Standard atmospheric pressure is 1·01325 bar.)

Temperature. The definition of temperature (T) is not simple. The human body is sensitive to "hotness" and "coldness," but this does not lead to a scientific definition and is, in any case, subjective—a bather may describe the water as warm while those about to enter it disagree.

All substances consist of large numbers of elementary particles or "molecules." These molecules are in motion, and from this point of view, the pressure of a fluid is the result of the forces developed when individual molecules rebound from the walls of their container, for example. The *kinetic theory* explains temperature as a measure of the molecular motion—the faster the motion, the higher the temperature. To be more precise, the temperature of a substance is a measure of the *mean translational kinetic energy* of its molecules.

Although this gives a valuable insight into the nature of temperature, it is of little use as a practical definition because the molecules of a substance cannot be observed and there is no simple way of measuring the speeds at which they move. A definition may, however, be based on the way in which heat transfer is brought about by temperature differences. Thus a simple test will determine whether two bodies are at the same temperature—if they are placed in contact, no heat transfer will take place. This can be recognised from the fact that no change in their properties is observed, and they are said to be in *thermal equilibrium*.

Temperature may thus be defined as "that property common to bodies in thermal equilibrium."

This definition involves the axiom that any number of bodies, all at the same temperature, will all be in thermal equilibrium. This is known as "the Zero'th law of thermodynamics" and is usually stated as follows: "If two bodies are in thermal equilibrium with a third body, they will be in thermal equilibrium with each other."

Variations of temperature produce several observable effects, any of which may be used for the construction of a thermometer. Such effects include differential expansion (as in the liquid-in-glass thermometer and the bimetal strip), variation of the liquid/vapour saturation pressure, variation in electrical resistance, the generation of an electric current in a circuit consisting of two dissimilar metals (known as a "thermocouple") and—at high temperatures—the emission of visible light.

The construction of a temperature scale involves (1) the choice of two "fixed points", and (2) the division of the interval between them into an arbitrary number of units. Early temperature scales had as their "fixed points" the melting point of ice and the boiling point of water at standard atmospheric pressure. On the Celsius scale, the interval between these points was divided into 100 "degrees," this being done by equal division of the scale on a mercury-in-glass thermometer. This method of division is unsatisfactory, for it depends on the properties of the substances of which the thermometer is made. If thermometers made of different materials or using different effects are calibrated in this way, they will agree at 0°C and 100°C but may disagree at intermediate points. Furthermore, the zero for this scale is arbitrary and does not correspond to any particularly significant physical condition. There are obvious advantages in choosing as zero the **lowest possible temperature.** This is called **absolute zero** and is about 273 degrees below 0°C.

Both these difficulties were overcome when in 1848 Lord Kelvin introduced the "absolute thermodynamic scale." The interval between the melting point of ice and the boiling point of water was still 100 units (thus 1 K = 1 °C) but the scale commenced at true zero and was divided so that the ratio of two temperatures was equal to the ratio of energies received and rejected by a perfectly reversible heat engine (see Chapter 27) working between them.

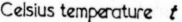

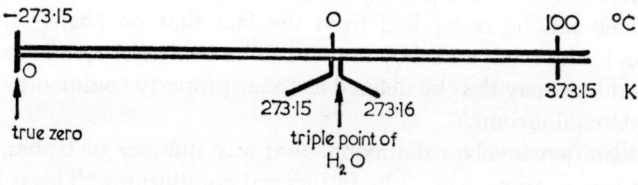

Fig. 25.1

In 1954 it was decided that the original fixed points should be replaced by (1) the true zero, and (2) the temperature of the triple point of water, or the temperature at which ice, water and water vapour will remain in equilibrium. This temperature is readily reproducible and careful measurements show that it corresponds to 0·01°C on the Celsius scale, and is 273·16 degrees above absolute zero. The *Kelvin thermodynamic temperature scale* is thus defined by the points 0 K = true zero and 273·16 K = triple point temperature of water, and the *Celsius scale* is defined from this by fixing its zero at 273·15 K (and the fact that 1 °C = 1 K).

This is shown diagrammatically in Fig. 25.1 and it is evident that Celsius scale temperature (t) is related to absolute temperature (T) by:

$$T = t + 273 \cdot 15 \text{ exactly}$$

For practical purposes, $T = t + 273$.

Internal energy. The total energy possessed by the molecules of a substance due to (1) motion, and (2) relative position is called the internal energy

(U). It is thus the sum of molecular kinetic energy (both translational and rotational) and potential energy (due to the positions of the molecules and the forces of attraction between them). Because we are always concerned with changes of internal energy, this may be calculated relative to an arbitrary zero. In tabulating the internal energy of steam, for example, "zero internal energy" is taken as that of liquid water at the triple point. The units of internal energy are joules. *Specific internal energy* (u) is the internal energy per unit mass and hence has the unit J/kg.

Enthalpy. The composite property (H) defined by $H = U + pV$ is called enthalpy. The reasons for this grouping of properties are discussed in Chapter 26. Since the units of U are joules and the units of pV are N/m² × m³ = Nm = J, the units of enthalpy are also joules. *Specific enthalpy* (h) is enthalpy per unit mass, and is given by $h = u + pv$ and has the unit J/kg.

Entropy. The important thermodynamic property entropy (S) is discussed in Chapter 27. A change of entropy is defined as follows: If, during a reversible process, a substance receives a quantity of heat δQ at temperature T, its entropy increases by an amount

$$\delta S = \delta Q / T$$

For a process in which temperature varies,

$$S_2 - S_1 = \int_1^2 \mathrm{d}Q_{\mathrm{rev}} / T$$

(The suffix "rev" indicates that the heat must be received in a reversible manner.) The quantity of heat $\mathrm{d}Q$ will be measured in joules and T in kelvins, so that the units of entropy are J/K and those of specific entropy (s) are J/kg K. As with internal energy, entropy may be referred to an arbitrary zero.

25.2 PROPERTIES OF GASES

It is convenient to divide fluids into two groups—gases whose properties are related by simple laws, and other fluids, vapours and liquids, in which there are in general no simple relationships.

The gas laws

In early experiments on gases such as oxygen and hydrogen, two relationships were found.

Boyle's law: If the temperature remains constant, the pressure and volume of a fixed mass of gas are inversely proportional to each other or, alternatively, pressure is inversely proportional to specific volume (see Fig. 25.2):

$$p \propto 1/v \quad \text{or} \quad pv = C$$

Charles' law. If the pressure remains constant, the volume of a fixed mass of gas is proportional to its absolute temperature (Fig. 25.3).

$$v \propto T \quad \text{or} \quad v = kT$$

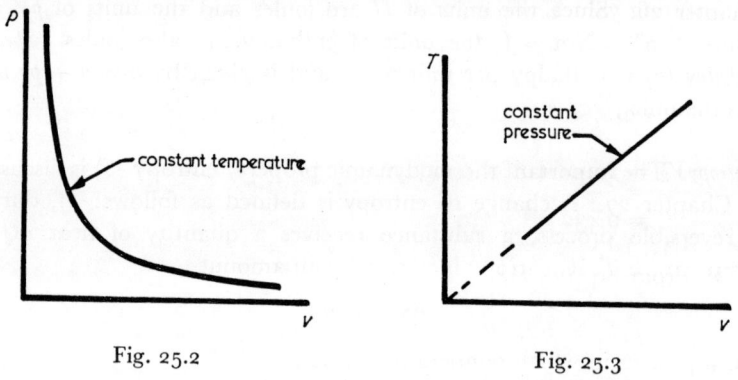

Fig. 25.2　　　　　　　　　　　　　Fig. 25.3

The characteristic equation. Combination of these two laws gives the relationship $pv = RT$ or, for a mass m of gas,

$$pV = mRT \tag{25.1}$$

This is the "characteristic gas equation" and the constant R (which will have different values for different gases) is called the "gas constant."

Example. *The receiver fitted to a small air compressor is a cylinder of internal diameter 200 mm and length 1·5 m, and contains air at a pressure of 1·2 MN/m²* (12 bar) *and temperature 80°C. Neglecting any curvature of the ends of the cylinder find (a) the volume the air would occupy at standard temperature and pressure, (b) the mass of air if R for air is 0·287 kJ/kg K.*

Note: When specifying a quantity of gas by volume, it is necessary to state the temperature and pressure at which this volume is to be measured, and a universally adopted practice is to state volumes at standard

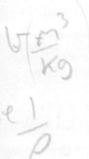

temperature and pressure which are, respectively, $0°C$ and $101·325 \text{ kN/m}^2$ ($1·013\ 25$ bar).

(a)   Volume of cylinder $= \frac{1}{4}\pi(0·2)^2 \times 1·5 = 0·04713 \text{ m}^3$

From Eq. (25.1),

$$p_1 V_1 = mRT_1 \quad \text{and} \quad p_2 V_2 = mRT_2$$

For a fixed mass of gas, mR is common to both equations, and

$$p_1 V_1 / T_1 = p_2 V_2 / T_2 \tag{25.2}$$

Substituting values, and putting

$$T = t + 273$$

$$1·2 \times 10^6 \times 0·04713/353 = 101·325 \times 10^3 \times V_2/273$$

$$V_2 = 1·2 \times 10^6 \times 0·04713 \times 273/(101·325 \times 10^3 \times 353)$$

$$= 0·432 \text{ m}^3$$

(b) From Eq. (25.1),

$$p_1 V_1 = mRT_1$$

$$\therefore\ m = p_1 V_1 / RT_1$$

$$= 1·2 \times 10^6 \times 0·04713/(0·287 \times 10^3 \times 353)$$

$$= 0·558 \text{ kg}$$

The mole and the universal gas constant. Although the gas constant R has different values for different gases, these values are not completely un-related. They are inversely proportional to the molecular weights of the gases. This is a consequence of Avogadro's law, which is that *equal volumes of all gases, under the same conditions of temperature and pressure, contain equal numbers of molecules.*

It has been estimated that 1 m^3 of *any* gas, at standard temperature and pressure (that is $0°C$ and $1·013\ 25$ bar), contains approximately 27×10^{24} molecules. The mass of 1 m^3 of any gas will thus be 27×10^{24} times the mass of an individual molcule. In other words, densities of gases (at S.T.P.) are proportional to their molecular weights.

An amount of gas with mass (in grams) numerically equal to the mole-cular weight is called the *mole*. Thus 1 kilomole (kmol) of a gas whose molecular weight is M would have a mass of M kg. It is found by experiment that 1 kmol of any gas (that is, 2 kg hydrogen, 32 kg oxygen, etc.) occupies a volume of 22·41 m³ at S.T.P. Applying Eq. (25·1) to the case of 1 kmol of any gas at S.T.P.,

$$pV_M = MRT$$

where V_M = Volume of 1 kmol of the gas,

$$MR = pV_M/T$$

$$= 1·013\ 25 \times 10^5 \times 22·41/273$$

$$= 8·314 \times 10^3\ \text{J/kmol K, or } 8·314\ \text{kJ/kmol K.}$$

This value applies to all gases and is called the "universal gas constant" R_0. Thus $pV_M = R_0 T$ and for a particular gas, the constant R may be found from $R = R_0/M$.

Example. *Taking the universal gas constant as* 8·314 kJ/kmol K, *find the specific volume of methane at a pressure of* 100 kN/m² *and temperature* 20°C.

Methane has the chemical formula CH_4, so that its molecular weight is $(12 + 4) = 16$.

$$R = R_0/M = 8·314/16$$

$$= 0·5196\ \text{kJ/kg K}$$

From Eq. (25.1),

$$pV = mRT$$

or

$$pv = RT$$

Hence

$$v = RT/p$$

$$= 0·5198 \times 10^3 \times 293/10^5$$

$$= 1·523\ \text{m}^3/\text{kg}$$

Joule's law of internal energy

The three basic properties of a fluid are pressure, volume and temperature, and internal energy must be a function of one or more of these. Joule's law states that for a gas, internal energy is a function only of temperature. Experimental proof of this law was first obtained using an apparatus consisting of two copper containers, A and B, connected by a valve and immersed in an insulated water bath (see Fig. 25.4). A contains air at a pressure of 22 atmospheres (that is, approximately 22 bar) and B is evacuated. On opening the valve and allowing conditions to stabilise, the thermometer shows that the apparatus has regained its original temperature.

In this experiment the air expands without doing any work, and since the rest of the apparatus returns to its original temperature it neither gains nor loses energy: thus the internal energy of the air must remain unchanged. Pressure and volume both change, hence internal energy must be independent of these and is thus a function of the only other property which does not change—the temperature.

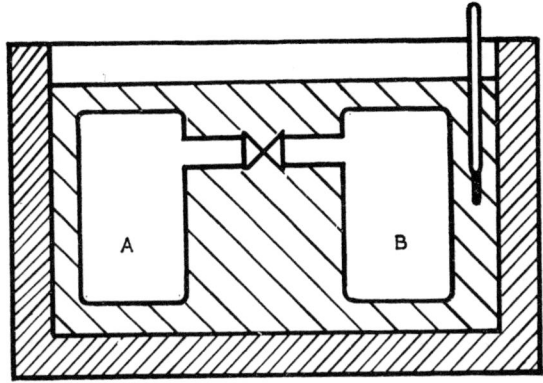

Fig. 25.4

Specific heats of a gas

The *specific heat* (or *specific heat capacity*) of a substance is defined as the heat transfer required to raise the temperature of unit mass of the substance by 1 kelvin. For copper, for example, a heat transfer of 0·4 kJ raises the temperature of 1 kg by 1 K. Thus the specific heat of copper $c =$ 0·4 kJ/kg K.

Example. *A 2 kW electric kettle is made of copper and has a mass of 900 g. It contains 1½ litres of water at a temperature of 20°C. How long after switching on will the temperature reach 100°C, neglecting heat losses? The specific heats of copper and water are respectively 0·4 kJ/kg K and 4·187 kJ/kg K. The mass of 1 litre of water is 1 kg.*

Heat transfer = Mass × Specific heat × Temperature change

∴ Total heat transfer = 0·9 × 0·4 × (100 − 20)
$$+ \ 1·5 × 4·187 × (100 − 20)$$
$$= 28·8 + 502·4 = 531·2 \text{ kJ}$$

Hence if rate of energy transfer is 2 kW, time taken is

$$531·2/2 = 265·6 \text{ seconds} \quad \text{or} \quad 4 \text{ min } 25·6 \text{ seconds}$$

For solids and (to a lesser extent) for liquids, the amount of heat required to raise the temperature of unit mass by 1 kelvin is substantially the same, irrespective of how the process is carried out. The heat capacity is only slightly affected by changes in pressure and volume. For a gas, however, the process may involve expansion during which the gas will do work. The heat transfer required will depend on the amount of this work (see Chapter 26) and the specific heat will have different values for different kinds of processes and may, in fact, have any value from zero to infinity. It is therefore usual to state specific heats of gases for two kinds of process—those carried out (1) at constant volume c_v, and (2) at constant pressure (c_p).

It will be shown in Chapter 26 that c_v represents the change in internal energy and c_p the change in enthalpy.

Vapours, gases and perfect gases

The gas laws were first deduced from experiments on gases such as oxygen (originally called the "permanent gases" because early investigators were unable to liquify them). Experiments with other gases show departures from these laws, particularly near the point at which liquifaction takes place. In such circumstances, the substance is called a "vapour." Obviously, there will be no sudden transition from gas to vapour but an arbitrary division between the two terms is based on the critical temperature, or the temperature above which a substance cannot exist as a liquid. Thus the term "gas" is reserved for substances at temperatures higher than the critical.

All real gases depart slightly from the gas laws and the specific heats c_p and c_v are not constant but increase with increasing temperature. A *perfect gas* is one having constant specific heats and obeying the gas laws exactly.

25.3 PROPERTIES OF LIQUIDS AND VAPOURS

All substances can exist in three different forms or *phases*—solid, liquid and vapour (or gas). At atmospheric pressure, for example, water is a solid below 0°C, a liquid between 0°C and 100°C, a vapour above 100°C and may be considered a gas if the temperature is sufficiently high. Raising the pressure lowers the freezing point (slightly) and raises the boiling point.

This is shown in Fig. 25.5, where the pressure scale has been expanded considerably at the lower end. Two points of particular importance are *T*, the triple point (at which all three phases are in equilibrium) and *C*, the critical point. For pressures above the critical, the division between liquid and vapour ceases to exist. In a supercritical steam boiler, for example, the operating pressure may be 25 MN/m² (250 bar) and the final temperature 600°C. Although this is referred to as a boiler, the transition of the water from initial to final states is a continuous smooth process. At no point does the phenomenon of boiling occur.

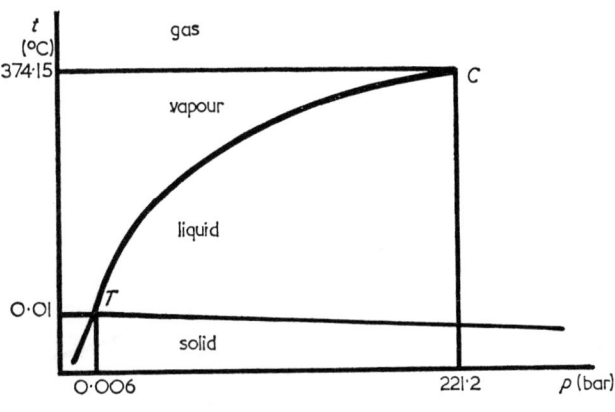

Fig. 25.5

For pressures below that at the triple point, there is no liquid phase and solids change directly into vapours—a process called *sublimation*. Some substances sublime at atmospheric pressure. A common example is solid carbon dioxide ("dry ice"), whose triple point pressure is approximately 5·2 bar.

If a change of phase at constant pressure takes place at constant temperature, diagrams like Fig. 25.5 would give no indication of the changes taking place in volume, enthalpy, etc. These may be shown on a *p − v* diagram. Fig. 26.6 shows, approximately to scale, the *p − v* diagram for water. (The solid phase is not shown.)

The line ABCD represents a constant pressure process in which heat is supplied to liquid water until it finally becomes superheated steam. At A the water is below the boiling point and would be described as a *compressed liquid*. As heat is supplied, its temperature rises until at B it reaches the boiling point or *saturation temperature*. "Saturation" refers to any situation in which two phases of a substance are in equilibrium, hence at B the water is a *saturated liquid*. From B to C the liquid changes to vapour at constant temperature, being called a *dry saturated vapour* at C. Between B and C there will be a mixture of liquid and vapour, and this is called a *wet vapour*. *The dryness fraction* of a wet vapour is defined as the ratio of the mass of vapour and the mass of mixture. Thus "1 kg of steam of dryness fraction 0·9" means 0·9 kg of vapour mixed with 0·1 kg of liquid. Further heating after C causes the temperature to rise above the saturation temperature and the vapour is now said to be *superheated*. At D the state of the steam may be defined by specifying its pressure and either (i) its temperature, or (ii) its *degree of superheat*, which is defined as the difference between the temperature and the saturation temperature.

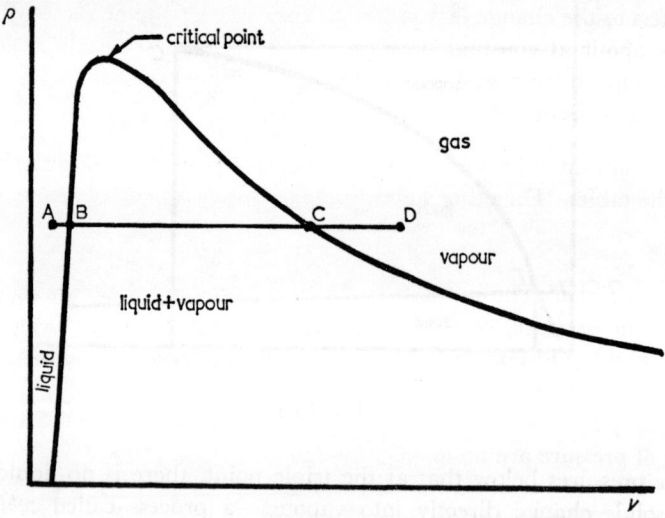

Fig. 25.6

At a pressure of 100 bar, for example, the saturation temperature is 311°C, so that steam at 100 bar and 400°C would be described as having 89 degrees of superheat.

Property tables

The properties of vapours and liquids are not related by simple laws, and must be found either (i) using diagrams constructed from experimental data, such as Figs. 25.5 and 25.6, or (ii) from property tables* which list the properties of water and other fluids used in heat engine cycles (including refrigerator cycles) under saturated and superheated conditions.

* Data used in this text has been obtained from "Thermodynamic and Transport Properties of Fluids in SI Units" by Y. R. Mayhew and G. F. C. Rogers, published by Basil Blackwell.

The properties listed are pressure p, Celsius temperature t, specific volume v, specific internal energy u, specific enthalpy h and specific entropy s. These symbols are qualified by suffixes s, f, g and fg, all of which refer to the condition of liquid-vapour saturation and have meanings as follows:

s refers to the saturation state, so that t_s means saturation temperature,

f refers to a property of a saturated liquid, so that h_f is the specific enthalpy of a liquid at the saturation temperature corresponding to its pressure,

g refers to a property of a saturated vapour, so that h_g is the specific enthalpy of a dry saturated vapour,

fg refers to the change in a property when its phase changes from liquid to vapour at constant pressure. Thus h_{fg} is the increase in specific enthalpy during evaporation at constant pressure, and (for a particular pressure) $h_{fg} = h_g - h_f$.

Data for saturated liquid and dry saturated vapour may be read directly from the tables. For other states, properties may be calculated as shown below.

Compressed liquid. The properties of a liquid are only slightly affected by changes in pressure, so that for practical purposes the properties of a compressed liquid may be considered equal to those of a saturated liquid at the same temperature. The tables give data relating to compressed liquid only for water, and chiefly at the higher pressures at which the effects of pressure are no longer negligible.

Example. *Water enters the drum of a boiler at* 200°C, *the working pressure being* 10 MN/m² (100 bar). *Find (a) its specific volume, (b) its specific enthalpy.*

(i) Approximate solution: taking the properties as equal to those of a saturated liquid at the same temperature: The saturation pressure for $t_s = 200$°C is 15·55 bar and for these conditions the values of v_f and h_f are, respectively, 0·001157 m³/kg and 853 kJ/kg.

(ii) Exact solution: the table referring to compressed water gives the following data:

$$(v - v_f) = -0.000009$$

and $$(h - h_f) = +4$$

hence $$v = 0.001157 - 0.000009 = 0.001148 \text{ m}^3/\text{kg}$$

and $$h = 853 + 4 = 857 \text{ kJ/kg}$$

For other temperatures and pressures, interpolation of this table may be necessary. Unless the pressure is high, corrections thus obtained will be of little practical significance and method (i) is sufficiently accurate for most purposes.

Wet vapour. If the dryness fraction is x, 1 kg of wet vapour will consist of x kg of saturated vapour and $(1 - x)$ kg of saturated liquid. Hence

$$u = xu_g + (1 - x)u_f \qquad (25.3)$$

$$h = xh_g + (1 - x)h_f \qquad (25.4)$$

$$s = xs_g + (1 - x)s_f \qquad (25.5)$$

$$v = xv_g + (1 - x)v_f \qquad (25.6)$$

The second term of (25.6) will usually be very much smaller than the first, hence for practical purposes the liquid volume may be neglected, giving

$$v = xv_g \qquad (25.7)$$

If values of h_{fg} or s_{fg} are available, Eqs. (25.4) and (25.5) may with advantage be transposed. Thus the first of them gives

$$h = xh_g + (1 - x)h_f$$

$$= h_f + x(h_g - h_f)$$

$$= h_f + xh_{fg} \qquad (25.8)$$

Similarly, $\qquad\qquad s = s_f + xs_{fg} \qquad (25.9)$

Example. *Find (a) the specific internal energy (b) the specific enthalpy of steam at a pressure of* 600 kN/m² (6 bar) *and of dryness fraction* 0·9.

(a) From Eq. (25.3),

$$u = xu_g + (1 - x)u_f$$

$$= 0·9 \times 2568 + 0·1 \times 669$$

$$= 2378 \text{ kJ/kg}$$

(b) From Eq. (25.8),

$$h = h_f + xh_{fg}$$

$$= 670 + 0·9 \times 2087$$

$$= 2548 \text{ kJ/kg}$$

Superheated vapour. For superheated vapours, properties may be stated in terms of actual temperature or of degree of superheat. Temperatures are listed at fairly wide intervals, and intermediate values are found by linear interpolation.

Example. *Find the specific enthalpy of (a) steam at a pressure of 2 MN/m²
(20 bar) and a temperature of 412°C, (b) ammonia at a pressure of 1·47 MN/m²
(14·7 bar) and a temperature of 60°C.*

(a) For a pressure of 20 bar, the tables give the following data:

$$t = 400, \; h = 3248; \quad t = 450, \; h = 3357$$

Thus for a temperature increase of 50 deg C, h increases by 109 kJ/kg.
Assuming the relationship between t and h to be linear, the value of h
corresponding to $t = 412°C$ will be

$$h = 3248 + 109 \times 12/50 = 3274 \text{ kJ/kg}$$

(b) For $p = 14·7$ bar, $t_s = 38°C$ hence the degree of superheat is
$(60 - 38) = 22$ deg.

Interpolating between h_g (which corresponds to zero superheat) and
h for 50 deg superheat,

$$h = 1472·6 + 147·5 \times 22/50 = 1537·5 \text{ kJ/kg}$$

Internal energy. Values of specific internal energy are stated in the tables
only for water. For other liquids, internal energies may be calculated
from other data. Since specific enthalpy is defined by

$$h = u + pv$$

it follows that $\qquad\qquad u = h - pv \qquad\qquad\qquad$ (25.10)

Example. *Find the specific internal energy of the refrigerant Freon-12 when
in the form of a wet vapour of dryness fraction 0·8 at a temperature of −10°C.*

From Eq. (25.4),

$$h = xh_g + (1 - x)h_f$$
$$= 0·8 \times 183·19 + 0·2 \times 26·87$$
$$= 151·93 \text{ kJ/kg}$$

From Eq. (25.7),

$$v = xv_g$$
$$= 0·8 \times 0·0766 \text{ m}^3/\text{kg}$$

For $\qquad\qquad t = -10°C,$

$$p_s = 2·191 \text{ bar}$$
$$\therefore \; pv = 2·191 \times 10^5 \times 0·8 \times 0·0766$$
$$= 13·43 \times 10^3 \text{ J/kg} \quad \text{or} \quad 13·43 \text{ kJ/kg}$$

From Eq. (25.10),

$$u = h - pv$$
$$= 151·93 - 13·43 = 138·50 \text{ kJ/kg}$$

Problems

For tutorials

1. A balloon is filled with 1000 m³ of helium, the pressure and temperature of both the helium and the surrounding atmosphere being 101 kN/m² (1·01 bar) and 15°C. Calculate its lifting force if the gas constants for air and helium are respectively 0·287 kJ/kg K and 2·079 kJ/kg K and if $g = 9·81$ m/s². What will happen to this lifting force as the balloon ascends into regions of lower pressure and temperature assuming that (i) the envelope is of large capacity so that the helium may expand without constraint, (ii) the ascent takes place slowly, so that there is thermal equilibrium at all times?

Ans. 10·33 kN

2. The specific volume of an unknown gas is found to be 12·47 m³/kg at a pressure of 100 kN/m² (1 bar) and temperature 27°C. If the universal gas constant is 8·314 kJ/kmol K, show that the gas is hydrogen. Could other gases be identified in this way?

3. Calculate the specific volume of H_2O under the following conditions, assuming it to obey the characteristic gas equation (and taking the universal gas constant as 8·314 kJ/kmol K). In each case use property tables to find the true value and calculate the percentage error (a) $p = 0·1$ bar, $t = 50°C$, (b) $p = 0·1$ bar, $t = 400°C$, (c) $p = 100$ bar, $t = 400°C$, (d) $p = 100$ bar, $t = 700°C$. Is the usual definition of a gas a guide as to conformity with the gas laws?

Ans. (a) +0·4 per cent; (b) +0·1 per cent; (c) +17·8 per cent;
(d) +3·3 per cent (based on $M = 18$)

4. (a) The boiling point of water is commonly stated to be 100°C. In what way is this statement incomplete and what are, in fact, the temperatures between which "the boiling point of water" may lie?

(b) High-altitude pilots and astronauts face the danger of "decompression" or exposure to very low atmospheric pressures. One of the effects of decompression is that body fluids boil if the pressure is sufficiently low. Treating body fluids as pure water and taking body temperature as 37°C, use property tables to estimate the altitude above which this could occur.

Ans. 19 200 m

General

1. A vessel of volume 2 m³ contains 10 kg of a gas at 500 kN/m² (5 bar) and 80°C. Find (a) the volume which the gas would occupy at standard temperature and pressure, (b) the value of the gas constant R.

Ans. (a) 7·64 m³; (b) 0·283 kJ/kg K

2. A cylinder for the storage of compressed gases has a volume of 0·05 m³ and is designed for a working pressure of 15 MN/m² (150 bar). What mass of (a) hydrogen, (b) oxygen could be stored at a temperature of 15°C? The universal gas constant is 8·314 kJ/kmol K.

Ans. (a) 0·627 kg; (b) 10·02 kg

3. Steam leaves a boiler at a pressure of 1·5 MN/m² (15 bar) and with dryness fraction 0·97. Find (a) its specific volume, (b) its specific internal energy, and (c) its specific enthalpy.

Ans. (a) 0·1277 m³/kg; (b) 2542 kJ/kg; (c) 2734 kJ/kg

4. Find the specific enthalpy of (a) steam at 13 MN/m² (130 bar) and 470 °C, (b) ammonia at 704·5 kN/m² (7·045 bar) and 75°C.

Ans. (a) 3249 kJ/kg; (b) 1615·3 kJ/kg

5. Find (a) the temperature, (b) the specific volume, (c) the specific enthalpy, and (d) the specific internal energy of wet mercury vapour (dryness fraction 0·95) at a pressure of 1 MN/m² (10 bar).

Ans. (a) 517·8°C; (b) 0·03103 m³/kg; (c) 344·64 kJ/kg;
(d) 313·61 kJ/kg

CHAPTER 26

The First Law of Thermodynamics

26.1 HEAT AND WORK

Systems and boundaries

To apply the laws of thermodynamics to a problem, the first step is to define the region to be considered. This is called the *system* and it is defined by *boundaries* which separate it from the *surroundings* (Fig. 26.1). These boundaries may be fixed or moving, and they may coincide with a real boundary (such as a cylinder wall) or they may be imaginary. The essential thing is that the boundaries completely enclose some volume in space, so that energy entering or leaving the system must cross the boundary and so can be accounted for.

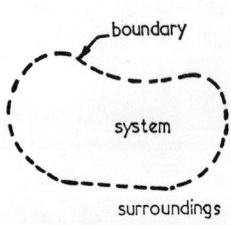

boundary

system

surroundings

Fig. 26.1

A system may consist simply of a quantity of fluid contained in a cylinder or it may be defined by placing an imaginary boundary around a complete power plant. In each case, the treatment is to investigate all energy transfers at the system boundary and to account for all forms of energy stored within the system. Systems may, however, be divided into two basic types—*closed systems*, in which matter does not cross the boundaries and in which *non-flow processes* may occur, and *open systems* in which matter flows across the boundaries and in which *flow processes* occur.

Heat, work and stored energy

Energy cannot be simply defined. The usual definition, "the ability to do work," must not be applied literally where heat is involved, since one major consequence of the *second* law of thermodynamics is that no heat engine can convert all the heat energy supplied to it into work. Thus "the possession of energy" does not always mean "the ability to perform an equivalent amount of work."

The various forms in which energy appears may, however, be readily recognised and it is important to distinguish between *stored energy* and energy *in transit*. The common forms of stored energy are potential energy V, kinetic energy T (see Chapter 1) and internal energy U (see Chapter 15). Heat Q and work W are both names given to the transfer of energy, and neither can be stored. If, say, a weight is being lowered so that it compresses a spring, the weight loses potential energy. This energy is transferred to the spring and is stored as strain energy, and work is the means by which the energy is transferred—the weight does work on the spring. Similarly, if a hot piece of metal is placed in a tank of cold water, the metal will lose internal energy and the water will gain internal energy. The energy transferred exists as heat only while it is in the process of transfer from the metal to the water. Thus heat is defined as "energy in transition from one body or system to another because of a temperature difference" and such terms as "heat content" and "stored heat" must be avoided: what is stored is not heat but internal energy.

Fig. 26.2

First law of thermodynamics

A classical statement of the first law of thermodynamics is that "heat and work are mutually convertible." An alternative is to say that "heat and work are both forms of energy," and the first law is now generally considered to include the principle of conservation of energy (Chapter 1). Thus if heat is a form of energy, the first law of thermodynamics becomes identical with the principle of conservation of energy—that energy may undergo conversion but can neither be created nor destroyed. Applying this principle to a system, the first law may be stated as "energy entering a system = increase of energy stored with the system + energy leaving the system."

26.2 CLOSED SYSTEMS

Fig. 26.2 shows a closed system in which a quantity of fluid is contained in a cylinder with a movable piston. The system is the fluid itself and only three kinds of energy are involved. The system may receive heat Q, its boundary may move doing work W on the piston, and the internal energy of the fluid may change from an initial value U_1 to a final value value U_2. The first law thus gives the equation

$$Q = (U_2 - U_1) + W \qquad (26.1)$$

As written, it refers to a process in which heat is received and work done by the system: but it may be applied generally by adopting the convention that heat leaving the system, and work done on the system, are negative.

Example. *A quantity of steam in an engine cylinder expands, doing* 750 J *of work. During this process there is a heat transfer of* 150 J *from the steam to the cylinder walls. What is the change in the internal energy of the steam?*

Work is done by the system, hence

$$W = +750 \text{ J}$$

Heat is transferred from the system, hence

$$Q = -150 \text{ J}$$
$$\therefore \ -150 = (U_2 - U_1) + 750$$
$$(U_2 - U_1) = -900 \quad \text{or} \quad U_2 = U_1 - 900$$

that is, the internal energy decreases by 900 J.

Work in a non-flow process

A quantity of fluid enclosed in a cylinder by a piston of area A (Fig. 26.3 (*a*)) undergoes an expansion process represented by 1–2 on the pressure-volume diagram (Fig. 26.3 (*b*)). The work done during the small piston displacement δx, during which the pressure is p, will be

$$W = \text{Force} \times \text{Distance} = (pA) \times \delta x$$
$$= p \times (A\delta x) = p\delta V$$
$$= \text{Area of shaded strip (Fig. 26.3 (*b*))}$$
$$\therefore \ \text{Total work done, } W = \int_{V_1}^{V_2} p \, dV$$
$$= \text{Total area under } p - V \text{ diagram}$$

This result is subject to the following conditions:

1. The expansion must be fully resisted. It is quite possible for a fluid to expand without doing any work at all—as in Joule's experiment on the internal energy of gases.

2. The operation must be carried out very slowly, so that the system is in equilibrium at all times. There must be no pressure gradients within the fluid nor any effects due to fluid friction. Such a process is said to be *quasistatic*.

All the effects excluded by the above conditions have something in common: they are *irreversible*. Thus $\int p \, dV$ represents the work done in a *reversible non-flow process*. A reversible process being defined as one in which reversal would return both the system and its surroundings to

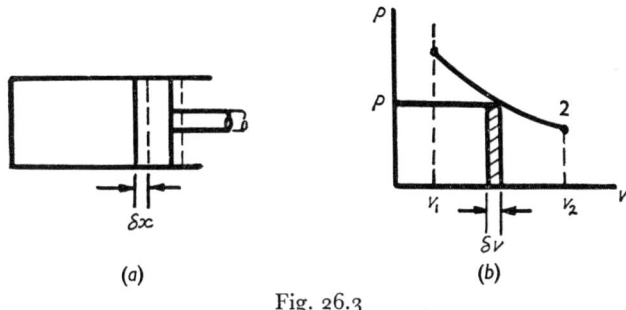

$$(a) \qquad\qquad (b)$$

Fig. 26.3

their original condition. Actual processes can never be perfectly reversible, but they are often assumed to be reversible. An irreversible process cannot in fact be represented by a line on a p–V diagram, since at any particular instant V will have a definite value but p may have different values in different parts of the system. If an irreversible process is to be shown on a p–V diagram (or any other property diagram), a dotted line is used to indicate the approximate path.

Non-flow processes for perfect gases

Reversible constant volume process. For reversible processes at constant volume, no work is done, so that

$$Q = (U_2 - U_1) + 0$$

The specific heat of a gas at constant volume c_v is defined as the heat transfer required to raise the temperature of 1 kg of the gas by 1 deg K, the volume being kept constant. Thus if the initial and final temperatures are T_1 and T_2, and the mass of gas is m,

$$Q = m c_v (T_2 - T_1)$$

Hence in this case,

$$(U_2 - U_1) = m c_v (T_2 - T_1)$$

Joule's law states that the internal energy of a perfect gas depends only on its temperature, so that the above expression must apply to any process in which the temperature changes from T_1 to T_2. Thus $mc_v(T_2 - T_1)$ is the change in internal energy of a gas during any process (including an irreversible process).

Reversible constant pressure process. For a reversible process at constant pressure,

$$W = \int_{V_1}^{V_2} p \, dV$$

$$W = p(V_2 - V_1) \qquad \text{(since } p \text{ is constant)} \qquad (26.2)$$

Hence $\qquad Q = mc_v(T_2 - T_1) + p(V_2 - V_1)$

Relationship between the specific heats of a gas. The specific heat of a gas at constant pressure c_p is defined as the heat transfer required to raise the temperature of 1 kg of the gas by 1 kelvin, the pressure being kept constant. Thus in this case

$$Q = mc_p(T_2 - T_1)$$

$$\therefore \ mc_p(T_2 - T_1) = mc_v(T_2 - T_1) + p(V_2 - V_1)$$

$$pV_2 = mRT_2 \quad \text{and} \quad pV_1 = mRT_1$$

Thus $\qquad mc_p(T_2 - T_1) = mc_v(T_2 - T_1) + mR(T_2 - T_1)$

$$\therefore \ c_p - c_v = R \qquad (26.3)$$

Reversible isothermal process. An isothermal process is one in which temperature is kept constant. Thus $T_2 = T_1$ and hence, for a gas, $U_2 = U_1$.

$$Q = (U_2 - U_1) + W$$

hence in this case

$$Q = 0 + W \quad \text{and} \quad W = \int_{V_1}^{V_2} p \, dV$$

From an isothermal process involving a gas, Boyle's law applies and

$$pV = C$$

$$\therefore \ W = C \int_{V_1}^{V_2} dV/V$$

$$= C(\log_e V_2 - \log_e V_1)$$

$$= pV \log_e (V_2/V_1) \qquad (26.4)$$

Thus $\qquad Q = pV \log_e (V_2/V_1)$

or $\qquad Q = mRT \log_e (V_2/V_1) \qquad (26.5)$

Reversible polytropic process. For practical non-flow processes, such as the expansions and compressions in engine cylinders, the relationship between pressure and volume often approximates to the law $pV^n =$ a constant, where the index n, usually referred to as "the index of expansion (or compression)" is a number between 1 and 2. Any process following the law $pV^n = C$ is termed *polytropic.*

$$Q = (U_2 - U_1) + W$$

For a gas,

$$(U_2 - U_1) = mc_v(T_2 - T_1)$$

For any fluid,

$$W = \int_{V_1}^{V_2} p \, dV$$

$$= C \int_{V_1}^{V_2} V^{-n} \, dV$$

$$= C[V^{1-n}/(1 - n)]_{V_1}^{V_2}$$

$$= C[(V_2^{1-n} - V_1^{1-n})/(1 - n)]$$

But

$$C = p_1 V_1^n = p_2 V_2^n$$

$$\therefore \ W = (p_2 V_2^n V_2^{1-n} - p_1 V_1^n V_1^{1-n})/(1 - n)$$

$$= (p_2 V_2 - p_1 V_1)/(1 - n)$$

Normally n will be greater than 1, and for an expansion $p_1 V_1$ will be greater than $p_2 V_2$, thus the expression is usually used in the form

$$W = (p_1 V_1 - p_2 V_2)/(n - 1) \tag{26.6}$$

Example. *A cylinder contains 0·1 m³ of air at 1·5 MN/m² (15 bar) and 200°C. The air is expanded (reversibly) according to the law $pV^{1·2} =$ constant to a final volume of 0·6 m³. Find (a) the mass of air, (b) the change in internal energy, (c) the work done, and (d) the heat transfer. For air, $c_p = 1·005$ kJ/kg K and $c_v = 0·718$ kJ/kg K.*

The arrangement is shown in Fig. 26.4 (a), the air being considered as the system. In Fig. 26.4 (b) the process is represented on a p–V diagram.

(a) $p_1 V_1 = mRT_1$,

$$\therefore \ m = p_1 V_1/RT_1$$

and, from Eq. (26.3),

$$R = c_p - c_v = 1·005 - 0·718 = 0·287 \text{ kJ/kg K}$$

$$\therefore \ m = 1·5 \times 10^6 \times 0·1/0·287 \times 10^3 \times 473 = 1·105 \text{ kg}$$

(*b*) Considering air to be a perfect gas,

$$(U_2 - U_1) = mc_v(T_2 - T_1)$$

In order to find T_2, a general relationship may be developed as follows:

$$p_1V_1^n = p_2V_2^n \tag{1}$$

But, for a gas,

$$p_1V_1/T_1 = p_2V_2/T_2 \tag{2}$$

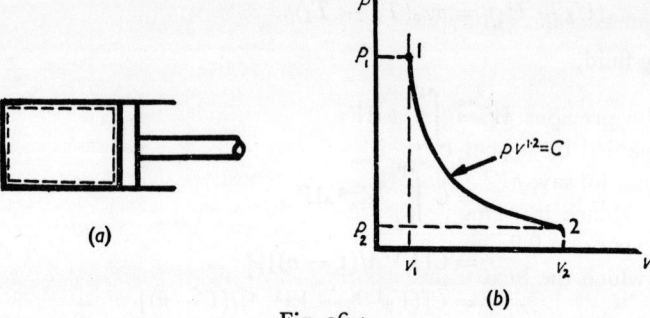

(*a*)

(*b*)

Fig. 26.4

Dividing (1) by (2),

$$V_1^{n-1} \times T_1 = V_2^{n-1} \times T_2$$

$$\therefore T_2 = T_1 \times (V_1/V_2)^{n-1} \tag{26.7}$$

Hence

$$T_2 = 473 \times (0\cdot1/0\cdot6)^{1\cdot2-1}$$

$$= 330\cdot6\,\text{K}$$

$$\therefore (U_2 - U_1) = 1\cdot105 \times 0\cdot718 \times (330\cdot6 - 473)$$

$$= -112\cdot9\,\text{kJ}$$

That is, internal energy decreases by $112\cdot9$ kJ.

(*c*) Work done = Area under $p - V$ diagram.
From Eq. (26.6),

$$W = (p_1V_1 - p_2V_2)/(n - 1)$$

$$p_1V_1^{1\cdot2} = p_2V_2^{1\cdot2}$$

hence

$$p_2 = p_1(V_1/V_2)^{1\cdot2} = 1\cdot5 \times 10^6 \times (0\cdot1/0\cdot6)^{1\cdot2}$$

$$= 1\cdot747 \times 10^5\,\text{N/m}^2$$

$$\therefore W = (1\cdot5 \times 10^6 \times 0\cdot1 - 1\cdot747 \times 10^5 \times 0\cdot6)/(1\cdot2 - 1)$$

$$= 2\cdot259 \times 10^5\,\text{J} \quad \text{or} \quad 225\cdot9\,\text{kJ}$$

(d) $\quad Q = (U_2 - U_1) + W$

$\qquad = -112\cdot9 + 225\cdot9$

$\qquad = +113 \text{ kJ}$

In other words, there is a heat transfer of 113 kJ to the air from the cylinder walls etc.

Reversible adiabatic process. A process in which no heat transfer takes place is termed *adiabatic*. The quantity $Q = 0$ and

$$-(U_2 - U_1) = W$$

In the previous example, an expansion according to $pV^{1\cdot2} = C$ was accompanied by a heat transfer *to* the system. For an expansion of air according to, say, $pV^{1\cdot6}$, however, a heat transfer *from* the system will be found. Hence it seems reasonable that somewhere between these will be an expansion following the same kind of law (i.e., a polytropic process) but in which the heat transfer is zero. Putting $pV^n = C$ in the equation above,

$$-mc_v(T_2 - T_1) = (p_1V_1 - p_2V_2)/(n - 1)$$
$$\therefore \ mc_v(T_1 - T_2) = mR(T_1 - T_2)/(n - 1)$$
$$\therefore \ c_v = R/(n - 1)$$
$$= (c_p - c_v)/(n - 1)$$
$$\therefore \ (n - 1) = (c_p - c_v)/c_v = (c_p/c_v) - 1$$
$$\therefore \ n = c_p/c_v$$

The ratio c_p/c_v is denoted by the symbol γ and, for a perfect gas, a reversible adiabatic process follows the law $pV^\gamma = C$.

Non-flow processes for steam and other vapours

The energy equation $Q = (U_2 - U_1) + W$ applies to any non-flow process. Provided the process is reversible, the work done

$$W = \int_{V_1}^{V_2} p \, dV$$

for vapours as well as gases. The change in internal energy of a vapour is not simply determined by the temperature change, and must be found from property tables.

Because steam and other vapours do not obey the gas laws, it also follows that a process obeying the law $pV = C$ will not be isothermal. (For a vapour, such a process is termed *hyperbolic* since $pV = C$ is the equation of a rectangular hyperbola.) Further, an adiabatic process will not follow the law $pV^\gamma = C$. For a steam, a reversible adiabatic expansion is found to follow, very approximately, the law $pV^n = C$ where $n = 1\cdot135$ if the steam is originally dry saturated; but this is only an approximation and a determination of the state of steam after an adiabatic expansion must be based on the methods of Chapter 27.

Example. *A closed vessel of volume* 2 m³ *contains steam at* 4 MN/m² (40 *bar*) *and* 0·9 *dry. The vessel is allowed to cool until the pressure falls to* 3 MN/m² (30 *bar*). *Find* (a) *the final dryness fraction*, (b) *the heat transfer.*

(a) Since V is constant, v is also constant, and neglecting liquid volumes,

$$x_1 v_{g1} = x_2 v_{g2}$$

∴ Final dryness fraction $x_2 = x_1 v_{g1}/v_{g2}$

$$= 0\cdot9 \times 4\cdot977 \times 10^{-2}/(6\cdot665 \times 10^{-2})$$

$$= 0\cdot672$$

(b) Here $W = 0$,

$$\therefore Q = (U_2 - U_1)$$

$$= m(u_2 - u_1)$$

Mass of steam $m = V/v_1 = 2/0\cdot9 \times 4\cdot977 \times 10^{-2}$

$$= 44\cdot65 \text{ kg}$$

$$\therefore (u_2 - u_1) = [x_2 u_{g2} + (1 - x_2)u_{f2}]$$
$$- [x_1 u_{g1} + (1 - x_1)u_{f1}]$$

$$= [0\cdot672 \times 2603 + 0\cdot328 \times 1004]$$
$$- [0\cdot9 \times 2602 - 0\cdot1 \times 1082]$$

$$= -371 \text{ kJ/kg}$$

∴ Heat transfer from vessel to surroundings

$$= 44\cdot65 \times 371 = 16570 \text{ kJ}$$

Example. *In a steam engine cylinder,* 0·2 kg *of steam at* 500 kN/m² (5 *bar*) *and* 0·8 *dry is expanded until the pressure falls to* 80 kN/m² (0·8 *bar*). *If the expansion is hyperbolic (i.e.,* pV = C), *find the heat transfer, stating its direction.*

Neglecting liquid volume,

$$v_1 = x_1 v_{g1} = 0\cdot8 \times 37\cdot48 \times 10^{-2} \text{ m}^3/\text{kg}$$

∴ Initial volume of steam,

$$V_1 = m v_1 = 0\cdot2 \times 0\cdot8 \times 37\cdot48 \times 10^{-2} = 0\cdot05997 \text{ m}^3$$

From Eq. (26.4),

$$W = pV \log_e (V_2/V_1)$$

and since
$$pV = C, \quad V_2/V_1 = p_1/p_2$$

$$\therefore \ W = 5 \times 10^5 \times 0.05997 \times \log_e (500/80)$$

$$= 5.494 \times 10^4 \text{ J} = 54.94 \text{ kJ}$$

Final volume of steam,

$$V_2 = V_1 \times p_1/p_2 = 0.05997 \times 500/80 = 0.3748 \text{ m}^3$$

$\therefore$ Final specific volume,

$$v_2 = 0.3748/0.2 = 1.874 \text{ m}^3/\text{kg}$$

For a pressure of 0.8 bar, $v_g = 2.087 \text{ m}^3/\text{kg}$. Hence the final condition of the steam is wet and, neglecting liquid volume,

$$\text{Final dryness fraction } x_2 = 1.874/2.087 = 0.8976$$

$$(u_2 - u_1) = [x_2 u_{g2} + (1 - x_2)u_{f2}] - [x_1 u_{g1} + (1 - x_1)u_{f1}]$$
$$= [0.8976 \times 2498 + 0.1024 \times 392]$$
$$- [0.8 \times 2562 + 0.2 \times 639]$$
$$= 106 \text{ kJ/kg}$$

Hence
$$(U_2 - U_1) = 0.2 \times 106 = 21.2 \text{ kJ}$$

$$\text{Heat transfer } Q = (U_2 - U_1) + W$$

$$= 21.2 + 54.94$$

$$= +76.14 \text{ kJ}$$

This means that there is a transfer of 76.14 kJ to the steam from its surroundings.

26.3 Open Systems

Where there is a continuous flow of fluid, as in a turbine or a nozzle it is convenient to consider as the system a region in space rather than a particular quantity of fluid. Such a region is termed an open system. For a simple open system, the boundary will be fixed so that the volume of fluid enclosed by it will be constant and there will be no work done due to movement of the system boundary.

Fluid entering or leaving the system will flow across the boundary and for devices such as turbines, the shaft will also cross the boundary, so that work may also enter or leave the system. To apply the first law of thermodynamics, it is necessary to list all the kinds of energy involved.

(i) *Energy stored within the system.* Within the system may be found not only the internal energy of the fluid but also the kinetic and potential energies of this fluid and of any mechanical moving parts. In a steady flow process, this energy remains constant.

(ii) *Simple energy transfers.* Energy may leave or enter the system without being associated with flow. There will be the shaft work W and the heat transfer Q. As for the closed systems, the convention is that heat transfer to the system, and work transfer from the system, are positive.

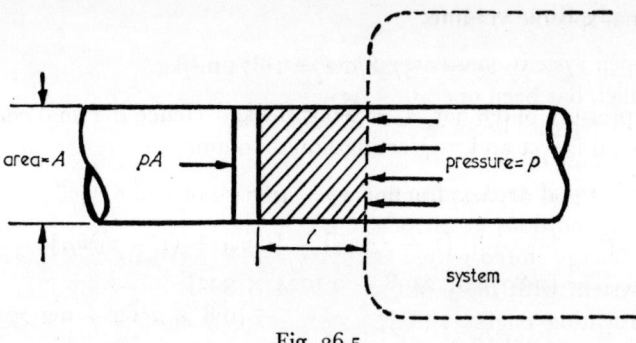

Fig. 26.5

(iii) *Energy transfers due to flow.* Fluid entering the system will possess internal energy U, potential energy V and kinetic energy T which will be transferred to the system as the fluid enters (similar energy transfers from the system taking place as fluid leaves). There is also another kind of energy transfer; the fluid within the system will be under pressure, and if a quantity of fluid is to enter, work must be done to overcome it. A practical example is the work done by a feed pump in forcing water into a boiler when the flow of fluid into the system is accompanied by a transfer of energy. This is not usually obvious, but in all cases work is done on the system by the fluid as it enters, and by the system on the fluid as it leaves. This kind of energy transfer is known as *flow work* or *flow energy* W_f.

This may be illustrated by considering a pipe of uniform cross-sectional area A along which fluid at pressure p is flowing into a system (Fig. 26.5). A small quantity of fluid, occupying a length l of the pipe, is about to cross the system boundary and an imaginary piston has been placed behind this fluid. Then

$$\text{Force exerted on piston} = pA$$

and $$\text{Work done by piston} = pAl = pV$$

where $V = $ Volume of fluid entering system.

This energy is transferred to the system as the fluid enters, hence

$$W_f = pV \qquad (26.8)$$

It should be noted that the "imaginary piston" is not itself a source of energy. The work is in fact done by the fluid following it. The piston is introduced only in order to make it apparent that work is done, and as an aid to calculation.

The steady flow system

If the open system considered is, say, a steam turbine forming part of a plant which has been operating for some time with a fixed power output, two characteristics become apparent:

1. The mass of fluid within the system remains constant, or the mass entering is equal to the mass leaving.

2. The conditions at all points within the system remain constant, so that the energy stored within the system does not change.

Any system with these characteristics is termed a steady flow system. A reciprocating engine running steadily may also be considered as a

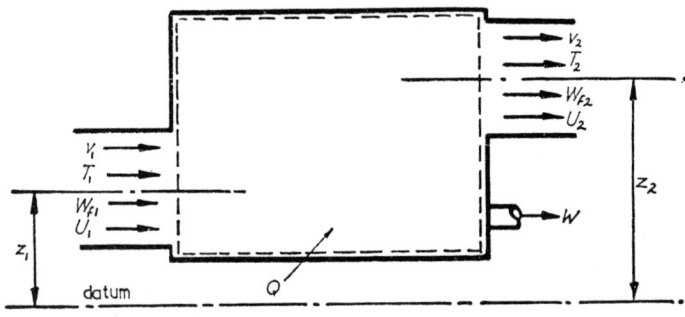

Fig. 26.6

steady flow system provided that a complete number of cycles occurs within the period of time considered.

The general steady flow system is shown diagrammatically in Fig. 26.6. Since stored energy is constant, the first law implies that

Energy entering the system = Energy leaving the system

$$\therefore\ V_1 + T_1 + U_1 + W_{f1} + Q = V_2 + T_2 + U_2 + W_{f2} + W \qquad (26.9)$$

From Eq. (1.10), potential energy $V = gmZ$ and from Eq. (1.11), kinetic energy $T = \frac{1}{2}mv^2$. From Eq. (26.8), it also follows that flow work

$W_f = pV$. Substituting these values,

$$gmZ_1 + \tfrac{1}{2}mv_1{}^2 + U_1 + p_1V_1 + Q = gmZ_2 + \tfrac{1}{2}mv_2{}^2 + U_2 + p_2v_2 + W$$
(26.10)

The flow equations may be simplified by substituting the enthalpy H_1 for $(U_1 + p_1V_1)$ and H_2 for $(U_2 + p_2V_2)$. It should be noted that although in flow processes, enthalpy represents the sum of two kinds of energy transfer, in non-flow processes it has no such meaning. Enthalpy itself is merely a convenient combination of properties and in a non-flow process the product pV does not represent energy. In these terms, Eq. (26.10) becomes

$$gmZ_1 + \tfrac{1}{2}mv_1{}^2 + H_1 + Q = gmZ_2 + \tfrac{1}{2}mv_2{}^2 + H_2 + W \quad (26.11)$$

or, for unit mass flow,

$$gZ_1 + \tfrac{1}{2}v_1{}^2 + h_1 + Q = gZ_2 + \tfrac{1}{2}v_2{}^2 + h_2 + W \quad (26.12)$$

Steady flow processes for liquids

An ideal liquid is assumed to be incompressible, so that reversible processes in which internal energy is converted to mechanical energy and vice versa (that is, reversible expansions and compressions) cannot take place. Fluid friction results in an irreversible conversion of mechanical energy to internal energy, but an ideal liquid is usually assumed to be frictionless. Thus for an incompressible frictionless liquid, energy conversions are of two completely separate kinds.

1. Those between heat transfer and internal energy.
2. Those between the various mechanical forms of energy.

For a steady flow process in which no external work is done, and neglecting terms relating to heat transfer and internal energy, Eq. (26.10) becomes

$$mgZ_1 + \tfrac{1}{2}mv_1{}^2 + p_1V_1 = mgZ_2 + \tfrac{1}{2}mV_2{}^2 + p_2V_2$$

or, since $V = m/\rho$ where ρ is the density of the liquid,

$$mgZ_1 + \tfrac{1}{2}mv_1{}^2 + mp_1/\rho = mgZ_2 + \tfrac{1}{2}mv_2{}^2 + mp_2/\rho$$

Hence for unit mass flow

$$gZ_1 + \tfrac{1}{2}v_1{}^2 + p_1/\rho = gZ_2 + \tfrac{1}{2}v_2{}^2 + p_2/\rho \quad (26.13)$$

This is Bernoulli's equation for the flow of an incompressible fluid (see Eq. (22.17)).

Steady flow processes for perfect gases

For a gas, the change in enthalpy during a process is given by

$$H_2 - H_1 = (U_2 + p_2 V_2) - (U_1 + p_1 V_1)$$
$$= (U_2 - U_1) + (p_2 V_2 - p_1 V_1)$$

For a gas,

$$(U_2 - U_1) = mc_v(T_2 - T_1)$$

also
$$p_2 V_2 = mRT_2 \quad \text{and} \quad p_1 V_1 = mRT_1$$
$$\therefore H_2 - H_1 = mc_v(T_2 - T_1) + mR(T_2 - T_1)$$
$$= m(c_v + R)(T_2 - T_1)$$
$$= mc_p(T_2 - T_1) \tag{26.14}$$

Reversible adiabatic flow processes. For a reversible adiabatic non-flow process, it has been shown that a perfect gas follows the law $pV^\gamma = C$ and this also applies to flow processes. To justify this statement, it is necessary to establish a connection between non-flow and flow processes, which may be done by considering for example, a duct through which air is being expanded adiabatically. Imagine a soap bubble to have been introduced into the air stream. The air within the bubble may now be regarded as a closed system which, as the bubble passes through the duct, undergoes reversible adiabatic expansion. Work is done by the expanding bubble on the surrounding air, not on a piston, but the process is nevertheless substantially the same as the non-flow adiabatic expansion previously dealt with. Considering air to be a perfect gas, it will follow the law $pV^\gamma = C$. For the flow process as a whole, the soap bubble merely separates a small quantity of air from the surrounding air stream, so that if the law $pV^\gamma = C$ applies to this particular quantity of air, it must also apply to the rest of the air stream.

Example. *Air enters a horizontal nozzle at a pressure of* 160 kN/m² (1·6 *bar*) *and* 150°C *with velocity* 150 m/s. *The air is expanded adiabatically and without friction to a final pressure of* 100 kN/m² (1 *bar*). *Find* (a) *the temperature, and* (b) *the velocity of the air leaving the nozzle. For air,* $c_p = 1·005$ kJ/kg K *and* $\gamma = 1·40$.

(a) A frictionless expansion is reversible, hence the air follows the law $pV^\gamma = C$.

For any polytropic expansion,

$$p_1 V_1^n = p_2 V_2^n \tag{1}$$

For a perfect gas,

$$p_1 V_1 / T_1 = p_2 V_2 / T_2 \qquad (2)$$

Raising (2) to power n,

$$p_1{}^n V_1{}^n / T_1{}^n = p_2{}^n V_2{}^n / T_2{}^n$$

Dividing by (1)

$$p_1{}^{n-1} / T_1{}^n = p_2{}^{n-1} / T_2{}^n$$

$$\therefore \ (T_2 / T_1)^n = (p_2 / p_1)^{n-1}$$

or

$$T_2 / T_1 = (p_2 / p_1)^{(n-1)/n} \qquad (26.15)$$

In this case $n = \gamma$, hence

$$T_2 = T_1 \times (p_2 / p_1)^{(\gamma - 1)/\gamma}$$

$$= 423(1/1 \cdot 6)^{0 \cdot 4 / 1 \cdot 4}$$

$$= 369 \cdot 8 \ \text{K} \quad \text{or} \quad 96 \cdot 8°\text{C}$$

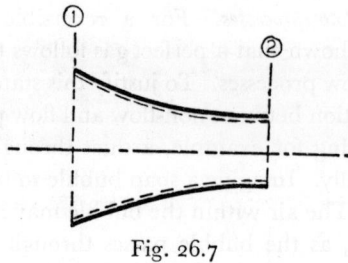

Fig. 26.7

(b) Considering the nozzle as a steady flow system (Fig. 26.7) and applying the energy equation (26.12) for unit mass flow:

$$gZ_1 + \tfrac{1}{2}v_1{}^2 + h_1 + Q = gZ_2 + \tfrac{1}{2}v_2{}^2 + h_2 + W$$

Here

$$Z_1 = Z_2 \quad \text{and} \quad Q = W = 0$$

Hence

$$\tfrac{1}{2}(v_2{}^2 - v_1{}^2) = h_1 - h_2$$

and, since air may be considered a perfect gas,

$$h_1 - h_2 = c_p(T_1 - T_2)$$

Hence

$$\tfrac{1}{2}(v_2{}^2 - 150^2) = 1 \cdot 005 \times 10^3 (423 - 369 \cdot 8)$$

$$v_2{}^2 = 106\,900 + 22\,500 = 129\,400$$

$$v_2 = 359 \cdot 8 \ \text{m/s}$$

Throttling. If a fluid flows from a higher to a lower pressure through a restriction such as a small orifice or a partially closed valve, it is said to be throttled. Heat transfer will be negligible, so that the process is adiabatic. But the process is also irreversible, and must not be confused with expansion in a correctly shaped nozzle (as in the previous example), which would produce a high-velocity jet of fluid. Small jets of fluid may in fact be formed at a throttling orifice, but their kinetic energy is quickly dissipated in turbulence and, apart from a small region close to the orifice, kinetic energy may usually be considered negligible on both sides. Potential energy change is also negligible and no external work is done, so that Eq. (26.12) gives

$$h_1 = h_2 \qquad (26.16)$$

In other words, specific enthalpy before throttling = specific enthalpy after throttling. For a perfect gas, Eq. (26.14) shows that temperatures before and after throttling must be equal. Actual gases do not obey Eq. (26.14) exactly. Their enthalpies depend on pressure as well as on temperature and experiments show that temperature after throttling may be slightly higher or lower than temperature before throttling. (This is known as the Joule-Thomson effect.)

Example. *A centrifugal air compressor takes in 2 kg/s of air at atmospheric pressure* (101·3 kN/m² *or* 1·013 bar) *and* 15°C. *The delivery pressure is* 190 kN/m² (1·9 bar). *Heat transfer through the casing is negligible so that compression is adiabatic but, due to friction follows the law* $pV^{1·6} = C$. *Find* (a) *the temperature of air leaving the compressor,* (b) *the power required to drive the compressor. Assume that differences in level and differences in kinetic energy between intake and delivery are negligible. For air,* $c_p = 1·005$ kJ/kg K.

(a) It should be noted that since this is an irreversible process, the relationship $pV^{1·6} = C$ applies only to the initial and final conditions. From Eq. (26.15),

$$T_2 = T_1 \times (p_2/p_1)^{(n-1)/n}$$
$$= 288 \times (1·9/1·013)^{0·6/1·6}$$
$$= 364·7 \text{ K or } 91·7°C$$

(b) In this case $Q = 0$, and changes in potential and kinetic energies are negligible, so that Eq. (26.11) gives

$$-W = H_2 - H_1$$

(The negative sign indicating that work is done on the system.) Hence work input per second is

$$H_2 - H_1 = mc_p(T_2 - T_1)$$
$$= 2 \times 1·005 \times 10^3(364·7 - 288)$$
$$= 154·2 \times 10^3 \text{ N}$$

Hence power input is $154·2 \times 10^3$ W or 154·2 kW.

Steady flow processes for steam and other vapours

The steady flow energy equation applies to all fluids, but (1) enthalpy changes are not related to temperature changes but must be found from property tables, and (2) relationships developed for gases do not apply—for example, a reversible adiabatic expansion does not follows the law $pV^\gamma = C$.

Throttling of a vapour. From Eq. (26.16), $h_1 = h_2$ and this may be applied to the use of a throttling process in the determination of the dryness fraction of steam. Property tables show that (for pressures below approximately 30 bar) h_g decreases with decreasing pressure which implies that if wet steam is throttled, it may become dry or superheated. If the steam is superheated, its enthalpy may be found by measuring its temperature and pressure and the original dryness fraction calculated as in the following example.

Example. *The steam supplied by a boiler is known to be wet, and a thermometer in the main steam pipe shows its temperature to be 195°C. Steam collected by a sampling pipe is throttled to atmospheric pressure (101·3 kN/m² or 1·013 bar) and the temperature after throttling is found to be 105°C. Find the dryness fraction of the steam supply.*

From Eq. (26.16), $h_1 = h_2$. At atmospheric pressure $t_s = 100°C$, so that the steam is superheated after throttling. By interpolation,

$$h_2 = 2686 \text{ kJ/kg}$$

Hence $$2686 = h_1 = h_f + xh_{fg}$$

The temperature of wet steam is the saturation temperature, and the pressure before throttling must be that corresponding to $t_s = 195°C$. From tables, this is 14 bar.

$$\therefore\ 2656 = 830 + 1960x$$

$$\therefore\ x = 1856/1960 = 0·947$$

The following example shows how the same method may be applied to other processes.

Example. *In a steam power plant, the boiler supplies 18000 kg/h of steam to the turbine via a pipe of 20 cm internal diameter, the pressure and temperature of the steam leaving the boiler stop valve being 4 MN/m² (40 bar) and 400°C. The turbine exhausts to a condenser at a pressure of 5 kN/m² (0·05 bar) and at this point the steam has a dryness fraction of 0·84. The cross-sectional area at the* *condenser inlet flange is 2·5 m² and the boiler stop valve is 25 m above the level of this flange. Calculate the power output of the turbine, assuming an hourly heat loss from steam pipe and turbine casing of 50 MJ. What is the effect of ignoring potential and kinetic energies in this calculation?*

The turbine and its supply pipe may be considered as a steady flow system, boundaries being placed (1) at a point just after the boiler stop valve, and (2) at the condenser flange. The steady flow energy equation may now be applied, and from Eq. (26.12) which refers to a mass transfer of 1 kg,

$$gZ_1 + \tfrac{1}{2}v_1{}^2 + h_1 + Q = gZ_2 + \tfrac{1}{2}v_2{}^2 + h_2 + W$$

Considering the condenser flange as datum level,

$$Z_1 = 25 \text{ m} \quad \text{and} \quad Z_2 = 0$$

Velocities v_1 and v_2 may be found from the relationship

Mass flow × Specific volume = Velocity × Cross-sectional area

Here   Mass flow = 18 000/3600 = 5 kg/s

From property tables, specific volume at (1) is 0·0733 m³/kg

$$\therefore \text{ Initial velocity } v_1 = 5 \times 0 \cdot 0733 / \tfrac{1}{4}\pi (0 \cdot 2)^2 = 11 \cdot 66 \text{ m/s}$$

Neglecting liquid volume, specific volume at (2) is xv_g, that is (0·84 × 28·20) m³/kg,

$$\therefore \text{ Final velocity } v_2 = 5 \times 0 \cdot 84 \times 28 \cdot 20 / 2 \cdot 5 = 47 \cdot 38 \text{ m/s}$$

From property tables, specific enthalpy $h_1 = 3214$ kJ/kg,

$$\text{Specific enthalpy } h_2 = h_f + xh_{fg}$$
$$= 138 + 0 \cdot 84 \times 2423 = 2173 \text{ kJ/kg}$$

Heat transfer $Q = 50$ MJ/h, so that heat transfer for unit mass flow of steam is $50 \times 10^6 / 18000 = 2778$ J. (Since the heat transfer is from system to surroundings, this will appear as a negative quantity.) Substituting in Eq. (26.12),

$$9 \cdot 81 \times 25 + \tfrac{1}{2}(11 \cdot 66)^2 + 3214 \times 10^3 - 2778$$
$$= 0 + \tfrac{1}{2}(47 \cdot 38)^2 + 2173 \times 10^3 + W$$

$$245 + 68 + 3214000 - 2778 = 0 + 1123 + 2173000 + W$$
$$\therefore W = 1037412$$

That means that a mass flow of 1 kg through the system results in a work output of 1 037 412 J or 1·037 412 MJ. (Since this result depends on values of specific enthalpy and, in this case, property tables give these to the nearest 1 kJ, the implied accuracy of 1 J is unjustified and a more reasonable statement would be $W = 1 \cdot 037$ MJ.)

$$\text{Mass flow of steam} = 5 \text{ kg/s}$$

so that

$$\text{Power output} = 5 \times 1 \cdot 037 = 5 \cdot 185 \text{ MJ/s}$$
$$= 5 \cdot 185 \text{ MW}$$

If potential and kinetic energies are ignored, Eq. (26.12) becomes:

$$h_1 + Q = h_2 + W$$

Substituting values,

$$3214 \times 10^3 - 2778 = 2173 \times 10^3 + W$$
$$W = 1\,038\,222 \text{ J}$$

Hence in this case,

$$\text{Power output} = 5 \times 1 \cdot 038 = 5 \cdot 19 \text{ MW}$$

If potential and kinetic energies are ignored, an error of approximately 0·1 per cent is introduced. Since actual measurements of pressures, etc., will be much less accurate, this error is negligible.

Example. *In a domestic refrigerator the refrigerant, Freon-12, passes through the compressor at the rate of 0·2 kg per minute. The refrigerant enters the compressor as a wet vapour of dryness fraction 0·95 at a temperature of −20°C and leaves at a pressure of 847·7 kN/m² (8·477 bar). The power input to the compressor is 130 W. What will be the state of the refrigerant after compression, assuming this to be adiabatic?*

Considering the compressor as a steady flow system, and neglecting potential and kinetic energies, Eq. (26.12) gives

$$h_1 + Q = h_2 + W$$

Here $Q = 0$ and for unit mass flow,

$$W = 130 \times 60/0 \cdot 2 = 39 \times 10^3 \text{ J}$$

(Since work is done on the system, this will appear as a negative quantity in Eq. (26.12).)

$$\therefore \ h_1 + 0 = h_2 - 39 \times 10^3$$

Expressing all quantities in kJ/kg,

$$h_2 = h_1 + 39$$

Since vapour is wet at (1),

$$h_1 = xh_g + (1 - x)h_f$$

Substituting values from property tables (for saturation temperature $t = -20°\text{C}$)

$$h_1 = 0.95 \times 178.73 + 0.05 \times 17.82$$
$$= 170.68 \text{ kJ/kg}$$

Hence $\qquad h_2 = h_1 + 39 = 209.68 \text{ kJ/kg}$

For $\qquad p_s = 8.477 \text{ bar}, \quad h_g = 201.45 \text{ kJ/kg}$

Hence after compression the refrigerant is superheated and, by interpolation, the degree of superheat is $11.0°\text{C}$: that is, the temperature is $(35 + 11) = 46°\text{C}$.

Application of the first law to a non-steady flow process

Examples of non-steady flow processes are those involving the components of, say, a steam turbine plant during periods of "warming-up" or of changing output and processes such as the filling of a compressed air receiver in which the flow is in one direction only. Such processes differ from the steady flow process in two essentials:

1. The mass of fluid within the system does not remain constant.

2. The energy content of the system varies, and this variation must be calculated. (If the system is, say, a steam turbine, this may be extremely difficult since it will require knowledge of the state of the steam, and the temperatures of the blading, etc., in all parts of the turbine.)

Applying the first law in its basic form,

(Energy entering system) = (Increase of energy stored)
$$\qquad\qquad\qquad + \text{(Energy leaving system)}$$

Hence if ΔE represents the increase in the energy stored by the system,

$$V_1 + T_1 + H_1 + Q = \Delta E + V_2 + T_2 + H_2 + W \qquad (26.17)$$

or $\quad gm_1Z_1 + \tfrac{1}{2}m_1v_1{}^2 + m_1h_1 + Q$
$$= \Delta E + gm_2Z_2 + \tfrac{1}{2}m_2v_2{}^2 + m_2h_2 + W \qquad (26.18)$$

Eq. (26.18) will be applicable only if v_1, h_1, v_2 and h_2 remain constant during the process. In the general non-steady flow process, these will vary and the energy equation must be applied to an infinitely short period of time:

$$dm_1(gZ_1 + \tfrac{1}{2}v_1{}^2 + h_1) + dQ = dE + dm_2(gZ_2 + \tfrac{1}{2}v_2{}^2 + h_2) + dW$$
$$(26.19)$$

Example. *An air tank has a volume of 2 m³ and contains air at atmospheric pressure* (101·3 kN/m² *or* 1·013 bar) *and* 10°C. *It is connected by a valve to a compressed air supply pipe in which the pressure and temperature are maintained at* 1·5 MN/m² (15 bar) *and* 25°C. *The valve is opened for a short period of time, allowing* 10 kg *of air to flow into the tank. What will then be the pressure and temperature of the air in the tank, assuming negligible heat transfer? For air* $R = 0·287$ kJ/kg K *and* $c_v = 0·718$ kJ/kg K.

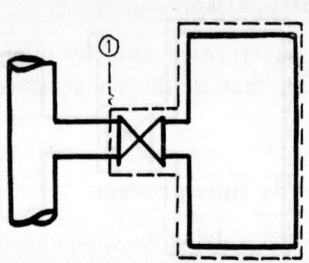

Fig. 26.8

The system boundary may be placed as shown in Fig. 26.8 so that h_1 remains constant and Eq. (26.18) may be used. Neglecting kinetic and potential energies, and noting that $Q = W = 0$,

$$m_1 h_1 = \Delta E$$

Since the only form of stored energy is the inernal energy of the fluid,

$$\Delta E = U' - U$$

where U and U' are the initial and final internal energies.

In order to calculate the values of h and U, an arbitrary zero for internal energy must be assumed. For a gas, internal energy depends only on temperature, and a convenient datum is 10°C (283 K).

From this datum,

$$U = m c_v (T - 283)$$

$$h_1 = u_1 + p_1 v_1$$

$$u_1 = 0·718(298 - 283) = 10·77 \text{ kJ/kg}$$

$$p_1 v_1 = RT_1 = 0·287 \times 298 = 85·53 \text{ kJ/kg}$$

$$\therefore \ h_1 = 10·77 + 85·53 = 96·30 \text{ kJ/kg}$$

$$\therefore \ m_1 h_1 = 10 \times 96·30 = 963·0 \text{ kJ}$$

Original mass of air in tank,

$$m = pV/RT = 101·3 \times 10^3 \times 2/0·287 \times 10^3 \times 283 = 2·495 \text{ kg}$$

Original internal energy,

$$U = 2·495 \times 0·718 \times (0) = 0$$

Final mass of air in tank,

$$m = 2·495 + 10 = 12·495 \text{ kg}$$

Final internal energy,

$$U' = 12 \cdot 495 \times 0 \cdot 718(T' - 283) \text{ kJ}$$

where T' is the final temperature.

$$m_1 h_1 = U' - U$$
$$963 \cdot 0 = 12 \cdot 495 \times 0 \cdot 718(T' - 283)$$
$$(T' - 283) = 107 \cdot 3$$

$\therefore$ Final temperature $T' = 390 \cdot 3 \text{ K}$ or $117 \cdot 3 °\text{C}$

$$p'V = m'RT'$$

$\therefore$ Final pressure $p' = m'RT'/V$

$$= 12 \cdot 495 \times 0 \cdot 287 \times 10^3 \times 390 \cdot 3/2$$
$$= 7 \cdot 00 \times 10^5 \text{ N/m}^2 = 700 \text{ kN/m}^2 \text{ or } 7 \text{ bar}$$

PROBLEMS

For tutorials

1. Investigate the throttling of dry saturated steam from pressures of (a) 10 bar, (b) 50 bar, (c) 180 bar, (d) 221·2 bar (i.e. the critical point) to progressively lower final pressures. What conclusions may be drawn regarding the determination of dryness fraction by throttling?

2. Is it possible to regard the extrusion of a metal as a steady flow process, and what, if any, is the difference between this process and throttling of a fluid? What assumptions are necessary if Eq. 26.16 is to be applied?

In an experiment on the extrusion of lead, a compressive stress of 120 MN/m² is found to be necessary. If the initial temperature of the lead is 20°C, estimate the temperature at which it will emerge from the die, assuming no heat transfer. Assume that the internal energy of lead is a function of temperature only and that $c_v = 0 \cdot 129 \text{ kJ/kg K}$. The density of lead is 11 340 kg/m³.

(102°C)

3. It is a matter of experience that if a steam boiler is operating steadily with a fixed output, failure of its feed pump will cause the steam pressure to rise: that is, removal of an energy input tends to *increase* output. Explain this with reference to steady and non-steady flow processes.

4. The question of whether compressed air contains energy has been the source of much argument. Investigate the energy transfers which take place when (a) air is compressed isothermally, (b) compressed air is expanded adiabatically. In (b), what is the source of the energy transferred in the form of work?

5. If compressed air (at normal room temperature) is throttled to normal atmospheric pressure, what happens to (a) its temperature, (b) its energy, (c) its "ability to do work"?

(*Note:* For a full explanation of the questions raised by (4) and (5) the second law of thermodynamics must be applied: see Chapter 27.)

General

1. A fluid, initially at a pressure of 2 MN/m^2 (20 bar) undergoes a reversible non-flow process in which its volume increases from $0 \cdot 5$ m^3 to 2 m^3. Find the work transfer if the process is (a) at constant pressure, (b) according to the law $pV = C$ and (c) according to the law $pV^{1 \cdot 3} = C$.

Ans. (a) 3 MJ; (b) $1 \cdot 386$ MJ; (c) $1 \cdot 134$ MJ

2. In a reversible process, $0 \cdot 2$ kg of CO_2 is compressed according to the law $pV^{1 \cdot 2} = C$ from initial pressure and temperature 100 kN/m^2 (1 bar) and $20°C$ to a final pressure of 1 MN/m^2 (10 bar). Considering CO_2 to be a perfect gas for which $c_v = 0 \cdot 65$ kJ/kg K and $R = 0 \cdot 189$ kJ/kg K, find (a) the final temperature, (b) the work done, (c) the change of internal energy and (d) the heat transfer, stating its direction.

Ans. (a) $157 \cdot 1°C$; (b) $25 \cdot 89$ kJ; (c) $17 \cdot 81$ kJ (increase);
(d) $8 \cdot 08$ kJ (from gas to surroundings)

3. A cylinder fitted with a frictionless piston contains $0 \cdot 1$ kg of a gas. The gas is expanded reversibly and adiabatically to twice its original volume and during this process the temperature of the gas falls from $208°C$ to $115°C$. The work done by the piston is 15 kJ. Calculate the two specific heats of the gas.

Ans. $c_v = 1 \cdot 613$ kJ/kg K, $C_p = 2 \cdot 113$ kJ/kg K

4. A vessel of volume $0 \cdot 1$ m^3 is filled with ammonia vapour and sealed. Its temperature is lowered to $-12°C$ and at this temperature the composition by mass is 48 per cent liquid, 52 per cent vapour. The vessel is now allowed to warm up slowly. At what temperature will the ammonia be dry saturated, and what heat transfer will have taken place? (Neglect liquid volume.)

Ans. $6°C$; 248 kJ

5. A "hot air blower" consists of a fan, electrical heating elements and a nozzle. The power input to the fan is 50 W and the heating elements take a current of 6 A at 240 V. Air enters the fan at a pressure of 100 kN/m^2 (1 bar) and temperature $20°C$ at the rate of 1 m^3 every 2 minutes: after passing over the heating elements, it emerges from the nozzle with a velocity of 25 m/s. What will then be its temperature? For air, $c_p = 1 \cdot 005$ kJ/kg K and $R = 0 \cdot 287$ kJ/kg K.

Ans. $169 \cdot 3°C$

6. A centrifugal compressor delivers air at a pressure of 250 kN/m^2 ($2 \cdot 5$ bar) and a temperature of $150°C$ into a horizontal diffuser (that is, a smoothly tapering duct of increasing cross-section) in which its velocity is reduced from 250 to 30 m/s. If the air flows through the diffuser adiabatically and without friction, what will be its final pressure?
(For air $c_p = 1 \cdot 005$ kJ/kg K and $\gamma = 1 \cdot 4$.)

Ans. 321 kN/m^2 ($3 \cdot 21$ bar)

7. Steam enters a nozzle with negligible velocity at a pressure of $1 \cdot 5$ MN/m^2 (15 bar) and $0 \cdot 95$ dry. At exit from the nozzle the steam is $0 \cdot 94$ dry and its pressure is 1 MN/m^2 (10 bar). Find the exit velocity, assuming negligible heat transfer.

Ans. 274 m/s

CHAPTER 27

The Second Law of Thermodynamics

27.1 REVERSIBILITY

Of all the physical laws, the second law of thermodynamics has, possibly, the widest field of application. It is also the least understood. One difficulty is that the law may be stated in many different ways, and it is not always obvious that these are connected. Thus the statements "heat will not flow up a temperature gradient" and "disorder tends to increase" seem, at first sight, to be statements of two completely different laws.

Both the first and second laws deal with the conversion of energy. The first law states that energy may be converted from one form to another, but that in all such conversions the total amount of energy remains constant. The second law points out that the converse of this statement is *not* true: it does *not* follow that, provided the total amount of energy remains constant, it may be freely converted into any desired form. To paraphrase George Orwell, "all forms of energy are equal, but some are more equal than others."

Two different kinds of energy conversion are compared in Fig. 27.1 (*a*) and (*b*). Fig. 27.1 (*a*) shows a simple lever and fulcrum arrangement acted on by forces of 100 kN and 1 MN. Fig. 27.1 (*b*) shows two tanks containing, respectively, 0·263 kg water at 100°C and 2·63 kg water at 0°C. If, in Fig. 27.1 (*a*), a slight impulse is given to the lever, the 100 kN force will fall and the 1 MN force rise. It is evident that when the lever is level, an energy transfer of 100 kJ will have occurred. If, in Fig. 27.1 (*b*), the tanks are brought into thermal contact, heat transfer will take place until both are at the same temperature. A simple calculation shows that the final temperature is 9·09°C and the heat transferred 100 kJ.

So far, there is a strong similarity between the two cases: but this ceases when an attempt is made to reverse the process. In Fig. 27.1 (*a*), all that needs to be done is to apply a small impulse to the lever, but in

Fig. 27.1 (*b*) nothing can be done at all. The mechanical conversion is
reversible and the thermal one *irreversible*. The second law is much concerned with *reversibility*: the fact that the energy conversion shown in
Fig. 27.1 (*b*) is irreversible will later be recognised as one of the ways in
which the law may be stated.

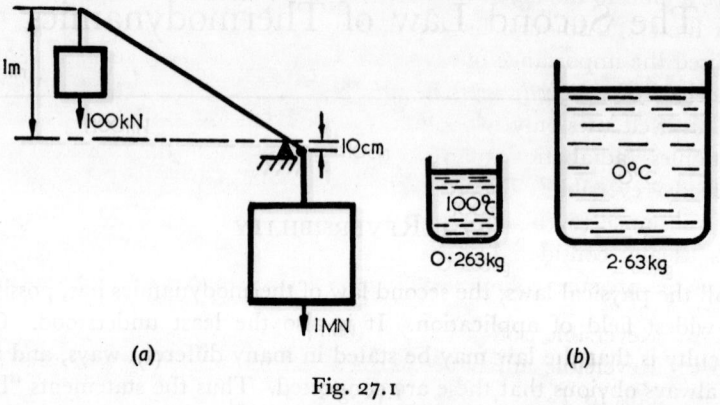

(*a*)   (*b*)

Fig. 27.1

Reversibility of a non-flow process. It has already been stated (in Chapter 26)
that reversibility is a necessary condition if the work done in a nonflow
process is to be equal to $\int p\,dV$. Irreversible factors are, in general,
friction (both mechanical and molecular) and heat transfer across a
finite temperature difference, and the effect of both is to *decrease* the amount
of energy converted into work. It is often convenient to separate these
factors into those concerned with the fluid itself (such as turbulence)
which determine *internal reversibility* and those concerned with the fluid's
surroundings (such as mechanical friction or heat flow through cylinder
walls) which determine *external reversibility*.

Reversibility of a heat engine. A "heat engine" is a device for the conversion
of heat into work and this is normally done by carrying out a series of
processes (the "cycle") on a "working fluid." Various kinds of cycle are
dealt with in Chapter 28, but all have three basic features. During part
of the cycle heat is taken in, during another part heat is given out or
"rejected" (at a lower temperature), and over the whole cycle there is a
net work output. Theoretically, any of these cycles may be operated
'in reverse' so that heat is taken in at low temperature and given out at
a higher temperature, the cycle receiving a net work input. Practical
engines working in this way are known (according to the purpose for
which they are used) as "heat pumps" or "refrigerators."

The Carnot cycle

The foundations of heat engine design—and possibly the foundations of thermodynamics as well—were laid in 1824 by N. L. S. Carnot. In an essay *Reflections on the motive power of heat* he considered the question "Is there a limit to the efficiency of a heat engine and, if so, what determines that limit?" Although he was not aware of the true nature of heat, he realised the importance of reversibility and, in order to test his theories, conceived of a *perfectly reversible cycle*.

This cycle uses only two kinds of process, isothermal heat transfer and frictionless adiabatic expansion (or compression), both of which are perfectly reversible. Theoretically, this cycle could be carried out using any substance (even, for example, a bar of metal), but for convenience it is usually considered to be carried out on a perfect gas. Fig. 27.2 shows the pressure-volume diagram and the processes are as follows:

a–b: Reversible isothermal heating at temperature T_1.

b–c: Reversible adiabatic expansion, during which the temperature falls to T_2.

c–d: Reversible isothermal cooling at temperature T_2.

d–a: Reversible adiabatic compression, during which the temperature rises to T_1.

As with all cycles, the fluid must be returned to its original state at the end of one complete cycle, that is, the point d must be chosen so that an adiabatic process will return the gas to point a. The condition for this may be found by consideration of the two adiabatic processes.

For a reversible adiabatic process with a perfect gas, $pV^\gamma = C$. Applying Eq. (26.7) to both adiabatic processes,

$$T_1/T_2 = (V_c/V_b)^{\gamma-1} = (V_d/V_a)^{\gamma-1}$$

Hence $$V_c/V_b = V_d/V_a$$

from which it follows that

$$V_b/V_a = V_c/V_d$$

Thus the required condition is that both isothermal processes shall be carried out over the same volume ratio: let this ratio be r.

Efficiency of the Carnot cycle. The efficiency of a heat engine is defined as the ratio of net work done and heat supplied. Since, for a complete cycle, the internal energy of the working fluid returns to its original value, it follows from the first law that

$$\text{Net work done} = \text{Heat supplied} - \text{Heat rejected}$$

Hence an alternative expression for cycle efficiency is

$$\text{Efficiency } \eta = (\text{Heat supplied}-\text{Heat rejected})/\text{Heat supplied} \quad (27.1)$$

$$= 1 - \text{Heat rejected/Heat supplied} \quad (27.2)$$

For the Carnot cycle, applying Eq. (26.5),

$$\text{Heat supplied (a to b)} = mRT_1 \log_e r \quad (27.3)$$

$$\text{Heat rejected (c to d)} = mRT_2 \log_e r \quad (27.4)$$

Hence, from Eq. (27.2),

$$\eta = 1 - (mRT_2 \log_e r / mRT_1 \log_e r)$$

$$= 1 - T_2/T_1 \quad (27.5)$$

Thus the efficiency of this cycle depends only on the temperatures between which it works—usually thought of as being the temperatures of a "heat source" and a "heat sink."

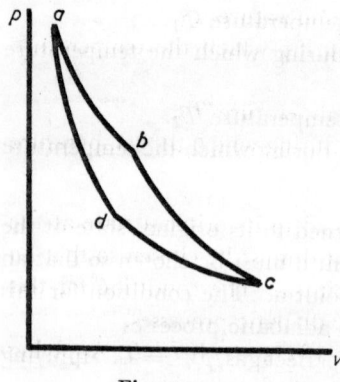

Fig. 27.2

Reversibility of the Carnot cycle. Since all processes in this cycle are, by definition, reversible, it follows that the cycle is itself perfectly reversible. In other words, although it is described (in Fig. 27.2) as $a \to b \to c \to d \to a$, it may just as readily be carried out as $d \to c \to b \to a \to d$. The engine then becomes a "heat pump," taking in heat at T_2 and giving it out at the higher temperature T_1. The amounts of energy involved are the same as before, but their directions are reversed: thus

$$\text{Heat taken in (at } T_2) = mRT_2 \log_e r \quad (27.6)$$

$$\text{Heat given out (at } T_1) = mRT_1 \log_e r \quad (27.7)$$

and
$$\text{Work } input = mR(T_1 - T_2) \log_e r \quad (27.8)$$

Carnot's principle

The important generalisation supplied by Carnot was that, for engines working between a "heat source" and a "heat sink" (the temperatures of both being fixed), *no engine can be constructed which will be more efficient than a perfectly reversible engine.* (Although there are many other "perfectly reversible engines" besides the one described by Carnot, because the efficiency of the Carnot cycle is readily deduced, most discussions are usually based on that.)

The proof of this principle rests on a demonstration that, if a more efficient engine could be constructed, it could be coupled to a "Carnot engine" in such a way as to produce either (i) a continuous transfer of energy from sink to source, or (ii) a continuous output of work accompanied only by a net transfer of heat from the source. Experience indicates that both are impossibilities. They are examples of the kind of energy conversion which would not violate the first law, but would be contrary to the second.

Condition (i) above may be demonstrated by considering the specific situation of a heat source at 600 K and a heat sink at 300 K. The efficiency of a Carnot cycle working between these temperatures would be

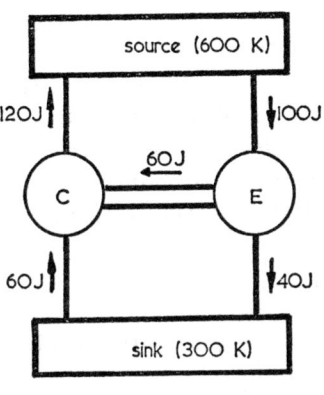

$$\eta = 1 - T_2/T_1 = 1 - 300/600 = 0.5$$

or   50 per cent

Fig. 27.3

Suppose there is some engine E with an efficiency of 60 per cent. This engine may be mechanically coupled to a "Carnot engine" C in the way shown by Fig. 27.3 so that E operates as a heat engine and C as a reversed heat engine or heat pump. Let engine E receive 100 J of heat from the source. Its efficiency is 60 per cent, hence it will convert 60 J into work and reject the remaining 40 J to the sink. The 60 J of work is transferred to the reversed heat engine C. From Eqs. (27.8), (27.6) and (27.7), the work input, heat taken in and heat given out are in the ratio

$$(T_1 - T_2) : T_2 : T_1$$

which in this case will be (600 − 300):300:600. Thus C will take 60 J from the sink and deliver 120 J to the source. These energy transfers are shown in Fig. 27.3, and it is seen that their net effect is a transfer of 20 J from sink to source. But since this process could be repeated indefinitely, it is impossible.

This demonstration can be carried out in general terms, and the same impossible situation will arise whenever E is more efficient than C. Thus a "Carnot engine" (or any other perfectly reversible engine) has the greatest possible efficiency which is, (Eq. (27.5)), $(1 - T_2/T_1)$.

The Carnot cycle cannot be achieved in practice since it involves (i) heat transfer across an infinitely small temperature difference;

(ii) adiabatic processes (perfectly insulating cylinders exist only in the imagination); and (iii) reversible processes (friction of all kinds will be present in an actual engine). Furthermore, an internal combustion engine based on this cycle would be an unusually large machine because the "work ratio" (see Chapter 28) is very small when the working fluid is a gas. Two fundamental properties do, however, arise from Carnot's principle and apply to all practical heat engines. Reversibility, although unattainable in practice, is still to be striven for; the elimination of a source of irreversibility will result in increased efficiency. It is even more important that $(1 - T_2/T_1)$ represents the maximum possible efficiency, and may be increased by raising T_1 or lowering T_2. Usually the "sink" into which a practical engine rejects heat is the surrounding atmosphere, so that T_2 cannot be varied. Thus, in heat engine design, much effort is directed towards the raising of T_1, the temperature at which heat is received. In actual heat engines, there is an obvious connection between maximum cycle temperature and engine efficiency.

27.2 The Second Law of Thermodynamics

Engineering is often concerned with the production of "power" (i.e., work or its equivalent) from natural sources which, directly or indirectly, provide energy in the form of heat, so that it is natural that the second law of thermodynamics should be stated in forms which apply to heat engines. Each of the two situations recognised as impossible is a separate statement of the law.

1. *It is impossible to construct a device which, operating in a cycle, produces no effect other than the transfer of heat from a reservoir at a low temperature to one at a higher temperature* or, *Heat cannot, of itself, pass from a lower to a higher temperature.*

2. *It is impossible to construct a device which, operating in a cycle, produces no effects except to do work and exchange heat with a single reservoir* or, *Heat cannot be completely and continuously converted into work.*

If one of these statements is true, so is the other. This can be illustrated by the arrangement shown by Fig. 27.3. It has been shown that this violates statement (1) but if, of the 60 J of work produced by E, only 40 J is transferred to C, the arrangement will produce a net output of 20 J of work while the source loses 20 J of heat—in effect, a violation of statement (2). A machine which violates statement (2) is said to produce "perpetual motion of the second kind." (Perpetual motion of the "first kind" involves a violation of the first law—the creation of energy.)

A simple (and, of course, impossible) example is that of a ship which takes in sea water, cools it, converts the heat extracted into work to drive its propellor and then discharges the cooled water back into the sea.

Probability and the second law

All these statements of the law are based on the observed behaviour of heat engines and of substances in general, but it is also possible to approach the second law by a consideration of the behaviour of individual molecules. The observed properties of substances are the averages of certain properties of the molecules of which they are composed. At the molecular level, motion is continuous and, furthermore, *random*. The molecules of a gas, for example, move in all directions with a variety of speeds, colliding frequently with each other and with the walls of their container. Thus the internal energy of a perfect gas (in which there are considered to be no forces of attraction between molecules) consists of the kinetic energies of its molecules. The only difference between this kinetic energy and that of a moving body is that molecular kinetic energy is *random* while that of the body is *ordered*.

From this point of view, the conversion of heat to work means the conversion of *disordered energy* into *ordered energy*. Experience indicates that ordered arrangements tend to become disordered, and that the creation of order from chaos requires considerable effort and never occurs naturally. At the molecular level, this is the second law in its most basic form: disorder always tends to increase and will never, in any natural process, decrease.

Order and disorder are not precise terms, which is why the principle is usually stated in terms of *probability*. The "probability" of any event is usually taken as the ratio of the number of ways in which it may occur to the total number of possibilities. Thus the probability of throwing a particular number using a single dice would be regarded as 1/6 since the throw has six possible results, only one of which satisfies the required condition. From this it follows that the greater the degree of order in an arrangement of objects (molecules included), the less its probability. For example, in the case of a pack of cards, there is only one completely ordered arrangement, but a great number of random ones so that the probability of the ordered arrangement is small. In these terms, the second law becomes:

3. *All spontaneous processes result in a more probable state.*

The calculation of such probabilities is far beyond the scope of this text, if only because very large numbers of molecules are involved (one milligramme of hydrogen contains about 3×10^{20}). The branch of science concerned is known as "statistical thermodynamics." Certain deductions are, however, possible even without precise calculations. An electric current, for example, is an essentially ordered phenomenon—all electrons move along the same path in the same direction, which is why it is to be expected that electrical and mechanical energy will be completely convertible. Experience shows this to be the case.

In some kinds of energy conversion, the fact that a more disordered state results is obvious and in others it may (with caution) be deduced. Consider, for example, the case of heat transfer between two bodies (Fig. 27.1 (*b*)). A completely disordered system is characterised by its uniformity. If different parts of a system can be distinguished, this implies a degree of order. In particular, two bodies at different temperatures constitute a more ordered arrangement than two similar bodies at the same temperature, so that statement (1) of the second law is seen to be a particular consequence of statement (3). (It should be emphasised that this argument is not a *proof*: that would require a rigorous examination of thermodynamic probabilities.)

Applications of the second law

The laws of thermodynamics do not apply simply to devices which convert heat into work and vice versa, but have a wide field of application. In chemistry, for example, the first law is used to calculate the energy released or taken in during a chemical reaction, but it is the second law which determines whether or not the reaction will take place at all. The second law is also relevant to situations unconnected with energy conversion. Thus in the branch of cybernetics (the study of control, data processing and automatic devices) dealing with "information theory," there is an axiom that "no operation on a message can, on the average, result in a gain in information." This is based on statistical arguments similar to those which are the foundation of the second law of thermodynamics, and has been encountered in Chapter 10 in the form "in any control system, the output can never be completely faithful to the input."

27.3 Entropy

The first law is based on the observation that in many different phenomena "something" appears to be conserved, leading to the concept of energy. The second law is, in general, the observation that many phenomena cannot be reversed—that "something" can move in one direction only—

and this has led to the concept of *entropy*. Entropy is usually defined mathematically, but, as with energy, it is also a physical concept. Indeed, the concepts of "thermodynamic probability" and "entropy" are essentially identical, so that entropy is sometimes loosely described as "the amount of disorder in a system." Thus substituting "entropy" for "probability," the second law becomes:

4. *All spontaneous processes result in an overall increase of entropy* or, *The entropy of an isolated system cannot decrease.*

A completely non-mathematical definition is that entropy is *a measure of the capacity of a system to undergo spontaneous change.* All systems tend towards a state of complete equilibrium, and when this has been reached the entropy of the system is considered to have reached its maximum value.

Entropy and available energy

A system which has "reached a state of complete equilibrium" is one in which all mechanical energy has been converted (by friction) to heat and distributed between the various bodies in the system as internal energy, and all bodies have subsequently exchanged heat until the whole system is at a uniform temperature. Since a heat engine can operate only if temperature differences exist, it is impossible to convert energy into work in such a system. The system contains energy, but it is "unavailable." The *available energy* of a system is that already in the form of, or convertible to, mechanical energy (or possibly, electrical energy).

For a system comprising two bodies at different temperatures, the available energy would be the total amount of work which could be produced by a reversible heat engine connected between them and allowed to run until the temperatures become equal. If, before doing this, the bodies were simply brought into contact for a time so that some irreversible heat transfer took place, it is obvious that less work would be produced. Thus irreversible processes reduce the available energy of a system and will, eventually, reduce it to zero. The energy in the system has not been *destroyed* but has been *degraded*, which is why the second law is often called "the law of degradation of energy."

It is now clear that an irreversible process is inevitably accompanied by a loss of available energy—a "degradation of energy"—and an increase in the entropy of the system. Thus the statement that the entropy of an isolated system cannot decrease also means that its available energy cannot increase. Alternatively, for a *reversible* process, both available energy and entropy remain constant, and this is why the concept of entropy is useful in practical calculations. Considering a fluid which undergoes a reversible adiabatic expansion, and taking as an "isolated system" the fluid itself and the device (purely mechanical) on which it does work, it follows that in such a process the entropy of the fluid remains constant: a reversible adiabatic process is thus said to be *isentropic*. Many calculations concern such processes and, for vapours, the fact that entropy remains constant provides the only convenient method of solution.

Entropy as a physical property

Consider a Carnot cycle carried out with a perfect gas so that, over one complete cycle, heat Q_1 is received at temperature T_1, heat Q_2 being rejected at T_2. From Eqs. (27.3) and (27.4), it follows that

$$Q_1/T_1 = Q_2/T_2$$

and, for a complete cycle,

$$\Sigma Q/T = 0$$

This applies for any working fluid, and may be extended to reversible cycles in which the temperatures of heat reception and rejection are not constant, so that for any reversible cycle,

$$\int dQ/T = 0$$

Fig. 27.4 shows a reversible cycle in which a fluid proceeds from state A to state B by a process represented by path 1, and returns from state B to state A by a different process, path 2. Then

$$\int dQ/T \text{ for path } 1 = -\int dQ/T \text{ for path } 2$$

Hence

$$\int dQ/T \text{ for path } 1 = \int dQ/T \text{ for a process from A to B by path } 2$$

Since 1 and 2 may be any reversible paths, it follows that the quantity $\int dQ/T$ is the same for any reversible process during which the state of the fluid changes from A to B, which means that $\int dQ/T$ represents the change in some *property* of the fluid. This property remains constant during a reversible adiabatic process, increases during an irreversible adiabatic process and, in fact, corresponds in every way to "entropy" as previously described, so that entropy S is defined by

$$(S_1 - S_2) = \int_1^2 dQ_{\text{Rev}}/T$$

This definition applies only to a reversible process, and if the process is not reversible, $\int dQ/T$ is meaningless. It should also be noted that the definition refers to a *change* in entropy—no attempt is made to define *total* entropy and, as in the case of internal energy, entropies must be referred to some chosen datum. Since, for a *small* reversible process, the increase in entropy is given by the ratio (heat received)/(absolute temperature), its units are kJ/K. The units of specific entropy (s) are kJ/kg K.

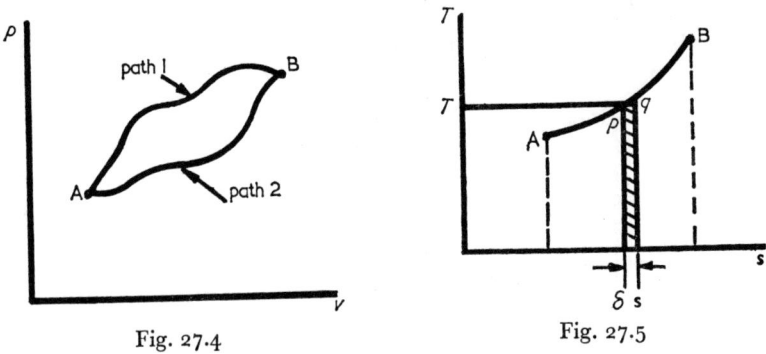

Fig. 27.4   Fig. 27.5

Temperature—entropy diagrams. Since specific entropy is an independent intensive property, it may be plotted on a diagram against other properties. Temperature-entropy diagrams are particularly informative for the areas beneath them may be shown to be related to heat transfer.

Referring to the reversible process represented by path *AB* in Fig. 27.5, let pq represent a small part of the process during which heat δQ is received at temperature T. Change in specific entropy

$$\delta s = \delta Q/T$$
$$\therefore \delta Q = T\delta s$$
$$= \text{Area of strip (shaded)}$$

Thus, for the whole process *A–B*, heat received per kg is represented by the area under the line *AB*. This applies only to reversible processes. That is, the area below a *dotted* line representing an *irreversible* process is meaningless. A reversible isothermal process will be represented by a horizontal line on the *T–s* diagram, and a reversible adiabatic or, as it

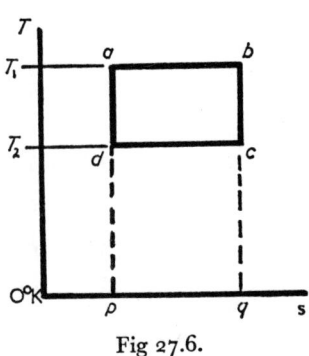

Fig 27.6.

may now be termed, isentropic process by a vertical line. Thus the *T–s* diagram for a Carnot cycle is simply a rectangle (Fig. 27.6). Since areas represent heat transfers, abqp represents heat received, cdpq represents heat rejected and the difference abcd represents work done. Comparing these areas, the cycle efficiency may be seen to be $(T_1 - T_2)/T_1$ or $1 - (T_2/T_1)$.

Example. *A cylinder contains* 0·5 kg *of a perfect gas at a temperature of* 20°C. *Calculate the change in entropy when* 250 kJ *of heat is received* (a) *during an isothermal expansion* (b) *at constant volume,* (c) *at constant pressure, and show each process on a T–s diagram. Consider all processes to be reversible and take* $c_p = 1·0$ kJ/kg K, $c_v = 0·7$ kJ/kg K.

(a) Since T is constant,

$$(S_2 - S_1) = Q_{Rev}/T$$

∴ Increase in entropy is

$$250/293 = 0·853 \text{ kJ/K}$$

The corresponding increase in specific entropy,

$$(s_2 - s_1) = (S_2 - S_1)/m$$
$$= 0·853/0·5 = 1·706 \text{ kJ/kg K}$$

(b) $$(S_2 - S_1) = \int_1^2 dQ_{Rev}/T$$

and, for a constant volume process,

$$dQ_{Rev} = mc_v \, dT$$

∴ $$(S_2 - S_1) = mc_v \int_{T_1}^{T_2} dT/T$$

or $$(s_2 - s_1) = c_v \int_{T_1}^{T_2} dT/T$$
$$= c_v \log_e (T_2/T_1) \tag{27.9}$$

For a constant volume process,

$$Q = mc_v(T_2 - T_1)$$

or $$250 = 0·5 \times 0·7 \times (T_2 - 293)$$

from which $$T_2 = 1007 \text{ K}$$

∴ $(s_2 - s_1) = 0·7 \log_e (1007/293) = 0·864 \text{ kJ/kg K}$

∴ Increase in entropy

$$(S_2 - S_1) = 0·864 \times 0·5 = 0·432 \text{ kJ/K}$$

(c) For a constant pressure process,

$$dQ_{Rev} = mc_p \, dT$$

so that $$(s_2 - s_1) = c_p \log_e (T_2/T_1) \tag{29.10}$$
$$Q = mc_p(T_2 - T_1)$$

from which $$T_2 = 793 \text{ K}$$

∴ $(s_2 - s_1) = (1·0) \log_e (793/293) = 0·996 \text{ kJ/kg K}$

and $$(S_2 - S_1) = 0·498 \text{ kJ/K}$$

Fig. 27.7 shows the T–s diagram and, since in all cases the gas received the same heat transfer, the areas under oa, ob and oc are seen to be equal.

Irreversible processes. As previously stated, for an irreversible process $\int dQ/T$ does not represent the entropy change. Entropy is, however, a *property*, so that the change in entropy depends only on the initial and final states and may be found by evaluating

$$\int_1^2 dQ_{\text{Rev}}/T$$

for *any* reversible process by which the substance could have been brought from state 1 to state 2.

Example. *A block of lead at 15°C is dropped from a height of 500 m on to ground which may be assumed to be unyielding and non-conducting. What is the change in its specific entropy? For lead,* $c_p = 0{\cdot}126$ *kJ/kg K.*

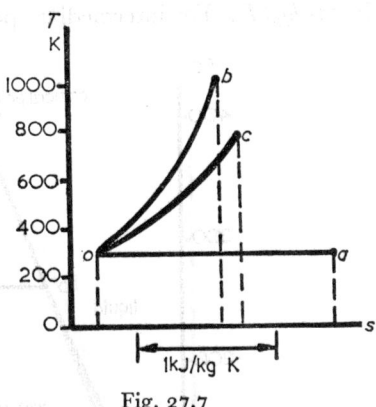

Fig. 27.7

This is a completely irreversible process in which the original potential energy of the lead becomes, finally, an addition to its internal energy. Applying the first law,

$$mgh = mc_p(T_2 - T_1)$$

$$\therefore gh = c_p(T_2 - T_1)$$

or

$$9{\cdot}81 \times 500 \times 10^{-3} = 0{\cdot}126(T_2 - 288)$$

$$\therefore T_2 = 327 \text{ K}$$

Thus the initial and final states of the lead are (1) atmospheric pressure and 288 K, and (2) atmospheric pressure and 327 K. An obvious reversible process connecting these states would be heat transfer at constant pressure, for which, from Eq. (29.10),

$$(s_2 - s_1) = c_p \log_e (T_2/T_1)$$

$$= 0{\cdot}126 \log_e 327/288 = 0{\cdot}016 \text{ kJ/kg K}$$

This is the change of specific entropy during the irreversible process and is positive—the entropy of the lead increases. This is expected, for an irreversible process must result in an overall increase in entropy and, in this case, no entropy changes take place in the surroundings.

Steam and other vapours. As with other properties, the entropy of a vapour is not simply related to temperature, so that entropy changes must be

found by using property diagrams or property tables. The T–s diagram
for steam is shown by Fig. 27.8, in which a–b–c–d represents a constant
pressure process. The "compressed liquid" at point a becomes, finally,
"superheated vapour" at d. From a to b, the process closely follows the
line separating liquid from wet vapour, which is a result of the fact that
liquids are almost incompressible. From b to c, the liquid evaporates
at constant temperature and, since the heat transfer is h_{fg}, the increase
in s is h_{fg}/T_s. For intermediate points (i.e., wet states), the heat transfer

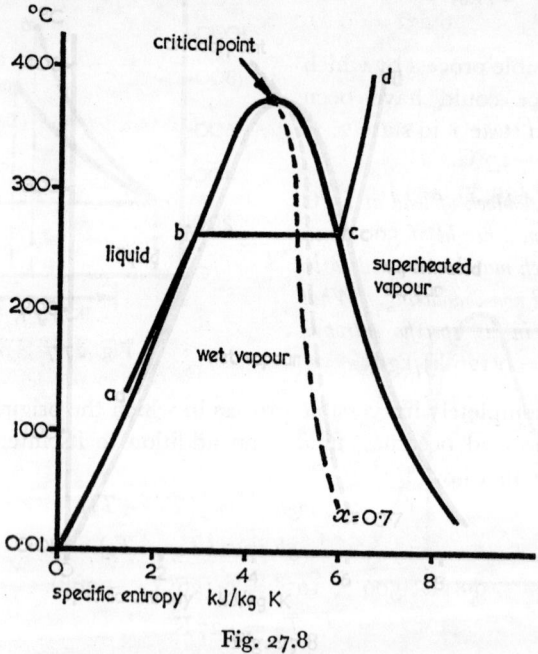

Fig. 27.8

would be xh_{fg}, so that s increases in direct proportion to x. In the final
stage c–d, the vapour is superheated and further heat transfer increases
both T and s.

Calculations may be based on readings taken directly from a T–s
chart or, more usually, a h–s chart (sometimes referred to as a "Mollier
diagram"). A more accurate method is to use the T–s diagram in order
to "visualise" a process and to suggest its approximate result, and to
obtain numerical values from property tables.

Example. *Steam enters a turbine at* 7 MN/m² *(70 bar) and* 350°C, *and the
condenser pressure is* 10 kN/m² *(0·1 bar). Assuming isentropic expansion,
find the state of the steam leaving the turbine.*

Fig. 27.9 represents the *T–s* diagram, on which the final state (2) is in the region of wet vapour. Using property tables,

$$s_1 = 6\cdot231 \text{ kJ/kg K}$$

From Eq. (25.9),

$$s_2 = s_f + x_2 s_{fg}$$
$$= 0\cdot649 + x_2(7\cdot500)$$

Equating entropies,

$$6\cdot231 = 0\cdot649 + x_2(7\cdot500)$$
$$\therefore x_2 = 5\cdot582/7\cdot500 = 0\cdot744$$

Example. *"Freon-12" enters the compressor of a refrigerator as a wet vapour* $(x = 0\cdot9)$ *at* $-15°C$. *What will be its state after isentropic compression to* $3\cdot979 \text{ MN/m}^2$ *(39·79 bar)?*

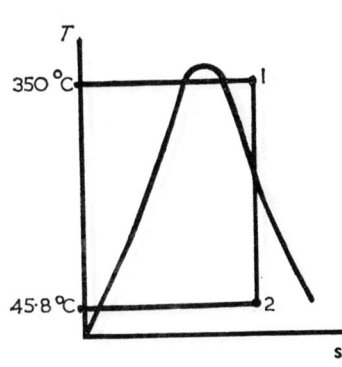

Fig. 27.9

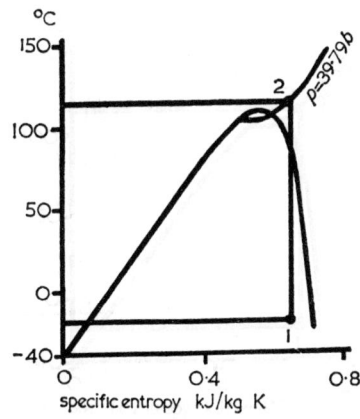

Fig. 27.10

Fig. 27.10 shows the *T–s* diagram for "Freon-12," and from this it is evident that the refrigerant will leave the compressor as a superheated vapour. From Eq. (25.5),

$$s_1 = x s_g + (1 - x) s_f$$
$$= 0\cdot9 \times 0\cdot7051 + 0\cdot1 \times 0\cdot0906$$
$$= 0\cdot6437 \text{ kJ/kg K}$$

For $p_s = 39\cdot79$ bar,

$$s_g = 0\cdot6076 \text{ kJ/kg K}$$

and, for 15°C of superheat $s = 0.6964$ kJ/kg K. Hence, by interpolation, the degree of superheat is

$$15 \times 0.0361/0.0888 = 6.1°C$$

or the final temperature is $(110 + 6.1) = 116.1°C$.

Example. *Dry saturated steam at* 200 kN/m² (*2 bar*) *enters a horizontal nozzle with negligible velocity. With what velocity will it leave after isentropic expansion to atmospheric pressure?*

From the *T–s* diagram, or by comparison of values of s_g, it is evident that the steam will be wet after expansion.

Using property tables, and equating entropies,

$$7.127 = 1.307 + x_2(6.048)$$

$$\therefore x_2 = 5.820/6.048 = 0.962$$

Applying Eq. (26.12),

$$\tfrac{1}{2}v_2{}^2 = h_1 - h_2$$

$$= 2707 - (417 + 0.962 \times 2256.7)$$

$$= 2707 - 2591 = 116 \text{ kJ/kg}$$

$$\therefore v_2 = \sqrt{232\,000}$$

$$= 482 \text{ m/s}$$

Example. *Dry saturated steam at* 200 kN/m² (*2 bar*) *is throttled to atmospheric pressure. Find (a) its final state, (b) the change in its specific entropy. Compare these results with those of the previous example.*

(a) From Eq. (26.16),

$$h_1 = h_2$$

$$\therefore 2707 = h_2$$

For atmospheric pressure, $h_g = 2676$ kJ/kg, so that the final state is *superheated vapour*.

By interpolation, the final temperature is 115°C.

(b) Using property tables and interpolating in the case of s_2,

$$s_2 - s_1 = 7.433 - 7.127$$

$$= +0.306 \text{ kJ/kg K}$$

PROBLEMS

For tutorials

1. What happens to the specific entropy of air when it is (a) compressed isothermally, (b) expanded adiabatically, (c) throttled? Reconsider questions (4) and (5) of the previous chapter, with particular reference to the available energy in an isolated system.

2. One natural source of energy is steam formed deep underground which finds its way to the surface. It has been suggested that greater efforts should be made to exploit "geothermal power" and that a considerable energy output could be obtained at the expense of a very small reduction in the temperature of the earth's core. Another suggestion is that, if the temperature of the oceans could be reduced by a fraction of a degree, the energy released would be sufficient for all our power requirements. What do you think of suggestion (1) and what is wrong with suggestion (2)?

3. Two devices which appear to violate statement (2) of the second law are the self-winding clock and the Crookes radiometer. Show that both are, in fact, heat engines by identifying "sources" and "sinks" and, in each case, describe a situation in which the device could not function (excluding, of course, complete isolation).

4. A large building has an estimated daily heat loss of 15×10^6 kJ. What will be its electrical power requirement if (a) it is electrically heated by radiators, etc.; (b) a heat pump, driven by an electric motor, circulates water at 70°C through a conventional central heating system, its heat source being a nearby river at 5°C? Assume that a perfectly reversible engine and a perfectly efficient motor are used in (b).
Practical heat pumps will not be as effective as this but (b) will always be much less than (a). Does this mean that the first law can be violated?

Ans. (a) 173·6 kW; (b) 32·9 kW

5. A promising field of research is concerned with the magnetohydrodynamic (MHD) generator. In this device a stream of ionised gas at high temperature is passed through a strong magnetic field, resulting in the generation of a direct current by the "Faraday effect." The device may be regarded as a heat engine with high temperatures of heat reception and rejection, and hence it is used as a "topper"—that is, on leaving it the gas enters a heat exchanger producing steam for a conventional power plant. Investigate its use in the following hypothetical arrangement: gas enters the MHD generator at 2600°C and leaves at 2000°C. The heat exchanger produces steam at 600°C and this passes through a turbine, leaving it at 35°C. Considering both as reversible heat engines, and assuming that all heat rejected by the MHD generator is received by the steam plant, find the efficiencies of (a) the MHD generator alone, (b) the steam plant alone, (c) both combined to form a single power plant, and (d) a reversible engine working between 2600°C and 35°C. Discuss the practical implications of (a), (b) and (c). Do you think the practical equivalent of (d) will ever be constructed?

Ans. (a) 20·9 per cent; (b) 64·7 per cent; (c) 72·0 per cent; (d) 89·3 per cent.

General

1. Give a proof of "Carnot's principle" based on the following assumptions: a heat source at 500°C and a heat sink at 0°C are available, also an engine which is thought to be capable, when working between these temperatures, of 70 per cent efficiency.

2. An inventor claims to have perfected a device which will increase the efficiency of a steam power plant to 60 per cent. The maximum and minimum cycle temperatures are 450°C and 40°C. Is his claim possible?

Ans. No; maximum possible efficiency is 56·7 per cent

3. Calculate the maximum possible efficiencies of:

(*a*) A steam engine exhausting at atmospheric pressure and supplied with steam at 150°C.

(*b*) The above engine, exhausting to a condenser in which the pressure is 10 kN/m² (0·1 bar).

(*c*) A steam power plant, maximum steam temperature 600°C with condensation at 25°C.

(*d*) An internal combustion engine: maximum and minimum cycle temperatures 1600°C and 60°C.

Ans. (*a*) 11·8 per cent; (*b*) 24·6 per cent; (*c*) 65·9 per cent; (*d*) 82·2 per cent.

4. Calculate the change in the entropy of 5 kg air, originally at 0°C, when 900 kJ of heat are received reversibly and at constant pressure. For air, $c_p = 1·005$ kJ/kg K.

Ans. 2·54 kJ/K, increase

5. Referring to Fig. 27.1 (*b*): when the tanks are brought into thermal contact find the change in entropy of (*a*) the smaller tank, (*b*) the larger tank, (*c*) the system comprising both tanks. Take c_p for water as 4·187 kJ/kg K and assume the containers to be of negligible heat capacity.

Ans. (*a*) 0·308 kJ/K, decrease; (*b*) 0·361 kJ/K, increase;
(*c*) 0·053 kJ/K, increase

6. Find the final state of steam when expanded isentropically from 15 MN/m² (150 bar) and 500°C to a final pressure of 20 kN/m² (0·2 bar).

Ans. Wet vapour of dryness fraction 0·78

CHAPTER 28

Gas and Vapour Power Cycles

28.1 ESSENTIALS OF A POWER CYCLE

A *heat engine* is a device which converts heat into work and it has been found in the last chapter that this conversion can never be complete. All heat engines have three essential features—the reception of heat at a high temperature, the conversion of some of this into work and the rejection of the remainder of the heat at a lower temperature. The concept of work may include any form of energy directly convertible to mechanical work, so that devices which convert heat into electrical energy (such as MHD and thermionic generators) may be regarded as heat engines. One of the simplest devices is the thermocouple which has all the characteristics of a heat engine. Heat is received at the hot junction and rejected at the cold junction. The process is theoretically reversible, and is indeed reversed in the "Peltier refrigerator." But there are in practice two irreversible processes—the dissipation of energy by electrical resistance and heat transfer by conduction between the two junctions. If these could be eliminated, the efficiency would be that of the Carnot cycle $(1 - T_2/T_1)$.

The term "heat engine" may therefore be applied not only to power plants but to a variety of mechanisms, one of which is the earth's atmosphere. This receives heat by radiation from the sun, rejects heat by radiation to outer space, and produces mechanical energy of various kinds and in vast quantities—in hurricanes, for example. A very small fraction of this energy is utilised as wind and water power.

For the production of power, heat energy is usually produced by the combustion of a fuel and converted into work by carrying out a series of processes on a gas or vapour, called the *working fluid*. This series of processes is termed the *cycle* and, since it is to be repeated indefinitely so as to obtain continuous production of energy, it must be *complete*, which means that the fluid must return to its original state after one complete cycle.

Theoretical and actual cycles

As in many other branches of design, the design of a power plant must involve a compromise between the ideal efficiency of a Carnot cycle and the practical problems of construction. For example, in the construction of cylinders and pistons there are many materials with excellent mechanical properties which are also excellent conductors of heat, so that truly adiabatic processes will not be possible in power plants in which they are used.

Thus there will be inevitable differences between a theoretically desirable process and its practical counterpart—in particular, no practical process is completely reversible. At first sight, this would seem a barrier to further investigation, for irreversible processes cannot be represented on a property diagram and the work done cannot be calculated. The difficulty is overcome by formulating, for an actual engine, the corresponding *theoretical cycle* in which all processes are considered to be reversible and in which practical imperfections, such as departures from truly adiabatic processes, are ignored. The theoretical cycle is, furthermore, always assumed to be a *closed cycle*, which means that the working fluid remains within the engine and passes repeatedly through the cycle. In practice, however, many engines have an *open cycle*—an internal combustion engine, for example, must discharge exhaust gases and take in fresh air and fuel. Such a process can be represented in a theoretical cycle by cooling at constant volume, and it may be seen that this is a true equivalent. Apart from the chemical changes involved, it is possible to suppose a quantity of gas being cooled by the surrounding atmosphere after leaving the engine and, later, re-entering it to undergo a further cycle.

Cycle efficiency and work ratio

The efficiency of a cycle is defined by

$$\eta = \frac{\text{Net work done}}{\text{Heat supplied}}$$

This may be calculated either from the fact that the net work is equal to the area enclosed by the p–V diagram and also to the area enclosed by T–S diagram, or by using Eq. (27.2).

The work ratio is defined as

$$r_w = \frac{\text{Net work done}}{\text{Positive work done}}$$

In all cycles work is done on the fluid during compressions and by it during expansions. The positive or gross work is the total work done by the fluid, and the net work is this quantity minus the work done *on* the fluid.

Although the work ratio is usually used in connection with gas turbines, it can be applied to any cycle. In practice, the work ratio is usually an indicator of the relative size of an engine—the smaller the ratio, the bigger the engine. (For a Carnot cycle with gas as the working fluid, the work ratio is very small.)

Non-flow cycles and flow cycles

Actual engines may be classified in two ways, (1) heat may be transferred to the working fluid after being produced in a separate combustion chamber, or it may be produced by *internal combustion*, and (2) the cycle may consist of non-flow processes or of flow processes. In a theoretical cycle, the way in which heat is received is unimportant so that the classification into *non-flow cycles* (as in internal combustion engines) and *flow cycles* (as in gas turbines and steam plant) is a useful one.

28.2 NON-FLOW CYCLES

The Otto cycle

The engine constructed by N. A. Otto in 1876 may be considered the first successful internal combustion engine. Its cycle, used by all modern petrol engines, was proposed in 1862 by Beau de Rochas and later, independently, by Otto, who in 1876 patented an engine based on it. The actual engine cycle is completed during four strokes of a piston:

Stroke (1): A mixture of air and fuel is drawn into the cylinder.

Stroke (2): The mixture is compressed, and, at the end of the stroke, ignited.

Stroke (3): After combustion (which takes place rapidly, so that for practical purposes it is complete when stroke (3) begins) the gases expand.

Stroke (4): The gases (which now include the products of combustion) are expelled from the cylinder.

In 1881, Sir Dugald Clerk produced an engine which eliminated strokes (1) and (4), replacing them by a "transfer" operation, and thus obtained a two-stroke cycle. His object was, primarily, to avoid infringing the patent held by N. A. Otto, but in doing so he produced an engine which is now widely used, for the elimination of intake and exhaust strokes results in a machine of light weight and simple construction. A modern development of the four-stroke cycle is the Wankel engine, in which the traditional piston and cylinder are replaced by a rotor/stator assembly, the "stroke" taking place in the spaces formed between these two components.

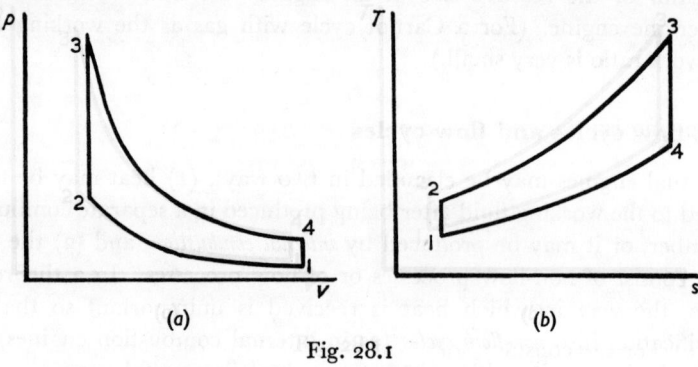

Fig. 28.1

The theoretical Otto cycle corresponds to strokes (2) and (3) only. Figs. 28.1 (*a*) and (*b*) show the corresponding p–v and T–s diagrams, and the processes (all reversible) are as follows:

1–2: Adiabatic compression.
2–3: Heat transfer at constant volume (corresponding, in the practical engine, to heat release by combustion).
3–4: Adiabatic expansion.
4–1: Heat transfer at constant volume (i.e., heat rejection).

Example. *In a theoretical Otto cycle, the compression ratio (that is, v_1/v_2) is 8 and the heat received at constant volume 1500 kJ per kg of working fluid. Assuming this working fluid to be air (for which $c_v = 0.718$ kJ/kg K and $\gamma = 1.40$) and considering air to be a perfect gas, find (a) the temperature at points 2, 3 and 4 if the temperature at point 1 is 27°C, (b) the heat rejected per kg of working fluid, and hence (c) the cycle efficiency.*

(*a*) For the reversible adiabatic process 1–2,

$$pv^\gamma = C$$

Applying Eq. (26.7),

$$T_2 = T_1(v_1/v_2)^{\gamma-1}$$

or, substantiating r (the "compression ratio") for (v_1/v_2),

$$T_2 = T_1 r^{\gamma-1}$$
$$= 300 \times 8^{(1\cdot40-1)}$$
$$= 689 \text{ K} \quad \text{or} \quad 416°\text{C}$$

For the constant volume process 2–3,

$$Q = mc_v(T_3 - T_2)$$
$$\therefore 1,500 = 1 \times 0\cdot718(T_3 - 689)$$
$$\therefore T_3 = 2778 \text{ K} \quad \text{or} \quad 2505°\text{C}$$

Applying Eq. (26.7) to process 3–4, and noting that

$$r = (v_1/v_2) = (v_4/v_3)$$
$$T_4 = T_3(1/r)^{\gamma-1}$$
$$= 2778 \times (1/8)^{(1\cdot40-1)}$$
$$= 1210 \text{ K} \quad \text{or} \quad 937°\text{C}$$

(b) For the constant volume process 4–1, heat rejected per kg,

$$Q = mc_v(T_4 - T_1)$$
$$= 1 \times 0\cdot718(1210 - 300)$$
$$= 653 \text{ kJ}$$

(c) From Eq. (27.2),

$$\eta = 1 - (\text{Heat rejected/Heat received})$$
$$= 1 - (653/1500)$$
$$= 0\cdot5647 \quad \text{or} \quad 56\cdot47 \text{ per cent}$$

Efficiency and compression ratio. Expressing the above results in general terms,

$$\eta = 1 - \frac{mc_v(T_4 - T_1)}{mc_v(T_3 - T_2)}$$
$$= 1 - (T_4 - T_1)/(T_3 - T_2)$$
$$T_4 = T_3(1/r)^{\gamma-1}$$

and
$$T_1 = T_2(1/r)^{\gamma-1}$$

hence
$$\eta = 1 - (1/r)^{\gamma-1} \tag{28.1}$$

Thus the efficiency of a theoretical Otto cycle depends only on its compression ratio r and will be increased if r is increased. This also applies to real cycles and, in practice, r is made as high as possible. It is, however, limited by (i) *pre-ignition*, and (ii) *detonation*. Pre-ignition is the spontaneous combustion of the air/fuel mixture before the end of the compression stroke and, in practice, is usually due to a "hot spot" developing on the cylinder head or piston crown. Detonation or "knock" occurs when unburned mixture ignites spontaneously ahead of the normal flame (which should spread outwards from the sparking plug quickly but smoothly). This gives rise to a rapid local increase of pressure, setting up pressure waves which strike the cylinder walls and piston. A distinctive noise results (aptly described by the term "pinking") and damage may be caused. This will occur, for an untreated fuel, when the compression ratio is greater than about 5. Higher compression ratios (up to 10 for normal purposes) are made possible by the addition to the fuel of "anti-knock" agents such as lead tetra-ethyl, $Pb(C_2H_5)_4$. These substances suppress detonation but many other factors (such as the shape of the combustion chamber) are involved.

It should also be noted that in practice the temperature T_3 after combustion will be much lower than that calculated for the theoretical cycle because of the increase in the specific heat c_v with rising temperature, chemical dissociation at high temperatures and heat transfer through cylinder walls, etc. For this reason the efficiency of a practical engine with $r = 8$ will usually be between 25 per cent and 30 per cent (the corresponding theoretical value being 56·47 per cent)*.

The Diesel cycle

Rudolf Diesel set out to design an internal combustion engine based on the Carnot cycle and, in addition, to use powdered coal as fuel—objectives which proved to be impracticable. The basic principle—that of injecting fuel at a controlled rate directly into the combustion chamber—was retained and, in 1893, a successful engine was constructed using oil as fuel

---

* The theoretical efficiency, $1 - (1/r)^{\gamma-1}$, corresponding to $\gamma = 1·40$, (i.e., taking the working fluid as air) is often referred to as the "Air Standard Efficiency". The Otto cycle is the most efficient of all the theoretical internal combustion engine cycles and it has been recommended that, for any engine, $1 - (1/r)^{0·4}$ should be taken as a standard of comparison and termed "the air standard efficiency of the engine" and that the ratio of this actual efficiency to the air standard efficiency should be termed the "Relative Efficiency" or "Efficiency Ratio". Some authorities, however, use the term "Air Standard Efficiency" to mean "the efficiency of a theoretical cycle, with air as working fluid, corresponding to the practical cycle of the engine concerned": and the term "Efficiency ratio" is occasionally used when comparing actual and Carnot cycles. To avoid confusion, these terms will not be used in this text.

and injecting it by a blast of compressed air. The resulting cycle differs from the actual Otto cycle in that (i) air only is drawn into the cylinder and compressed, (ii) no sparking plug or other ignition device is required, the temperature after compression being sufficient to ignite the fuel as it enters, and (iii) the combustion process continues over the whole of the injection period, so that a constant pressure process may result, and this is assumed in the theoretical cycle.

Most modern engines do not use air-blast injection (and for this reason are more correctly referred to as "compression-ignition engines" although commonly called "diesels"). The fuel is injected at high pressure through atomising nozzles, this system being known as *direct* or *solid injection*. As with the Otto cycle, both four-stroke and two-stroke engines may be made.

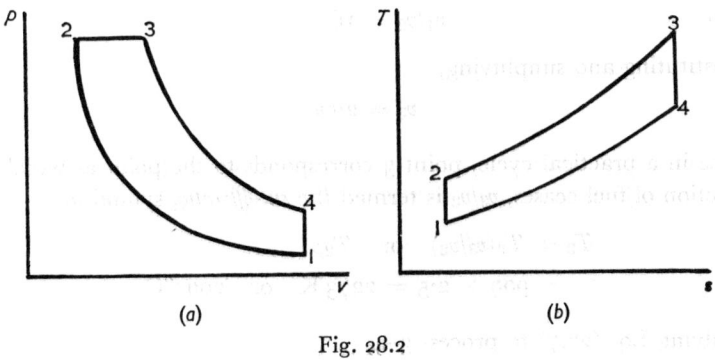

Fig. 28.2

The theoretical Diesel cycle. The *p–v* and *T–s* diagrams for the Diesel cycle are shown in Figs. 28.2 (*a*) and (*b*), and the (reversible) processes are:

1–2: Adiabatic compression.
2–3: Heat transfer at constant pressure (corresponding, in the practical engine, to controlled injection of the fuel, heat being released at such a rate as to keep pressure constant).
3–4: Adiabatic expansion.
4–1: Heat transfer at constant volume (i.e., heat rejection).

Example. *In a theoretical Diesel cycle the compression ratio v_1/v_2 is 16 and heat is supplied at constant pressure for $1/10$ stroke. The working fluid is air (which may be considered a perfect gas having $c_p = 1·005$ kJ/kg K and $c_v = 0·718$ kJ/kg K) and at the commencement of the cycle its temperature is $27°C$. Find (a) the cycle temperatures, (b) the heat received and the heat rejected per kg of working fluid and hence (c) the cycle efficiency.*

(a) $T_1 = 27°C = 300 K$

Applying Eq. (26.7) to process 1–2,

$$T_2 = T_1 r^{\gamma-1}$$

Here $\gamma = c_p/c_v = 1.005/0.718 = 1.40$

∴ $T_2 = 300 \times 16^{(1.40-1)}$

$= 909 K$ or $636°C$

Since process 2–3 is at constant pressure, Charles' law applies and

$$T_3/T_2 = v_3/v_2$$

The process occupies 1/10 stroke, that is,

$$(v_3 - v_2) = 1/10(v_1 - v_2)$$

Also $v_1/v_2 = 16$

Substituting and simplifying,

$$v_3 = 2.5v_2$$

Since in a practical cycle, point 3 corresponds to the point at which the injection of fuel ceases, v_3/v_2 is termed the *cut-off ratio*, symbol ρ

$$T_3 = T_2(v_3/v_2) \quad \text{or} \quad T_2\rho$$

$$= 909 \times 2.5 = 2273 K \quad \text{or} \quad 2000°C$$

Applying Eq. (26.7) to process 3–4,

$$T_4 = T_3(v_3/v_4)^{\gamma-1}$$

It should be noted that the expansion ratio (v_3/v_4) is *not* equal to the compression ratio r.

$$v_3/v_4 = (v_3/v_2) \times (v_2/v_4)$$

$$= \rho/r \quad \text{since } v_4 = v_1$$

$$= 2.5/16 \quad \text{or} \quad 0.1563$$

∴ $T_4 = 2273 \times 0.1563^{(1.40-1)}$

$$= 1081 K \quad \text{or} \quad 808°C$$

(b) Heat is received at constant pressure from 2 to 3, therefore, heat received per kg,

$$Q = mc_p(T_3 - T_2)$$

$$= 1 \times 1.005(2273 - 909)$$

$$= 1371 \text{ kJ}$$

Heat is rejected at constant volume from 4 to 1, therefore heat rejected per kg,

$$Q = mc_v(T_4 - T_2)$$
$$= 1 \times 0{\cdot}718(1081 - 300)$$
$$= 561 \text{ kJ}$$

(c) Applying Eq. (27.2),

$$\eta = 1 - (561/1371)$$
$$= 0{\cdot}591 = 59{\cdot}1 \text{ per cent}$$

Factors affecting efficiency. If the above calculations are carried out in general terms, the efficiency will be found to be

$$\eta = 1 - (1/r)^{\gamma-1} \frac{\rho^\gamma - 1}{\gamma(\rho - 1)} \qquad (28.2)$$

This is seen to differ from the expression derived for the Otto cycle only in that it includes the factor $(\rho^\gamma - 1)/\gamma(\rho - 1)$, which has a minimum value of 1 and which increases with increasing values of ρ, the cut-off ratio. Thus the Diesel cycle efficiency is increased by increasing the compression ratio r, but will always be less than that of an Otto cycle having a similar compression ratio.

In practice, compression-ignition engines achieve higher efficiencies than engines working the Otto cycle because their compression ratios are higher. Pre-ignition and detonation are associated only with mixtures of fuel and air, and hence do not occur. The minimum compression ratio which will ensure ignition of the fuel is about 12, and values of 14 to 18 are commonly used. As with the Otto cycle, the maximum temperature and efficiency of the practical cycle are reduced by variations of specific heats with temperature, dissociation and heat transfer. It must also be pointed out that only slow-speed engines have cycles approximating to the theoretical Diesel cycle. In most cases heat is released partly at constant volume and partly at constant pressure, and the corresponding theoretical cycle is referred to as the *dual-combustion cycle*.

28.3 FLOW CYCLES

The steam plant cycle

Fig. 28.3 (*a*) shows the arrangement of a typical steam power plant and Fig. 28.3(*b*) the corresponding theoretical *T–s* diagram. The cycle operated by this plant is known as the *Rankine cycle* and consists of the following processes:

1–2: Liquid enters a feed pump at (theoretically) condenser pressure and leaves at boiler pressure after isentropic compression. (In practice, two pumps would be used: an *extraction pump* delivering condensate at atmospheric pressure to a hotwell, and a *feed pump* transferring liquid from hotwell to boiler.)

2–3–4: In the boiler, heat transfer occurs at constant pressure: the steam leaves the boiler at 4 as a superheated vapour.

4–5: Steam expands isentropically in a turbine (or in a reciprocating steam engine).

5–1: Steam is condensed at constant pressure (and hence at constant temperature).

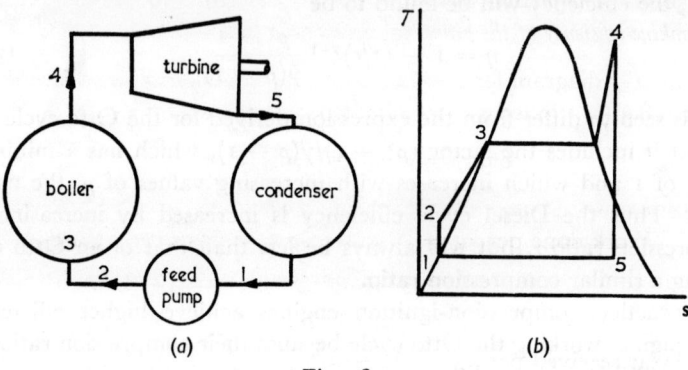

Fig. 28.3

Efficiency of the Rankine cycle. Considering 1 kg of working fluid (i.e. water/steam).

$$\text{Heat received in boiler} = h_4 - h_2$$

Neglecting kinetic energies at entry and exit,

$$\text{Work done in turbine} = h_4 - h_5$$

and this is known as the *isentropic enthalpy drop*.

Work done by feed pump $W_p = h_2 - h_1$

$$\therefore \text{Cycle efficiency } \eta = \frac{(h_4 - h_5) - W_p}{(h_4 - h_2)}$$

Since,

$$h_2 = h_1 + W_p$$

$$\eta = \frac{(h_4 - h_5) - W_p}{(h_4 - h_1) - W_p} \qquad (28.3)$$

The feed pump work is small and may usually be neglected (in Fig. 28.3 (*b*) the line 1–2–3 would, if drawn correctly, be almost coincident with the phase boundary) so that for practical purposes

$$\eta = (h_4 - h_5)/(h_4 - h_1) \qquad (28.4)$$

Rankine efficiency. For an actual steam power plant, the efficiency of a theoretical Rankine cycle in which point 4 corresponds to actual boiler output conditions and point 5 to actual condenser pressure is termed "the Rankine efficiency of the plant."

Example. *In a steam plant, steam leaves the boiler at* 2 MN/m² (*20 bar*) *and* 350°C. *The pressure in the condenser is* 12 kN/m² (*0·12 bar*). *Calculate the Rankine efficiency of the plant, neglecting feed pump work.*

The *T–s* diagram for the corresponding Rankine cycle will be similar to Fig. 28.3 (*b*). Since 4–5 is an isentropic process, $s_4 = s_5$, or

$$6{\cdot}957 = 0{\cdot}696 + x_5 \cdot 7{\cdot}389$$

$$\therefore\ x_5 = 0{\cdot}847$$

Work done per kg = "Isentropic enthalpy drop" $(h_4 - h_5)$

$$= 3138 - (207 + 0{\cdot}847 \times 2383)$$

$$= 912 \text{ kJ/kg}$$

Heat received per kg = $(h_4 - h_1)$

$$= 3138 - 207 = 2931 \text{ kJ/kg}$$

$\therefore$ Ranking efficiency = $(h_4 - h_5)/(h_4 - h_1)$

$$= 912/2931$$

$$= 0{\cdot}311 \quad \text{or} \quad 31{\cdot}1 \text{ per cent}$$

Gas turbine cycles

The gas turbine is not a recent invention. In 1791 a patent was obtained for a device which was never constructed but which could only be described as a primitive gas turbine. Another design, patented in 1861, was basically similar to the present gas turbine, and the first successful machine was constructed in 1905 by the *Societé des Turbomotors* of Paris. The chief reason for the relatively slow development of the gas turbine is that, unlike the internal combustion engine, it contains mechanical parts (i.e., the turbine blades) which are at the same temperature as the working fluid. If reasonable efficiency is to be obtained, high maximum temperatures must be used and, in the absence of materials capable of operating at high temperatures, efficiency is low.

The gas turbine may have a closed or an open cycle, as shown by Figs. 28.4 (*a*) and (*b*). In both cases, heat is received and rejected at constant pressure, so that the theoretical cycle is the *constant pressure cycle*. This cycle is also known as the Joule or Brayton cycle (both names being associated with reciprocating machines using constant pressure cycles) and the theoretical *T–s* diagram is shown by Fig. 28.5.

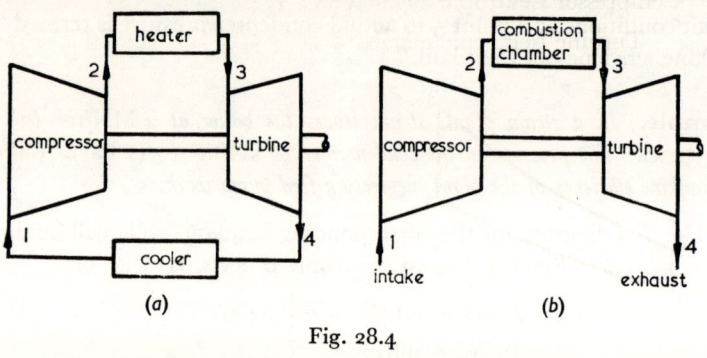

Fig. 28.4

Efficiency of the constant pressure cycle. From Eq. (27.2),

$$\eta = 1 - \frac{mc_p(T_4 - T_1)}{mc_p(T_3 - T_2)}$$

From Eq. (26.15),

$$T_2 = T_1(p_2/p_1)^{(\gamma-1)/\gamma}$$

and

$$T_4 = T_3(p_4/p_3)^{(\gamma-1)/\gamma}$$

But

$$p_2/p_1 = p_3/p_4$$

$$\therefore \ T_4/T_3 = T_1/T_2 = 1/r_p^{(\gamma-1)/\gamma}$$

where r_p is the *pressure ratio*

$$p_2/p_1 = p_3/p_4$$

Substituting and simplifying,

$$\eta = 1 - 1/r_p^{(\gamma-1)/\gamma} \qquad\qquad (28.5)$$

Irreversible compression and expansion. The cycle shown by Fig. 28.5 assumes isentropic compression and expansion. Actual rotary compressors and turbines cannot achieve this and the effect is to modify the cycle, as shown by Fig. 28.6, from 1–2–3–4 to 1–2′–3–4′.

Isentropic efficiencies. The ratio of the actual work to the corresponding isentropic work for a compressor or a turbine is called its *isentropic efficiency*. Since both operate under conditions of steady flow, and kinetic energies at entry and exit may usually be neglected, the work done is, in all cases, equal to the enthalpy change. For a gas, this is proportional to the temperature change, so that

Compressor isentropic efficiency $\eta_c = (T_2 - T_1)/(T_2' - T_1)$

and   Turbine isentropic efficiency $\eta_t = (T_3 - T_4')/(T_3 - T_4)$

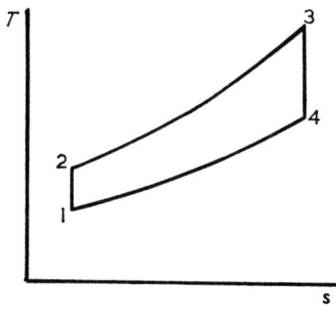
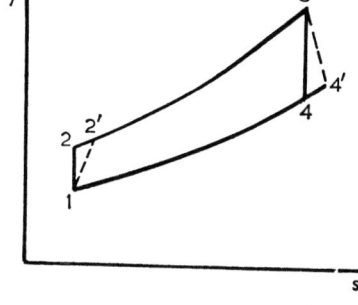

Fig. 28.5   Fig. 28.6

Example. *In a simple gas turbine plant, air at 20°C is taken in at 100 kN/m²* (1 bar) *and compressed to 500 kN/m²* (5 bar). *It then passes through a combustion chamber, the pressure in which may be assumed to be uniform, and in which its temperature is raised to 800°C. After expansion in the turbine it is exhausted at the original pressure of 100 kN/m². If the isentropic efficiencies of compressor and turbine are, respectively, 0·82 and 0·85 find* (a) *the cycle efficiency* (b) *the work ratio and* (c) *the mass flow required for an output of 1500 kW. Neglect the additional mass flow due to fuel and assume that* $\gamma = 1·40$ *and* $c_p = 1·005$ *kJ/kg K.*

The *T–s* diagram for this cycle will be similar to Fig. 28.6.

From Eq. (26.15),

$$T_2 = T_1 r_p^{(\gamma-1)/\gamma}$$

Here $r_p = 500/100 = 5$

$$\therefore \; T_2 = 293 \times 5^{0·4/1·4}$$

$$= 464 \text{ K}$$

$$\eta_c = (T_2 - T_1)/(T_2' - T_1)$$

$$\therefore \; 0·82 = (464 - 293)/(T_2' - 293)$$

$$T_2' = 501 \text{ K}$$

Similarly, $\qquad T_4 = 1073/5^{0.4/1.4}$

$\qquad\qquad\qquad = 678\,\text{K}$

and $\qquad\qquad 0.85 = (1073 - T_4')/(1073 - 678)$

$\qquad\qquad \therefore T_4' = 737\,\text{K}$

Considering a mass flow of 1 kg,

$$\text{Heat received} = c_p(T_3 - T_2') = c_p \times 572\,\text{kJ}$$
$$\text{Turbine work} = c_p(T_3 - T_4') = c_p \times 336\,\text{kJ}$$
$$\text{Compressor work} = c_p(T_2' - T_1) = c_p \times 208\,\text{kJ}$$
$$\therefore \text{Net work output} = c_p \times 336 - c_p \times 208 = c_p \times 128\,\text{kJ}$$

(a) $\qquad$ Cycle efficiency = Net work output/Heat received

$$= c_p \times 128/(c_p \times 572) = 0.224$$

$\qquad\qquad$ or 22·4 per cent.

(b) $\qquad\qquad$ Work ratio = Net work/Positive work

$$= c_p \times 128/(c_p \times 336) = 0.381$$

(c) A mass flow of 1 kg gives a net output of (1.005×128) kJ.

$\therefore$ Mass flow for 1500 kW output is

$$1500/(1.005 \times 128) = 11.7\,\text{kg/s}$$

Gas turbine modifications

Regneration. In the previous example, T_4' the exhaust temperature is greater than T_2', the temperature after compression. Thus a heat exchanger could recover some of the heat rejected, reducing the amount of heat to be supplied and increasing plant efficiency. The use of such a heat exchanger is termed *regeneration* and the arrangement is shown in Fig. 28.7.

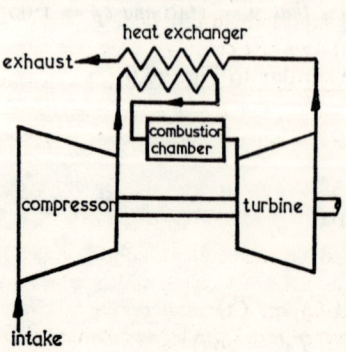

Fig. 28.7

Jet propulsion. The energy output of a gas turbine plant may be obtained as the kinetic energy of a jet, and this is the form used for aircraft propulsion. The jet propulsion unit differs from the usual gas turbine plant in that the turbine produces only the power required to drive the compressor. Gases leaving it pass into a nozzle, where their remaining available energy is converted into kinetic energy. When the engine is in motion, the air entering the unit also has kinetic energy which produces a "ram effect"—that is, an addition to the enthalpy of the air entering the compressor.

PROBLEMS

For tutorials

1. A theoretical Otto cycle has a compression ratio of 7 and the conditions at commencement of compression are: pressure 100 kN/m² (1 bar), temperature 15°C and volume 0·1 m³. Heat addition at constant volume results in a doubling of pressure. Taking $c_v = 0·718$ kJ/kg K and $c_p = 1·005$ kJ/kg K, find the cyclic work done (i) as the difference between heat received and heat rejected, and (ii) as the area enclosed by the $p - V$ diagram: show that these are identical.

2. A theoretical Otto cycle has a compression ratio of 9. Its lowest temperature is 10°C and the heat supplied at constant volume is 600 kJ per kg of working fluid, which may be assumed to have $c_v = 0·718$ kJ/kg K and $\gamma = 1·40$. Compare the efficiency of this cycle with that of a Carnot cycle (a) working between the same maximum and minimum temperatures, (b) having the same adiabatic compression ratio. Comment on these comparisons, especially (b).

3. A theoretical Diesel cycle has a compression ratio r and a cut-off ratio ρ. By expressing all cycle temperatures in terms of that at the commencement of compression, develop an expression for the cycle efficiency in terms of r, ρ and γ.

4. On the same T–s diagram show (i) a Rankine cycle for steam, (ii) a Carnot cycle operating between the same pressures. In what ways do these differ, for what practical reasons, and with what effects on cycle efficiency and work ratio?

5. The efficiency of a steam plant may be improved by increasing boiler pressure, or by superheating.
Compare these by calculating the efficiency of a Rankine cycle in which steam leaves the boiler (a) at 5 MN/m² (50 bar) and dry saturated; (b) at 10 MN/m² (100 bar) and dry saturated; (c) at 5 MN/m² (50 bar) and 311°C. The condenser pressure in all cases is 10 kN/m² (0·1 bar). Why does (b) represent a greater improvement than (c), and why is this not always done in practice?
Ans. (a) 34·7 per cent; (b) 37·5 per cent; (c) 35·2 per cent

6. Thermal efficiencies of gas turbine plants are, in general, much lower than those of internal combustion engines. Discuss (i) the reasons for this, (ii) the advantages of the gas turbine which make low efficiencies acceptable, (iii) ways of improving gas turbine plant efficiency.

General

1. In an ideal Otto cycle having a compression ratio of 6, the conditions at the beginning of compression are: pressure, 100 kN/m² (1 bar) and temperature 30°C. The heat received during the cycle is 800 kJ/kg. Calculate (a) the maximum pressure reached during the cycle, (b) the heat rejected per kg and hence (c) the cycle efficiency. Take $c_v = 0·718$ kJ/kg K and $\gamma = 1·40$.
Ans. (a) 3·43 MN/m² (34·3 bar); (b) 390 kJ/kg; (c) 51·2 per cent

2. A gas engine works on the four-stroke Otto cycle. It has a bore of 160 mm, a stroke of 280 mm and the clearance volume is 1 litre. The air/fuel ratio is 6:1 by volume and the calorific value of the gas used is 18 000 kJ/m³ (measured at standard atmospheric pressure and 20°C). Conditions in the engine cylinder at commencement of the compression stroke are: temperature 80°C and pressure 90 kN/m² (0·9 bar). It is estimated that only 75 per cent of the cylinder contents are fresh mixture, the remaining 25 per cent being exhaust gases from the previous cycle. If the engine operated the theoretical Otto cycle, what would be (a) the compression ratio, (b) the cycle efficiency, and (c) the indicated power output when running at 300 rev/min? Take $\gamma = 1·40$.

Ans. (a) 6·63; (b) 53·1 per cent; (c) 12·5 kW

3. In an ideal Diesel cycle, the compression ratio is 17. The temperature at commencement of compression is 40°C and the maximum cycle temperature is 1671°C. Find (a) the temperature at the end of compression, (b) the "cut-off ratio," (c) the temperature at the end of expansion, and (d) the cycle efficiency. Take $\gamma = 1·40$.

Ans. (a) 699°C; (b) 2·0; (c) 553°C; (d) 62·3 per cent

4. In a steam power plant, the boiler and condenser pressures are respectively, 2 MN/m² (20 bar) and 7 kN/m² (0·07 bar). Steam leaves the boiler with a dryness fraction of 0·97. Find (a) the Rankine efficiency of the plant, neglecting the feed pump work, (b) the efficiency of a Carnot cycle working between the same pressures.

Ans. (a) 31·4 per cent; (b) 35·7 per cent

5. In a theoretical constant pressure cycle, the pressure ratio is 6 and the maximum and minimum temperatures are, respectively, 500°C and 20°C. Considering the working fluid to be air, for which $c_p = 1·005$ kJ/kg K and $\gamma = 1·40$, find (a) the heat received, and (b) the heat rejected per kg of working fluid, also (c) the cycle efficiency.

Ans. (a) 286 kJ/kg; (b) 171 kJ/kg; (c) 40·1 per cent

6. A simple gas turbine plant burns oil of calorific value 43 000 kJ/kg and has a power output of 2 MW. The pressure at both intake and exhaust is 100 kN/m² (1 bar) and the pressure in the combustion chamber is 500 kN/m² (5 bar). Air enters the compressor at 15°C and leaves the combustion chamber at 700°C. The isentropic efficiencies of compressor and turbine are, respectively, 80 per cent and 84 per cent. Neglecting the effect of additional mass flow due to fuel, and assuming that for both air and combustion products $c_p = 1·005$ kJ/kg K and $\gamma = 1·40$, find (a) the mass flow of air, (b) the hourly fuel consumption, (c) the thermal efficiency, and (d) the work ratio.

Ans. (a) 21·8 kg/s; (b) 874 kg; (c) 19·2 per cent; (d) 0·302

CHAPTER 29

Heat Transfer

29.1 Conduction, Convection and Radiation

Heat has been defined as "energy transferred from one body to another because of a difference in temperature" and there are three mechanisms by which this transfer may take place—conduction, convection (either natural or forced) and radiation.

Conduction

Conduction consists of a direct transfer of energy from fast moving molecules at high temperature to slower molecules at lower temperatures. It is found by experiment that the heat Q transferred through a material is proportional to (i) the temperature gradient dT/dx, (ii) the cross-sectional area A, and (iii) the time t. It also depends on the material used, so that the relationship may be stated in the form of "Fourier's equation"

$$Q = -kAt \frac{dT}{dx} \tag{29.1}$$

Alternatively,

Rate of heat transfer $\dot{Q} = -kA \, dT/dx$ $\tag{29.2}$

where the constant k is known as the *thermal conductivity* of the material. The negative sign is necessary since heat transfer always takes place in the direction of *decreasing* temperature, so that if positive heat transfer is to occur dT/dx must be negative. Heat transfer by conduction is dealt with more fully later in this chapter.

Convection

The mechanism involved in heat transfer to liquids, vapours and gases is called convection. Most fluids are poor heat conductors, so that heat transfer by direct conduction is small. Thus, if a hot object is suspended in still air, what happens is as follows. A thin layer of air surrounding the object is heated by conduction, and the resulting decrease of density causes the air to rise and to be replaced by cool air. Thus a "convection current" of air rises continuously from the object. A similar current would fall from an object cooler than the surrounding air. This process is termed *natural convection*. In many cases the movement of the fluid is not the result of differences in density but of, say, a fan or pump, and this is termed *forced convection*.

The calculation of heat transferred by natural or forced convection involves several quantities including density, viscosity, specific heat and thermal conductivity of the fluid, its coefficient of expansion, the velocity of flow and its nature (that is, streamline or turbulent) and the nature of the surface. Experiments are therefore essential and the results are best analysed by the technique of *dimensional analysis*. Three important dimensionless quantities are known as the Nusselt number, the Prandtl number and the Grashof number. The use of such dimensionless groups is illustrated in Chapter 20 but, in general, the solution of problems involving convection is beyond the scope of this text. A more limited approach is, however, possible. In any particular case, the rate of heat transfer per unit area from a body to a surrounding fluid is (for practical purposes) directly proportional to the temperature difference between the body and the bulk of the fluid—that is, excluding the fluid in the immediate vicinity of the body.

Thus the rate of heat transfer is

$$Q = hA(T_1 - T_2) \qquad (29.3)$$

where T_1 and T_2 are the temperatures of the body and the fluid, A is the surface area of the body and h is a constant usually known as the *surface heat transfer coefficient*.

Radiation

All bodies emit radiation—energy in the form of electromagnetic waves. No conducting medium is required and this, for example, is how the earth receives heat from the sun. Electromagnetic waves are propagated, at a speed of 300000 km/s, by the interaction of magnetic and electrostatic fields and, according to their wavelengths, produce widely differing effects. The longer wavelengths (up to say, 2 km) are radio waves, the extremely short wavelengths (about 10^{-11} m) are known as gamma rays. About halfway between these two extremes is a small section (0.4 micron to 0.75 micron) to which the human eye is sensitive and between this "visible light" and the "ultra high frequency" end of the region of radio waves lies the range of wavelengths by which heat is transmitted. Since this radiation occupies a range of frequencies just below that of red light, it is often referred to as "infrared." The temperature of a body determines (i) the amount of energy radiated, and (ii) the range of wavelengths over which the radiation takes place. This may be clearly seen in, say, the heating of a steel billet. At 100°C, the amount of radiation is very small and is entirely in the infrared region. At 600°C, the amount of radiation has increased considerably and the range of wavelengths now extends into the red end of the region of visible light— the billet is "red-hot." At 1350°C the radiation is intense and has extended over the whole of the visible spectrum, so that the billet emits "white light." (In fact, most of the energy radiated is in the invisible or infrared region.) This applies also to higher temperatures: an electric arc (about 3500°C) emits light which is predominantly blue, together with a considerable amount of ultraviolet light and the sun (surface about 6000°C) emits a spectrum extending to the region of X-rays.

Black body emission. A *black body* is one which absorbs all the radiated heat falling on its surface, and experiments have shown that the energy radiated by such a body is proportional to the fourth power of its (absolute) temperature. This is expressed by the Stefan–Boltzmann law:

$$\text{Heat radiated } Q = \sigma A T^4 \tag{29.4}$$

where A is the area of the emitting surface, T its (absolute) temperature and σ the Stefan–Boltzmann constant, found by experiment to have the value

$$56.7 \times 10^{-12} \text{ kW/m}^2 \text{ K}^4$$

If two bodies having temperatures T_1 and T_2 are in such a position that each receives the radiation of the other, the heat transfer may be found using the principle that each body radiates energy according to the above law, and the net heat transfer is the difference between energy received and energy radiated. Thus, if both are at the same temperature, both are considered to radiate energy but the net heat transfer is zero.

In practice, bodies are not "black." A polished metal surface, for example, reflects most of the radiation falling on it. Hence the equation must be modified by the inclusion of a factor which takes account of this and is termed the *emissivity* (symbol ε). It may be shown that this factor is the same for emission and absorption (so that a polished surface is also a poor radiator) and the Stefan–Boltzmann equation becomes:

$$\text{Heat transfer } Q = \varepsilon \sigma A (T_1^4 - T_2^4) \tag{29.5}$$

A further factor is, however, necessary. A condition of Eq. (29.5) is that each body should receive *all* the radiation from the other and, unless one body completely encloses the other, this will not be the case. Thus account must be taken of the extent to which one object "sees" the other. The calculation of such factors (sometimes called "view factors") can be complicated and calculations of heat transfer by radiation are beyond the scope of this text: with the exception that the "surface heat transfer coefficients" already mentioned as a means of calculating heat transfer by convection (and used only for small temperature differences) may include the small amount of heat transfer by radiation in such cases.

29.2 Heat Transfer by Conduction

Plane layers

The conduction of heat through plane layers follows the law previously stated in the form of Eqs. (29.1) and (29.2), the only qualification being that the temperature gradient shall be uniform—that is, that the layer is of *uniform thickness* and has *isothermal boundaries*.

Example. *One wall of a refrigerated room is 3 m long and 2 m high. It is insulated by a layer of cork 200 mm thick and the temperatures at the inner and outer surfaces of this layer are −15°C and 20°C. If the thermal conductivity of cork is 0·04 W/m K, find the hourly heat leakage through the wall.*

From Eq. (30.2),

$$\dot{Q} = -kA\,dT/dx$$

Here
$$dT/dx = (T_2 - T_1)/x$$
$$= (-35)/0\cdot 2 = -175\ \text{K/m}$$
$$A = 3 \times 2 = 6\ \text{m}^2$$
$$\therefore \dot{Q} = -(0\cdot04) \times 6 \times (-175)$$
$$= 42\ \text{W}$$
$$\therefore \text{Hourly leakage} = 42 \times 60 \times 60$$
$$= 151\,200\ \text{J} \quad \text{or} \quad 151\cdot2\ \text{kJ}$$

Composite plane layers

Although the "electrical analogy" provides the most convenient method of solution, composite plane layers may be dealt with by the application of Eq. (29.2) as in the following example:

Example. *A furnace wall consists of a layer of firebrick* 200 mm *thick, to which is attached an outer layer of asbestos board* 10 mm *thick. The inner and outer surfaces of the composite wall are at temperatures of* 600°C *and* 50°C *respectively, and the thermal conductivities of firebrick and asbestos are respectively* 0·7 *and* 0·12 W/m K. *Find* (a) *temperature at the junction of the two materials,* (b) *the rate of heat loss per* m² *of the composite wall.*

For any complete plane layer, the same amount of heat flows through each material in turn: that is, Q is common to all layers. Considering 1 m² of the composite wall, let the temperature of the interface be t°C.

Eq. (30.2) may be written

$$Q = kA(T_1 - T_2)/x$$

where T_1 and T_2 are the temperatures on each side of a layer of thickness x. Applying this equation, and noting that $A = 1$,

$$Q = 0·7(600 - t)/0·2 = 0·12(t - 50)/0·01$$

or $$Q = 3·5(600 - t) = 12(t - 50)$$

(a) Solving for t:

$$2100 - 3·5t = 12t - 600$$
$$2700 = 15·5t$$
$$\therefore t = 174·2°C$$

(b) Q may be found by substituting in *either* expression for heat transfer. Choosing the expression for firebrick,

$$Q = 0·7(600 - 174·2)/0·2$$
$$= 1490 \text{ W}$$

Thus the rate of heat transfer through the composite wall is 1·49 kW/m².

The electrical analogy

The solution above follows the same pattern as the solution, from first principles, of an electrical circuit consisting of two resistances in series. Fig. 29.1 shows this comparison. In each case, there is the same kind of relationship between "driving force" and "resultant flow" and in each case the flow is common to both elements. The only difference is that, as yet, the term "resistance" in the electrical circuit has no thermal counterpart,

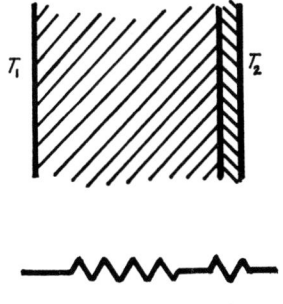

Fig. 29.1

Comparing electrical and thermal laws,

$$I = V/R \quad \text{and} \quad \dot{Q} = (T_1 - T_2) \times (kA/x)$$

It is obvious that $\dot{Q}$ corresponds to I and $(T_1 - T_2)$ to V. Hence (x/kA) is the equivalent of R and may be regarded as the *thermal resistance* of a plane layer. As in an electrical circuit, the total resistance of a composite layer is the sum of the resistances of each individual layer, so that the laborious calculation of interface temperatures by solving simultaneous equations is unnecessary. Furthermore, the similarity between electrical and thermal circuits means that complex cases of heat transfer by conduction may be dealt with using the techniques of the "analogue computer." That is, an arrangement of thermal resistances may be represented by a similar arrangement of electrical resistances, as in Fig. 29.2 (which includes "surface heat transfer coefficients"). An electrical circuit may thus be set up and a voltage (representing the overall temperature difference) applied. The currents flowing in the various parts of the circuit will then represent heat flows, and the potential difference between any two points will represent the corresponding temperature difference.

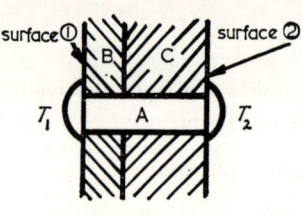

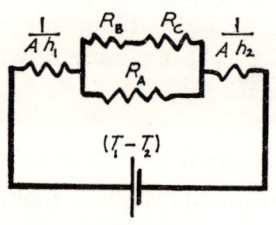

Fig. 29.2

The previous example may be solved by this method. Considering 1 m² of composite wall, the thermal resistances are:

Firebrick: $x/kA = 0{\cdot}2/0{\cdot}7 = 0{\cdot}2857$ K/W

Asbestos: $x/kA = 0{\cdot}1/0{\cdot}12 = 0{\cdot}0833$ K/W

Total resistance $R = 0{\cdot}3690$ K/W

$$\therefore \text{ Heat flow } \dot{Q} = (T_1 - T_2)/R$$

$$= (600 - 50)/0{\cdot}3690 = 1490 \text{ W}$$

The interface temperature may be found by considering, say, the layer of asbestos.

$$\dot{Q} = (T_1 - T_2)/R$$

$$1490 = (t - 50)/0{\cdot}0833$$

from which $t = 174{\cdot}2°C$

Overall heat transfer coefficients

It is often convenient to quote a single coefficient for a composite plane layer rather than to state the individual conductivities and thicknesses. Thus in general terms,

$$\dot{Q} = UA(T_1 - T_2) \tag{29.6}$$

where U is known as the *overall heat transfer coefficient* and, for the composite wall in the above example, has the value $2 \cdot 71$ W/m^2 K. Such coefficients are much used in the calculation of the heat loss from buildings.

Surface heat transfer coefficients

Heat transfer to a fluid by convection may be dealt with using Eq. (29.3)

$$\dot{Q} = hA(T_1 - T_2)$$

where h is the *surface heat transfer coefficient*. Many overall heat transfer coefficients (U) refer to cases in which a composite wall separates two fluids, which means that the coefficient applies to heat transfer between the fluids—that is, U includes both surface coefficients.

Example. *The wall of a house consists of two brick walls, each* 100 mm *thick, separated by a cavity. The inner wall is coated with a* 15 mm *layer of plaster. One room has an external wall measuring* 5 m × 2·5 m *containing a window* 2 m × 1 m *made of glass* 5 mm *thick. (a) Calculate the rate of heat loss from this wall when the outside temperature is* 2°C *and the room is being maintained at* 20°C. *(b) If the dew point of the air in the room is* 10°C, *will condensation occur on the window? The thermal conductivities of brick, plaster and glass are respectively* 0·7, 0·4 *and* 0·9 W/m K *and the overall heat transfer coefficient for the cavity is* 6 W/m^2 K. *Surface heat transfer coefficients for both wall and window may be taken as: inside,* 10 W/m^2 K *and outside,* 20 W/m^2 K.

(a) From Eq. (29.2), it follows that

$$R = x/kA \tag{29.7}$$

From Eq. (29.6), it follows that

$$R = 1/UA \tag{29.8}$$

From Eq. (29.3), it follows that

$$R = 1/hA \tag{29.9}$$

Area of wall $= (5 \times 2\cdot5) - (2 \times 1) = 10\cdot5$ m^2

Thermal resistances:

$$\text{For inner surface, } 1/(10 \times 10 \cdot 5) = 0 \cdot 009524 \text{ K/W}$$
$$\text{For } 2 \times 100 \text{ mm brick, } 0 \cdot 2/(0 \cdot 7 \times 10 \cdot 5) = 0 \cdot 02721 \text{ K/W}$$
$$\text{For cavity, } 1/(6 \times 10 \cdot 5) = 0 \cdot 01587 \text{ K/W}$$
$$\text{For } 15 \text{ mm plaster, } 0 \cdot 015/(0 \cdot 4 \times 10 \cdot 5) = 0 \cdot 003571 \text{ K/W}$$
$$\text{For outer surface, } 1/(20 \times 10 \cdot 5) = 0 \cdot 004762 \text{ K/W}$$
$$\text{Total resistance} = 0 \cdot 060937 \text{ K/W}$$
$$\therefore \text{ Heat flow through wall } \dot{Q} = (T_1 - T_2)/R$$
$$= (20 - 2)/0 \cdot 060937$$
$$= 295 \cdot 4 \text{ W}$$
$$\text{Area of window} = (2 \times 1) = 2 \text{ m}^2$$

Thermal resistances:

$$\text{For inner surface, } 1/(10 \times 2) = 0 \cdot 05 \text{ K/W}$$
$$\text{For 5 mm glass, } 0 \cdot 005/(0 \cdot 9 \times 2) = 0 \cdot 00278 \text{ K/W}$$
$$\text{For outer surface, } 1/(20 \times 2) = 0 \cdot 025 \text{ K/W}$$
$$\text{Total resistance} = 0 \cdot 07778 \text{ K/W}$$
$$\therefore \text{ Heat flow through window } \dot{Q} = (T_1 - T_2)/R$$
$$= (20 - 2)/0 \cdot 07778$$
$$= 231 \cdot 6 \text{ W}$$
$$\therefore \text{ Total heat flow} = 295 \cdot 4 + 231 \cdot 6$$
$$= 527 \text{ W}$$

(b) Applying the relationship $\dot{Q} = (T_1 - T_2)/R$ to the inner surface of the window,

$$231 \cdot 6 = (20 - t)/0 \cdot 05$$

from which the temperature t of the glass at its inner surface is 8·42°C. This is below the stated dew point, hence moisture will condense—that is, the window will "mist."

Heat transfer through cylindrical layers

For cylindrical structures, outward conduction occurs through an increasing area, and Eq. (29.2) may be applied only to an element with an area which is substantially the same on both sides—that is, to a thin cylindrical element. Fig. 29.3 shows the cross-section of a cylinder (considered to be of considerable length so that no heat flows *along* the cylinder at this point) having internal and external radii r_1 and r_2 and internal and external surface temperatures T_1 and T_2 respectively. Consider a thin cylindrical element of radius r, thickness δr and axial length l, the temperature difference across the layer being δT.

Applying Eq. (29.2),

$$\dot{Q} = -kA \, dT/dx$$

$$= -k(2\pi rl)\delta T/\delta r$$

The heat flow (assuming it to be completely radial) must be the same for all such layers, thus over the whole cylinder

$$\dot{Q} = -k(2\pi rl) \, dT/dr$$

or

$$\dot{Q} \, dr/r = -k(2\pi l) \, dT$$

Integrating both sides from the inner to the outer surface,

$$\dot{Q} \int_{r_1}^{r_2} (1/r) \, dr = -k(2\pi l) \int_{T_1}^{T_2} dT$$

$$\dot{Q} \log_e (r_2/r_1) = -k(2\pi l)(T_2 - T_1)$$

$$= k(2\pi l)(T_1 - T_2)$$

$$\therefore \dot{Q} = k(2\pi l)(T_1 - T_2)/\log_e (r_2/r_1) \qquad (29.10)$$

Example. *A steam pipe is 20 m long and of external diameter 150 mm. Its surface temperature is 200°C and it is lagged to an overall diameter of 250 mm. A thermometer shows the surface temperature of the lagging to be 50°C. If the thermal conductivity of the lagging is 0·09 W/m K, find the rate of heat loss.*

From Eq. (29.10),

$$\dot{Q} = k(2\pi l)(T_1 - T_2)/\log_e (r_2/r_1)$$

Here

$$(r_2/r_1) = (d_2/d_1) = (250/150)$$

$$\therefore \text{Heat flow } \dot{Q} = 0·09 \times 2\pi \times 20(200 - 50)/\log_e (250/150)$$

$$= 3320 \text{ W} = 3·32 \text{ kW}$$

Heat transfer through spherical layers

For a spherical geometry with isothermal boundaries, the heat flow will be completely radial. Fig. 30.3 could thus refer to a hollow sphere, the only modification being that the thin element shown now represents a thin spherical layer. The surface area of this layer is $4\pi r^2$, hence the expression for heat flow becomes

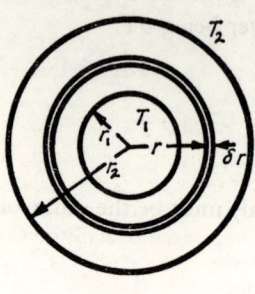

$$\dot{Q} = -k(4\pi r^2)\delta T/\delta r$$

from which $\quad \dot{Q}\,\mathrm{d}r/r^2 = -k(4\pi)\,\mathrm{d}T$

Integrating, as before, from the inner to the outer surface,

$$\dot{Q}(1/r_2 - 1/r_1) = -k(4\pi)(T_2 - T_1)$$

or $\quad \dot{Q}(1/r_1 - 1/r_2) = k(4\pi)(T_1 - T_2)$

$$\therefore Q = k(4\pi)(T_1 - T_2)/(1/r_1 - 1/r_2) \quad (29.11)$$

Fig. 29.3

Example. *A chemical reaction vessel is spherical and has a working temperature of 150°C (which may be assumed to be the temperature of its outer surface). The diameter of the vessel is 1 m and it is covered by a layer of lagging 200 mm thick. Taking the thermal conductivity of the lagging as 0·09 W/m K and the temperature of its outer surface as 25°C, find the rate of heat leakage from the vessel.*

Putting $r_1 = 0\cdot5$ m and $r_2 = (0\cdot5 + 0\cdot2) = 0\cdot7$ m in Eq. (29.11),

$$\dot{Q} = 0\cdot09 \times 4\pi(150 - 25)/(1/0\cdot5 - 1/0\cdot7)$$

$$= 45\pi/(2 - 1\cdot429)$$

$$= 248 \text{ W}$$

Problems

For tutorials

1. Glass is not a particularly bad conductor of heat, yet glass fibre is extensively used in heat insulating. What, in this case is the real "insulator" and what is the function of the glass fibre? How many other heat insulators can you name and, of these, which use the same principle as glass fibre?

2. The most perfect heat insulator in common use is the "vacuum flask." How does this deal with (*a*) conduction, (*b*) convection, and (*c*) radiation? In spite of everything, a small amount of heat transfer still takes place. In what ways does this occur and can you suggest any means by which it might be reduced?

3. It is well known that a carbon steel bar of, say, 100 mm diameter cannot be hardened by simple heating and quenching. Explain the reasons for this and give

a rough proof based on the assumptions that (i) the steel is a perfect conductor, (ii) the surface heat transfer coefficient (for a hot bar immersed in an ordinary quenching bath) is unlikely to exceed 10 kW/m² K. Take the density and specific heat of steel as, respectively, 7800 kg/m³ and 0·48 kJ/kg K.

4. The internal and external radii of a hollow sphere are in the ratio 1:1·2. What will be the percentage error introduced by treating this as a plane surface (based on mean radius) in heat transfer calculations?

Ans. Calculated heat flow will be 0·833 per cent high

5. An orbiting capsule is said to be under "zero *g*." Does this affect heat transfer within the capsule (assuming it to be pressurised to normal atmospheric conditions) and what special cooling arrangements for, say, electronic apparatus would you suggest? The passenger in such a capsule is given a box of matches and instructed to strike one while in orbit. Assuming that the air in the capsule is of normal atmospheric composition, what do you think will happen?

General

1. The "cold chamber" of a large refrigerator is rectangular in shape and measures 2 m × 1 m × 1·2 m. It is insulated by a layer of cork board 150 mm thick applied to all surfaces. The refrigerator is capable of extracting heat at the rate of 150 W. Neglecting the effect of "corners" i.e., treating the cork board as a plane surface whose area is that of the cold chamber, and assuming its outer surface temperature to be 20°C, what is the lowest possible internal temperature? The thermal conductivity of cork board is 0·043 W/m K.

Ans. −26·7°C

2. The wall of a house consists of a 150 mm thick layer of concrete, to the inner surface of which is attached a 10 mm thick insulating board, fixed on battens so as to leave a 15 mm space. The thermal conductivities of concrete and of insulating board are respectively 1·0 and 0·06 W/m K, and the overall heat transfer coefficient for the air space (including battens) may be taken as 7 W/m² K. Using surface heat transfer coefficients of 9 W/m² K for the inner surface and 22 W/m² K for the outer surface, find (*a*) the rate of heat leakage per m² if the room temperature is 19°C and the outside temperature −2°C, (*b*) the value, for this wall, of the overall heat transfer coefficient *U*.

Ans. (*a*) 34·1 W/m²; (*b*) $U = 1·62$ W/m² K

3. A composite panel is made by riveting a 10 mm thick insulating sheet to a 20 mm plywood backing using 5 mm aluminium rivets. The arrangement is similar to that shown in Fig. 29.2, and rivets are spaced so that their distribution is 10 per m². Taking the thermal conductivities of insulating sheet, plywood and aluminium as 0·03, 0·14 and 200 W/m K respectively, assuming surface heat transfer coefficients of 10 W/m² K for both sides of the panel and neglecting rivet heads, find the rate of heat transfer for 0·1 m² of the panel (i.e., the surface supported by one rivet) if the temperature difference is 25°C.

Ans. 5·07 W

4. Liquid methane is stored in a spherical container of diameter 2 m, covered with insulation of uniform thickness 120 mm, having a thermal conductivity of 0·025 W/m K. The temperature of the outer surface of this insulation is 9°C and the temperature of its inner surface may be taken as that of the liquid, −162°C. At what rate will heat leak into the vessel?

Ans. 502 W

5. A steam pipe 8 m long is of 100 mm external diameter and is covered by a 50 mm thick layer of lagging. The pipe conveys wet steam at a pressure of 1·5 MN/m² (15 bar) and the ambient temperature is 22°C. Taking the thermal conductivity of the lagging as 0·15 W/m K, assuming its surface heat transfer coefficient to be 10 W/m² K, and considering the pipe and its inner surface to have negligible thermal resistance, estimate the mass of steam condensed per hour.

Ans. 2·92 kg

Index